Geophysical Monograph Series

Including

**IUGG Volumes
Maurice Ewing Volumes
Mineral Physics Volumes**

Geophysical Monograph Series

120 **GeoComplexity and the Physics of Earthquakes** *John B. Rundle, Donald L. Turcotte, and William Klein (Eds.)*

121 **The History and Dynamics of Global Plate Motions** *Mark A. Richards, Richard G. Gordon, and Rob D. van der Hilst (Eds.)*

122 **Dynamics of Fluids in Fractured Rock** *Boris Faybishenko, Paul A. Witherspoon, and Sally M. Benson (Eds.)*

123 **Atmospheric Science Across the Stratopause** *David E. Siskind, Stephen D. Eckerman, and Michael E. Summers (Eds.)*

124 **Natural Gas Hydrates: Occurrence, Distribution, and Detection** *Charles K. Paull and William P. Dillon (Eds.)*

125 **Space Weather** *Paul Song, Howard J. Singer, and George L. Siscoe (Eds.)*

126 **The Oceans and Rapid Climate Change: Past, Present, and Future** *Dan Seidov, Bernd J. Haupt, and Mark Maslin (Eds.)*

127 **Gas Transfer at Water Surfaces** *M. A. Donelan, W. M. Drennan, E. S. Saltzman, and R. Wanninkhof (Eds.)*

128 **Hawaiian Volcanoes: Deep Underwater Perspectives** *Eiichi Takahashi, Peter W. Lipman, Michael O. Garcia, Jiro Naka, and Shigeo Aramaki (Eds.)*

129 **Environmental Mechanics: Water, Mass and Energy Transfer in the Biosphere** *Peter A. C. Raats, David Smiles, and Arthur W. Warrick (Eds.)*

130 **Atmospheres in the Solar System: Comparative Aeronomy** *Michael Mendillo, Andrew Nagy, and J. H. Waite (Eds.)*

131 **The Ostracoda: Applications in Quaternary Research** *Jonathan A. Holmes and Allan R. Chivas (Eds.)*

132 **Mountain Building in the Uralides: Pangea to the Present** *Dennis Brown, Christopher Juhlin, and Victor Puchkov (Eds.)*

133 **Earth's Low-Latitude Boundary Layer** *Patrick T. Newell and Terry Onsage (Eds.)*

134 **The North Atlantic Oscillation: Climatic Significance and Environmental Impact** *James W. Hurrell, Yochanan Kushnir, Geir Ottersen, and Martin Visbeck (Eds.)*

135 **Prediction in Geomorphology** *Peter R. Wilcock and Richard M. Iverson (Eds.)*

136 **The Central Atlantic Magmatic Province: Insights from Fragments of Pangea** *W. Hames, J. G. McHone, P. Renne, and C. Ruppel (Eds.)*

137 **Earth's Climate and Orbital Eccentricity: The Marine Isotope Stage 11 Question** *André W. Droxler, Richard Z. Poore, and Lloyd H. Burckle (Eds.)*

138 **Inside the Subduction Factory** *John Eiler (Ed.)*

139 **Volcanism and the Earth's Atmosphere** *Alan Robock and Clive Oppenheimer (Eds.)*

140 **Explosive Subaqueous Volcanism** *James D. L. White, John L. Smellie, and David A. Clague (Eds.)*

141 **Solar Variability and Its Effects on Climate** *Judit M. Pap and Peter Fox (Eds.)*

142 **Disturbances in Geospace: The Storm-Substorm Relationship** *A. Surjalal Sharma, Yohsuke Kamide, and Gurbax S. Lakhima (Eds.)*

143 **Mt. Etna: Volcano Laboratory** *Allessandro Bonaccorso, Sonia Calvari, Mauro Coltelli, Ciro Del Negro, and Susanna Falsaperla (Eds.)*

144 **The Subseafloor Biosphere at Mid-Ocean Ridges** *William S. D. Wilcock, Edward F. DeLong, Deborah S. Kelley, John A. Baross, and S. Craig Cary (Eds.)*

145 **Timescales of the Paleomagnetic Field** *James E. T. Channell, Dennis V. Kent, William Lowrie, and Joseph G. Meert (Eds.)*

146 **The Extreme Proterozoic: Geology, Geochemistry, and Climate** *Gregory S. Jenkins, Mark A. S. McMenamin, Christopher P. McKay, and Linda Sohl (Eds.)*

147 **Earth's Climate: The Ocean–Atmosphere Interaction** *Chuzai Wang, Shang-Ping Xie, and James A. Carton (Eds.)*

148 **Mid-Ocean Ridges: Hydrothermal Interactions Between the Lithosphere and Oceans** *Christopher German, Jian Lin, and Lindsay Parson (Eds.)*

149 **Continent-Ocean Interactions Within East Asian Marginal Seas** *Peter Clift, Wolfgang Kuhnt, Pinxian Wang, and Dennis Hayes (Eds.)*

150 **The State of the Planet: Frontiers and Challenges in Geophysics** *Robert Stephen John Sparks and Christopher John Hawkesworth (Eds.)*

151 **The Cenozoic Southern Ocean: Tectonics, Sedimentation, and Climate Change Between Australia and Antarctica** *Neville Exon, James P. Kennett, and Mitchell Malone (Eds.)*

152 **Sea Salt Aerosol Production: Mechanisms, Methods, Measurements, and Models** *Ernie R. Lewis and Stephen E. Schwartz*

153 **Ecosystems and Land Use Change** *Ruth DeFries, Gregory Asner, and Richard Houghton (Eds.)*

154 **The Rocky Mountain Region—An Evolving Lithosphere: Tectonics, Geochemistry, and Geophysics** *Karl E. Karlstrom and G. Randy Keller (Eds.)*

155 **The Inner Magnetosphere: Physics and Modeling** *Tuija I. Pulkkinen, Nikolai A. Tsyganenko, and Reiner H. W. Friedel (Eds.)*

Geophysical Monograph 156

Particle Acceleration in Astrophysical Plasmas: Geospace and Beyond

Dennis Gallagher
J. L. Horwitz
J. D. Perez
R. D. Preece
J. Quenby
Editors

American Geophysical Union
Washington, DC

Published under the aegis of the AGU Books Board

Jean-Louis Bougeret, Chair; Gray E. Bebout, Carl T. Friedrichs, James L. Horwitz, Lisa A. Levin, W. Berry Lyons, Kenneth R. Minschwaner, Andy Nyblade, Darrell Strobel, and William R. Young, members.

Library of Congress Cataloging-in-Publication Data
Particle acceleration in astrophysical plasmas : geospace and beyond / Dennis Gallagher ... [et al.], editors.
      p. cm. -- (Geophysical monograph, ISSN 0065-8448 ; 156)
   Includes bibliographical references.
   ISBN-13: 978-0-87590-421-4
   ISBN-10: 0-87590-421-1
      1. Particle acceleration. 2. Plasma astrophysics. 3. Space plasmas. I. Gallagher, Dennis L. II. Series.

QB464.15.P369 2005
523.01'973--dc22
                                      2005017995

ISBN-13: 978-087590-421-4
ISBN-10: 0-87590-421-1
ISSN 0065-8448

Copyright 2005 by the American Geophysical Union
2000 Florida Avenue, NW
Washington, DC 20009

Front Cover: A composite image of Chandra X-ray (blue) and radioastronomy (red) observations showing the inner 4,000 light years of a magnetized jet in Centaurus A. Purple regions are bright in both radiowave frequencies and X-rays. This jet originates from the vicinity of the supermassive black hole at the center of the galaxy. The radioastronomy observations were taken between 1991 and 2002 at the NRAO/VLA (National Radioastronomy Observatory/Very Large Area) by M. J. Hardcastle, Bristol University, Bristol, U.K. The X-rays were recorded with the National Aeronautics and Space Administration Chandra X-ray telescope, under the aegis of Chandra X-Ray Center/ Smithsonian Astrophysical Observatory (NASA/CXC/SAO), also by M. J. Hardcastle.

Back Cover: View of the Aurora Australis, or Southern Lights, from Space Shuttle Discovery (STS-39) in 1991. Source: NASA.

Figures, tables, and short excerpts may be reprinted in scientific books and journals if the source is properly cited.

Authorization to photocopy items for internal or personal use, or the internal or personal use of specific clients, is granted by the American Geophysical Union for libraries and other users registered with the Copyright Clearance Center (CCC) Transactional Reporting Service, provided that the base fee of $1.50 per copy plus $0.35 per page is paid directly to CCC, 222 Rosewood Dr., Danvers, MA 01923. 0065-8448/05/$1.50+0.35.

This consent does not extend to other kinds of copying, such as copying for creating new collective works or for resale. The reproduction of multiple copies and the use of full articles or the use of extracts, including figures and tables, for commercial purposes requires permission from the American Geophysical Union.

Printed in the United States of America

CONTENTS

Preface
D. L. Gallagher, J. L. Horwitz, J. D. Perez, R. D. Preece, and J. J. Quenby vii

Introduction to Particle Acceleration in the Cosmos
D. L. Gallagher, J. L. Horwitz, J. D. Perez, and J. J. Quenby 1

Shock Acceleration

Diffusive Shock Acceleration in Astrophysics
J. J. Quenby and A. Meli 9

On Bow Shock Source of Cusp Energetic Ions
Shen-Wu Chang and Karlheinz J. Trattner 21

Generation of Diamagnetic Cavities at the Bow Shock by Ion Kinetic Effects
Yu Lin 31

Diffusive Compression Acceleration
Joe Giacalone, Jack R. Jokipii, and Jozsef Kóta 41

Particle Acceleration and Transport at CME-Driven Shocks: A Case Study
Gang Li, G. P. Zank, M. I. Desai, G. M. Mason, and W.K.M. Rice 51

Cosmic Ray Acceleration at Relativistic Shock Waves With a "Realistic" Magnetic Field Structure
Jacek Niemiec and Michal Ostrowski 59

Kinetics of Particles in Relativistic Collisionless Shocks
Mikhail V. Medvedev, Luis O. Silva, Ricardo A. Fonseca, J. W. Tonge, and Warren B. Mori 65

Studies of Relativistic Shock Acceleration
A. Meli and J. J. Quenby 71

Energy Spectra of Energetic Ions Around Quasi-Parallel Shocks
T. Sugijama, M. Fujimoto, and H. Matsumoto 87

Particle Acceleration in Shell Supernova Remnants: Observational Evidence
Stephen P. Reynolds 97

Waves and Turbulence Acceleration

Overview: Particle Acceleration by Waves and Turbulence
Robert Y. Lysak 107

The Connection Between Parallel Electric Fields and Ion Acceleration in Astrophysical Plasmas
E. J. Lund 109

Simulation Study of Beam–Plasma Interaction and Associated Acceleration of Background Ions
X. Y. Wang and Y. Lin 117

CONTENTS

Particle Acceleration and Current Disruption From the Cross-Field Current Instability
Anthony T. Y. Lui 125

Magnetic Energy Storage and Stochastic Acceleration

The Role of Electron Acceleration in Quick Reconnection Triggering
Masaki Fujimoto and Iku Shinohara 139

The Hall Current System for Magnetic Reconnection in the Magnetotail
T. Nagai and M. Fujimoto 149

Plasma Acceleration due to Transition Region Reconnection
J. Büchner, B. Nikutowski, and A. Otto 161

RHESSI Observations of Particle Acceleration in Solar Flares
R. P. Lin 171

Magnetohydrodynamic Analysis of the January 20, 2001, CME–CME Interaction Effect
A. H. Wang, S. T. Wu, and N. Gopalswamy 185

The Quadrupole as a Source of Cusp Energetic Particles: I. General Considerations
Robert B. Sheldon, Theodore A. Fritz, and Jiasheng Chen 197

Adiabatic, Diffusive, and Double Layer Acceleration

Electron Phasespace Density Analysis Based on Test-Particle Simulations of Magnetospheric Compression Events
Jennifer L. Gannon and Xinlin Li 205

Parameterization of Ring Current Adiabatic Energization
M. W. Liemohn and G. V. Khazanov 215

Interrelation Among Double Layers, Parallel Electric Fields, and Density Depletions
Nagendra Singh 231

Acceleration and Outflow of Matter From Celestial Objects
Rickard Lundin, Stanislav Barabash, and Anatol Guglielmi 249

Studies of Exotic Acceleration Processes and Fundamental Physics

Exotic Acceleration Processes and Fundamental Physics
Giovanni Amelino-Camelia 265

Improving Limits on Planck-Scale Lorentz Symmetry Test Theories
Giovanni Amelino-Camelia 269

Spectral Evolution of Two High-Energy Gamma-Ray Bursts
Yuki Kaneko, Robert D. Preece, María Magdalena González, Brenda L. Dingus, and Michael S. Briggs 275

PREFACE

Space is dominated by plasma and a myriad of physical processes through which it has been evolving for billions of years. In our effort to understand what the universe has been and what it will become, we turn to the nature of matter, space, and the various interactions that change this environment. Our contact with the remote regions of the universe is dominated by the photons and high-energy particles produced there long ago, with light often the result of physical processes that involve particles accelerated to high energy. Developing our knowledge of particle acceleration across all spatial scales and energies is one of the keys to understanding the evolving universe. It is also the subject of this book.

Our ability to advance knowledge about the universe springs from what we gain by experiments in Earth-based plasma laboratories and our attempt to apply that knowledge in space near Earth and in the solar system where available to direct measurement. A key feature is the experimental ability to test theoretical knowledge of astrophysical plasma at least partially through in situ measurement. We must know not only what can happen if we manufacture a specific set of circumstances in the laboratory but also which of these circumstances occur naturally in space and the role they play in evolving plasma systems.

We hope that the current monograph—which draws from research in magnetospheric physics, solar physics, and astrophysics—will guide the reader in considering the full significance of what we have learned, and continue to learn, in the laboratory, in Earth's magnetosphere, in the solar wind, at the Sun, and beyond in the astrophysical plasma. Linking the physical processes in all these places is the fundamental nature of matter and energy and their evolving properties. The acceleration of particles is one of the unifying phenomena fundamental to the exploration of the universe. Whether working the details of how cold 0.1 eV ions and electrons in the Earth's ionosphere escape outward and possibly become accelerated more than 10,000 times to later rain down to form the auroras or how Exa-eV (10^{18} eV) particles can be created in cosmic events deep in space, the same basic physical processes that exchange energy between particles and electric and magnetic fields must be considered. Although we believe we have gained at least a partial understanding of a number of these processes, we cannot say the same for many other processes, which remain largely unknown and untested.

Researchers from the three fields of study mentioned above have made this book possible. Some participated directly in the founding workshop for the monograph, "Astrophysical Particle Acceleration in Geospace and Beyond" (the fall 2002 Huntsville Workshop in Chattanooga, Tennessee) by testing the idea that we could work together toward a common goal. Many of their names and affiliations can be found on the papers that follow. Many others contributed through their thoughtful evaluation of the papers in this book. The workshop participants, contributing authors, and reviewers have made immensely valuable contributions to this book and to the proposition that we should be involved in carrying hard-won knowledge about the evolution of plasma systems across all the spatial scales and energies that make up our universe.

D. L. Gallagher
J. L. Horwitz
J. D. Perez
R. D. Preece
J. J. Quenby
Editors

Introduction to Particle Acceleration in the Cosmos

D. L. Gallagher,[1] J. L. Horwitz,[2] J. D. Perez,[3] and J. J. Quenby[4]

Accelerated charged particles have been used on Earth since 1930 to explore the very essence of matter, for industrial applications, and for medical treatments. Throughout the universe, nature employs a dizzying array of acceleration processes to produce particles that span 20 orders of magnitude in energy range, while shaping our cosmic environment. Here, we introduce and review the basic physical processes causing particle acceleration in astrophysical plasmas, from geospace to the outer reaches of the cosmos. These processes are chiefly divided into four categories: adiabatic and other forms of nonstochastic acceleration, magnetic energy storage and stochastic acceleration, shock acceleration, and plasma wave and turbulent acceleration. The purpose of this introduction is to set the stage and context for the individual papers included in this monograph.

1. INTRODUCTION

Particle acceleration is ubiquitous in space. Energy gains of even less than 1 eV can cause atmospheric outflow from planetary ionospheres illuminated by stellar light. Indications of extremely high-energy (~10^{20} eV) cosmic rays challenge our basic understanding of matter and space. Across all energies, accelerated particles carry energy and information concerning the physical processes that produce these accelerations. Accelerated particles are therefore diagnostic of the flow of mass and energy as stellar, planetary, and galactic systems evolve. Our understanding of the behavior of these systems is dependent on our understanding of the processes that accelerate particles and our resulting ability to interpret those processes through remote measurement.

Much of what we know about acceleration processes has been derived from laboratory experiments and associated theoretical studies. Early experiments in the development of atomic theory led Rutherford to first speculate on the value of laboratory acceleration devices as tools for probing matter [*Rutherford*, 1919, 1928]. R. Wideröe proposed the first accelerator design in 1928, soon followed by D. H. Sloan and E. O. Lawrence in 1931 [*Persico et al.*, 1968]; the first operating model was constructed at Cambridge University by John Cockroft and E. T. S. Walton [*Gamov*, 1961]. The promise of new insight into the fundamental structure of matter fueled the rapid development of more and more energetic particle accelerators, which continues today. Toward the end of the 19th century, electromagnetic diagnostics of accelerated particles began to be appreciated. In 1895, Wilhelm Konrad Roentgen noticed the production of electromagnetic radiation by bombarding a target with electrons [*Gamov*, 1961]. This radiation later became known as Bremsstrahlung radiation. Although anticipated for many years [*Liénard*, 1898], the first experimental evidence of synchrotron radiation from accelerated electrons was obtained with a 70 MeV accelerator assembled by Robert Langmuir and Herbert Pollock in 1948 [*Pollock*, 1982]. The leap into space and astrophysics came with the first cosmic radiation measurements by Hess in 1912 in a balloonborne experiment [*Sekido and Elliot*, 1985] and with the launch of the space program on the LUNIK 1 and 2 [*Gringzuz et al.*, 1960] and Explorer 1 and 3 [*Van Allen*, 1959] satellites [see account in *Lemaire and Gringauz*, 1998].

The study of acceleration processes and accelerated particles is intended to help us understand the operating and evolutionary processes taking place in the cosmos. The study

---

[1] Space Science Branch, Marshall Space Flight Center, National Aeronautics and Space Administration, Huntsville, Alabama.
[2] Department of Physics, University of Texas in Arlington, Arlington, Texas.
[3] Department of Physics, Auburn University, Auburn, Alabama.
[4] Astrophysics Group, Blackett Laboratory, Imperial College of Science, Technology and Medicine, London, United Kingdom.

of these processes in regions of space accessible to us enables a highly detailed diagnostic of the physical properties of space plasma undergoing change, which in turn leads to much enhanced constraints on the physical descriptions of those processes we develop. The attributes of accelerated particles and the circumstances of their existence directly reflect the basic physics of the cosmos and must be consistent with any theory we may develop to explain the nature of the universe. Therefore each measurement reflects an attempt to temper or strengthen our collective understanding of the cosmos.

The study of acceleration processes within accessible regions such as the solar system, particularly geospace, offers the opportunity to understand at least some processes for acceleration of charged particles in detail and to subject concepts to some level of rigorous experimental testing. Astrophysical observations, on the other hand, offer indications of particle acceleration levels and types that appear quantitatively different from those in the solar system in terms of parameter regimes. We can thus learn from the full range of apparent particle acceleration mechanisms associated with regions extending from geospace to the furthest reaches of the universe. The purpose of this monograph is to bring together observations and theories on particle acceleration for regions throughout the cosmos to help foster greater understanding from multiple perspectives.

Particles are accelerated through a wide range of physical processes. Here we discuss these processes in the following four sections, organized by physical mechanism. Our objective is to provide an overall perspective on particle acceleration physics and the dynamic processes occurring in space where acceleration takes place.

2. ADIABATIC AND OTHER FORMS OF NONSTOCHASTIC ACCELERATION

One general category for acceleration of charged particles in geospace and beyond could be those types of particle acceleration that are relatively smooth. Acceleration by quasi-DC electric fields parallel to the magnetic field can be of this type. Such acceleration is believed to occur under some circumstances in auroras. Another type of acceleration that may often be relatively smooth occurs when adiabatic invariants are conserved during motion in nonuniform magnetic fields. For example, if charged particles in the middle terrestrial magnetosphere are moved Earthward toward regions of higher magnetic fields by convection electric fields, they may conserve their first adiabatic invariants [e.g., *Chen*, 1984, p. 44], the ratio of their energies perpendicular to the magnetic field to the strength of the magnetic field, thereby increasing their perpendicular energies in this process. This process is thought to be at least partially operative in the formation of the ring current from inward injections or motions of the inner plasma sheet of the magnetosphere. In this volume, *Liemohn and Khazanov* determine, through simulations of a magnetospheric storm, that a good parameter for determining the net adiabatic energy gain is the instantaneous value of the product of the maximum westward electric field at the outer boundary with the nightside plasma sheet density.

For particles mirroring along a magnetic field line, the integral of the parallel velocity between mirror points gives the second invariant [e.g., *Chen*, 1984, p. 46]. When such mirroring particles are transported onto shorter field lines, their parallel energies increase. This process may be important for energizing cosmic ray particles mirroring between approaching magnetic clouds, and for increasing parallel energies of particles convected Earthward from distended field lines in the deep magnetotail toward more dipolar-shaped field lines., The acceleration that *Sheldon et al.* suggest occurs in the magnetospheric cusp [this volume] is partially related to Fermi acceleration in magnetic geometries.

A third invariant [e.g., *Chen*, 1984, p. 49] is the total magnetic flux encompassed by a complete drift path of a charged particle, e.g., an energetic particle gradient drifting around the Earth, if changes in external agents do not occur more rapidly than the full periods for such drifts. *Gannon and Li* [this volume], considering test-particle simulations of magnetospheric compression events, argue that the radiation belt particles involved tend to conserve their third adiabatic invariants.

Another form of smooth, nonstochastic particle acceleration is "centrifugal" acceleration [*Cladis*, 1986; *Horwitz et al.*, 1994] in which convection of magnetic field lines with changing directions "whips" the particles outward along the lines. This energization may be relevant to acceleration of outflows from planetary ionospheres (and perhaps other objects) to substorm-associated dipolarization events in the magnetosphere of the Earth [e.g., *Liu et al.*, 1994], and perhaps elsewhere, and to large, rapidly rotating planetary magnetospheres such as that of Jupiter. A characteristic of this type of acceleration is that the velocity gains tend to be constant for different ion species, so that ions with different masses receive different energy boosts.

Another nonstochastic parallel electric field acceleration of ionospheric plasma transport may result from the effects of photoelectrons. In this phenomenon, to limit the net current and maintain quasi-neutrality in ionospheric polar wind plasma transport in the presence of upward-moving photoelectrons, an upward electric field develops to suppress the upward photoelectron flow; consequently, ionospheric ions are accelerated upward. *Su et al.* [1998] used a combination fluid–kinetic steady-state treatment to indicate that the photoelectron-produced electric potential should concentrate into a thin double layer 2-4 R_E in altitude, with drops of 10s of volts to accelerate polar ions to energies of a few 10s of electron volts. *Lundin and Barabash* [this volume] discuss aspects of acceleration and outflow from celestial objects. Though these authors do not specifically mention centrifugal and photoelectron-driven acceleration processes, both are examples of nonstochastic acceleration that could be important in acceleration and outflow from planetary ionospheres and other cosmic objects.

Many papers in auroral physics investigations have treated the parallel electric potential distribution as approximately stationary for electron acceleration, driven variously by imposed field-aligned electric currents, imposed large-scale drops in potential, or differential anisotropy in hot plasma distributions [e.g., *Knight*, 1973; *Evans*, 1974; *Block*, 1978; *Ergun et al.*, 2002]. *Singh* [this volume] presents simulations of electric double layers driven by imposed voltage drops and currents. Such electric double layers can be important in quasi-smooth versions of the observed parallel acceleration of electrons [e.g., *Burch et al.*, 1976] and ions [e.g., *Steinbach-Nielsen et al.*, 1984] in the auroral regions.

Downward versions of such parallel electric fields can also play an indirect role in energizing ions in the auroral regions through the so-called pressure-cooker effect [e.g., *Gorney et al.*, 1985; *Wu et al.*, 2002], in which downward electric fields trap in or slow the transit of ions through regions of transverse wave heating, thereby increasing the wave-driven heating itself. In this case, the main energization agent (the transverse ion heating) is stochastic in nature, but the important contributing factor of the downward electric field may comparatively smooth. *Lund* [this volume] considers a version of this process through test-particle simulations and also suggests that similar processes may be operative in the solar corona and perhaps in other astrophysical processes.

3. MAGNETIC ENERGY STORAGE AND STOCHASTIC ACCELERATION

A fundamental process in plasmas is energy storage in magnetic fields and its subsequent conversion to kinetic energy of particles. The idea that what we now call magnetic reconnection plays a key role in this process was first advanced by *Giovanelli* [1947] to explain particle acceleration in solar flares. *Hoyle* [1941] and *Dungey* [1961] applied the idea to the Earth's magnetosphere. *Priest and Forbes* [2000] have reviewed magnetic connection phenomena from laboratory machines, the Earth's magnetosphere, and the Sun's atmosphere to flare stars and astrophysical accretion disks.

Magnetic reconnection involves a breakdown of the ideal magnetohydrodynamic frozen-field approximation in which charged particles that are connected by a magnetic field line at one instant of time must remain connected by a field line for all time. This process, which is mathematically singular and usually highly nonlinear, induces electric fields parallel to the magnetic field and thus accelerates the plasma particles. For a review of early work on the theory of and evidence for magnetic reconnection, see *Hones* [1984]. For a description of the mathematical theory for the structure of magnetic reconnection layers, see *Lin and Lee* [1993].

In this volume, *Fujimoto and Shinohara* describe a rapid, spontaneous triggering mechanism for magnetic reconnection in an ion-scale current sheet. *Nagai and Fujimoto* present direct evidence of the Hall current system that plays such a vital role for reconnection in the Earth's magnetotail. *A.-H. Wang et al.* report the results of numerical simulations of the interaction of two coronal mass ejections and the associated reconnection.

Acceleration from reconnection sites is the result of electric fields parallel to the local magnetic field. Likewise, diffusive acceleration at shocks, referred to as first-order Fermi acceleration and discussed elsewhere in this volume, occurs when the electric field generated by the changing magnetic field is parallel to the magnetic field. In contrast, stochastic acceleration results from randomly oriented electric fields associated with magnetohydrodynamic waves or turbulence. This mechanism has been used to address the origin and transport of solar energetic particles and interstellar pickup ions [*Committee on Solar and Space Physics*, 2004].

In this volume, *Sheldon et al.* propose stochastic acceleration in a magnetic quadrupole trap in the Earth's cusp as the source of the observed cusp energetic particles. They suggest that this mechanism may be important elsewhere, e.g., magnetic cusp geometries in the heliosphere and the galaxy. Also, *R. P. Lin* [this volume] presents exciting new data from the NASA Reuven Ramaty High Energy Solar Spectroscopic Imager (RHESSI) small explorer spacecraft launched 5 February 2002. The capability of using the data from RHESSI along with detailed quantitative analysis to probe the particle acceleration mechanism(s) in flares and related phenomena was foreshadowed in *Lin et al.* [2002].

4. SHOCK ACCELERATION

Another important category of cosmic particle acceleration is energization by shock structures that may occur in the Earth's bow shock, within the solar wind as driven by coronal mass ejections, solar flares, supernovae remnants (SNR), jets of Active Galactic Nuclei (AGN), relativistic shocks associated with gamma-ray bursts and intercluster gas. For a shock to form on Earth, collisions in neutral gases are required, but in space, collisionless shocks are formed in quasi-neutral plasmas consisting of positively and negatively charged particles and electric and magnetic fields. They provide an arena for the acceleration of particles to unexpectedly high energies. Observations suggest power-law spectra for accelerated particles, particularly at the bow shock, in SNR and in AGN jets. Moreover, the cosmic ray spectrum is now known to obey a power law in energy with exponent of about −2.65 from 10 GeV to 10^6 GeV, slightly steepening at higher energies to 10^{10} GeV until another flatter portion and a likely cutoff at ~10^{11} GeV are encountered. *Reynolds* [this volume] presents experimental evidence of acceleration of cosmic ray electrons to a few thousand TeV. *Meli and Quenby* [this issue] present simulations illustrating the acceleration that forms ultra-high-energy cosmic rays.

It was *Fermi* [1959] who proposed a mechanism for cosmic ray acceleration in the interstellar medium that could produce a power spectrum; this was, in fact, the prototype of the second order or stochastic mechanisms already mentioned in the Section 3. This elegant approach—to predicting the outcome of random energetic particle collisions with tangled magnetic fields carried by moving gas clouds—treats the collision in the cloud frame. Thus electric field effects do not have to be followed in detail and the energy change finally depends only on the particle direction change and the cloud velocity. Fermi found the power-law exponent to be specified by the ratio of the particle loss rate in the medium to the cloud velocity, an unsatisfactory result because the numerical coincidence required to explain observations seemed too great. The break-through came when it was realized that collisionless, magnetohydrodynamical (MHD) shock waves provide an ideal situation to exploit a first-order variant of the Fermi mechanism, originally conceived as the action of advancing magnetic "jaws". This diffusive shock acceleration mechanism, proposed by several authors in 1977–1978 [see *Quenby and Meli*, this volume] envisaged the successive energization of an energetic particle bouncing between magnetic scattering centers upstream and downstream of an MHD shock. Although some derivations of the resulting power-law slope followed Fermi's basic approach, the new result was that this exponent depends only on the shock compression ratio, not on details of the scattering or plasma flows, unless extreme relativistic conditions were encountered. In fact, for strong shocks, the power law in momentum p is p^{-2}, not too dissimilar to the observed cosmic ray spectrum, especially when particle escape is added to the model.

In this volume, *Chang and Trattner* associate particles accelerated by this same mechanism at the quasi-parallel, Earth's bow shock with energetic particles observed in the Earth's magnetic cusp. *Sugiyama et al.* find that the presence of He^{2+} at the Earth's bow shock enhances this acceleration. *Li et al.* present a model in which this mechanism explains solar energetic particles produced in shocks associated with coronal mass ejections. *Niemiec and Ostrowski* find that the same mechanism is responsible for the generation of cosmic rays in mildly relativistic shocks.

One attraction of diffusive shock acceleration is the promise of a universal mechanism accounting for the energization of diverse cosmic particle populations. Particle energies range between 10^4 and 10^{20} eV and are encountered in shocked flows with velocities ranging from 4×10^7 cm/s to those with relativistic Lorentz Γ factors of 10^3. Current theoretical interests in astrophysics deal with the relativistic flow regime and the back-reaction or loading on the plasma as a result of the accelerated particle spectrum and hence consider the dynamical importance of these energetic particles in cosmic physics. *Medvedev et al.* [this volume] describe a mechanism in which a two-stream, nonresonant (Weibel) instability produces small-scale magnetic fluctuations that accelerate electrons and produce radiation having a signature different from synchrotron radiation. They use three-dimensional particle-in-cell (PIC) simulations to relate this process to gamma ray bursts. Recent PIC simulations using counter-streaming relativistic jets also show that acceleration is provided locally in the downstream jet, rather than by scattering particles back and forth across the shock as in Fermi acceleration [*Silva et al.*, 2003; *Frederiksen et al.*, 2003, 2004]. *Medvedev and Loeb* [1999], *Brainerd* [2000], *Pruet et al.* [2001], and *Gruzinov* [2001] have found that the Weibel instability generates current filaments and associated magnetic fields as dominant structures at relativistic shocks. Particle acceleration perpendicular and parallel to the jet propagation direction accompanies the nonlinear development of filamentary structures that cannot be characterized as Fermi acceleration.

Giacolone et al. [this volume] show that a process called diffusive compression acceleration can also accelerate particles. This process occurs not at shocks but when plasma is being gradually compressed; moreover, the acceleration can be both diffusive and nondiffusive. They illustrate the process in a number of geophysical processes, e.g., acceleration of particles by corotating solar wind compressions, acceleration in a sinusoidally varying fluid flow, the formation of high-energy tails in pickup-ion distributions, and acceleration of galactic cosmic rays by turbulent flows in the interstellar medium. The interaction of particles with the formation of a shock and with other structures in the plasma can lead to numerous interesting and unexpected phenomena. *Y. Lin* [this volume] shows that accelerated ion beams interacting with the Earth's bow shock and tangential discontinuities in the solar wind lead to the formation of diamagnetic cavities and the production of hot flow anomalies.

5. PLASMA WAVE AND TURBULENT ACCELERATION

In this section, we discuss particle acceleration in the context of plasma waves. Weak interactions, or interactions where wave amplitudes remain small, can be treated by linear approximation of the relevant equations. Under these conditions, energy in excited waves remains small relative to the thermal energy in the plasma [*Sagdeev and Galeev*, 1969, p. 1]. Interactions that proceed to large wave amplitude violate the conditions for weak perturbation analysis and can be expected to lead to strong nonlinear coupling between waves and a cascade of energy from long to short wavelengths in plasma [*Chen*, 1984, p. 288]. Such turbulent systems may support nonresonant particle acceleration and must be treated with a statistical approach. In either case, plasma systems that are not in thermal equilibrium will be unstable to plasma waves, and the resulting exchange of energy can accelerate particles to very high energy.

Weak interactions are considered to be resonant interactions where coupling is achieved between particles and waves through approximate matching of particle velocity and wave phase velocity. The relationship for wave–particle interactions is expressed through resonant matching conditions. Without a background magnetic field or along an ambient magnetic field, the resonance matching conditions can be expressed as follows [*Davidson*, 1972, p. 134]:

(a) $w_k - \vec{k} \cdot \vec{v} = 0$ and

(b) $\sum_{i=1}^{n}(w_{k_i} - \vec{k}_i \cdot \vec{v}) = 0, n \geq 2.$

where w_k is wave frequency, $\vec{k}$ is wave vector, $\vec{v}$ is particle velocity, and i designates one of n possible waves when multiple waves participate in a resonant interaction with particles. Condition (a) holds for a single wave where the component of the particles' velocity along the wave vector matches the wave phase velocity. This has been referred to as linear wave–particle processes. Nonlinear wave–particle processes, described by the second resonance condition (b), involve two or more waves interacting with particles. For particle motions transverse to an ambient magnetic field, the resonance-matching condition can be written as follows [*Gurnett and Bhattacharjee*, 2005, p. 341]:

$$w' = w - k_{||}v_{||} = nw_c.$$

In this case n is an integer corresponding to harmonics of the particle gyration frequency (w_c). The resonance condition takes into account the Doppler shift experienced by a particle as it interacts with a wave. The Doppler shift is given by the product of the parallel wave vector ($w_{||}$) and the parallel particle velocity ($v_{||}$). Ion cyclotron damping is an example of this type of interaction, in which a particle gyrating in either direction along a magnetic field can gain energy from a linearly or elliptically polarized wave. Whistler waves in the terrestrial magnetosphere are thought to be a source of energetic electrons in the Van Allen radiation belts attributable to this resonant wave–particle interaction [see, e.g., *Summers and Ma*, 2000]. In this case, electrons of a few hundred electron volts injected earthward from the terrestrial plasmasheet are accelerated to more than 1 MeV.

Linearly unstable conditions in plasma can generally be classified into four categories: (1) streaming instabilities, (2) Rayleigh–Taylor instabilities, (3) universal instabilities, and (4) kinetic instabilities [*Chen*, 1983, p. 208]. Streaming instabilities arise from the flow of charged particles through background plasma. The resulting interaction leads to the growth of plasma waves at the expense of the streaming particles. Type III solar radio bursts are an example of this instability: Energetic electrons streaming outward from a solar flare through the background solar wind will generate Langmuir waves at the local electron plasma frequency [*Gurnett and Bhattacharjee*, 2005, p. 329]. In this case, particle energy is lost to electrostatic, nonpropagating plasma waves that nonlinearly couple to electromagnetic waves at twice the electron plasma frequency [*Ginzburg and Zheleznakov*, 1958]. These waves display a characteristic downward sweep in frequency with time, as observed from Earth orbit, when the energetic electrons pass outward through decreasing solar wind density. This deceleration of energetic plasma and generation of remotely observable waves can be expected at any stellar object where the impulsive release of energy sends energetic particles through background plasma. The same process takes place in the upstream region of the Earth's bow shock where solar wind ions are reflected. X. Wang and Lin [this volume], studying the consequence of wave generation and energy coupling between the streaming and background ions, found strong heating (factor ~100) and acceleration of the background ions, in addition to strong heating (factor ~900) of the initial streaming ion distribution.

The Rayleigh–Taylor instability results from opposition of a nonelectromagnetic force, such as gravity or centripetal acceleration, and a density gradient in plasma, otherwise maintained by a magnetic field. This instability is thought to play a role in accreting neutron stars, where an outward plasma density gradient exists against the star's gravitational attraction [see *Wang and Nepveu*, 1983, and cited references]. In this case, the magnetosphere of a neutron star strongly influences, but cannot prevent, the inward fall of plasma. A full understanding requires the treatment of this instability as it grows to become strongly nonlinear. The Rayleigh–Taylor instability also acts in the Earth's equatorial ionosphere to produce density irregularities that affect radio communication and navigation systems [see, e.g., *Basu*, 2005].

In universal instabilities, plasma with a density gradient will expand without any other contributing force. Because all plasma systems are finite, density gradients will always exist at system boundaries and thus lead to instability and particle acceleration. An application of this type of instability to expanding magnetized relativistic plasma is given by *Liang et al.* [2003]. In their study hot, magnetized, collisionless plasma is released into a vacuum to expand perpendicular to the ambient magnetic field. The strong diamagnetic current resulting from the surface gradient confines the surface field to a thin layer. In a relativistic expanding cloud, an $\vec{E} \times \vec{B}$ drift is induced at the surface. Particles that stay in phase with the drift field are continuously accelerated to higher energies. They found that nonrelativistic particles could be accelerated to relativistic energies.

In a more general sense, deviations away from thermal equilibrium in plasma are characterized as deviations away from a Maxwellian velocity distribution. Anisotropy in plasma temperature parallel and perpendicular to an ambient magnetic field is one example of a kinetic instability. *Gary et al.* [1994] used the proton temperature anisotropy developed in compression of the terrestrial magnetosheath by the solar wind to drive the proton cyclotron instability, which in turn heated cold sheath plasma. The expectation is that this process will be active in other planetary magnetospheres.

Nishikawa et al. [2003, 2005, and references cited therein] find both electron and ion acceleration in the downstream region of relativistic astrophysical jets due to the kinetic Weibel instability. The Weibel instability results from small transverse perturbations in the ambient magnetic field that grow as a result of changes in electron trajectories, which enhance filamentary current structures and increase the field perturbation. The importance of advancing our knowledge of relativistic shock properties was given in *Nishikawa et al.* [2003], who stated that such studies are critical to development of a "proper understanding of the prompt gamma-ray and afterglow emission in gamma-ray bursts, and also to an understanding of the particle reacceleration processes and emission from the shocked regions in relativistic AGN jets". The Weibel and two-steam instabilities are also suspected of strong particle acceleration in thin current sheets in astrophysical plasma as a class of instabilities called the cross-field current instabilities [see *Lui*, this volume]. Thin current sheets with strong current densities can be found in the terrestrial magnetotail and elsewhere in astrophysical space plasmas. Disruption of the terrestrial magnetotail, possibly through cross-field current instabilities, is responsible for the impulsive release of energy in the form of accelerated particles.

Particle acceleration often appears as a result of several processes operating in concert. For example, ion cyclotron heating in the terrestrial ionosphere accelerates particles perpendicular to the ambient magnetic field. Divergence of the Earth's magnetic field with altitude leads to a mirror force that accelerates particles outward. When combined with the presence of a downward-directed parallel electric field, ions remain trapped until their energies have been pumped to values greater than the trapping potential at high altitudes. On Earth this process accelerates ionospheric ions on the order of 0.1 eV to many hundreds of electron volts in energy. *Lund* [this volume] discusses the process in the context of similar magnetic topology in the solar corona, suggesting this may be common in astrophysical plasma systems.

6. SUMMARY

As we have discussed, there are many types of charged particle acceleration processes known or suspected to take place within the cosmos. As we noted in the Introduction section, these acceleration processes range widely in complexity and accessibility to measurement. In all cases, the acceleration of charged particles is fundamental to the evolution of cosmic plasma systems. The basic physical processes have been and are being characterized in laboratories on the ground and in solar system space, where natural processes are directly sampled. Our local experiences guide our interpretation of remote plasma systems that may involve similar acceleration processes, albeit involving higher energies and other significant scaling differences. The brief descriptions and references provided here are intended to familiarize the reader with the diverse venues of particle acceleration, which share common physical mechanisms.

Acknowledgments. The authors thank Ken-Ichi Nishikawa and S.-T. Wu for their thoughtful contributions in reviewing this manuscript.

REFERENCES

Basu, B., Characteristics of electromagnetic Rayleigh-Taylor modes in nighttime equatorial plasma, *J. Geophys. Res.*, **110**, A02303, doi:10.1029/2004JA010659, 2005.

Block, L.P., A double layer, *Rev. Astrophys. Space Sci.*, **55**, 59, 1978.

Brainerd, J. J., A plasma instability theory of gamma-ray burst emission, *The Astrophysical Journal*, **538**, 628-637, 2000.

Burch, J. L., S. A. Fields, W. B. Hanson, R. A. Heelis, R. A. Hoffman, and R. W. Janetzke, Characteristics of auroral electron acceleration regions observed by Atmospheric Explorer-C, *J. Geophys. Res., 81,* 2223, 1976.

Chen, F. F., *Introduction to Plasma Physics and Controlled Fusion, volume 1, Plasma Physics,* 2nd ed., Plenum Press, New York and London, 1984.

Cladis, J. B., Parallel acceleration and transport of ions from polar ionosphere to plasma sheet, *Geophys. Res. Lett., 13,* 893, 1986.

Committee on Solar and Space Physics, *Plasma Physics of the Local Cosmos,* National Research Council, Washington, DC, 2004.

Davidson, Ronald C., *Methods in Nonlinear Plasma Theory,* Academic Press, New York, 1972.

Dungey, J. W., Interplanetary magnetic fields and auroral zones, *Phys. Rev. Lett., 6,* 47, 1961.

Ergun, R. E., L. Anderson, C. W. Carlson, M.V. Goldman, D. S. Main, J. P. McFadden, F. S. Mozer, D. L. Newman, and Y.-J. Su, Parallel electric fields in the upward current region of the aurora: Numerical solutions, *Physics of Plasmas, 9,* 3695, 2002.

Evans, D. S., Precipitating electron fluxes formed by a magnetic field-aligned potential difference, *J. Geophys. Res., 79,* 2853, 1974.

Fermi, E., On the origin of cosmic radiation, *Phys. Rev., 75,* 1169-1174, 1949

Frederiksen, J. T., C. B. Hededal, T. Haugbølle, and Å. Nordlund, Collisionless shocks—Magnetic field generation and particle acceleration, *Proc. From 1st NBSI on Beams and Jets in Gamma Ray Bursts* (held at NBIfAFG/NORDITA, Copenhagen, Denmark, August, 2002, astro-ph/0303360), 2003.

Frederiksen, J. T., C. B. Hededal, T. Haugbølle, and Å. Nordlund, Magnetic field generation in collisionless shocks: Pattern growth and transport, *The Astrophysical Journal, 608,* pt. 2, L13, doi:10.1086/421262, 2004.

Gamov, G., *Biography of Physics,* Harper & Row, New York & Evanston, IL, 1961.

Gary, S, M. Moldwin, M. Thomsen, D. Winske, and D. McComas, Hot proton anisotropies and cool proton temperatures in the outer magnetosphere, *J. Geophys. Res., 99* (A12), 23603-23616, 1994.

Ginzburg, V. L., and V. V. Zhelezniakov, On the possible mechanism of sporadic solar radio emission (radiation in an isotropic plasma). *Sov. Astron.* AJ *2,* 653-666, 1958.

Giovanelli, R. G., Magnetic and electric phenomena in the sun's atmosphere associated with sunspots, *Mon. Notices Roy. Astron. Soc., 107,* 338, 1947.

Gorney, D. J., Y. T. Chiu, and D. R. Croley, Jr., Trapping of ion conics by downward electric fields, *J. Geophys. Res., 90,* 4205, 1985.

Gringauz, K. I., V. V. Bezrukikh, V. D. Ozerov, and R. E. Rybchinsky, The study of the interplanetary ionized gas, high-energy electrons and corpuscular radiation of the Sun, employing three-electrode charged particle traps on the second Soviet space rocket. *Doklady Akademiya Nauk SSSR, 131,* 1302-1304, 1960; translated in *Soviet Physics Doklady, 5,* 361-364, 1960; published again in *Planetary and Space Science, 9,* 103-107, 1962.

Gruzinov, A., Gamma-ray burst phenomenology, shock dynamo, and the first magnetic fields, *The Astrophysical Journal, 563,* pt. 2, L15, doi:10.1086/324223, 2001.

Gurnett, D. A., and A. Bhattacharjee, *Introduction to Plasma Physics, With Space and Laboratory Applications,* Cambridge University Press, London, 2005.

Hones, E. W., Jr., ed., *Magnetic Reconnection in Space and Laboratory Plasmas,* American Geophysical Union Monograph **30**, 1984.

Horwitz, J. L., C. W. Ho, H. D. Scarbro, G. R. Wilson, and T. E. Moore, Centrifugal acceleration of the polar wind, *J. Geophys. Res., 99,* 5051, 1994.

Hoyle, F., *Some Recent Researches in Solar Physics,* Cambridge University Press, London, 1949.

Kirk, J. G., and P. Duffy, Particle acceleration and relativistic shocks, *J. Phys. G: Nucl. Part. Phys., 25,* R163-R194, 1999.

Knight, L., Parallel electric fields, *Planet. Space Sci., 21,* 741, 1973.

Lemaire, J. F., and K. I. Gringauz, *The Earth's Plasmasphere,* Cambridge University Press, London, 1998.

Liang, E., K. Nishimura, J. Li, and S. P. Gary, Particle energization in an expanding magnetized relativistic plasma, *Phys. Rev. Lett., 90,* 85001, 2003.

Liénard, A., Champ electrique et magnetique produit par une charge concentrée en un point et animée d'un mouvement quelconque, *L'Eclairage Elec., 16,* 5, 1898.

Lin, R.P. et al., The Reuven Ramaty High-Energy Solar Spectroscopic Imager (RHESSI), *Sol. Phys., 210,* 3, 2002.

Lin, Y. and L. C. Lee, Structure of reconnection layers in the magnetosphere, *Space Science Reviews, 0[?],* 1, 1993.

Liu, C., J. D. Perez, T. E. Moore, and C. R. Chappell, Low-energy particle signature of substorm depolarization, *Geophys. Res. Lett., 21,* 229, 1994.

Medvedev, M. V., and A. Loeb, 1999, Generation of magnetic fields in the relativistic shock of gamma-ray burst sources, *The Astrophysical Journal, 526,* pt. 1, 697, doi:10.1086/308038, 1999.

Nishikawa, K.-I., Particle acceleration in relativistic jets due to Weibel instability, *Astrophys. J., 595,* 555-563, 2003.

Nishikawa, I.-I., P. Hardee, G. Richardson, R. Preece, H. Sol, and J. J. Fishman, Particle acceleration and magnetic field generation in electron-positron relativistic shocks, *Astrophys. J., 623,* 049702, 2005 (in press)

Persico, E., E. Ferrari, and S. E. Segre, *Principles of Particle Accelerators,* W. A. Benjamin, Inc., New York & Amsterdam, 1968.

Pollock, H. C., The discovery of synchrotron radiation, *Am. J. Phys.* **51,** 278-280, 1982.

Priest, E., and T. Forbes, *Magnetic Reconnection: MHD Theory and Applications,* Cambridge University Press, London, 2000.

Pruet, J., K. Abazajian, G. M. Fuller, New connection between central engine weak physics and the dynamics of gamma-ray burst fireballs, *Phys. Rev. D, 64,* 063002-1, 2001.

Rutherford, E., Collision of α particles with light atoms, IV. An anomalous effect in nitrogen, *Phil. Mag. and J. of Sci., 37,* 581-587, 1919.

Rutherford, E., *Royal Soc. of London, Series A*, 117, 1928.

Sagdeev, R. Z., and A. A. Galeev, *Nonlinear Plasma Theory*, revised and edited by T. M. O'Neil and D. L. Book, W. A. Benjamin, Inc., 1969.

Sagdeev, R. Z., and C. F. Kennel, Collisionless shock waves in interstellar matter, *Scientific American*, p. 40, April 1991.

Sekido, Y., and H. Elliot (eds.), *Early History of Cosmic Ray Studies*, D. Reidel Publishing Co., 17-31, 1985.

Silva, L. O., R. A. Fonseca, J. W. Tonge, J. M. Dawson, W. B. Mori, and M. V. Medvedev, Interpenetrating plasma shells: Near-equipartition magnetic field generation and nonthermal particle acceleration, *The Astrophysical Journal*, **596**, pt. 2, L121, doi:10.1086/379156, 2003,.

Steinbach-Nielsen, H. C., T. J. Hallinan, and E. M. Wescott, Acceleration of barium ions near 8,000 km above an aurora, *J. Geophys. Res.*, **89**, 19788, 1984.

Su, Y.-J., J. L. Horwitz, G. R. Wilson, P. G. Richards, D. G. Brown, and C. W. Ho, Self-consistent simulation of the photoelectron-driven polar wind from 120 km to 9 R_E altitude, *J. Geophys. Res.*, **103**, 2279, 1998.

Wu, X.-Y., J. L. Horwitz, and J.-N. Tu, Dynamic fluid kinetic (DyFK) simulation of auroral ion transport: Synergistic effects of parallel potentials, transverse ion heating, and soft electron precipitation, *J. Geophys. Res.*, **107**, A10, 1283, doi:10.1029/2000JA000190, 2002.

Van Allen, J. A., The geomagnetically trapped corpuscular radiation, *J. Geophys. Res.*, **64**, 1683-1689, 1959.

Wang, Y.-M., and M. Nepveu, A numerical study of the nonlinear Rayleigh–Taylor instability, with application to accreting X-ray sources, *Astron. Astrophys.*, **118**, 267-274, 1983.

D. L. Gallagher, Space Science Branch, National Aeronautics and Space Administration Marshall Space Flight Center, 320 Sparkman Drive, Huntsville, Alabama 35805, USA. (dennis.l.gallagher @nasa.gov)

J. L. Horwitz, Department of Physics, Box 19059, University of Texas in Arlington, Arlington, Texas 76019, USA. (horwitz@ uta.edu)

J. D. Perez, Department of Physics, Allison Laboratory, Room 206, Auburn University, Auburn, Alabama 36849, USA. (perez@ physics. auburn.edu)

J. J. Quenby, Astrophysics Group, Blackett Laboratory, Imperial College of Science, Technology and Medicine, Prince Consort Road, SW7 2BW, London, United Kingdom. (j.quenby@imperial. ac.uk)

Diffusive Shock Acceleration in Astrophysics

J. J. Quenby

Imperial College of Science, Technology and Medicine, London, UK

A. Meli

Max Planck Institut für Radioastronomie, Bonn, Germany

Diffusive shock acceleration is invoked to explain non-thermal particle acceleration at the Bow shock, in the Interplanetary Medium and Solar Flares, in Supernova Remnants, the Interstellar Medium, Active Galactic Nuclei Jets and in large-scale cosmic structures. The basic test particle mechanism to describe the spectrum and acceleration time is detailed for non-relativistic flows, and the non-linear regime where the accelerated particles react back on the flow is mentioned. Extension of the acceleration mechanism to relativistic flows is outlined, with emphasis on Monte Carlo results. The importance of explaining the time constant for achieving the highest observed energies in a given situation is a recurring theme. Current problems range from predicting the high energy output of Gamma-Ray Bursts to establishing the injection mechanism in all situations.

1. INTRODUCTION

Detection of 10^6–10^{20} eV nuclei within the solar system, implying their presence through much of the observable universe, has stimulated the development of the diffusive shock acceleration model whereby particles are repeatedly accelerated in multiple crossings of a shock interface, due to collisions with upstream and downstream magnetic scattering centres. Early theoretical work on compressional acceleration includes that of *Jokipii* (1966a) for the earth's bow shock, but it was the four late-'70s papers of *Krymsky* (1977), *Bell* (1978), *Axford et al.* (1978), and *Blandford and Ostriker* (1978) that established the basic mechanism for diffusive particle acceleration in non-relativistic flows. However, extension of the model to oblique shocks (fields not parallel to the shock normal), relativistic plasma flows, and the incorporation of non-linear effects has not yet resulted in a emergence of a consensus physical picture. Active Galactic Nuclei (AGN) jet flow Lorentz factors of $\Gamma = 10$, estimated Gamma Ray Bursts (GRB) outflows of $\Gamma = 10$–10^3, together with gamma-ray observations from AGN and GRB, suggest the need for extensive study of the diffusive shock mechanism for extremely relativistic plasma flows. Suggestions that GRB could be the source of ultra-high-energy cosmic rays and neutrinos (*Vietri*, 1998; *Waxman*, 1995) are based on the potentially rapid acceleration that may occur in the $\Gamma = 10^3$ flows.

Non-linear theory take into account the back-reaction of the accelerated particles on the shock which can modify the shock to produce a finite width and can feed energy back into the scattering waves, hence modifying the energetic particle spectrum and shock compression ratio. A full theory needs to include injection with account taken of the behaviour of individual ions and the electron plasma gas, non-linear interactions, and the following of individual high-energy particles. Much work is limited to a test particle regime where the accelerated particle "gas" energy is small compared to that in plasma waves. A recent general review in the relativistic flow regime, including the more difficult non-linear problem, has been published by *Kirk and Duffy* (1999).

Particle Acceleration in Astrophysical Plasmas
Geophysical Monograph Series 156
Copyright 2005 by the American Geophysical Union
10.1029/156GM03

2. SCATTERING AND PROPAGATION MODEL

We start by reviewing assumptions in the diffusive shock model, especially in relation to Heliospheric propagation studies but including both relativistic and non-relativistic flows.

2.1. Reference Frames, Guiding Centre Model

Analytical, or Monte Carlo, test particle modelling considers two-dimensional (2-D) motion in a plane containing the magnetic fields and shock normal with the particle's guiding centre following field lines between scattering centres. Frames of reference to be used are the local fluid frames, namely, upstream and downstream frames denoted henceforth as "1" and "2", respectively; the shock normal frame, where all flows are along the normal; and the de Hoffmann–Teller frame, which is only possible for sub-luminal flows where a transformation is made by boosting along the shock front to make $\mathbf{E} = 0$. In general, flows are not along the shock normal, but a transformation into the shock normal frame considerably simplifies the model (e.g. *Begelman and Kirk*, 1990). The model sets out to solve the time-independent Boltzman equation for the distribution function, f,

$$\Gamma (V + v\mu) \frac{\partial f}{\partial x} = \frac{\partial f}{\partial t}\bigg|_c \qquad (1)$$

for particle velocity v at cosine pitch angle μ and with a collision operator $\partial f/\partial t|_c$. Here pitch angle is measured in the local fluid frame and position along the field, x, in the shock rest frame. The neglect of cross-field diffusion will be justified and only elastic scattering in the fluid frame is included, the stochastic non-elastic effect being neglected in relation to the shock acceleration. For large angle scattering, the collision operator is

$$\frac{\partial f}{\partial t}\bigg|_c = \frac{<f>_\mu - f}{T} \qquad (2)$$

where T is the mean time between collision, related to the mean free path by $\lambda = vT$. For small angle scattering, resulting in pitch angle diffusion,

$$\frac{\partial f}{\partial t}\bigg|_c = \frac{\partial}{\partial \mu}\left(D_{\mu\mu}\frac{\partial f}{\partial \mu}\right) \qquad (3)$$

where $D_{\mu\mu}$ is the pitch angle diffusion coefficient.

2.2. Shock Model, Field Directions

Plane fast magnetohydrodynamic (MHD) shocks with a transition from supersonic to subsonic flow will be dealt with in the following, the Alfvén speed, practically the fast MHD mode speed, usually defining the Mach number.

Adopting a planar geometry (Figure 1) with the normal flows, V_1, V_2 in the x–y plane, directed along x positive with the discontinuity at $x = 0$, x negative upstream, we place the fields $\mathbf{B_1}, \mathbf{B_2}$ in the negative x,y directions at shock normal angles ψ_1, ψ_2. $\mathbf{E} = -\mathbf{V} \times \mathbf{B}$ yields only E_z finite and positive. In $c = 1$ units (where c is the speed of light), the fields transform according to

$$\mathbf{E}' = \Gamma(\mathbf{E} + \mathbf{V} \times \mathbf{B}), \quad \mathbf{B}' = \Gamma(\mathbf{B} - \mathbf{V} \times \mathbf{E}) \qquad (4)$$

where Γ is the Lorentz transformation factor and unprimed and primed quantities refer to the shock and a moving frame, respectively. The only electric field component present is in the z direction, $E_z' = \Gamma(VB\sin\psi - V_T B\cos\psi)$ and this is reduced to zero under a frame transformation by adding a velocity, $V_T = V_{DH} = V\tan\psi$ in the negative y direction, where V, ψ are measured either upstream or downstream. This de Hoffmann–Teller frame corresponds to sub-luminal shocks, but if $V_T > c$, we have the super-luminal case. A boost along the negative y axis with a velocity V_T produces $B_x' = \Gamma(-B\cos\psi + V_T VB\sin\psi)$. Hence if in this later, super-luminal case we put $V_T = cot\psi/V$, the magnetic fields are parallel to the shock surface. Applying the general transform between shock and plasma frames to the actual field directions of the shock geometry yields $B_x' = B_x$ and $B_y' = \Gamma(-B\sin\psi + V^2 B\sin\psi)$ and $E_z' = 0$. Also $\tan\psi' = \Gamma^{-1} \tan\psi$. Apart from a shock frame situation where the x component of magnetic field is close to zero, we note from these transformations that for high upstream Γ, the induced electric field

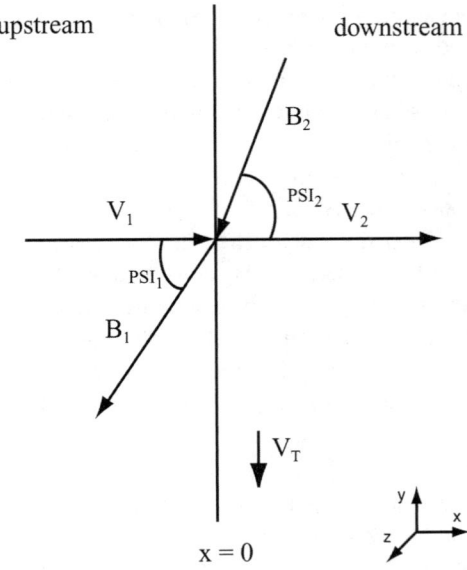

Figure 1. A planar shock geometry.

in turn produces a back-induction to suppress the y component of magnetic field in the upstream frame, thus swinging the magnetic field lines towards the x axis. Also $tan\psi_1' \sim \Gamma_1^{-1} \sim \psi_1'$. If the relativistic plasma is originally accelerated from an isotropic field distribution, extreme flow frame anisotropy followed by relative isotropy in the shock frame can be reasonable.

2.3. Parameter Regimes for Small and Large Angle Scattering

A small angle scattering description of the particle propagation is achieved by allowing random change in pitch angle $\delta\theta$ up to a prescribed limit so that the new pitch angle is given by

$$cos\theta_n = cos\theta \sqrt{1 - sin^2\delta\theta} + sin\delta\theta \sqrt{1 - cos^2\theta cos\phi} \quad (5)$$

at phase ϕ (*Ellison et al.*, 1990). To find the maximum allowed $\delta\theta$, we note that a particle entering the upstream region from downstream will move furthest upstream if it starts parallel to the shock normal. In terms of upstream quantities (we neglect primes in this subsection), **B**, **V**, **n**, υ (magnetic field, plasma velocity, shock normal unit vector and particle velocity, respectively) are all nearly parallel if Γ is large. If the particle moves at $\upsilon \sim c$ in the shock frame, the upstream frame particle speed is $\upsilon = (c - V_1)/(1 - cV_1/c^2)$; the shock in this frame moving in the $-x$ direction with V_1 and $V_1 = c - \delta$, $\upsilon \sim c$. The particle can no longer outrun the shock after a deflection through an angle α with respect to **n** so that $c \times cos\alpha = V_1$. That is, $\alpha = cos^{-1}(V_1/c) = cos^{-1}(1 - 1/2\Gamma_1^2)$ or $\alpha \sim 1/\Gamma_1$. Hence a way to ensure the particle to multi-scatter upstream is that each scatter is small angle with $\delta\theta < 1/\Gamma_1$.

To allow large angle scatter, we need room to change the particle's position by a Larmor radius within the scattering region, $r_{g,1}$ (r_g is the particle gyroradius), smaller than the distance of the particle to the shock. If it takes a time Δt_1 to reach the scatter centre at λ_1, then $\Delta t_1 = \lambda_1/c$, and meanwhile, the shock moves in the upstream frame $(c - \delta)\Delta t_1$. Hence we require $r_{g,1} < (\lambda_1 - (c - \delta)\Delta t_1)$ or $r_{g,1} < \lambda_1/2\Gamma_1^2$ as the condition allowing large angle scatter. To show this criterion is compatible with motion in oblique shocks, note that in one-half a cyclotron period, the angle α has doubled and the shock is catching up the particle with maximum speed possible at the starting pitch angle, even if there is no scattering. If we compare the time to go one-half cycle at angular frequency ω_b, which is π/ω_b, with the time to reach the large angle scatter centre, Δt_1, we find we need $(\pi/N)(\lambda_1/2\Gamma_1^2 > r_{g,1}$ for scatter before "catch-up" where we allow $\lambda_1 = Nr_b$, where ω_b and r_b are the angular frequency and Larmor radius in the ambient field, respectively. Since as we will see shortly, $N \geq 10$, the criteria are compatible.

2.4. Possible Scattering Regimes and Diffusion Coefficients

It is in the interplanetary medium that some of the most comprehensive studies of energetic particle scattering have been performed. Here it has been possible to use the measured propagation characteristics of solar-accelerated particles to compare with theoretical predictions of the diffusion coefficients based on actual interplanetary field data. The theoretical description of the parallel mean free path derives from the pitch angle scattering of particles performing helical motion around a field line due to gyro-resonance. Here a particle receives maximum deflection if in one gyro-period it moves over one field perturbation wavelength. This quasi-linear theory leads to a parallel diffusion coefficient, $K_\parallel$, due to a spectrum of fluctuations perpendicular to the mean field, $P(k) = Ak^{-n}$ of wavenumber k where according to *Jokipii* (1966b)

$$K_\parallel = \frac{\upsilon B^2 r_g^{2-n}}{\pi(2 - n)(4 - n)A} \quad (6)$$

This expression effectively ignores the problem of scattering through 90° pitch angles where there are no resonant waves, but this is overcome by invoking longitudinal waves or waves with large angular changes. It is known to overestimate the scattering by a factor of a few, probably due to the finite amplitude of the actual field fluctuations and their 3-D nature. Detailed numerical simulation of scattering of 100 MeV particles in a 3-D field model based on actual spacecraft magnetometer data suggests that $\lambda_\parallel = Nr_g$, where $N \geq 10$ (*Moussas et al.*, 1992, and references therein). Since interplanetary field fluctuations typically contain up to one-third of the ambient field power, it is unlikely that situations involving non-linear wave generation can produce greater scattering than this value for N under guiding centre propagation conditions. The most extreme scatter situation is likely to be where a particle encounters an intense "blob" of plasma containing a rather homogeneous field of dimension $\sim r_g$ which can contain the particle sufficiently for significant scattering. A further experimental indication from interplanetary propagation studies relevant to the diffusion coefficient is that in the relativistic regime $\lambda_\parallel \propto \gamma$, where γ is the particle Lorentz factor, a dependency is automatically achieved if $\lambda \propto r_g$.

The most favourable site to find high-Γ, large angle scattering is in GRB arising from collapse resulting in a rapidly rotating black-hole torus system (*van Putten*, 2001). Rotation produces a $1/r$ field falloff and typically the torus would be at one Schwarzchild radius for a 20 M/M_o hypernova collapse. Although the equipartition field strength on collapse could be as high as $\sim 10^{18}$ gauss, the extra quantum-mechanical dissipation processes occurring at $\Gamma B > 10^{14}$ gauss will rapidly reduce the field strength (*Lerche and Schramm*, 1977) and therefore 10^{14} gauss is more reasonable for B_o, the torus field. Blobs of plasma emission, rather than a steady wind, arise in some models of the flow from the hypernova collapse (e.g. *Blackman et al.*, 1996). Hence the comoving frame "bullet" fields in the transverse direction, which are the fields most likely to be effective for scattering, are estimated by $B'_s = B_o r_o / r\Gamma$. *Meszaros et al.* (1993) suggest a fireball size in comoving coordinates during the main GRB acceleration phase of $r'_b \sim 10^{14} \rightarrow 10^{16}$ cm and a mean fireball field in comoving coordinates due to internal shock equipartition of $\sim 10^2$ gauss. We put $\lambda'_a = N r'_{g,a}$. An additional constraint is that this mean free path is significantly smaller than the dimensions of the accelerating region. Hence we require $\lambda'_a = min[r'_b/10, N r'_{g,a}]$. Then inserting this expression in the large angle scatter limit requires either $B'_a/B'_s < N/2\Gamma^2$ independent of γ, i.e., particle momentum, or p' (eV).

$$p'(\text{eV}) < \frac{300 B'_s r'_b}{10 \times 2\Gamma^2} \text{ (gauss,cm)} \quad (7)$$

where the particle momentum p' and the field B'_s are measured near the fireball boundary. Taking extreme values of $\Gamma = 10^3$, $N = 40$, the first inequality is marginally satisfied if $r'_b < 10^{14}$ cm, while the second inequality is satisfied by a comoving frame momentum of $< 10^{16}$ eV or a rest frame momentum of $< 10^{19}$ eV. Thus it is possible to consider GRB models with $\Gamma < 10^3$ where large angle scattering is occurring, especially if a larger torus or torus field is allowed.

Gallant and Achterberg (1999) pointed out an important consequence of the limitation to small angle scattering for certain field models. Putting all particle velocity vectors positive in the positive x direction, the ratio of energies crossing upstream to downstream measured in the respective fluid frames is $\Gamma(1 + \beta_r \mu'_{\rightarrow d})$ where β_r is the relative velocity of the two streams as a fraction of c, and $\mu'_{\rightarrow d}$ is the cosine of the upstream pitch angle at the moment of crossing. If the particle is hardly deviated on crossing into the upstream flow before recrossing, $\mu'_{\rightarrow d} \sim -1 + 1/2\Gamma^2$. Meanwhile the final energy in the upstream frame on recrossing with $\mu^*_{\rightarrow u}$ as cosine of the downstream frame pitch angle, measured in that frame, on crossing, is

$$E'_f / E'_i = \Gamma^2 (1 + \beta_r \mu'_{\rightarrow d})(1 - \beta_r \mu^*_{\rightarrow u}) \quad (8)$$

Now for $\beta_r \sim (1 - 1/2\Gamma^2)$, the cosine of the downstream pitch angle only needs to be less than the inverse compression ratio for recrossing to be possible, but for small angle scatter, $1 + \mu'_{\rightarrow d} \sim 1/2\Gamma^2$. Hence the ratio of energies is close to unity on subsequent cycles, whereas on the first complete cycle with arbitrary upstream pitch angle, Γ^2 factors are achieved. This last energy gain factor, first highlighted by *Quenby and Lieu* (1989), is therefore likely to be very dependent on the scattering model and the actual fluctuations around the mean behaviour. *Achterberg et al.* (2001) found a variety of upstream field models, excluding the situation where the field measured in the upstream frame is nearly along the shock normal, which suggest use of a small angle deflection model with the deflection limited to $0.1/\Gamma$. *Protheroe et al.* (2002, personal communication) investigated the relation of a small angle limit to the spatial diffusion.

2.5. Validity of Guiding Centre Approximation

Test particle acceleration theory is usually performed under the guiding centre approximation, with adjustment made for cross-field scattering. There are three conditions for the approximation to be fulfilled relating to the ratio $K_\perp/K_\parallel$, the probability of hitting the shock at the last scattering, including the action of field line wandering and a dependence on the magnitude of cross-field scattering jumps. These will now be considered, using only plasma frame variables.

If $K_\perp/K_\parallel \leq 0.1$, diffusion away from contact with the shock is clearly predominantly along the field line. Using the quasi-linear expression for $K_\parallel$ with $n = 1$ to obtain the usually assumed $\lambda_\parallel \propto r_g$ and with $P(k)$ limited between k_o and k_1 to prevent divergence of fluctuation power, we obtain

$$\lambda_\parallel = \frac{r_g}{\pi} \frac{\ln(k_1/k_o)}{\delta b^2} \quad (9)$$

where δb^2 is the power in transverse fluctuations normalized to the mean field. As is found in the interplanetary medium, field line wandering is at least a major cause of perpendicular diffusion and we adopt the *Jokipii and Parker* (1969) expression,

$$K_\perp = \frac{v}{4B^2} P_\parallel(k \rightarrow 0) = \frac{v}{4k_o} \frac{\delta b_\parallel^2}{\ln(k_1/k_0)} \quad (10)$$

where $P_\parallel(k \rightarrow 0)$ is the power in the parallel field fluctuations at very low wave number. Even if the fluctuations are isotropic and $k_o^{-1} = L_b/2\pi$, where L_b is the scale size of the acceleration region, allowing $\delta b \sim 0.3$ and $k_1/k_0 \sim 10$ and demanding $r_g \leq L_b/10$, we expect $K_\perp/K_\parallel \sim 0.05$. Hence at high r_g, a guiding centre motion seems reasonable. At low r_g, perpendicular motion is entirely determined by the wandering of the field line bundles to which the particle is "attached" and it is the local value of ψ close to the shock which determines the acceleration details.

A second criterion considers where the gyrating particle on average scatters at a distance of one $\lambda_\parallel$ from the shock front. It must not jump away from the original field line bundle far enough so as to intersect the front. This means $\tan\psi \leq \lambda_\parallel/r_g$. The reason for scattering, large angle or multiple small angle, is not crucial. With the above parameters, the requirement becomes $\psi \leq 84°$. However, with these same parameters, the field wander rms is ~18° and hence the guiding centre approximation in strong turbulence is thus probably limited to field-shock normal angles of ≤66° at all r_g.

As explained by *Baring et al.* (1995), the third criterion depends on the jump across the field by each scatter. This jump changes the distance to the shock by $\leq r_g$ and the change can be incorporated in Monte Carlo codes which are basically guiding centre approximations to correct for perpendicular diffusion. A numerical regime where $\lambda_\parallel \geq 10 r_g$ is a situation where cross-field diffusion may be neglected.

3. OUTLINE OF ANALYTICAL TEST PARTICLE MODEL

Physical insight into diffusive shock acceleration is perhaps best gained by outlining the analytical approach to the mechanism, confining the description mainly to non-relativistic flow regimes. Much of the material here is derived from *Drury* (1983), where a more rigorous exposition can be found.

3.1. Adiabatic Invariant Conservation

The first adiabatic invariant, $p_\perp^2/B$, is approximately or rigorously conserved in a variety of models and is therefore extremely useful in analytic and computational approaches. It is best seen for a super-luminal shock in a frame where **B** is rendered parallel to the interface but with finite **E**. This **E** causes the guiding centres of the energetic test particles to drift towards and across the interface with the plasma velocity under the **E** × **B** action with little additional effect on energy, essentially assuming gyrophase independence in the limit, $v \sim c \gg V$. The momentum change on shock crossing can be obtained from Liouville's theorem, which results from the invariance of the 6-D distribution function between side "1" and side "2",

$$2\pi p_{\perp,2} dp_{\perp,2} dp_{\parallel,2} dx_2 dy_2 dz_2 = 2\pi p_{\perp,1} dp_{\perp,1} dp_{\parallel,1} dx_1 dy_1 dz_1$$

Since for the perpendicular shock, $p_{\parallel,1} = p_{\parallel,2}$, then with compression ratio r,

$$r = \frac{dx_1 dy_1 dz_1}{dx_2 dy_2 dz_2} = \frac{p_{\perp,2} dp_{\perp,2}}{p_{\perp,1} dp_{\perp,1}} = \frac{B_2}{B_1} \quad (11)$$

Integrating yields the constancy of the first invariant, $p_{\perp,2}^2/p_{\perp,1}^2 = B_2/B_1$. Hence while all strong to weak field transitions are possible, there is a reflection coefficient, weak to strong. Numerical integration for non-relativistic flow in the **E** = 0 frame shows that the adiabatic approximation gets worse as the inclination of the field-shock normal angle decreases (*Hudson*, 1965), and the maximum acceleration occurs for particles just reflected at the transmission/reflection boundary at high inclinations (*Armstrong et al.*, 1977, *Webb et al*, 1983). These last authors investigate the equivalent field gradient and curvature drift treatment in the shock frame, finding energy loss at the lowest pitch angles. *Lieu and Quenby* (1990) both find factors ~2 breaking of invariant conservation at intermediate inclinations and a "resonance acceleration" when the particle and de Hoffman-transformation speeds coincide.

3.2. Non-Relativistic Flow Linear Theory

To illustrate the derivation of the standard shock spectrum, we follow the *Bell* (1978) microscopic approach. Simplifying to parallel (**B** aligned with normal) shocks, an upstream particle of momentum p_1' in the upstream rest frame approaching the shock has a momentum $p_1'[1 + V_1\cos\theta/v]$ in the shock frame, all small changes of θ being neglected. No reflection is possible. Transforming to the downstream frame gives $p_1'[1 + (V_1 - V_2)\cos\theta/v]$. Applying solid angle and particle current weighting, the average momentum change is

$$2\pi p_1' \int_0^{\pi/2} \frac{(V_1 - V_2)}{v} \cos\theta \sin\theta \cos\theta \, d\theta \quad (12)$$

Using a similar procedure after elastic scatter downstream and return upstream, the mean total momentum change becomes

$$\langle \Delta p_1' \rangle \geq \frac{4}{3} p_1' \frac{(V_1 - V_2)}{v} \quad (13)$$

Let $n (\geq p)$ be the number density of particles reaching p or more where it takes l complete cycles to reach p. Then $\ln(p_1/p_o) = (4l/3v)(V_1 - V_2)$. The probability of downstream escape per cycle is the ratio of convection current to isotropic current for the particles, $nV_2/(nc/4)$. Hence the probability of experiencing l cycles, $Prob_l \propto n$, is given by

$$\ln(Prob_l) = l\ln(1 - \frac{4V_2}{v}) \approx -l\frac{4V_2}{v} \quad (14)$$

Combining these equations yields $dn/dp \propto p^{-s}$, where $s = (r+2)/(r-1)$ and $s = 2$ for strong, dominantly hydrodynamic shocks. *Axford et al.* (1978) and *Bell* (1978a) generalize this result for inclined shocks.

A time constant for acceleration may be obtained using the 1-D diffusion–convection balance equation of Cosmic Ray Modulation theory in the absence of adiabatic deceleration since there is no flow divergence (*Lagage and Cesarsky*, 1981). Upstream, this balance of diffusive flow away from and convection towards the shock is given by $nV_1 = K_1/(\partial n/\partial x)$. Solving and integrating, we find the total number of particles upstream is $\int_{-\infty}^{0} n_o exp(V_1 x/K_1) dx = K_1 n_o/V_1$. Since the flux across the shock is $n_o v/4$, the mean residence time is the ratio of total number to this time, or $4K_1/V_1 v$. Downstream, only that subset of particles which can return to the shock against convection must be counted and they obey $nV_2 = K_2/(\partial n/\partial x)$. Using an argument similar to that for the upstream case, the downstream mean residence time becomes $4K_2/V_2 v$. Adding the residence times and dividing into the total momentum increase per cycle yields a time constant, τ

$$\tau = \frac{3}{V_1 - V_2}\left(\frac{K_1}{V_1} + \frac{K_2}{V_2}\right) \quad (15)$$

Lieu and Quenby (1990) explicitly generalize this last expression for oblique shocks using K_{norm} appropriate to the shock normal direction. Because this equation is clearly an inadequate measure of the acceleration time in the case of relativistic plasma flows, it is worth noting that it may be written as $\tau = [4 \to (8/3)](Nr_g/c)\cos^2\psi$ for $\lambda_\parallel = Nr_g$ with a compression ratio 4 and a downstream reduction in $\lambda_\parallel$ by a factor 4. Hence time to move over r_g is a measure of the acceleration time constant.

4. NON-LINEAR MODIFICATIONS

The two basic non-linear modifications to the test particle acceleration theory for sharp discontinuities will now be outlined in the non-relativistic flow limit.

4.1. Self-Excited Waves

Bell (1978) suggested self-excited waves upstream of the shock could provide the scattering necessary for the acceleration, strong downstream turbulence being expected anyway. If the scattering centres all travel away from the shock with the Alfvén speed, V_a, the 1-D transport equations for the energetic particle distribution function and the wave energy density, $E_w(x,p)$ per unit log bandwidth, resonant with p, for a quasilinear diffusion coefficient, K, related to our previous formula but written in terms of gyroradius, are, with no wave damping,

$$\frac{\partial f}{\partial t} + (V - V_a)\frac{\partial f}{\partial x} = \frac{\partial}{\partial x}\left(K\frac{\partial f}{\partial x}\right)$$
$$K(x,p) = \frac{4}{3\pi} r_g v \frac{B^2/8\pi}{E_w} \quad (16)$$
$$\frac{\partial E_w}{\partial t} + (V - V_w)\frac{\partial E_w}{\partial x} = \frac{4\pi}{3} V p^4 v \frac{\partial f}{\partial x}$$

Physically, the energy source term in the wave transport equation is the VdP work done by a particle pressure per unit log bandwidth, $(4\pi/3)p^4 v\,f(p)$, on the waves. Bell's steady-state solution is

$$f(x,p) = f_1(p) + [f_2(p) - f_1(p)](1 - x/x_o)^{-1} \quad (17)$$
$$x_o = \frac{1}{\pi^2}\frac{B^2/8\pi}{p^4 v[f_2(p) - f_1(p)]}\frac{v}{V}r_g$$

This solution shows a buildup of distribution function upstream, and equivalently of wave amplitude, over a typical distance vr_g/V multiplied by the ratio of magnetic to energetic particle pressure. It is perhaps not surprising that the first factor is equivalent to the e-folding length of the solution of the convection-diffusion modulation equation with $\lambda_\parallel \sim r_g$. In the solar wind, the particle pressure is usually small compared with the magnetic pressure, except at the boundary of the Heliosphere and hence this non-linear modification to shock structure is difficult to find. However, in astrophysical situations where magnetic and particle equipartition is assumed, the necessary upstream wave spectrum should develop, regardless of the initial conditions.

4.2. Self-Consistent Shock Model

The non-linear reaction between the plasma and energetic particle (cosmic ray, usually) fluids modifies the shock structure, in particular the relationship between upstream and downstream states previously described by plasma Rankine-Hugoniot relations and smoothing the structure while allowing a region of sharp or sub-shock transition. In the non-relativistic flow limit, the 1-D steady-state conservation laws of mass, momentum, and energy flows are intuitively as follows (see *Drury*, 1983, for a rigorous derivation):

$$\rho V = A$$
$$AV + P_C + P_G = B$$
$$\frac{1}{2}AV^2 + \frac{\gamma_G}{\gamma_G - 1}VP_G + \frac{\gamma_C}{\gamma_C - 1}VP_C = C + \frac{\overline{K}}{\gamma_C - 1}\frac{\partial P_C}{\partial x}$$
$$P_G V^{\gamma_G} = D$$

Here P_G, P_C are gas and cosmic ray pressures, respectively, with appropriate ratios of specific heats, $\gamma = 5/3$ non-relativistic or $4/3$ relativistic, and the energy density $E_G = P_G/(\gamma_G - 1)$. The right-hand side of the energy flow equation is the cosmic ray diffusive energy flux averaged over the whole spectrum. The gas entropy is assumed to be constant as stated by the last, adiabatic law, except at a sub-shock. Elsewhere, cosmic rays take little energy from the flow but affect the momentum flow. As shown by *Axford et al.* (1982), these equations can be reduced to a single first-order ordinary differential equation in one unknown, which yields the shock structure. *Drury* (1983) gives an illustrative example of a plasma with $P_G = 0$. The above equations under this approximation, introducing the steady upstream and downstream states, V_1, V_2, the elimination of P_C, and integration yield

$$\overline{K}\frac{\partial V}{\partial x} = \frac{\gamma_C + 1}{2}(V - V_1)(V - V_2)$$

$$V(x) = \frac{V_1 + V_2}{2} - \frac{V_1 - V_2}{2}\tanh\left(\frac{(1+\gamma_C)(V_1 - V_2)}{4}\int_{xo}^{x}\frac{dx'}{\overline{K}(x')}\right)$$

Again we have an e-folding length scale $Vx/\overline{K}$ similar to diffusion-convection modulation for the smoothed transition. If the wave energy density reaches equipartition with that of the cosmic rays, a similar solution is expected for the buildup of wave energy.

An upstream to downstream transition may be smooth, following the plasma gas adiabatic law, provided the sonic line where the plasma velocity equals the sound speed is not crossed. Alternatively the flow may first follow the adiabatic law until it reaches a point where a sub-shock takes the flow to downstream conditions, obeying the pure gas shock Rankine–Hugoniot conditions. *Drury and Volk* (1981) show for strong shocks with weak cosmic ray pressure that three downstream solutions are possible for some upstream states. Within the limitations of the above model, they further find that moderate Mach number ($M \sim 3$) shocks with dominant upstream gas pressure yield sub-shocks. However, for $M \sim 10$ shocks, the transition is smooth and most of the energy goes into cosmic rays. The resulting downstream compression ratio of 7 provides additional acceleration, which can more than compensate for the loss of acceleration due to the smoothed transition. Wave growth and damping need to be incorporated into the above conservation laws (e.g. *McKenzie and Volk*, 1982).

4.3. Relativistic Shock Jump Conditions

A full relativistic shock treatment with MHD was provided by *Akhiezer and Polovin* (1959) but here we briefly give the conclusions of the accessible approach of *Kirk and Duffy* (1999). Proper pressure, P, energy density, E, density, ρ, and enthalpy, $W = E + P$, measured in the plasma flow rest frame, are introduced. A Synge equation of state is used which correctly relates W with the temperature of the relativistic plasma particles. In the shock rest frame, the conservation of mass, momentum density, and energy density flow becomes

$$\Gamma_1 \rho_1 V_1 = \Gamma_2 \rho_2 V_2$$
$$\Gamma_1^2 W_1 V_1^2 + P_1 = \Gamma_2^2 W_2 V_2^2 + P_2$$
$$\Gamma_1^2 W_1 V_1 = \Gamma_2^2 W_2 V_2$$

Γ factors arise because of the Lorentz transformation of volume element and of mass to the observing frames. Introducing MHD requires use of a source-free Maxwell equation where $\mathbf{E} = -\mathbf{V}_{1,2} \times \mathbf{B}$, the idealized Ohm's law; in general, displacement current cannot be neglected, $\nabla_\mu(B^\mu V^\nu - V^\mu B^\nu) = 0$. Magnetic energy density, $B^2/4\pi$, and pressure, $B^2/8\pi$, must be included and the Alfvén speed becomes

$$V_A = \left(\frac{B^2/4\pi}{W + B^2/4\pi}\right)^{0.5} \cos\phi \qquad (18)$$

for an angle between field and propagation direction, ϕ. *Kirk and Duffy* (1999) present (Figure 2) numerical solutions based on the above for a strong, oblique, fast-mode shock front ($\phi = 45°$) propagating into a magnetized plasma. The figure shows the proper compression ratio as a function of upstream proper velocity for various Alfvénic Mach numbers, together with an ultra-relativistic hydromagnetic approximation. For a large region of flow conditions, the compression ratio decreases from 4 to 3 as the flow becomes relativistic.

5. DEVELOPMENT OF ACCELERATION MODELS

Some notable steps in the emerging theory of particle acceleration include the oblique shock drift approach, treating the problem as particle motion along the shock front

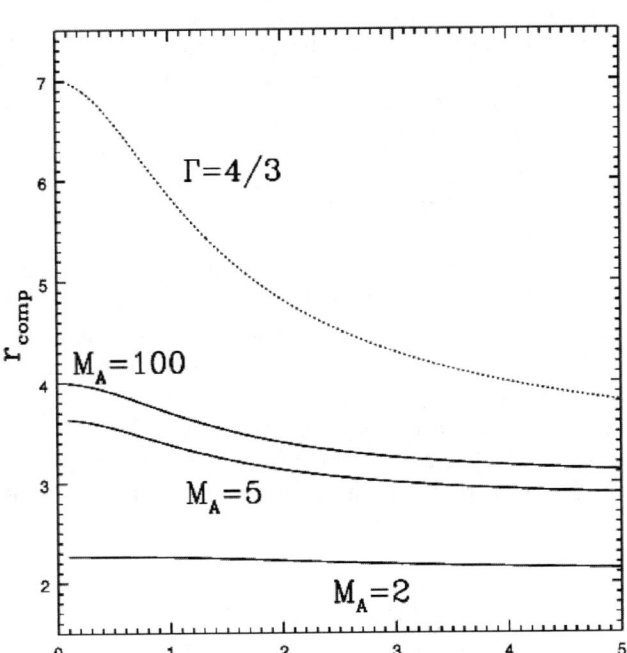

Figure 2. Proper compression ratio as a function of upstream proper velocity for various Alfvénic Mach numbers, together with an ultra-relativistic hydromagnetic approximation. Figure taken from *Kirk and Duffy* (1999); used with permission.

driven by the magnetic field gradient and seeing the shock frame electric field (*Armstrong and Decker*, 1979). This provides results equivalent to the above treatment transforming to the $E = 0$ frame but also implies a shock drift acceleration for one shock passage in the absence of scattering. *Kirk and Schneider* (1987) showed with an analytical solution that as the flow becomes relativistic, the spectrum flattens with respect to the non-relativistic limit. *Schneider and Kirk* (1989) incorporated a prescribed velocity profile for a finite, parallel, shock thickness in the non-relativistic limit. *Heavens and Drury* (1988) emphasized a result of analytical solution of mildly relativistic cases that, for a variety of circumstances, the differential energy spectral index tended to −2.2 as Γ increased.

Because of analytical difficulties with the developing anisotropies as Γ increases, both *Kirk and Heavens* (1989) and *Ostrowski* (1991) used numerical methods to show spectral flattening and increased anisotropy in the angular distribution of particles as the relativistic flow speed increased for inclined shocks. *Quenby and Lieu* (1989) were the first to find a relativistic "speed-up" in the acceleration whereby the acceleration time constant is reduced progressively below its non-relativistic theory value as Γ increases for parallel shocks. These last authors confined their work to the large angle scattering case, but *Ellison et al.* (1990) extended the computation of a speed-up to include a small-angle scattering model; *Lieu et al.* (1994) demonstrated this effect for inclined, sub-luminal shocks under large angle scattering. Even more general was the work of *Bednarz and Ostrowski* (1996) in considering time constant reduction for both large and small angle scattering models in inclined sub- and super-luminal shocks, taking some account of cross-field diffusion. High turbulence appears to reduce the acceleration time to the upstream gyroperiod in some situations. The computations of these last authors and of *Baring* (1999) for the small angle case also yielded a spectral index tending to −2.2 at high Γ. *Meli and Quenby* (2003a,b; and this volume) find and confirm a speed-up effect also for parallel and oblique highly relativistic shocks. Strong hints of structure in the particle spectrum appear in the results of *Baring et al.* (1994) for non-relativistic, inclined shocks. Regarding the appropriate scattering, *Begelman and Kirk* (1990) noted that for low-turbulence super-luminal shocks, the difficulty for energetic particles to return upstream means that only shock compression acceleration is allowed in this situation. However, *Lucek and Bell* (1994) find special trajectories in such a model if there are large-scale variations in the field vector that allow energy increases by a factor of hundreds. *Gallant and Achterberg* (1999) provided a severe constraint in that for several classes of magnetic field configuration in relativistic flows, only small angle scattering can take place upstream. *Gallant and Arons* (1994) provide a rare example of a hybrid model of relativistic particle acceleration, applied to the Crab Nebula, using a fluid approach for an $e^{\pm}$ pair in plasma together with a kinematic treatment of ions to compute the synchrotron radiation spectrum.

6. BRIEF SUMMARY OF APPLICATION OF SHOCK ACCELERATION MODELS IN ASTROPHYSICS

While the applications of the diffusive shock acceleration model cover a vast literature, there is space here only to list some original and recent papers to indicate the extent of the field, together with relevant contributions from authors of chapters in this volume.

6.1. Heliosphere

Scholer et al. (1981) observed ion intensity increases in the solar wind, upstream of the Earth's Bow shock on field lines connected to the shock. *Lee* (1982) related an expected wave growth to the particle acceleration. A review of the possible mechanisms for producing energetic ions in the polar cusps, including Fermi acceleration, is provided by Chen-Wu Chang (this volume). Magnetic connection between the bow shock and cusp is the likely key to the observations. Sugiyama et al. (this volume) apply to the Bow shock a 1-D hybrid computational code, considering ions as particles and electrons as a massless fluid, to investigate the excitation of turbulence. They find that increased α ion content hardens the non-thermal ion spectra.

Co-rotating interaction regions within the solar wind have traditionally provided the strongest in situ evidence for the operation of diffusive shock acceleration, starting from the work of *Barnes and Simpson* (1976) on the increases seen at the forward and reverse shocks of these plasma interactions. *Fisk and Lee* (1980) give the appropriate steady-state solution for the acceleration in an expanding medium. *Sanderson et al.* (1985) find anisotropy evidence for both statistical and shock drift acceleration in such plasma streams, but *Moussas et al.* (1987) are among those who have difficulty in matching the magnetic field data to an acceleration time constant fast enough to work in the interplanetary medium as compared with the available transit times and adiabatic deceleration rate. *Reames* (1999) emphasizes that for coronal mass injection (CME) events at least, acceleration must take place mainly close to the sun and it is the CME, rather than the associated solar flare, which is the basic signature of 100–1000-keV particle acceleration. *Rice et al.* (2003) provide a self-consistent model of CME acceleration of solar particles, incorporating non-linear effects with information on the time dependence of the diffusion coefficient necessary for significant particle flux amplification. However, Giacalone et al. (this volume) emphasise diffusive compressional acceleration at corotating interaction regions. Some particles are clearly accelerated in the impulsive phase of solar flares and shock acceleration can play a role (*Ellison and Ramaty*, 1985). A recent attempt at Monte Carlo modelling the combined effects of reconnection injection of a fast plasma and subsequent shock acceleration is due to *Naito* (2003). Meanwhile, Medvedev (also this volume) reviews the basic stucture of solar wind and relativistic plasma shocks.

The shocked termination of the solar wind at the interface with the interstellar medium appears to provide the source of the anomalous cosmic ray component, accelerating originally neutral interstellar atoms after their ionization in the solar wind (*Pesses et al.*, 1981). This could be an example of a shock with structure modified by cosmic ray pressure (*Donohue and Zank*, 1993).

6.2. Galactic Acceleration

It was the pioneering work of *Ginzburg and Syrovatskii* (e.g. 1964) that established type 2 supernova as the likely source, on energy grounds, of the bulk of galactic cosmic radiation. *Drury et al.* (1989) found up to 30% of the supernova remnant (SNR) energy is actually necessary to replenish the galactic flux, due to escape of particles, adiabatic deceleration in the expanding remnant, and earlier cooling of the SNR in the adiabatic, Sedov phase as the internal energy of the gas is removed. *Berezhko and Volk* (1997) include Alfvénic heating to yield a suitable high-efficiency model. Moreover, while observed SNR radio spectra fit a synchrotron plus inverse Compton source model following electron acceleration, the acceleration cuts off at $10^{13} \rightarrow 10^{14}$ eV (e.g. *Lipari and Morlino*, 2003). Reynolds (this volume) reviews the best evidence that SNRs actually yield electrons, at least up to 100 TeV.

To extend the ability of the Galaxy to acceleration to higher energies, *Jokipii and Morfill* (1985) considered the possibility that a terminal shock of a 500-km/s galactic wind could produce particles up to 10^{19} eV. Work on the production of such a wind, driven by cosmic rays and hot gas from the disk, includes that of *Breitschwerdt et al.* (1991). *Volk and Zirakashvili* (2003) investigate the effects of a spiral shock structure in a non-uniform galactic wind flow and predict extra cosmic ray acceleration up to 10^{17} eV.

6.3. Extra-Galactic Shock Acceleration

Relativistic jets from AGN terminate in shocks as they encounter the ambient medium, while variability in the central "engine" causes internal shocks to propagate down the plasma flow. *Begelmann et al.* (1984) consider non-thermal particle acceleration in the context of a comprehensive review of jets. *Quenby and Lieu* (1989) discuss the ability of $\Gamma = 10$ jet hot spots to produce 10^{19} cosmic rays by shock acceleration. *Dermer and Schlickeiser* (1993) are among the first to describe a homogeneous, internal shock model for blazar emission over a wide range of X-ray to γ-ray energies via radiation from electrons. One example of a description of the jet radio emission and the possible sources of jet production, including particle acceleration, is given in *Particles and Fields in Radio Galaxies*, edited by *Laing and Blundell* (2002).

GRB fireballs are believed to produce plasma flows with Γ up to 1,000, hence the concentration in this volume by Niemiec and Ostrowski and by Meli and Quenby, in providing test particle models of high Γ factor and with various realistic scattering regimes. The fireball plasma may initially be $e^{\pm}$ (*Cavallo and Rees*, 1978), but eventually high-energy electrons and protons are produced, leading to synchrotron inverse, Compton, and even ultra-high-energy cosmic ray and neutrino emission (*Vietri*, 1998; *Waxman*, 1995). The cannonball model of *Dar and Rujula* (2003), based on sudden collapse of part of an accretion disk around a supernova yields a jet of baryonic matter. These last authors question the observational existence of fast-flow, MHD Bow shocks with downstream thermalization, even for SNR velocities $\sim 4 \times 10^4$ km/s, although the fastest solar wind shocks are $\sim 2 \times 10^4$ km/s. Since radio synchrotron emission is seen in the structure attributed to cannon-balls on all jet scales, the essential upstream/downstream magnetic field structure with a sharp transition, necessary for test particle acceleration, is present, whether or not a thermalized downstream plasma develops.

There is now a growing realization of the likely importance of shock acceleration in the intracluster space, where large-scale structure formation, particularly cluster merger, is evident. The observations of excess extreme ultraviolet by *Lieu et al.* (1996) led to the finding of diffuse synchrotron radio halos, where large-scale dynamical behaviour, rather than cooling flows, is evident. N-body modelling of structure evolution confirms the likelihood of particle acceleration, revealed by the non-thermal photon spectra, taking a significant fraction of the thermal energy dissipated in cosmological shock waves (*Ryu et al.*, 2003; *Takizawa et al.*, 2003).

REFERENCES

Achterberg, A., Gallant, Y. A., Kirk, J.G., and Guthmann, A. W., M.N.R.A.S, 328, 393, 2001.
Akhiezer, I.A., and Polovin, R.V., Sov. Phys., JETP, 9, 1316, 1959.
Armstrong, T.P., Chen, G., Sarris, E.T., and Krimigis, S.M., in M.A. Shea et al. (eds), Study of Travelling Interplanetary Phenomena, Reidel, Dordrecht-Holland, 364-389, 1977.
Armstrong, T.P., and Decker, R.B., Particle Acceleration Mechanisms in Astrophysics, AIP Proc. No. 55, 101, 1979.
Axford, W.I., Leer, E., and Skadron, G., 15th ICRC, Plovdiv, Bulgaria, 11, 132, 1978.
Axford, W.I., Leer, E., and McKenzie, J.F., Astron. Astrophys., 111,317,1982.
Baring, M.G., Ellison, D.C., and Jones, F.C., Astrophys. J.S, 90, 574, 1994.
Baring, M.G., Ellison, D.C., and Jones, F.C., Adv. Space Res., 15, No. 8/9, 397, COSPAR, 1995.
Baring, M.G., Proc. 26th ICRC, Salt Lake City. 4, 5, 1999.
Barnes, C.W. and Simpson, J.A., Astrophys. J., 120, L91, 1976.
Bednarz, J., and Ostrowski, M., M.N.R.A.S., 283, 447, 1996.
Begelman, M.C., Blandford, R.D., and Rees, M.J., Rev. Mod. Phys. 56, 255, 1984.
Begelman, M.C., and Kirk, J.G., Astrophys. J.,353, 66, 1990.
Bell, A.R., M.N.R.A.S., 182, 147, 1978.
Berezhko, E.G., and Volk, H.J., Astroparticle Phys.,7, 183, 1997.
Blackman, E.G., Yi, I., and Field, G.B., Astrophys. J., 473, L79, 1996.
Blandford, R.D., and Ostriker, J.P., Astrophys. J. Lett., 221, L29.
Breitschwerdt, D., Volk, H.J., and McKenzie,J.F., Astron. and Astrophys., 245,79, 1991.
Cavallo, G., and Rees, M.J., M.N.R.A.S., 183, 359, 1978.
Dar, A., and De Rujula, A., Astro-ph/0308248, 2003
Dermer, C.D., and Schlickeiser, R., Astrophys. J., 416, 458, 1993.

Donohue, D.J., and Zank, G.P., J.G.R., 98, 19,005, 1993.
Drury, L. O'C., Rep. Prog. Phys., 1, 973, 1983.
Drury, L. O'C., and Volk, H.J., Astrophys. J., 248, 344, 1981.
Drury, L.O'C., Markeiwicz, W.J. and Volk, H.J., Astron. and Astrophys., 225, 179, 1989.
Ellison, D. C. and Ramaty, R., Astrophys. J., 298, 400, 1985.
Ellison, D.C., Jones, F.C., and Reynolds, S.P., Astrophys. J., 360, 702, 1990.
Fisk, L.A., and Lee, M.A., Astrophys. J., 237, 620, 1980..
Gallant, Y.A., and Arons, J., Astrophys. J. 435, 230, 1994.
Gallant, Y. A., and Achterberg, A., M.N.R.A.S. 305, 6, 1999.
Ginzburg, V.L., and Syrovatskii, S.I., The Origin of Cosmic Rays, Pergamon, Oxford, 1964.
Heavens, A.F., and Drury, L. O'C., M.N.R.A.S., 235, 997, 1988.
Hudson, P.D., M.N.R.A.S., 131, 23, 1965.
Jokipii, J.R., Astrophys J., 143, 961, 1966a.
Jokipii, J.R., Astrophys. J., 146, 480, 1966.
Jokipii, J.R., and Parker, E.N., Astrophys. J., 155, 177, 1969.
Jokipii, J.R., and Morfill, G.E., Astrophys. J. Lett. 290, L1, 1985.
Kirk, J.G., and Duffy, P., Astro-ph/9905069, 1999.
Kirk, J.G., and Schneider, P., Astrophys. J.,315,425, 1987.
Kirk, J.G., and Heavens, A.F., M.N.R.A.S., 239, 995, 1989.
Krymsky, G. F., Dokl. Acad. Nauk., SSSR, 234, 1306, 1977.
Lagage, P. O., Cesarsky, C. J., Proc. Int. School and Workshop on Plasma Astrophysics, Varenna, ESA, SP-161, 317, 1981.
Lee, M.A.,J.G.R.,87, 5036, 1982.
Lerche, I., and Schramm, D.N., *Astrophys. J.*, 216, 881, 1977.
Liang, R.A., and Blundell, K.M., Astr. Soc. Pacific Conf. Series, 250, 2002.
Lieu, R., Mittaz, J.P.D., Bowyer, S., Lockman, J.F., Hwang, C.-Y. and Schmitt, J.H.M.M., Astrophys. J. 458, L5, 1996.
Lieu, R., and Quenby, J.J. Astrophys. J., 350, 692, 1990.
Lieu, R., Quenby, J.J., Drolias, B., and Naidu, K. Astrophys. J., 421, 211, 1994.
Lipari, P., and Morlino, G., Proc 28th ICRC, Tsukuba, 2027, 2003.
Lucek, S.G., and Bell, A.R., M.N.R.A.S. 268, 581, 1994.
McKenzie, J. F., and Volk, H.J., Astron. and Astrophys., 116, 191, 1982.
Meli, A., and Quenby, J. J., 2003a, Ast.Part. Phys., 19..649
Meli, A., and Quenby, J. J., 2003b, Ast.Part. Phys., 19, 637
Meszaros, P., Laguna, P., and Rees, M.J., Astrophys. J. 415, 181, 1993.
Moussas, X., Quenby, J.J., and Valdes-Galicia, J.F., Solar Phys., 112, 395, 1987.
Moussas, X., Quenby, J.J., Theodossiou-Ekaterinidi, Z., Valdes-Galicia, J.F., Drilla, A.G., Roulias, D., and Smith, E.J., Solar Phys., 140, 161, 1992.
Naito, T., Proc 28th ICRC, Tsukuba, 3347, 2003.
Ostrowski, M., M.N.R.A.S., 249, 551, 1991.
Pesses, M.E., Jokipii, J.R., and Eichler, D., Astrophys. J., 246, L85, 1981.
Protheroe, R.J., Meli, A., and Donea, A.-C., Sp. Sc. Rev., 107, 369, 2002.
Quenby, J.J., and Lieu, R., Nature, 342, 654, 1989.
Reames, D.V., Space Sci. Rev., 90, 413, 1999.

Rice, W.K.M., Zank, G.P., and Li, G., J. Geophys. Res., 108, A10, 1369, 2003, doi:10.1029/2002JA009756.

Ryu, D., Kang, H., Hallman, E., and Jones, T.W., Proc. 28th ICRC, Tsukuba, 2055, 2003.

Sanderson, T.R., Reinhard, R., Van Ness, P., and Wentzel, K.P., J.G.R., 90, 119, 1985.

Schneider, P., and Kirk, J.G., Astron. and Astrophys., 217, 344, 1989.

Scholer, M., Hoverstadt, D., Iparvitch, F.M., and Bostrom, C.O., J.G.R, 86, 9040, 1981.

Takizawa, M., Naito, T., Ohno, H., and Shibata, S., Proc 28th ICRC, Tsukuba, 2059, 2003.

van Putten, M.H.P.M. Physics Reports, 345, 1, 2001.

Vietri, M. Ap J. Lett., 448, L105, 1988.

Volk, H.J., and Zirakashvili, V.N., Proc 28th ICRC, Tsukuba, 2031, 2003.

Waxman, E., Phys. Rev. Lett., 75, 386, 1995.

Webb, G.M., Axford, W.I., and Terasawa, T., Astrophys. J., 270, 537, 1983.

A. Meli, Max Planck Institut für Radioastronomie, Auf dem Hügel 69, D-53121 Bonn, Germany. (meli@physik.uni-dortmund.de)

J. J. Quenby, Astrophysics Group, Blackett Laboratory, Imperial College of Science, Technology and Medicine, Prince Consort Road, SW7 2BW, London, UK. (j.quenby@imperial.ac.uk)

On Bow Shock Source of Cusp Energetic Ions

Shen-Wu Chang

Center for Space Plasma and Aeronomic Research, University of Alabama in Huntsville, and Space Science Research Center, National Space Science and Technology Center, Huntsville, Alabama

Karlheinz J. Trattner

Lockheed Martin Advanced Technology Center, Palo Alto, California

Energetic ions from ~10 to 150 keV e^{-1} are frequently observed by the Polar spacecraft in the magnetic cusp of the Earth's magnetosphere. Composition measurements indicate that they are of solar origin. Their energy spectra resemble those of energetic diffuse ions observed upstream and downstream from the quasi-parallel bow shock where solar wind ions are accelerated via first order Fermi mechanism. These bow shock accelerated ions show characteristic abundance ratios, composition ratios, and spectral dependence on solar wind parameters. Similar features also appear in the cusp energetic ions. Results from a comparison of energetic ions detected in the cusp and in the quasi-parallel magnetosheath and foreshock suggest that the observed cusp ions are very likely accelerated at the quasi-parallel bow shock. When the cusp is magnetically connected to this shock region, bow shock energetic ions can enter the cusp along the interconnected magnetic flux tubes.

1. INTRODUCTION

Energetic particles of solar origin with energies up to several hundreds of keV e^{-1} are frequently observed by the Polar spacecraft in the magnetospheric cusp and its vicinity. Such events were originally named cusp energetic particle (CEP) events [*Chen et al.*, 1997]. It is well known that shocked solar wind ions can directly enter the cusp after magnetic merging occurs at the magnetopause [e.g., *Reiff et al.*, 1977; *Onsager et al.*, 1995]. Ion compositions measured in the cusp also confirm their solar wind origin [e.g., *Fuselier et al.*, 1998]. They gain energy up to ~1 keV while crossing the magnetopause current layer [e.g., *Cowley*, 1982], resulting in a typical energy of a few keVs. Thus, additional acceleration is required to account for the observed energies of CEPs.

Three sources for CEPs have been discussed and at least partially investigated in the literature: Fermi acceleration at the quasi-parallel ($Q_{\parallel}$) bow shock [*Chang et al.*, 1998; *Trattner et al.*, 1999, 2001], magnetospheric energetic particles [*Delcourt and Sauvaud*, 1999; *Blake*, 1999] and local acceleration [*Chen et al.*, 1998]. The flux, spectral slope, and composition of the lower-energy component (≤150 keV e^{-1}) of CEPs agree very well with those of energetic ions observed at the $Q_{\parallel}$ bow shock, where the angle (θ_{Bn}) between the upstream magnetic field line and shock normal is less than 45°. These results suggest bow shock acceleration. Energetic ions from the duskside plasma sheet can also drift into the cusp and constitute the higher-energy component of CEPs (>150 keV e^{-1}).

In contrast, local acceleration in the cusp does not require any direct connection between the cusp and other regions of energetic ions such as the $Q_{\parallel}$ shock and plasma sheet. Because an acceleration mechanism has yet to be identified, it is not clear how keV ions are locally accelerated to hundreds of keVs in the cusp. In this paper, we review current evidence of bow shock accelerated particles as a source for

---

Particle Acceleration in Astrophysical Plasmas
Geophysical Monograph Series 156
Copyright 2005 by the American Geophysical Union
10.1029/156GM04

CEPs. Following a brief introduction of bow shock acceleration in section 2, evidence for shock acceleration is subsequently presented as we move along in discussing CEP observations in sections 3 and 4. Section 5 provides discussions and summary. Our findings suggest that CEPs with energies from ~10 to 150 keV e^{-1} most likely come from the $Q_\parallel$ bow shock. The magnetosphere itself may account for CEPs above 150 keV e^{-1}.

2. BOW SHOCK ACCELERATION

Energetic ions up to ~150 keV e^{-1} have often been reported upstream and downstream from the Earth's bow shock [e.g., *Ipavich et al.*, 1981; *Gosling et al.*, 1989a; *Fuselier et al.*, 1995, and references therein]. Beginning with the earliest observations, it was realized that the connection of the interplanetary magnetic field (IMF) to the bow shock is a necessary condition for the presence of energetic particles on IMF field lines [e.g., *Asbridge et al.*, 1968; *Scudder et al.*, 1973; *Lin et al.*, 1974]. The fact, that the enhanced plasma and magnetic turbulence and their associated energetic ions are similar in both the upstream and downstream regions, led to the suggestion of a common bow shock source region [e.g., *West and Buck*, 1976; *Asbridge et al.*, 1978]. The occurrence rate of enhanced energetic ion events on either side of the bow shock increases as θ_{Bn} decreases [e.g., *Bonifazi and Moreno*, 1981; *Mitchell and Roelof*, 1983]. Furthermore, upstream energetic ion flux also grows with decreasing θ_{Bn} [*Kudela et al.*, 2002].

It is generally accepted that solar wind ions are accelerated up to ~150 keV e^{-1} via first-order Fermi mechanism in the turbulent regions upstream and downstream from the $Q_\parallel$ bow shock by scattering back and forth across the shock [e.g., *Lee*, 1982; *Ellison*, 1985]. These ions undergo substantial energy gain and spatial diffusion in the foreshock region and they are known as bow shock diffuse ions. This acceleration mechanism is more efficient for ions with a higher mass to charge ratio [e.g., *Ipavich et al.*, 1981; *Trattner and Scholer*, 1994]. Energy spectra of bow shock diffuse ions for different species have a similar exponential shape and they are ordered in energy per charge with similar *e*-folding energy over a range from 20 to 150 keV e^{-1}. Their spectral characteristics mainly depend on the solar wind density and velocity [e.g., *Trattner et al.*, 1994]. Higher solar wind velocity yields harder spectrum.

In addition to Fermi acceleration, magnetospheric energetic ions can also escape the magnetosphere to populate the upstream region [e.g., *Sarris et al.*, 1976; *Krimigis et al.*, 1978]. Fluxes from shock acceleration dominate in the diffuse population and the spectra of escaping ions extends to still higher energies (hundreds of keV e^{-1}).

3. CEP PROPERTIES

High charge-state energetic, heavy ions presumably of solar origin are commonly observed in the CEP events. Furthermore, typical CEP fluxes and $Q_\parallel$ bow shock energetic ion fluxes generally agree [*Chang et al.*, 1998, 2003a; *Trattner et al.*, 1999]. These results suggest a bow shock source for CEPs.

Trattner et al. [2001] compiled 53 CEP events from the Polar database and compared CEP characteristics with bow shock and magnetosheath observations and predictions from shock acceleration theory. During these CEP events, magnetosheath-like plasmas, i.e., intense keV ions and 100 eV electrons, observed with Toroidal Imaging Mass-Angle Spectrograph (TIMAS) [*Shelley et al.*, 1995] and Hydra [*Scudder et al.*, 1995] instruments coexisted with energetic ions observed with Magnetospheric Ion Composition Sensor (MICS) of the CAMMICE instrument (see *Wilken et al.* [1992] for a similar detector flown on the CRRES satellite). The presence of these thermal plasmas excludes energetic particle events in regions neighboring the cusp, such as the low-latitude boundary layer (LLBL) and boundary plasma sheet. Including events in these regions would contaminate CEP statistics.

3.1. September 22, 1996, Event

Figure 1 depicts energy spectra of H^+ and He^{2+} in energy per charge format for one of the 53 Polar CEP events, occurring on September 22, 1996. Both species contain an energetic component from ~10 to 200 keV e^{-1} and a thermal component below 10 keV e^{-1} (defined below as the transmitted and reflected populations). Spectral shapes for these two energetic components are very similar. They are ordered by energy per charge, not by total energy. Furthermore, a spectral break appears at about 20 keV e^{-1} for both ions as marked by the dashed line. In addition, energetic high charge-state oxygen ions are also observed during this event. Although their fluxes are close to the one-count level, their energy spectrum shows signatures in agreement with the other two species (see Figure 4 of *Trattner et al.* [2001]). All of the CEP characteristics described above are also common features of energetic ions at the $Q_\parallel$ shock [e.g., *Ipavich et al.*, 1981; *Lee*, 1982]. For comparison, a typical energetic He^{2+} spectrum observed by the AMPTE/IRM spacecraft downstream from the $Q_\parallel$ bow shock is also shown in Figure 1 [*Ellison et al.*, 1990]. CEP fluxes and $Q_\parallel$ magnetosheath energetic ion fluxes especially at higher energies agree quite well, implying that bow shock energetic ions can be a sufficient source of CEPs.

To compare CEP properties with $Q_\parallel$ magnetosheath ener-

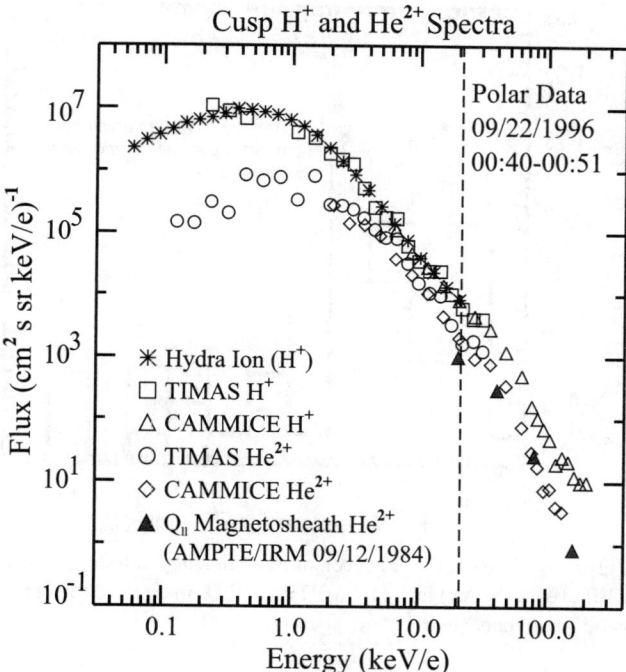

Figure 1. Cusp H^+ and He^{2+} spectra observed by Polar during the September 22, 1996, CEP event (after Figures 2 and 3 of *Trattner et al.* [2001]). A typical $Q_\parallel$ magnetosheath energetic He^{2+} spectrum observed by AMPTE/IRM is shown for reference [*Ellison et al.*, 1990].

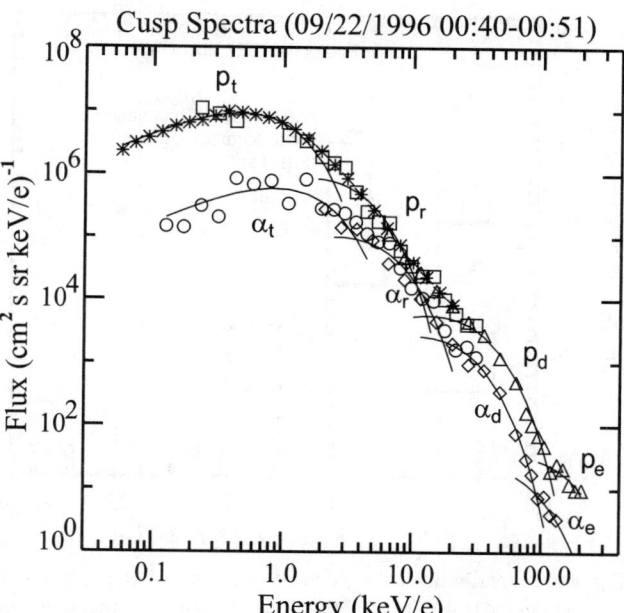

Figure 2. A repeat of Figure 1 with four Maxwellian fits for each spectrum.

getic ion observations and predictions from shock acceleration theory, CEP density and temperature are estimated by fitting cusp ion phase space density f with Maxwellian distributions. A least-squares fitting for $ln(f)$ and ion energy E is performed in each of the three or four contiguous energy intervals selected for each event. These intervals are chosen to minimize the total χ^2 in the fitting. Figure 2 presents the Maxwellian fits for the previously discussed H^+ and He^{2+} spectra. Each spectrum is divided into four different parts. From low to high energy, they represent the transmitted, reflected, diffuse, and energetic components of cusp ions. The first two components are the shocked solar wind ions that are directly transmitted (transmitted), or initially reflected at the shock but later transmitted through the shock (reflected). The reflected component is ~20% of the transmitted component for protons, consistent with early reports [e.g., *Gosling et al.*, 1989b; *Trattner and Scholer*, 1994]. The diffuse and energetic protons with energies above 10 keV e^{-1} are ~0.3% and 0.004% of the shocked solar wind protons, respectively. Because of the spectral break at ~150 keV e^{-1} indicated in Figure 2, different sources are most likely required for these two components.

The density ratio (0.3%) and temperature (1.4×10^8 K) estimated for the diffuse component of cusp protons are comparable to the average values for $Q_\parallel$ magnetosheath energetic ions [*Fuselier*, 1994]. The energetic component which represents ~1% of the total energetic protons may come from another region, such as the magnetosphere, as demonstrated in several numerical simulations [e.g. *Blake*, 1999; *Delcourt and Sauvaud*, 1999]. As for the cusp He^{2+} ions, the diffuse component derived from Figure 2 is ~1.3% of the shocked solar wind He^{2+} and has a temperature of 2.5×10^8 K that are also comparable to the density ratio and temperature of $Q_\parallel$ magnetosheath energetic He^{2+} ions [*Fuselier*, 1994]. These results further suggest that the diffuse component (~10–150 keV e^{-1}) of CEPs for this event originates from the $Q_\parallel$ bow shock.

3.2. Density Ratio and Average Temperature

Applying the same Maxwellian fitting technique to all the 53 CEP events yields several average quantities for the diffuse component of CEPs: H^+ density ratio of 0.26 ± 0.2%, H^+ temperature of $1.2 \pm 0.5 \times 10^8$ K, and He^{2+} temperature of $2.8 \pm 1.4 \times 10^8$ K. These quantities are consistent with the corresponding values (0.34%, $1.6 \pm 0.5 \times 10^8$ K, $2.5 \pm 0.9 \times 10^8$ K) for the 41 AMPTE/CCE $Q_\parallel$ magnetosheath energetic ion events reported by *Fuselier* [1994].

Although average values of the H^+ density ratio in the cusp and $Q_\parallel$ magnetosheath are comparable, distributions of the ratio in these regions are somewhat different as shown in Figure 3. Generally speaking, the occurrence rate for CEP events increases as the density ratio decreases. In contrast,

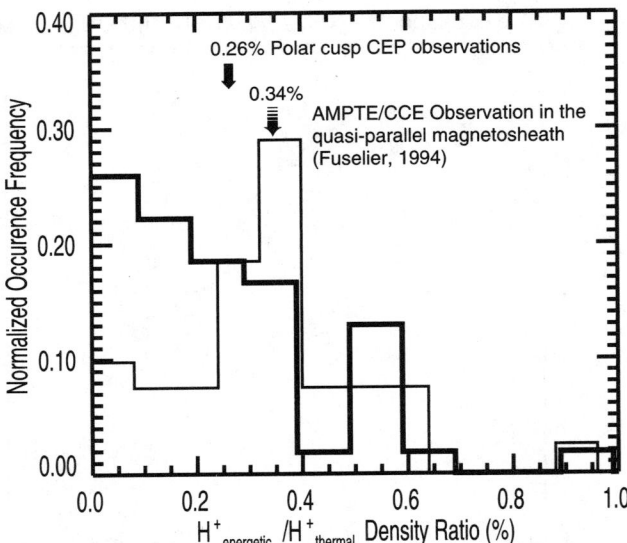

Figure 3. Distributions of ion density ratio for 53 Polar CEP events (heavy) and 41 AMPTE/CCE $Q_\parallel$ magnetosheath energetic ion events (light).

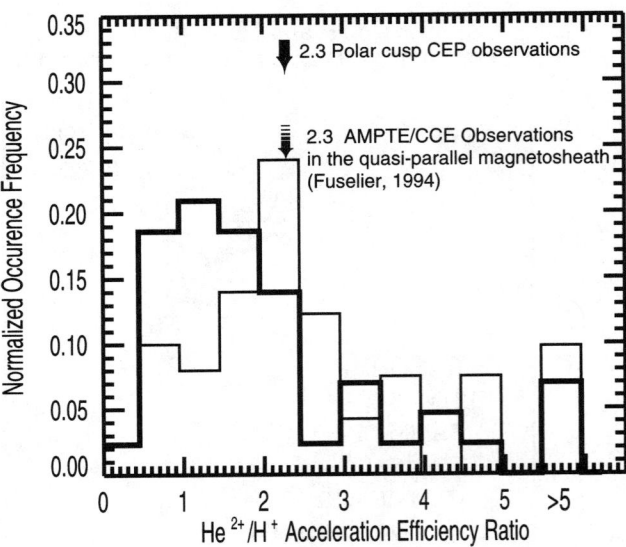

Figure 4. Distributions of acceleration efficiency ratio for 53 Polar CEP events (heavy) and 41 AMPTE/CCE $Q_\parallel$ magnetosheath energetic ion events (light).

the distribution for the $Q_\parallel$ magnetosheath events peaks around the average value. Possible reasons for the difference include an energy-dependent process for ions entering the cusp, and cusp ions with different energies originating from different parts of the magnetosheath in which ion distribution shows spatial structure. These subjects require further investigation and will not be discussed here.

3.3. Acceleration Efficiency

Monte Carlo simulations of shock acceleration have yielded an acceleration efficiency ratio of He^{2+} to H^+ of 2 [e.g., *Ellison*, 1985; *Ellison et al.*, 1990]. The ratio is defined as (see *Möbius et al.* [1987] and *Fuselier* [1994])

$$\frac{\eta_2}{\eta_1} = \frac{P_2 m_1 n_{sw1}}{P_1 m_2 n_{sw2}}$$

where subscripts 1 and 2 stand for H^+ and He^{2+}, respectively, P is the pressure from energetic ions, m is the ion mass, and n_{sw} is the density for each species in the solar wind. Similarly, the acceleration efficiency ratios for the 53 Polar CEP events are calculated using parameters derived from the Maxwellian fits described earlier. Their average value is 2.3 ± 1.9, which is consistent with the value obtained from the simulations for upstream diffuse ions. It is also consistent with the values for the 41 AMPTE/CCE $Q_\parallel$ magnetosheath events, which range between 2.0 and 2.5 [*Fuselier*, 1994].

Despite the agreement between the average efficiency ratio for CEP events and the average value for magnetosheath events, differences are found in their distributions as shown in Figure 4. There are a considerable number of CEP events with efficient ratio below 2 whereas the distribution for magnetosheath events peaks around 2. Nevertheless, reducing magnetosheath energetic He^{2+} to H^+ density ratio by 40% would move the peak distribution of magnetosheath events to the value of CEP events. This result implies a mass-dependent process for energetic ions entering the cusp. This conjecture is supported by a previous report in the literature that there is a 40% reduction in the He^{2+} to H^+ density ratio for solar wind ions entering the LLBL [*Fuselier et al.*, 1997].

3.4. Solar Wind Dependence

Bow shock diffuse ions have an exponential energy (E/q) spectrum with a spectral slope depending on the solar wind velocity [e.g., *Ipavich et al.*, 1981; *Trattner et al.*, 1994]. Similar relation should appear in CEP events if CEPs originate from the $Q_\parallel$ bow shock. Figure 5 presents spectral slope of CEP energy spectrum for the 53 CEP events and the associated solar wind velocity. There is a correlation between these two quantities with a correlation coefficient of 0.68 as shown in Figure 5. In general, spectral slope increases with solar wind velocity. Such a relation has been reported in the upstream diffuse ion events. Figure 6 depicts spectral slope and the component of solar wind velocity along the magnetic field direction for 382 AMPTE/IRM upstream events [*Trattner et al.*, 1994]. The dependence of spectral slope on solar wind velocity in this case is very similar to

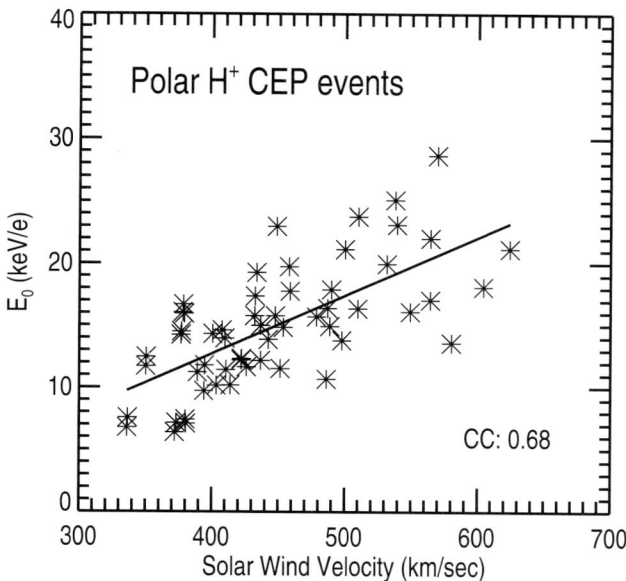

Figure 5. Spectral slope of CEP spectrum and associated solar wind velocity for 53 Polar CEP events.

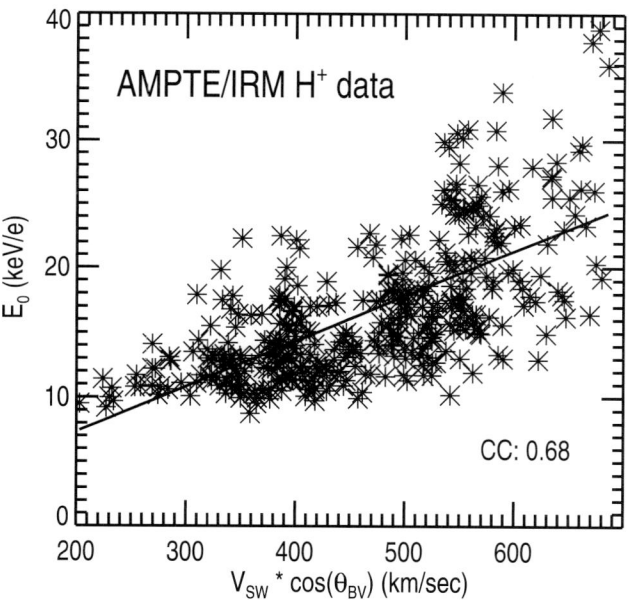

Figure 6. Exponential spectral slope of energetic ion spectrum and associated solar wind velocity component along magnetic field for 382 AMPTE/IRM upstream diffuse ion events [*Trattner et al.*, 1994].

the result for CEP events. This finding further supports the suggestion that CEPs and bow shock diffuse ions are of the same origin.

4. MAPPING

Statistically, CEPs demonstrate characteristics of bow shock diffuse ions as discussed in the previous section. To make a strong case of cusp energetic ions originating from the $Q_\parallel$ bow shock, mapping for these two regions has to be addressed. Figure 7 illustrates an example of magnetic topology for CEP events. Under this condition, $Q_\parallel$ bow shock and cusp are connected by magnetic field lines so that bow shock energetic ions can simply enter the cusp along interconnected magnetic flux tubes [*Chang et al.*, 1998]. Two cases of magnetic conjunction involving Interball Tail and Polar are discussed below, one for bow shock/magnetosheath conjunction and the other for bow shock/cusp conjunction.

4.1. May 4, 1998, Event

During the May 4, 1998, magnetic storm, Interball was upstream from the $Q_\parallel$ bow shock and Polar was just upstream from the cusp in the magnetosheath. Both regions are magnetically connected when very intense energetic ion fluxes (much greater than typical CEP fluxes) are detected simultaneously by these two spacecraft. From 0840 to 1200 UT, magnetosheath energetic ion fluxes of solar origin observed by two energetic particle instruments onboard Polar, CAMMICE and Comprehensive Energetic Particle and Pitch Angle Distribution Experiment (CEPPAD) [*Blake et al.*, 1995], show variations as large as 2 orders of magnitude. These temporal variations are caused by abrupt changes in the bow shock magnetic geometry as shown in the IMF data from the Wind spacecraft. A cross-correlation analysis was performed for this interval. Plate 1 presents correlation coefficient for magnetosheath ion flux (Polar) and IMF cone angle (Wind), the angle between upstream magnetic field and the Sun–Earth line which serves as a proxy for θ_{Bn}. For ions of solar origin with energies above ~41 keV e^{-1}, their fluxes are strongly anticorrelated with the IMF cone angle at a time lag of 37 min (Plate 1a through 1d). In contrast, literally no correlation is found between He^+ flux, presumably of ionospheric origin, and cone angle (Plate 1e). This result suggests that the above anticorrelation being consistent with the Fermi process at the $Q_\parallel$ shock is not incidental.

Moreover, energetic ions of solar origin show spectral characteristics of bow shock diffuse ions, i.e., their energy spectra are ordered by E/q, not total energy, with a common e-folding energy (see Figure 6 of *Chang et al.* [2000]). This e-folding energy for this case is about 41 keV e^{-1}, which is also the threshold energy for anticorrelation described above. These results imply that the observed magnetosheath energetic ions with energies above this threshold energy are accelerated at the $Q_\parallel$ bow shock.

In addition to the above evidence, the observed magnetosheath energetic ions of solar origin demonstrate a strong flow toward the magnetopause, consistent with an upstream

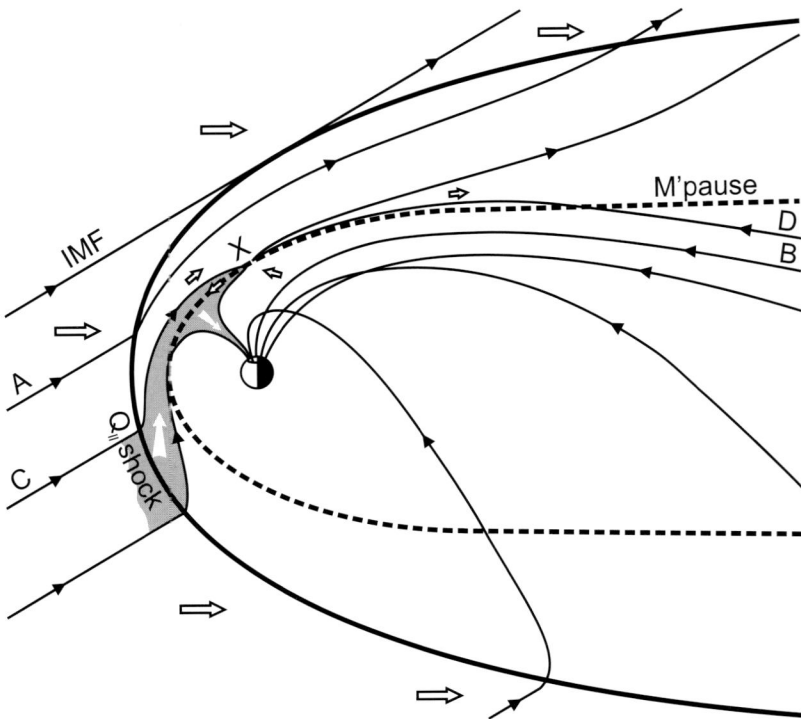

Figure 7. Magnetic field topology under northward and antisunward IMF condition showing the connection between the $Q_\parallel$ bow shock and cusp.

source for these ions. Furthermore, bow shock energetic ion fluxes derived from foreshock energetic ion observations from the energetic particle instrument DOK-2 [*Kudela et al.*, 1995] onboard Interball Tail match very well with the magnetosheath energetic ion fluxes from Polar as shown in Figure 8. This result further supports a bow shock source for the observed magnetosheath ions and it is consistent with the bow shock model for CEPs. Detailed discussions for this event can be found in *Chang et al.* [2000, 2001, 2003b] that demonstrate $Q_\parallel$ shock acceleration and transport of energetic ions from bow shock to the magnetosheath just upstream from the cusp. These studies achieve an important step in addressing the mapping problem.

4.2. April 16, 1999, Event

Next step for the mapping problem is to examine bow shock diffuse ions and CEPs that are simultaneously detected by two or more spacecraft. Such an example has been found in the Polar and Interball database [*Chang et al.*, 2003c]. Interball was upstream from the $Q_\parallel$ bow shock and Polar was in the cusp during a CEP event occurring on April 16, 1999. Both regions are interconnected by magnetic field lines as concluded from upstream magnetic field observations from Interball, ACE, and Wind. Bow shock energetic ion fluxes observed by Interball DOK-2 are anticorrelated with θ_{Bn} as expected for the Fermi process. Following the procedure to scale foreshock diffuse ion fluxes for *e*-folding distances as described in *Chang et al.* [2001], energetic ion fluxes at $Q_\parallel$ bow shock can be derived from Interball observations. Figure 9 presents the resulting bow shock ion spectrum and the observed CEP spectrum from Polar. Both energetic ion spectra are almost indistinguishable. This result strongly supports the bow shock model for CEPs.

4.3. LLBL Event

The magnetic configuration presented in Figure 7 is not the only way for bow shock energetic ions to access the cusp. Furthermore, bow shock energetic ions initially entering the cusp on open magnetic field lines may become trapped on closed field lines as magnetic field topology evolves. In a recent study, *Fuselier et al.* [2002] addressed such a mapping problem. These authors utilized global MHD simulations to demonstrate the whole process and interpret complex observations from Polar in the LLBL. As shown in their Figure 8, magnetic reconnection first occurs in one hemisphere to yield open magnetic field lines. They convect toward the magnetopause in the opposite hemisphere and then participate in another magnetic reconnection to become

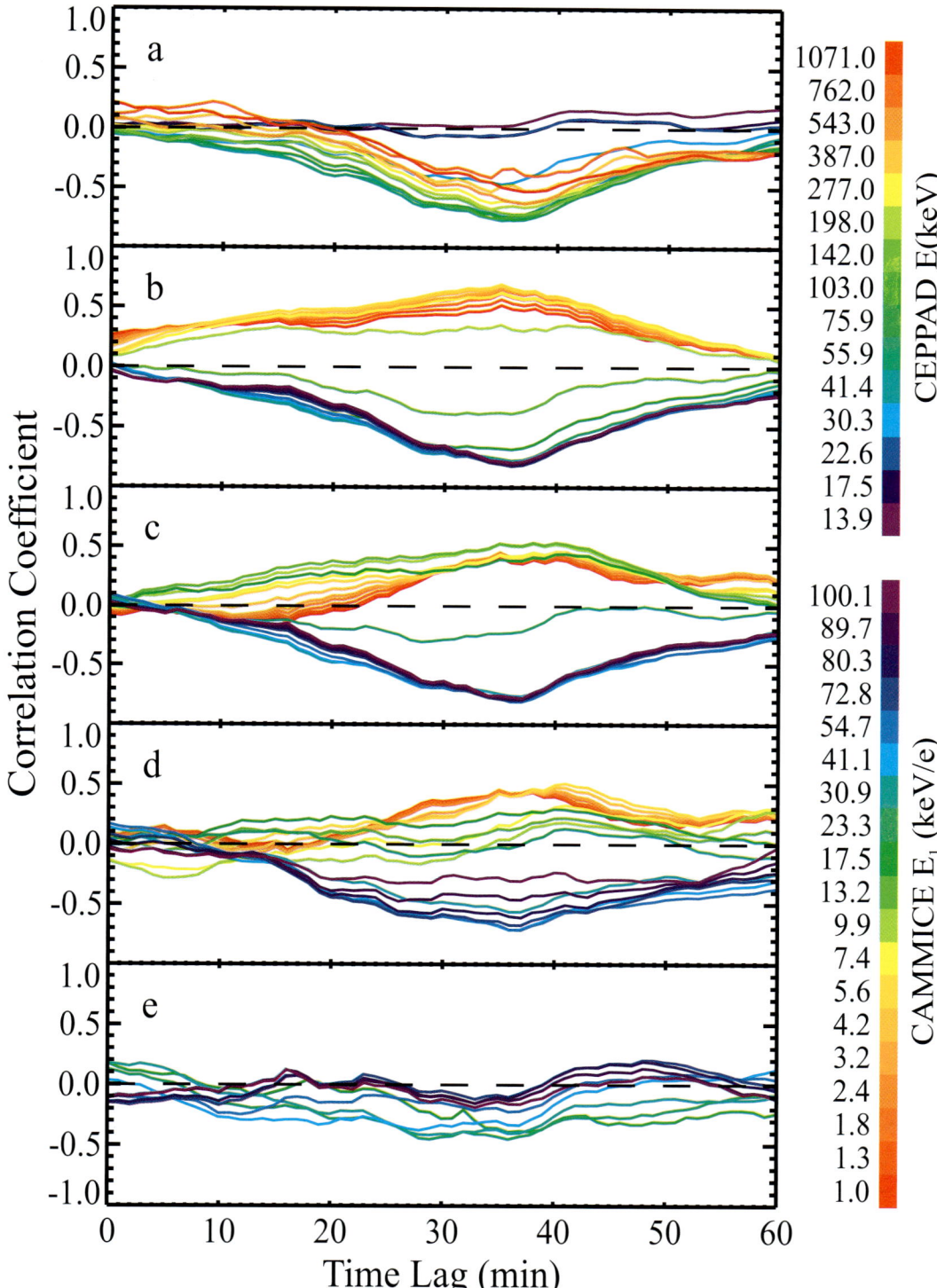

Plate 1. Correlation coefficient for ion flux and IMF cone angle at assumed time delays for (a) CEPPAD ion, (b) CAMMICE ion (assuming H^+), (c) CAMMICE He^{2+}, (d) CAMMICE $O^{>2+}$, and (e) CAMMICE He^+ at energies labeled by the color bars.

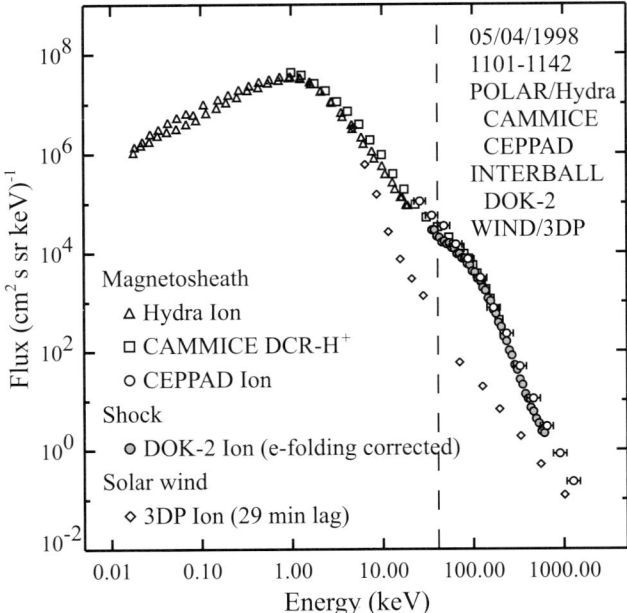

Figure 8. Magnetosheath ion spectrum observed by Polar and $Q_{\|}$ energetic ion spectrum estimated from Interball DOK-2 observations on May 4, 1998.

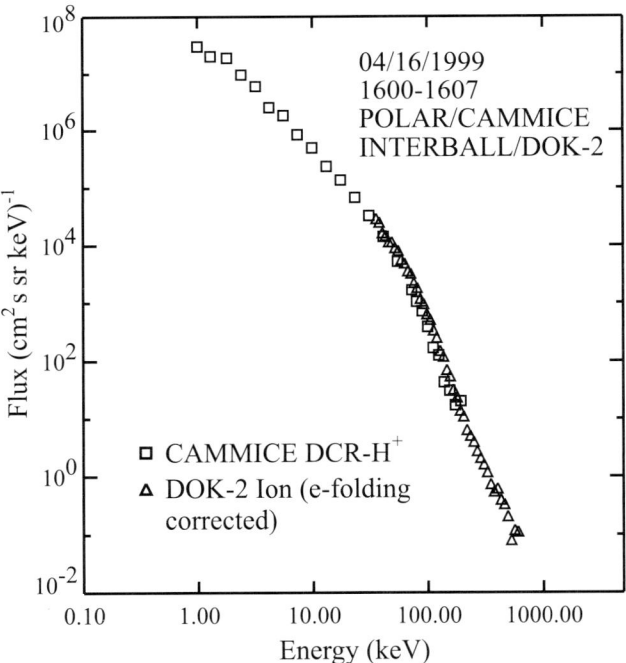

Figure 9. CEP spectrum (square) from Polar and $Q_{\|}$ energetic ion spectrum (triangle) estimated from Interball observations for April 16, 1999, CEP event.

closed field lines that thread the LLBL. As a result, magnetosheath plasmas and bow shock energetic ions entering magnetosphere on open field lines earlier are trapped on closed field lines. Energetic ions at the duskside magnetosphere with much higher energies also drift into this region of closed field lines. Therefore, three kinds of ion populations occupying different energies simultaneously exist on the same LLBL field lines, as observed by Polar.

5. DISCUSSION AND SUMMARY

As shown above, CEPs with energies from ~10 to 150 keV e^{-1} demonstrate characteristics of bow shock diffuse ions. Their fluxes, energy distribution, spectral slope, density, temperature, and acceleration efficiency agree very well with those at the $Q_{\|}$ bow shock and values predicted by shock acceleration theory and simulations. From several event studies, we conclude that bow shock diffuse ions propagate downstream into the magnetosheath and then the cusp, and on some occasion enter LLBL. These ions contribute ~99% of CEPs. The remaining 1% is the most energetic component that presumably comes from the magnetosphere. Thus, bow shock acceleration appears to be the dominant source of CEPs.

Upstream magnetic turbulence and waves that are associated with bow shock diffuse ions can propagate downstream into the magnetosheath and cusp [e.g. *Omidi et al.*, 1999]. Energetic particles may also produce waves at the magnetopause as they stream toward the current layer. Therefore, a good correlation between energetic ion fluxes and waves in the cusp is not unexpected in the bow shock model and does not necessarily imply local acceleration. After bow shock energetic ions enter the cusp, some wave-particle interactions may occur and modify the particle distribution. This subject requires further investigation.

Acknowledgments. The authors are indebted to many people from the Polar, Interball, Wind, and ACE instrument teams for their valuable data and advice provided for CEP research. The work at UAH was supported in part by NASA grant NAG5-12008 and NSF grant ATM-0242427. The work at Lockheed Martin was supported by NASA contracts NAS5-30302 and NAG5-9867.

REFERENCES

Asbridge, J. R., S. J. Bame, and J. B. Strong, Outward flow of protons from the Earth's bow shock, *J. Geophys. Res., 73*, 5777, 1968.

Asbridge, J. R., S. J. Bame, J. T. Gosling, G. Paschmann, and N. Sckopke, Energetic plasma ions within Earth's magnetosheath, *Geophys. Res. Lett., 5*, 953, 1978.

Blake, J. B., Comment on "Cusp: A new acceleration region of the magnetosphere" by J. Chen et al., *Czech. J. Phys., 49*, 675, 1999.

Blake, J. B., et al., CEPPAD comprehensive energetic particle and pitch angle distribution experiment on POLAR, *Space Sci. Rev., 71,* 531, 1995.

Bonifazi, C., and G. Moreno, Reflected and diffuse ions backstreaming from the Earth's bow shock, *J. Geophys. Res., 86,* 4405, 1981.

Chang, S.-W., J. D. Scudder, K. Kudela, H. E. Spence, J. F. Fennell, R. P. Lepping, R. P. Lin, and C. T. Russell, Mev magnetosheath ions energized at the bow shock, *J. Geophys. Res., 106,* 19,101, 2001.

Chang, S.-W., J. D. Scudder, K. Kudela, H. E. Spence, J. F. Fennell, R. P. Lepping, R. P. Lin, and C. T. Russell, Reply to comment on "MeV magnetosheath ions energized at the bow shock" by J. Chen, T. A. Fritz, and R. B. Sheldon, *J. Geophys. Res., 108,* 1312, doi:10.1029/2002JA009724, 2003b.

Chang, S.-W., K. Kudela, J. F. Fennell, and H. E. Spence, Energetic ion observations at the cusp and bow shock, *Eos Trans. AGU, 84* (46), 1149, Fall Meet. Suppl., Abstract SM32D-04, 2003c.

Chang, S.-W., et al., Cusp energetic ions: A bow shock source, *Geophys. Res. Lett., 25,* 3729, 1998.

Chang, S.-W., et al., Energetic magnetosheath ions connected to the Earth's bow shock: Possible source of cusp energetic ions, *J. Geophys. Res., 105,* 5471, 2000.

Chang, S.-W., et al., Correction to "Cusp energetic ions: A bow shock source" by S.-W. Chang, J. D. Scudder, S. A. Fuselier, J. F. Fennell, K. J. Trattner, J. S. Pickett, H. E. Spence, J. D. Menietti, W. K. Peterson, R. P. Lepping, and R. Friedel, *Geophys. Res. Lett., 30,* 1149, doi:10.1029/2002GL016613, 2003a.

Chen, J., T. A. Fritz, R. B. Sheldon, H. E. Spence, W. N. Spjeldvik, J. F. Fennell, and S. Livi, A new, temporarily confined population in the polar cap during the august 27, 1996 geomagnetic field distortion period, *Geophys. Res. Lett., 24,* 1447, 1997.

Chen, J., et al., Cusp energetic particle events: Implications for a major acceleration region of the magnetosphere, *J. Geophys. Res., 103,* 69, 1998.

Cowley, S. W. H., The causes of convection in the Earth's magnetosphere: A review of developments during the IMS, *Rev. Geophys., 20,* 531, 1982.

Delcourt, D. C., and J.-A. Sauvaud, Populating of the cusp and boundary layers by energetic (hundreds of keV) equatorial particles, *J. Geophys. Res., 104,* 22,635, 1999.

Ellison, D. C., Shock acceleration of diffuse ions at the Earth's bow shock: Acceleration efficiency and A/Z enhancement, *J. Geophys. Res., 90,* 29, 1985.

Ellison, D. C., E. Möbius, and G. Paschmann, Particle injection and acceleration at Earth's bow shock: Comparison of upstream and downstream events, *Astrophys. J., 352,* 376, 1990.

Fuselier, S. A., A comparison of energetic ions in the plasma depletion layer and the quasi-parallel magnetosheath, *J. Geophys. Res., 99,* 5855, 1994.

Fuselier, S. A., M. F. Thomsen, F. M. Ipavich, and W. K. H. Schmidt, Suprathermal He^{2+} in the Earth's foreshock region, *J. Geophys. Res., 100,* 17,107, 1995.

Fuselier, S. A., E. G. Shelley, and O. W. Lennartsson, Solar wind composition changes across the Earth's magnetopause, *J. Geophys. Res., 102,* 275, 1997.

Fuselier, S. A., E. G. Shelley, W. K. Peterson, and O. W. Lennartsson, Solar wind He^{2+} and He^+ distributions in the cusp for southward IMF, in *Polar Cap Boundary Phenomena,* edited by J. Moen, A. Egeland, and M. Lockwood, NATO ASI Ser., Ser. C, 509, p. 63, 1998.

Fuselier, S. A., J. Berchem, K. J. Trattner, and R. Friedel, Tracing ions in the cusp and low-latitude boundary layer using multi-spacecraft observations and a global MHD simulation, *J. Geophys. Res., 107,* 1226, doi:10.1029/2001JA000130, 2002.

Gosling, J. T., M. F. Thomsen, S. J. Bame, and C. T. Russell, On the source of diffuse, suprathermal ions observed in the vicinity of the Earth's bow shock, *J. Geophys. Res., 94,* 3555, 1989a.

Gosling, J. T., M. F. Thomsen, S. J. Bame, and C. T. Russell, Ion reflection and downstream thermalization at the quasi parallel bow shock, *J. Geophys. Res., 94,* 10,027, 1989b.

Ipavich, F. M., A. B. Galvin, G. Gloeckler, M. Scholer, and D. Hovestadt, A statistical survey of ions observed upstream of the Earth's bow shock: Energy spectra, composition, and spatial variation, *J. Geophys. Res., 86,* 4337, 1981.

Krimigis, S. M., D. Venkatesan, J. C. Barichello, and E. T. Sarris, Simultaneous measurements of energetic protons and electrons in the distant magnetosheath, magnetotail, and upstream in the solar wind, *Geophys. Res. Lett., 5,* 961, 1978.

Kudela, K., M. Slivka, J. Rojko, and V. N. Lutsenko, The apparatus DOK-2 (project Interball): Output data structure and modes of operation, *UEF 01-95,* p. 18, Inst. of Exp. Phys., Kosice, Slovakia, March, 1995.

Kudela, K., V. N. Lutsenko, D. G. Sibeck, and M. Slivka, Energetic ions upstream of the Earth's bow shock: Interball-1 survey, *Adv. Space Res., 30,* 2731 2736, 2002.

Lee, M. A., Coupled hydromagnetic wave excitation and ion acceleration upstream of the Earth's bow shock, *J. Geophys. Res., 87,* 5063, 1982.

Lin, R. P., C.-I. Meng, and K. A. Anderson, 30- to 100-kev protons upstream from the Earth's bow shock, *J. Geophys. Res., 79,* 489, 1974.

Mitchell, D. G., and E. C. Roelof, Dependence of 50-kev upstream ion events at IMP 7&8 upon magnetic field bow shock geometry, *J. Geophys. Res., 88,* 5623, 1983.

Möbius, E., M. Scholer, N. Sckopke, H. Lühr, G. Paschmann, and D. Hovestadt, The distribution function of diffuse ions and the magnetic field power spectrum upstream of Earth's bow shock, *Geophys. Res. Lett., 14,* 681, 1987.

Omidi, N., D. Krauss-Varban, S.-W. Chang, and J. D. Scudder, Connection between cusp energetic ions and the bow shock, *Eos Trans. AGU, 80* (46), Fall Meet. Suppl., F900, 1999.

Onsager, T. G., S.-W. Chang, J. D. Perez, J. B. Austin, and L. X. Janoo, Low-altitude observations and modeling of quasi-steady magnetopause reconnection, *J. Geophys. Res., 100,* 11,831, 1995.

Reiff, P. H., T. W. Hill, and J. L. Burch, Solar wind injection at the dayside magnetospheric cusp, *J. Geophys. Res., 82,* 479, 1977.

Sarris, E. T., S. M. Krimigis, and T. P. Armstrong, Observations of magnetospheric bursts of high-energy protons and electrons at ~35 R_E with IMP 7, *J. Geophys. Res., 81,* 2341, 1976.

Scudder, J. D., D. L. Lind, and K. W. Ogilvie, Electron observations in the solar wind and magnetosheath, *J. Geophys. Res., 78,* 6535, 1973.

Scudder, J. D., et al., Hydra: A 3-dimensional electron and ion hot plasma instrument for the Polar spacecraft of the GGS mission, *Space Sci. Rev., 71,* 459, 1995.

Shelley, E. G., et al., The toroidal imaging mass-angle spectrograph (TIMAS) for the POLAR mission, *Space Sci. Rev., 71,* 497, 1995.

Trattner, K. J., and M. Scholer, Diffuse minor ions upstream of simulated quasi-parallel shocks, *J. Geophys. Res., 99,* 6637, 1994.

Trattner, K. J., E. Möbius, M. Scholer, B. Klecker, M. Hilchenbach, and H. Lühr, Statistical analysis of diffuse ion events upstream of the Earth's bow shock, *J. Geophys. Res., 99,* 13,389, 1994.

Trattner, K. J., S. A. Fuselier, W. K. Peterson, and S.-W. Chang, Comment on "Correlation of cusp MeV helium with turbulent ULF power spectra and its implications" by J. Chen and T. A. Fritz, *Geophys. Res. Lett., 26,* 1361, 1999.

Trattner, K. J., S. A. Fuselier, W. K. Peterson, S.-W. Chang, R. Friedel, and M. R. Aellig, Origins of energetic ions in the cusp, *J. Geophys. Res., 106,* 5967, 2001.

West, H. I., Jr., and R. M. Buck, Observations of >100-kev protons in the Earth's magnetosheath, *J. Geophys. Res., 81,* 569, 1976.

Wilken, B., W. Weiss, D. Hall, M. Grande, F. Soraas, and J. F. Fennell, Magnetospheric ion composition spectrometer onboard the CRRES spacecraft, *J. Spacecr. Rockets, 29,* 585, 1992.

S.-W. Chang, National Space Science and Technology Center, SD50, 320 Sparkman Drive, Huntsville, Alabama 35805. (changs@cspar.uah.edu)

K. J. Trattner, Lockheed Martin Advanced Technology Center, 3251 Hanover Street, B255, ADCS, Palo Alto, California 94304. (trattner@mail.spasci.com)

Generation of Diamagnetic Cavities at the Bow Shock by Ion Kinetic Effects

Yu Lin

Physics Department, Auburn University, Auburn, Alabama

Two-dimensional global-scale hybrid simulations are carried out to study the generation of diamagnetic cavities at the bow shock due to accelerated ion beams in (1) intrinsic process of the shock and (2) the interaction of the shock with an interplanetary tangential discontinuity (TD). In the simulation of the bow shock alone, strong electromagnetic waves occur in the foreshock regions of the quasi-parallel shock due to the interaction between thin, hot backstreaming/reflected ions and the cold, dense solar wind plasma. Diamagnetic cavities form in these foreshock regions. The cavities are crater-like and have a width ~1–2R_E, with a low-density and low-magnetic field center bounded by a rim of high density and high magnetic field. The craters convect downstream with the solar wind. When the interplanetary magnetic field (IMF) lies nearly parallel to the solar wind flow, the craters develop into field-aligned structures in upstream and downstream. On the other hand, as a TD carrying IMF direction change interacts with the bow shock, a strong hot flow anomaly (HFA) can be generated by coherent, gyrating multiple beams of reflected ions that are accelerated toward the TD by motional electric field. In the center of the HFA, a low plasma density and a low magnetic field strength are present. The plasma is heated to a nearly isotropic temperature, different from that in the foreshock cavities generated by process (1). The hot HFA cavity and the magnetosheath behind it may bulge into the solar wind by several R_E. These transient cavities generated by the bow shock itself and a simple variation in the IMF direction can produce pressure pulses on the magnetopause.

1. INTRODUCTION

Ion kinetic physics plays an important role in the structure of the Earth's bow shock and its interaction with incoming waves and discontinuities from the solar wind. The presence of accelerated ion beams of reflected or backstreaming ions can lead to the generation of transient plasma structures at the bow shock through two important methods. (1) Pressure pulse and the associated plasma and magnetic field variations can be generated in the foreshock regions by intrinsic kinetic processes in the bow shock [e.g., *Sonnerup*, 1969; *Fairfield et al.*, 1990], and (2) significant transient waves can also be produced by accelerated ion beams as interplanetary discontinuities interacts with the bow shock [*Thomsen et al.*, 1986; *Hubert and Harvey*, 2000]. These transient structures can be transmitted through the magnetosheath and interact with the magnetosphere. In addition to the significant dayside transients caused by the arrival of interplanetary discontinuities [*Sibeck et al.*, 1999], observations have also shown that foreshock fluctuations generated in the foreshock of quasi-parallel shocks can be linked to ULF pulsations [*Engebretson et al.*, 1991], and traveling convection vortices [*Sibeck et al.*, 2003] in the magnetosphere and ionosphere.

In the quasi-parallel shocks, the solar wind ions can be reflected into upstream nearly along field lines, and backstreaming ions from the downstream can escape a large distance in the foreshock region. Near the shock transition, the instabilities due to interactions between ion beams contribute to the energy dissipation required by the shock [*Tanaka et al.*, 1983; *Sckopke et al.*, 1983]. It is believed that first-order Fermi acceleration of ions between upstream waves and the shock leads to ion heating and energetic ions at quasi-parallel shocks [e.g., *Lee*, 1982], yielding the diffuse ion distribution. Pressure pulses and the associated magnetic field variations can be generated in the foreshock regions by processes within the bow shock [*Fairfield et al.*, 1990], with positively correlated variations in magnetic field and plasma density.

Observations indicate that diamagnetic cavities are frequently present in foreshock regions of quasi-parallel shocks [*Wibberenz et al.*, 1985; *Sibeck et al.*, 2001, 2002; *Fairfield et al.*, 1990]. Pressures associated with the energetic ions are found to depress the foreshock magnetic field strength and plasma densities [*Sibeck et al.*, 2001]. These cavities occur preferentially during high-speed solar wind streams. In a further study based on Wind observations, *Sibeck et al.* [2002] have compared the observed foreshock cavities with a hybrid simulation by *Thomas and Brecht* [1988] for the interaction of a spatially limited beam of backstreaming ions with the incoming solar wind, and attributed the cavities to the diamagnetic effects of ions Fermi accelerated within the foreshock.

On the other hand, another type of stronger but less common diamagnetic cavities, the so-called hot flow anomalies (HFAs), is also observed near the Earth's bow shock [*Schwartz*, 1995]. These events usually contain a hot subsonic plasma embedded in the upstream wind, showing a low-density core flanked by narrow regions of high density as well as strong field. The isotropic distribution of heated ions in the HFA is quite different from that in the foreshock cavities mentioned above. The majority of the events are found to be associated with a gross rotation in the interplanetary magnetic field (IMF) [*Paschmann et al.*, 1988; *Thomsen et al.*, 1993]. *Schwartz et al.* [2000] find that most of HFAs have quasi-perpendicular shock conditions on at least one side of the HFA, although several HFAs correspond to quasi-parallel conditions. Their study also shows that HFAs tend to be associated with a "directional" tangential discontinuity (TD), which carries a change in the IMF orientation and has nearly no change in the density and field strength. In addition, HFAs are produced from TDs whose normals make a large cone angle with the sunward direction. Recent satellite observations show that the magnetopause can bulge into the upstream solar wind when an HFA affects the magnetosphere [*Sibeck et al.*, 1999], indicating a strong response of the magnetosphere to a simple variation in the IMF direction.

Hybrid simulations, which include full ion kinetics, have been carried out for the nonlinear structure and ion heating of the bow shock [e.g., *Leroy and Winske*, 1983; *Goodrich*, 1985; *Thomas et al.*, 1990; *Scholer and Terasawa*, 1990]. These simulations have revealed important kinetic structure and ion heating of collisionless fast Shocks. The interaction between backstreaming ion beams and the ambient plasmas is found to lead to the generation of large-amplitude electromagnetic waves in the upstream and downstream regions. A simulation addressing the interaction between the reflected ions and a current sheet [*Burgess and Schwartz*, 1988] shows that counterstreaming ions can lead to an ion thermalization in the low field region. Hybrid simulation of the interaction between a fast shock and a tangential discontinuity [*Thomas et al.*, 1991] indicates that a hot diamagnetic cavity can be generated under the "proper" type of electric field that focuses the reflected ions into the current sheet, consistent with observations of the HFA [*Schwartz*, 1995].

Recently, we carried out two-dimensional (2-D) global-scale hybrid simulations of the bow shock-magnetosheath-magnetosphere system for the structure of the curved bow shock [*Lin*, 2003] and its interaction with interplanetary TDs [*Lin*, 2002]. It was reported that significant multi-dimensional plasma and magnetic field structures can be generated at the curved bow shock, and transmitted through the magnetosheath. In this paper, a review is given for our recent investigations of the transient kinetic processes at the bow shock. The simulation results of the diamagnetic cavities generated in the foreshock of quasi-parallel shock are presented in section 2, and the results on the generation of HFAs during the interaction of the bow shock with a TD are shown in section 3. A summary is given in section 4.

2. GENERATION OF FORESHOCK CAVITIES IN THE QUASI-PARALLEL SHOCK

In this section, we present the simulation results of case 1 for the bow shock alone without external perturbations from the solar wind, with a focus on the quasi-parallel shock. The hybrid code used in this study was developed by *Swift* [1996]. The detailed description of the simulation model can be found in *Lin* [2002, 2003]. The simulation includes the entire system of the bow shock, magnetosheath, and magnetosphere on the dayside. The 2-D ($\partial/\partial y = 0$) simulation is carried out in the noon–midnight meridian plane, in which the x axis points along the Sun–Earth line and toward the Sun, and the z axis points from south to north. The Earth is located at the origin $(x,z) = (0,0)$. A polar coordinate system is used in the simulation; this system consists of the radial distance r in the xz plane and the polar angle $\theta = \tan^{-1}(x/z)$. The domain lies within the region with $5R_E < r < 30R_E$ and $0° < \theta < 180°$.

The ion gyrofrequency Ω_0 in the solar wind is chosen to be 0.6 s^{-1}. The solar wind ion inertial length λ_0 is chosen to be 0.04–0.08 R_E. Nonuniform grid sizes Δr are used in the r direction, with the grid size $\Delta r \sim 0.5\lambda_0$ for $r > 13R_E$. Physical quantities are normalized as follows. The magnetic field B is normalized to the IMF B_0, the ion number density N is normalized to the solar wind number density N_0, the flow velocity V is normalized to the solar wind Alfven speed V_{A0}, and the time t is normalized to Ω_0^{-1}. The thermal pressure is expressed in units of $P_{00} = B_0^2/\mu_0$, and the temperature is in units of P_{00}/N_0. The bow shock, magnetosheath, and magnetopause form by the interaction between the supersonic solar wind and the geomagnetic field.

In case 1, the IMF cone angle is equal to 0°, with the IMF $B_{x0} = B_0$ and $B_{y0} = B_{z0} = 0$. This choice of IMF direction results in the presence of quasi-parallel and parallel shocks in a large portion of the dayside domain. The solar wind ion plasma beta is chosen as $\beta_0 = 0.5$. The solar wind flow speed is assumed to be $V_0 = 5V_{A0}$, corresponding to a Mach number $M_A = 5$.

Figure 1 shows the simulation results at $t = 48\Omega_0^{-1}$. Shown from the top are magnetic field lines plotted against the field magnitude on the logarithmic scale, contours of the B_y component of magnetic field, contour plot of the ion number density N, flow vectors plotted on top of the flow speed, and contours of the ion parallel temperature $T_\parallel$. Only part of the domain ($r = 26R_E$) is shown. At this time, the bow shock is present at a standoff distance of $\sim 15R_E$ in front of the dayside magnetosphere, as seen from the enhanced ion density and magnetic field and the diverted flow vectors across the shock front. Some crater-like structures, as indicated in Figure 1 by the cross marks, are present in the foreshock region upstream of the parallel and quasi-parallel shocks. These craters exist in the region where backstreaming ions, often associated with a diffuse ion distribution in the foreshock, are present. In these structures, a low density and low magnetic field diamagnetic cavity is surrounded by high density and high magnetic field strength. The spatial variations in B and N are in phase and well correlated through these craters. For the three craters shown in Figure 1, at the rim of the craters B and N increase by a factor of 1.4 on average relative to those in the ambient solar wind. In the center of these craters they are $\sim 75\%$ of the average values in the solar wind. The flow speed slightly decreases, by $\sim 13\%$, in the diamagnetic cavities.

Large-amplitude electromagnetic waves are present in the vicinity of the shock due to the interaction between the background plasma and the counterstreaming ion beams. In addition, particle scattering between the upstream and downstream waves provides a mechanism for Fermi acceleration of the reflected/backstreaming solar wind ions, which

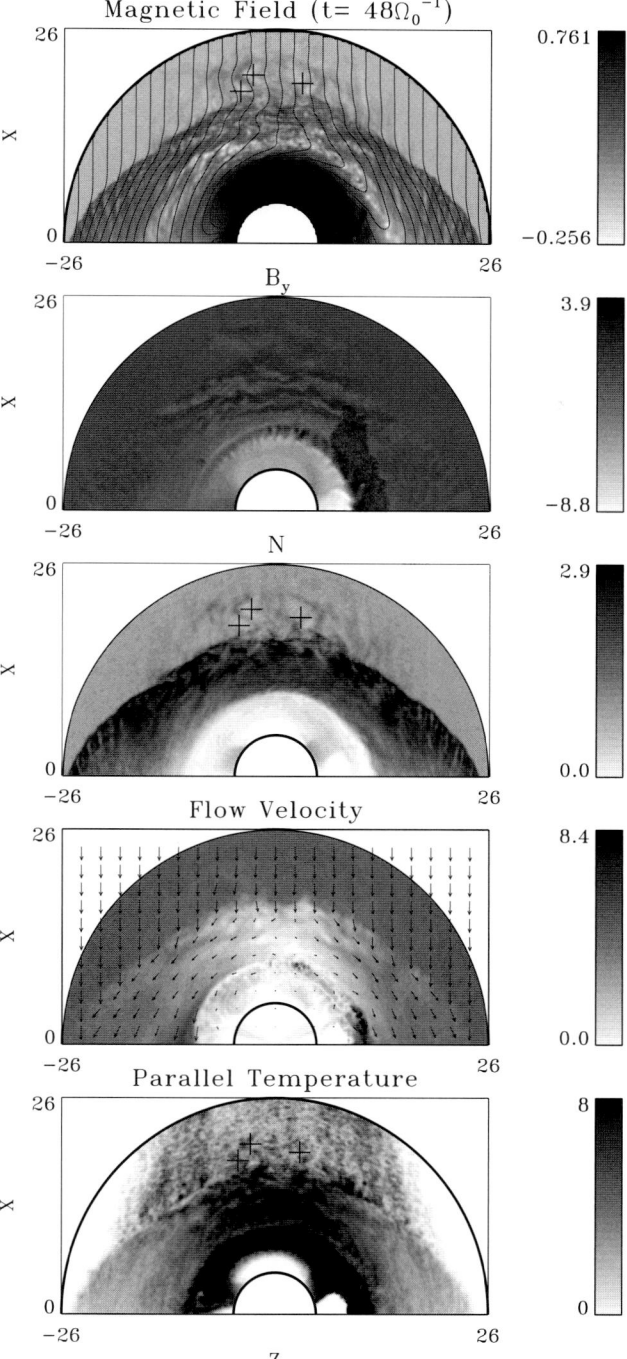

Figure 1. (From top) Magnetic field lines on top of the contours of field magnitude B on the logarithmic scale, contours of the B_y component of magnetic field, contours of ion number density N, flow vectors on top of the contours of flow speed, and contours of the ion parallel temperature at $t = 48\Omega_0^{-1}$ in case 1 for the foreshock structure of the bow shock. The cross marks indicate some of the diamagnetic cavities in the foreshock.

evolve to diffuse foreshock ion distribution in phase space. Overall, the sunward, hot ion beam can escape a large distance following field lines in the upstream. Enhanced parallel temperatures are present upstream of the quasi-parallel shock, as seen in Figure 1. The presence of the craters in the foreshock is associated with the interaction between the hot, tenuous, sunward ion beam and the dense, incoming solar wind plasma, with relative speeds between the two population >> local Alfven speed. In the foreshock region where the craters are seen, the density of the hot beam ions is ~6–20% of that of the background solar wind. In early times with $t < 60\Omega_0^{-1}$, fast compressional magnetosonic/ whistler waves, with right-hand polarized magnetic fields in the solar wind plasma frame of reference, are found in the regions upstream and downstream from the parallel and quasi-parallel shocks. They propagate with wave vectors k nearly parallel to the magnetic field B. Meanwhile, wave structures with in-phase fluctuations in B and N are also present with oblique k's in the 2-D plane, resulting in the crater-like structures, which saturate at sizes of $1-2R_E$.

The magnetic field and density structure of each diamagnetic cavity, in the foreshock region, obtained in our self-consistent simulation of the bow shock is very similar to that created by *Thomas and Brecht* [1988] in a simulation of interaction between an ion beam and a plasma. The magnetic field and plasma signatures in the foreshock cavities obtained in our simulation are similar to those observed by IMP-8 and WIND in the foreshock regions [*Sibeck et al.*, 2001, 2002]. The observations also indicated that the diamagnetic cavities are very common near the quasi-parallel bow shock.

The waves and the associated cavity-like structures are carried downstream by the earthward flow convection. Meanwhile, the foreshock fluctuations are evolving with time. In later times, they evolve into some structures elongated along field lines in the foreshock region of the quasi-parallel shock, with alternate field-aligned filaments (with enhanced density) and cavity-like areas (with decreased density). The filaments/cavities develop from the foreshock through the shock transition, following streamlines into the downstream magnetosheath. Figure 2 shows the magnetic field, density, and temperature $T_\parallel$ at $t = 132\Omega_0^{-1}$. The upstream parts of some of the cavity areas (filaments) are indicated by the crosses (heavy dots). The magnetic field varies with an in-phase correlation with the density throughout the field-aligned cavities/filaments. Similar to the foreshock craters at earlier times, the corresponding structures in the parallel temperature are in an anti-phase correlation with N or B, with an enhanced $T_\parallel$ in the cavity areas. These field-aligned structures are phase-standing in the plasma frame, and thus stationary in the simulation (Earth) frame in this case with flow velocities V nearly parallel to B. On

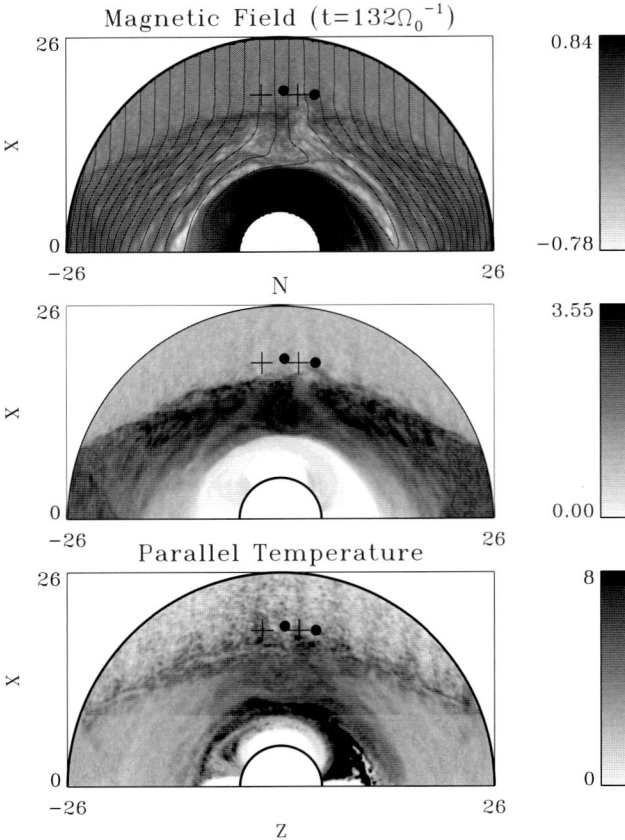

Figure 2. Simulation results at $t = 132\Omega_0^{-1}$ of case 1: Cavities/ filaments are seen from the foreshock to the downstream magnetosheath. The upstream part of some of the cavity areas is indicated by the crosses, while the upstream parts of some of the filaments are marked by heavy dots.

the other hand, waves with right-hand polarizations in B_y and B_z (not shown) remain in the downstream region. The presence of field-aligned density and magnetic field structures has also been found in the 2-D hybrid simulation of beam–plasma interaction by *Wang and Lin* [2003, and this volume]. Our spectrum analysis for the foreshock regions indicates that in early times, fluctuations in B and N are at wave vectors k in nearly all the directions, corresponding to the crater-like structures. Later, only modes with k nearly $\perp B_0$ are dominant in the foreshock regions of the quasi-parallel shock. In general cases with the magnetic field B oblique to V, the field-aligned cavities/filaments can convect as transient structures oblique to the flow. A case with the IMF oblique to the solar wind flow can be seen in *Lin* [2003]. The foreshock cavities can convect to the magnetopause as transient compressional structures.

The increase in $T_\parallel$ in most of these foreshock cavities indicates the presence of energetic ion particles, as shown in the ion scatter plots in *Lin* [2003]. The perpendicular ion temperature $T_\perp$, however, does not always increase as much as $T_\parallel$ in the foreshock cavities. Throughout the foreshock cavities, the hot, tenuous ion beam population is well distinguished from the cold, dense solar wind plasma population. This feature of ion distribution is very different from that in HFAs, in which the interaction between the specularly reflected, multiple ion beams causes an isotropic, single-component ion distribution in the HFA, as shown in the next section.

3. SIMULATION OF INTERACTION OF THE BOW SHOCK WITH INTERPLANETARY TD—GENERATION OF HFA

When an interplanetary TD arrives at and interacts with the bow shock, a diamagnetic cavity that is much stronger than the cavities shown in section 2 can be generated. The interaction between the bow shock and a directional TD is performed by allowing an interplanetary TD to propagate into the simulation domain from the upstream boundary at $30R_E$. The initial TD is assumed to be a planar 1-D structure that lies perpendicular to the simulation (xz) plane, across which the physical quantities experience a jump only along the discontinuity normal n, which is also the direction of wave vector k of the discontinuity. For the directional TD, we assume that the direction of the tangential magnetic field can change by an arbitrary angle $\Delta\Phi$ across the initial TD, but plasma density, pressure, velocity, and field strength remain constant. The half width of the initial TD is assumed to be about $1.5\lambda_0$.

Case 2, shown in this section, corresponds to one of the strongest HFAs obtained in our simulations. In this case, the component of the initial IMF in the xz plane makes an angle of $10°$ with the $-x$ direction, while the IMF has an angle of $-40°$ relative to the xz plane, with $B_{x0} = -0.75B_0$, $B_{y0} = -0.64B_0$, and a southward $B_{z0} = -0.13B_0$. The solar wind is again assumed to have a speed of $V_0 = 5V_{A0}$, and the upstream ion beta $\beta_0 = 0.5$. Across the TD that convects into the domain, the magnetic field changes direction by $\Delta\Phi = 80°$, so that behind the TD the IMF B_y is equal to $-B_{y0}$. Note that the TD front is aligned with the magnetic field direction. Its normal direction n thus makes an angle of $\gamma = 80°$ relative to the $-x$ axis, pointing earthward and northward. The motional electric field is symmetric and its normal component, $E_n = E_0 \cdot n = -(V_0 \times B_0) \cdot n$ points inward from both sides of the TD.

The left column of Figure 3 shows magnetic field lines and the magnitude of the magnetic field on the logarithmic scale for $t = 60, 120, 168, 216,$ and $264\Omega_0^{-1}$. The right column of Figure 3 shows contours of the ion number density N on a linear scale at the corresponding times. At $t = 60\Omega_0^{-1}$, the bow shock is present at a standoff distance of $r \sim 18R_E$. The bow shock in the southern hemisphere and around the equator is a quasi-perpendicular shock, whereas the quasi-parallel shock exists on the north $z > 0$. The incident TD, which carries no changes in the field strength or the ion density, has convected into the domain and reached the shock front at $z \sim -17R_E$. The bow shock, however, is nearly unaffected by the interaction. The significant ion kinetic effects in the shock-TD interaction first show up at $t = 120\Omega_0^{-1}$. An HFA cavity forms at the intersection between the TD and the shock front around $z = -9R_E$, in the quasi-perpendicular shock region, as seen in the second row of Figure 3. A low magnetic field and a low density are present in the center of the cavity. The cavity has an extended structure along the incident and the transmitted TD. At $t = 168\Omega_0^{-1}$, the cavity bulges into the upstream solar wind. Two boundaries with an enhanced magnetic field and density flank the central low-density bulge. At $t = 216\Omega_0^{-1}$, the HFA region has extended to about $26R_E$ in the solar wind, and has passed the equator into the northern hemisphere. Some wavy structures in the magnetic field occur in the center of the HFA. The plasma region just earthward of the HFA cavity is also seen to expand sunward along the much wider current sheet. By $t = 264\Omega_0^{-1}$, the core of the upstream HFA has expanded to a width of $\sim 10R_E$, and the magnetosheath has expanded into the upstream region. Note that the magnetopause has eroded earthward by about $2R_E$ due to the removal of the magnetic flux in a magnetic reconnection at the magnetopause [*Lin*, 2002]. It is found that the magnetopause reconnection has little effect on the formation of the HFA. A significant sunward expansion of the magnetosheath and magnetopause due to the HFA has also been found by satellite observations of *Sibeck et al.* [1999].

In the HFA throughout its upstream and magnetosheath parts, the ion bulk flow speed is significantly reduced. The solar wind inflow velocity is greatly deflected sunward. The ion temperature, on the other hand, is greatly enhanced in the core of the HFA. Notice the presence of field-aligned cavity/filament structures in the quasi-parallel shock, north of the equator, with alternate increases and decreases in N (and B), as seen in Figure 3. These structures are associated with the foreshock cavities due to backstreaming ions in the quasi-parallel shock, as discussed in section 2, and are not due to the interplanetary TD.

The generation of the HFA is due to the specular reflection of ions at the quasi-perpendicular bow shock. These ions are coherent and much denser than the scattered backstreaming ions that cause the foreshock cavities at the quasi-parallel shock. Since in case 2 the electric field on both sides of the TD points toward the TD, the reflected ions that gyrate

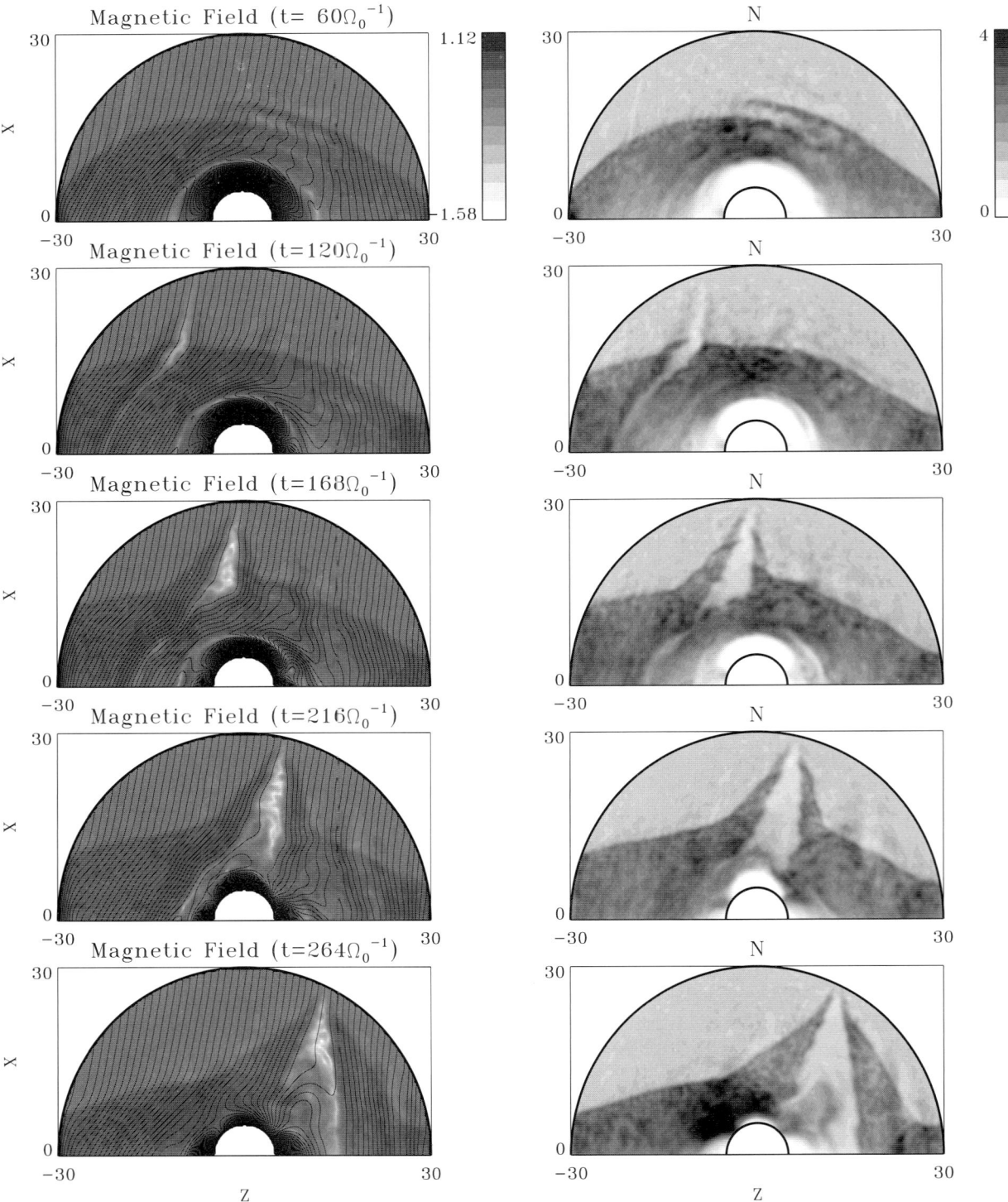

Figure 3. (Left) Magnetic field lines plotted against the field strength in the logarithm scale and (right) contours of the ion number density at various times in case 2 during the bow shock–TD interaction. The formation of an HFA is seen in both the upstream region and the magnetosheath.

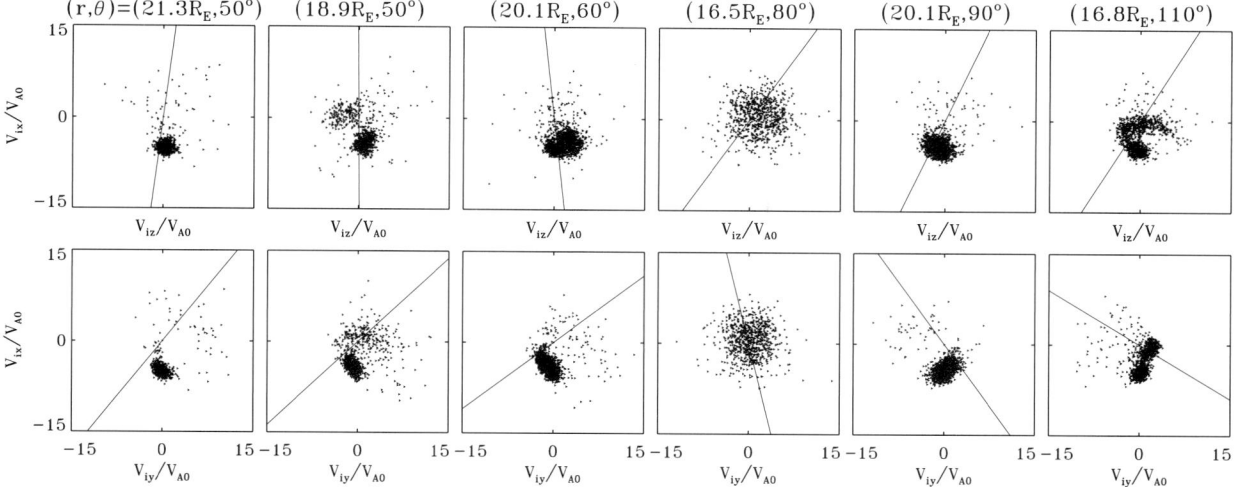

Figure 4. Ion velocity distributions obtained from case 2 at $t = 216\Omega_0^{-1}$ at six locations. From left to right: Location 1 is in the upstream ahead of the TD, location 2 at the leading edge of the HFA, location 3 further in the leading boundary of the HFA, location 4 in the core of the HFA cavity, location 5 in the trailing boundary of the HFA, and location 6 at the trailing edge of the HFA. The solid lines point to the local magnetic field direction.

around the magnetic field experience the strong electric force toward the TD from both sides. This force results in an accumulation and self-interaction between the reflected ion beams in the current sheet. Figure 4 shows the ion velocity (v_i) distributions in the v_{iz}–v_{ix} plane and the v_{iy}–v_{ix} plane at $t = 216\Omega_0^{-1}$. Shown in Figure 4 from the left column are the phase space distributions at location 1 in the upstream ahead of the TD, location 2 in the shock ramp at the leading edge of the HFA boundary, location 3 sunward of the shock front but inside the leading boundary of the HFA, location 4 in the core of the HFA cavity, location 5 in the trailing boundary of the HFA, and location 6 at the trailing edge of the HFA boundary. The solid line in each plot indicates the local magnetic field direction. The cold upstream ions are seen with some hot, reflected and backstreaming ions at location 1. Unlike case 1, however, the strong HFA cavity is not caused by the low-density backstreaming ions, but rather by the dense, specularly reflected ions around the quasi-perpendicular shock. At locations 2 and 6, the solar wind ions are just being shocked, and a significant number of the ions are being reflected back and gyrate around the magnetic field. The incident solar wind interacts with the reflected, coherent ion beam. At locations 3 and 5 in the boundary of the HFA, the plasma is dense. The boundary of the HFA contains the ions being reflected and arriving at the TD. At location 4 in the center of the HFA, the temperature becomes nearly isotropic and much hotter than the shocked solar wind in the magnetosheath because of the interaction among the ion beams.

The spatial profiles of various physical quantities through the upstream part of the HFA, along the θ-coordinate line at $r = 22R_E$ at $t = 216\Omega_0^{-1}$, are shown in Figure 5. The significant HFA is embedded in the upstream solar wind from $\theta \sim 62°$ to $90°$, where B_y gradually changes sign. In the HFA, the flow component V_x is greatly reduced to nearly zero from the upstream inflow speed. The V_z velocity, which is nearly zero upstream, has a positive (northward) perturbation in the HFA, and a southward perturbation on the trailing (southward) boundary of the HFA. The magnetic field and density increase by factors of 2 in the leading and trailing boundaries, while they are comparable to their upstream values in the core. While the dynamic pressure decreases in the core of the HFA, the sum of the thermal and magnetic pressure, P_{tot}, increases greatly, leading to the expansion of the cavity into the ambient solar wind. As mentioned earlier, large amplitude HFAs occur when the normal component of the electric field $E_n \sim E_z$ points inward toward the TD, with $E_z < 0$ ($E_z > 0$) on the leading (trailing) side of the TD, as seen in the fifth row of Figure 5. The structure of the HFA obtained from the simulation is very similar to that from the satellite observation [e.g., *Paschmann et al.*, 1988].

The formation of the HFA is closely related to the strength of the ambient electric field that leads to the accumulation of the reflected ion beams into the incident current sheet. The simulation has also been carried out for cases similar to case 2, with a symmetric B_y across the interplanetary TD, but with various field rotation angles $\Delta\Phi$ across the TD. The size of the HFA increases with positive $\Delta\Phi$, which corresponds to the proper electric field component pointing into the TD. Negative $\Delta\Phi$, on the other hand, does not produce an HFA. The dependence of HFA characteristics on orientation of the TD front is also studied, and the size of the HFA cavity is found to increase with the angle γ, but the maximum HFA occurs when the angle between the TD normal and the x-axis is ~80°.

4. SUMMARY

In summary, our 2-D hybrid simulation shows that significant pressure pulses can be generated at the bow shock due to ion kinetic effects while there exists no plasma perturbations from the solar wind. The main results are:

(1) As the bow shock forms under certain IMF and solar wind condition, crater-like diamagnetic cavities develop self-consistently in the foreshock of the parallel and quasi-parallel shocks due to the interaction between the accelerated backstreaming ions and the incoming solar wind. The center of the craters, with a low density and low magnetic field strength, is bounded by a rim with enhanced density and magnetic field. The bulk flow speed slightly decreases in the center of the cavity, while the total ion temperature often increases. In the cases in which the IMF is nearly parallel to the solar wind flow, the craters develop into field-aligned density and magnetic field structures in both upstream and the magnetosheath, phase-standing in the plasma flow, with alternate decrease and increase in plasma density. The corresponding variations in magnetic field are in phase with those of the density. In the general cases with the IMF oblique to the solar wind, craters and field-aligned cavities/filaments convect with the flow toward the Earth as transient structures, some may impinge on the magnetopause.

(2) HFAs can be generated while the bow shock interacts with an interplanetary TD that carries a simple change in the IMF direction. The formation of strong HFAs is mainly due to the presence of a normal component of the electric field pointing from the two sides toward the TD, and thus the specularly reflected, gyrating ion beams at the quasi-perpendicular bow shock are accelerated toward the TD by the electric field. The interaction of the ion beams results in the hot diamagnetic cavity along the TD near the intersection between the TD and the BS. In the part of the HFA upstream of the bow shock, the core of the diamagnetic cavity contains a hot, thin plasma. The total pressure of the core plasma

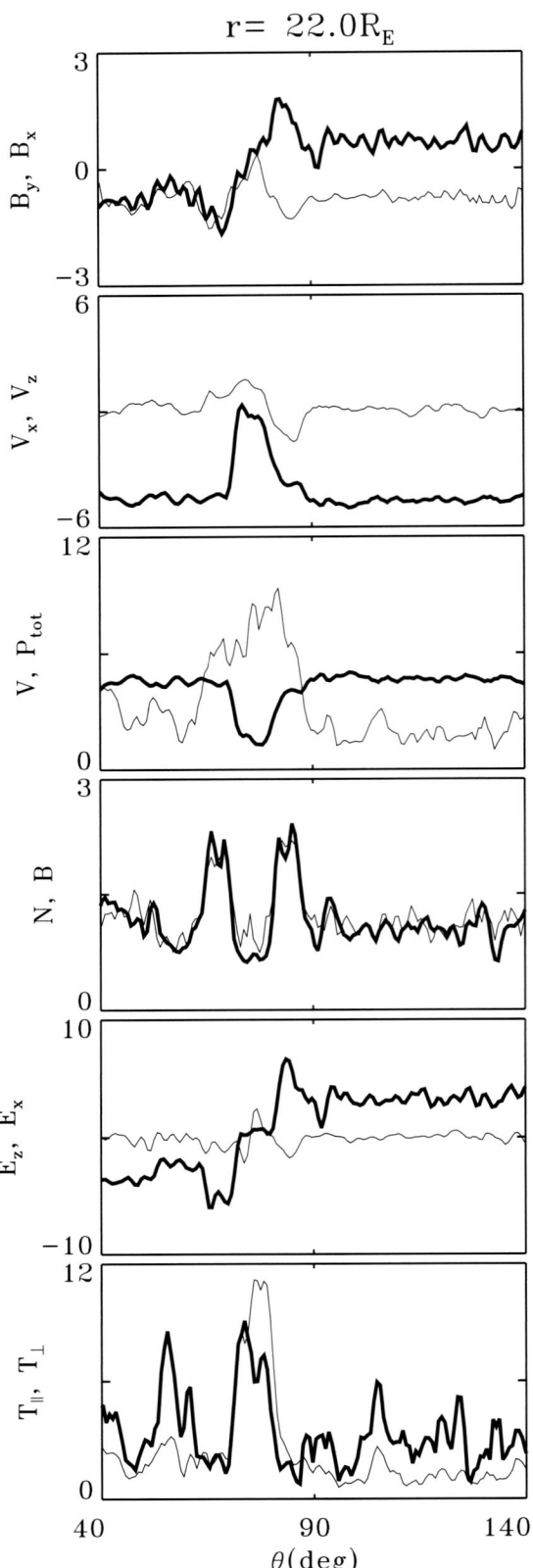

Figure 5. Spatial profiles of various quantities at $t = 216\Omega_0^{-1}$ in case 2, along the azimuthal coordinate line (constant r) through the upstream part of HFA. The heavy curves show, from the top, B_y, V_x, ion flow speed V, N, electric field E_z, and ion temperature $T_\parallel$. The light curves show the B_x, V_z, pressure P_{tot} ($= P_\perp + P_B$), field strength B, electric field E_x, and $T_\perp$, with $P_\perp$ and P_B being the perpendicular and magnetic pressures, respectively.

perpendicular to the TD is much larger than that of the ambient solar wind, and thus the HFA expands into the ambient solar wind. The leading and trailing boundaries of the HFA contain a high density plasma and high magnetic field strengths. In the HFA, the solar wind inflow is strongly deflected sunward, and the flow speed is greatly reduced. The local magnetopause region just earthward of the HFA expands sunward locally as the HFA passes by.

Acknowledgments. This work was supported by NASA grant NAG5-12899 and NSF grant ATM-0213931 to Auburn University. Computer resources were provided by the National Partnership for Advanced Computational Infrastructure and the Arctic Region Supercomputer Center.

REFERENCES

Burgess, D., and S. J. Schwartz, Colliding plasma structures: Current sheet and perpendicular shock, *J. Geophys. Res., 93*, 11,327, 1988.

Engebretson, M. J., N. G. Lin, W. Baumjohann, H. Luehr, B. J. Anderson, L. J. Zanetii, T. A. Potemra, R. L. McPherron, and M. G. Kivelson, A comparison of ULF fluctuations in the solar wind, magnetosheath, and dayside magnetosphere, 1. Magnetosheath morphology, *J. Geophys. Res., 96*, 3441, 1991.

Fairfield, D. H., W. Baumjohann, G. Paschmann, H. Luhr, and D. G. Sibeck, Upstream pressure variations associated with the bow shock and their effects on the magnetosphere, *J. Geophys. Res., 95*, 3773, 1990.

Goodrich, C. C., Numerical simulations of quasi-perpendicular collisionless shocks, in *Collisionless Shocks in the Heliosphere: Reviews of Current Research*, Geophys. Monogr. Ser., Vol. 35, edited by B. T. Tsurutani and R. G. Stone, p. 153, AGU, Washington, D. C., 1985.

Hubert, D., and C. C. Harvey, Interplanetary rotational discontinuities: from the solar wind to the magnetosphere through the magnetosheath, *Geophys. Res. Lett., 27*, 3149, 2000.

Lee, M. A., Coupled hydromagnetic wave excitation and ion acceleration upstream of the Earth's bow shock, *J. Geophys. Res., 87*, 5063, 1982.

Leroy, M. M., and D. Winske, Backstreaming ions from oblique Earth bow shocks, *Ann. Geophys., 1*, 527, 1983.

Lin, Y., Generation of anomalous flows near the bow shock by its interaction with interplanetary discontinuities, *J. Geophys. Res., 102*, 24,265, 1997.

Lin, Y., Global hybrid simulation of hot flow anomalies near the bow shock and in the magnetosheath, *Planet. Space Sci., 50*, 577, 2002.

Lin, Y., Global-scale simulation of foreshock structures at the quasi-parallel bow shock, *J. Geophys. Res., 108* (A11), SMP 3, 2003.

Paschmann, G., G. Haerendel, N. Sckopke, E. Mobius, H. Luhr, and C. W. Carlson, Three-dimensional plasma structures with anomalous flow directions near the Earth's bow shock, *J. Geophys. Res., 93*, 11,279, 1988.

Scholer, M., and T. Terasawa, Ion reflection and dissipation at quasi-parallel collisionless shocks, *Geophys. Res. Lett., 17*, 119, 1990.

Schwartz, S. J., Hot flow anomalies near the Earth's bow shock, *Adv. Space Res., 15(8/9)*, 107, 1995.

Schwartz, S. J., G. Paschmann, N. Sckopke, T. M. Bauer, M. Dunlop, A. N. Fazakerley, and M. F. Thomsen, Condition for the formation of hot flow anomalies at Earth's bow shock, *J. Geophys. Res., 105*, 12,639, 2000.

Sckopke, N., G. Paschmann, S. J. Bame, J. Gosling, and C. T. Russell, Evolution of ion distributions across the nearly perpendicular bow shock: Specularly and non-specularly reflected ions, *J. Geophys. Res., 88*, 6121, 1983.

Sibeck, D. G. et al., Comprehensive study of the magnetospheric response to a hot flow anomaly, *J. Geophys. Res., 104*, 4577, 1999.

Sibeck, D. G., R. B. Decker, D. G. Mitchell, A. J. Lazarus, R. P. Lepping, K. W. Ogilvie, and A. Szaba, Solar wind preconditioning in the flank foreshock: IMP-8 observations, *J. Geophys. Res., 106*, 21,675, 2001.

Sibeck, D. G., T.-D. Phan, R. Lin, R. P. Lepping, and A. Szabo, Wind observations of foreshock cavities: A case study, *J. Geophys. Res. (A10), 107*, SMP 4, 2002.

Sibeck. D. G., N. B. Trivedi, E. Zesta, R. B. Decker, H. J. Singer, A. Szabo, H. Tachihara, and J. Watermann, Pressure-pulse interaction with the magnetosphere and ionosphere, *J. Geophys. Res., 108*, 2003.

Sonnerup, B. U. O., Acceleration of particles reflected at a shock front, *J. Geophys. Res., 74*, 1301, 1969.

Swift, D. W., Use of a hybrid code for a global-scale plasma simulation, *J. Comput. Phys., 126*, 109, 1996.

Tanaka, M., C. C. Goodrich, D. Winske, and K. Papadopoulos, A source of backstreaming ion beams in the foreshock region, *J. Geophys. Res., 88*, 3046, 1983.

Thomas, V. A., and S. H. Brecht, Evolution of diamagnetic cavities in the solar wind, *J. Geophys. Res., 93*, 11,341, 1988.

Thomas, V. A., D. Winske, and N. Omidi, Reforming supercritical quasi-parallel shocks, 1, One and two dimensional simulations, *J. Geophys. Res., 95*, 18,809, 1990.

Thomas, V. A., D. Winske, M. F. Thomsen, and T. G. Onsager, Hybrid simulation of the formation of a hot flow anomaly, *J. Geophys. Res., 96*, 11,625, 1991.

Thomsen, M. F., J. T., Gosling, S. A. Fuselier, S. J. Bame, and C. T. Russell, Hot, diamagnetic cavities upstream from the Earth's bow shock, *J. Geophys. Res., 91*, 2961, 1986.

Thomsen, M. F., V. A. Thomas, D. Winske, J. T. Gosling, M. H. Farris, and C. T. Russell, Observational test of hot flow anomaly formation by the interaction of a magnetic discontinuity with the bow shock, *J. Geophys. Res., 98*, 15,319, 1993.

Wang, X. Y., and Y. Lin, Generation of nonlinear Alfven and magnetosonic waves by beam plasma interaction, *Phys. Plasmas, 10*, 3528, 2003.

Wibberenz, G., F. Zollich, H. M. Fischer, and K. Keppler, Dynamics of intense upstream ion events, *J. Geophys. Res., 90,* 283, 1985.

Yu Lin, Physics Department, Auburn University, 206 Allison Laboratory, Auburn, Alabama 36849-5311. (ylin@physics.auburn.edu)

Diffusive Compression Acceleration

Joe Giacalone, Jack R. Jokipii, and Jozsef Kóta

Lunar and Planetary Laboratory, University of Arizona, Tucson, Arizona

We discuss the physics of diffusive compression acceleration of charged particles. This mechanism is similar to diffusive shock acceleration, which is well known, except that it applies to a gradual compression of the plasma, rather than a shock. It also applies to fluctuations in velocity, such as turbulence. We present the results from theoretical and numerical calculations for a variety of applications of this mechanism. These include: acceleration of particles by corotating solar-wind compressions, acceleration in a sinusoidally varying fluid flow, the formation of high-energy tails in pickup-ion distributions, and the acceleration of Galactic cosmic rays by turbulent flows in the interstellar medium. We also show that this mechanism naturally favors acceleration of particles with larger gyroradii. This may help explain recent observations of some solar-energetic particle observations showing abundance enhancements that increase with particle gyroradius.

1. INTRODUCTION

Understanding the acceleration of cosmic rays or energetic charged particles is one of the most fundamental goals in astrophysics. A number of mechanisms for accelerating particles to high energies have been suggested. *Swann* [1933] showed that the electric field associated with the compression of a magnetic field could lead to particle energization. The model was incomplete because it did not discuss what would happen when the field eventually decreased. *Alfvén* [1950] demonstrated that by including pitch-angle scattering there is a net acceleration. A similar idea was proposed by *Fermi* [1949]. More recently, the acceleration by collisionless shocks has received considerable attention. In its modern form (e.g., *Axford et al.* [1978]; *Bell* [1978]; *Krymsky* [1977]; *Blandford and Ostriker* [1978]; and reviews by *Drury* [1983] and *Jones and Ellison* [1991]), magnetic irregularities scatter the charged particles and keep them nearly isotropic in the fluid frame. In this limit the particle transport is diffusive and is described by Parker's transport equation [*Parker*, 1965]. Diffusive shock acceleration is simply the application of this equation to a shock. This mechanism produces a near-universal energy spectrum over a broad range of parameters, which is close to that observed. It seems likely that most cosmic rays are produced by this mechanism.

However, observations suggest that there may be important situations where energetic particles are accelerated where no shocks are present. One recent and particularly clear example consists of the energetic ions observed in interplanetary corotating interaction regions near 1 AU, well inside the radius where the associated co-rotating shocks form [*Mason*, 2000]. *Schwadron et al.* [1996] have suggested that wave-particle acceleration may play a role in producing these particles. More recently, *Giacalone et al.* [2002] demonstrated that these particles could instead have resulted from acceleration in the regions of compression between fast and slow wind, inside the point at which the shocks form. They found that the charged-particle scattering mean-free paths were comparable to the scale of the compression, so diffusion was not a good approximation, and direct particle integrations were performed. We note, that such compressive acceleration might also occur where diffusion is a good approximation. This has been discussed briefly by *Jokipii et al.* [2003] and from a different point of view by *Webb et al.* [2003].

It is the purpose of this paper to review the topic of particle acceleration by gradual plasma compressions. We define such compressions as regions where, in contradistinction to shocks, the spatial scale is set up by large-scale plasma dynamics, and not by the plasma microphysics. Several applications are discussed. We present quantitative calculations for three specific applications. These include the acceleration of particles by corotating compression regions, acceleration by velocity compressions that vary sinusoidally, and acceleration of particles with different gyroradii by a one-dimensional, planar, gradual plasma compression. Other applications are also discussed with qualitative estimates of the acceleration time scale and efficiency.

2. ANALYTIC CONSIDERATIONS: THE DIFFUSION APPROXIMATION

In the diffusion approximation, the (nearly-isotropic) cosmic-ray distribution function $f(x_i,p,t)$, as a function of position x_i, momentum magnitude p and time t satisfies the Parker equation [*Parker*, 1965]

$$\frac{\partial f}{\partial t} = \frac{\partial}{\partial x_i}\left[\kappa_{ij}\frac{\partial f}{\partial x_j}\right] - U_i\frac{\partial f}{\partial x_i} + \frac{1}{3}\frac{\partial U_i}{\partial x_i}\frac{\partial f}{\partial \ln(p)} + Q - L \quad (1)$$

where κ_{ij} is the diffusion tensor, U_i is the flow velocity of the background plasma, and Q and L represent any additional sources and losses. This equation is remarkably general and applies if there is enough scattering that the distribution function remains nearly isotropic, even at discontinuities such as current sheets and shock waves. Although Equation 1 does not contain the electric field explicitly, it nonetheless describes particle acceleration which is contained in the term $\partial U_i/\partial x_i$.

If Equation 1 is applied to a one-dimensional system having a flow velocity $U_x(x,t)$ which contains a planar shock, where the flow velocity changes discontinuously, the term $\partial U_x/\partial x$ becomes a δ-function. The resulting equation can be solved quite easily, and one obtains the standard results of diffusive shock acceleration mentioned above. If the shock wave is not a discontinuity, but instead is a more-gradual compression having a characteristic length scale L_c, one can show that in the limit where the ratio of the diffusive skin depth $L_d = \kappa_{xx}/U_x$ to the length scale L_c is large, or, equivalently,

$$\xi = \frac{L_d}{L_c} = \frac{\kappa_{xx}}{U_x L_c} \gg 1 \quad (2)$$

the solution for the cosmic-ray distribution f goes over to the standard diffusive shock solution (for an earlier discussion see *Gombosi et al.*, 1989). Because there is no shock, there must be some other source of seed particles to be accelerated. In the opposite limit $\xi \ll 1$, the cosmic rays are closely tied to the convecting fluid, and simply compress adiabatically by the compression factor with the rest of the gas, and the acceleration negligible.

This scenario is similar to that involved in the theory of particle acceleration at cosmic-ray modified shocks (e.g. *Ellison et al.* [1981]; *Drury et al.* [1982]). If the acceleration process is extremely efficient, the accelerated particles contain a significant amount of energy density to modify the shock structure itself and cause it to broaden. In this case, the upstream plasma gradually decelerates ahead of a thin subshock. While this is similar to what we describe in this paper, our mechanism does not, in general, involve a shock.

3. APPLICATIONS

3.1. Compressions Corotating With the Sun

Gradual compressions of the solar wind plasma can occur in a number of ways. One of the clearest examples is the case of a corotating interaction region (CIR) at times around solar minimum. At distances beyond a few AU, the CIR produces two shock waves: a forward shock, which moves away from the Sun in the solar-wind frame of reference, and a reverse shock, which moves towards the Sun in this frame. At distances inside where the shocks form (<~2 AU), the transition from slow to fast wind occurs through a gradual compression of the plasma.

We consider the acceleration of interstellar pickup ions by such a solar wind compression region. Pickup ions are formed as interstellar neutrals enter the solar system, due to the Sun's motion relative to the local interstellar medium, and are ionized. Neutral He is ionized predominantly by photoionization and can penetrate inside of 1 AU before being ionized. Thus, there is a significant abundance of pickup He at Earth's orbit. The dominant process for ionizing neutral hydrogen atoms is charge exchange with the solar wind. There are relatively few pickup protons at 1 AU (the characteristic ionization distance for neutral protons is about 5 AU).

We consider a spherical coordinate system. The relevant diffusion coefficient describing the spatial diffusion of the pickup ions to be used in Equation 2 is κ_{rr}, and the flow speed is U. κ_{rr} can be related to the diffusion coefficients parallel and perpendicular to the average magnetic field $\kappa_\parallel$ and $\kappa_\perp$, respectively. Assuming $\kappa_\parallel \gg \kappa_\perp$ and a nominal Parker spiral magnetic field at 1 AU, we obtain $\kappa_{rr} \approx (1/2)\kappa_\parallel$. Thus, Equation 2 leads to $(1/6)(w/U)\lambda_\parallel > L_c$, where w is the particle speed. Taking $\lambda_\parallel = 0.25-2$ AU, as suggested by observations [*Gloeckler et al.*, 1995; *Nemeth et al.*, 2000] problem and the environment. As a consequence, it is necessary to identify specific events, which are relatively "clean" when performing a comparison with a model calculation. In this work, we consider an event of April 21, 2002. This event exhibited the following basic characteristics:

1. It occurred in active region 9906 (western hemisphere);
2. There was a CME eruption at 0100 on April 21 (day 111);
3. GOES-8 showed significant proton increases up to their >100 MeV channel; and
4. No clear signatures of other events from day 109 to day 115 were observed.

We chose this event for its clean time-intensity profile, which is relatively free of contamination from other events. The parent shock is thought to be driven by a CME associated with active region 9906 (S03 W56) and was magnetically well connected to Earth. Velocity dispersion observed during the onset of the event is a clear signature of particles being accelerated near the Sun and their subsequent transport through the interplanetary medium to 1 AU. We concentrate on modeling the time intensity profiles. Our model results are compared with those obtained over a broad range of energies between 0.08 keV/amu through 100 MeV/amu from instruments on ACE, Wind, and SAMPEX. We also study the particle energy spectrum, and in particular, the time evolution of the spectrum.

We model the shock by fitting the arrival time of the observed shock and the shock velocity at 1 AU. The CME-driven shock is introduced at 0.1 AU, where the solar wind is supersonic, by temporarily increasing the number density and solar wind velocity by a factor of 3.25 for half an hour. This corresponds to an energy injection of order $\sim 10^{32}$ erg, a typical value for a coronal mass ejection. We assume $\gamma = 5/3$ to model the solar wind. The shock has an initial velocity of 950 km/s and drops to about 620 km/s at 1 AU, taking ~ 51 hours to reach 1 AU. Figure 1 plots the shock properties. The upper panel plots the shock velocity as a function of heliodistance. The lower panel is the compression ratio s. The shock velocity and the compression ratio observed at 1 AU are shown as the dot with the error bar. The model predicts a compression ratio that is larger than that observed. This is probably because we are using a 1-D model, and so we are always connected to the nose of the shock. The observation, however, is likely made at the western flank of the shock by the time the shock reaches 1 AU.

Plate 1 is a plot of the time intensity profile for CNO

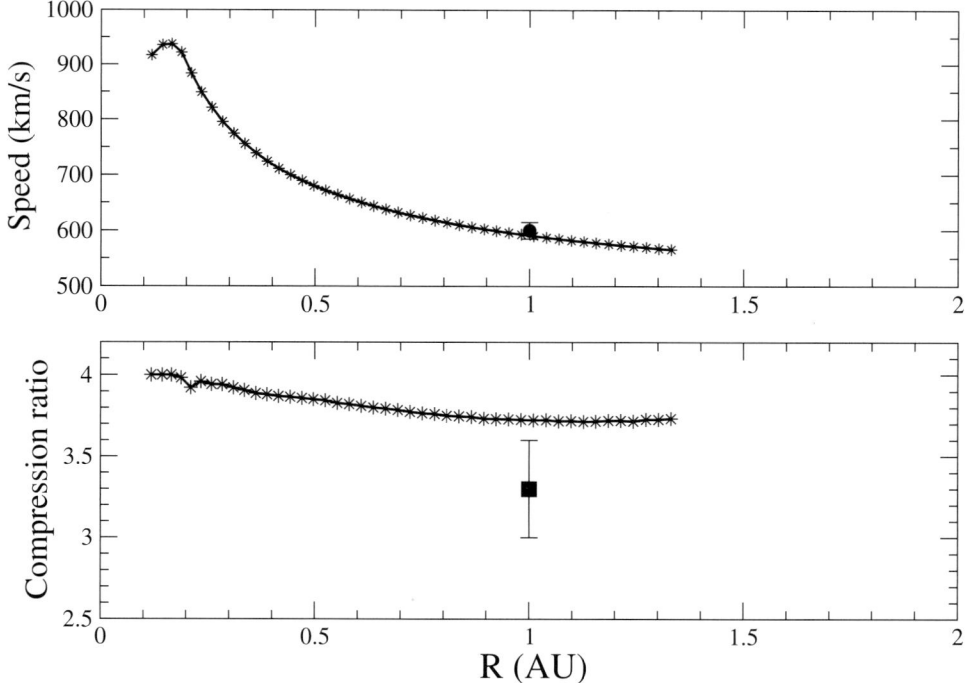

Figure 1. The shock properties. The upper panel shows the shock velocity as a function of heliodistance. The lower panel shows the compression ratio s. The shock is initially strong with $s = 4$ and a speed of ~950 km/s, and weakens to $s = 3.7$ and $v \sim 620$ km/s at 1 AU. Typical values of solar wind speed and compression ratio observed at 1 AU are also indicated in the error bars. See text for details.

56 PARTICLE ACCELERATION AND TRANSPORT AT CME-DRIVEN SHOCKS

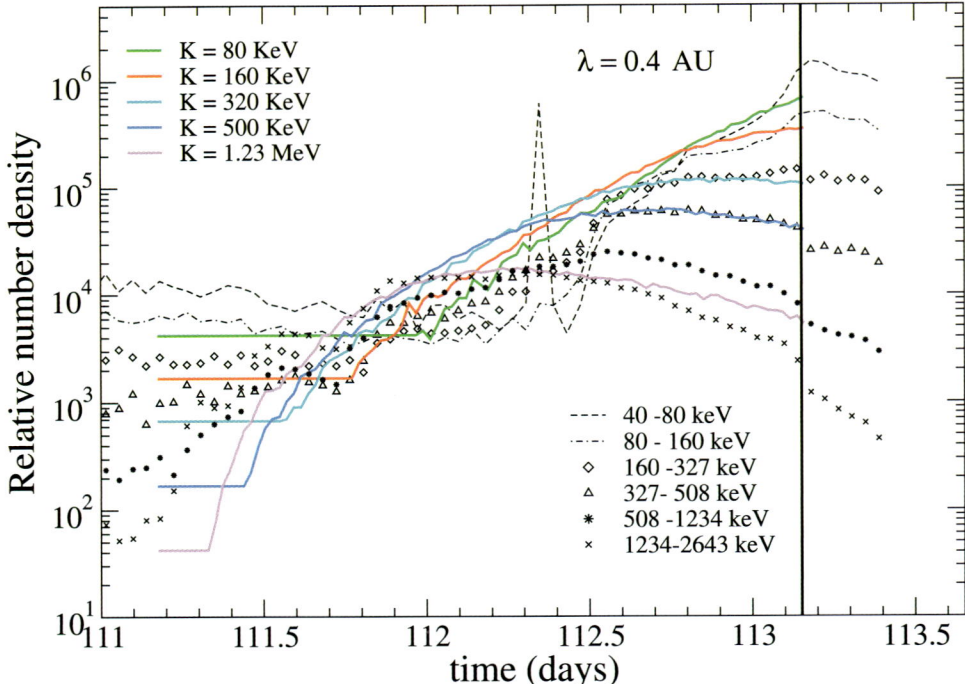

Plate 1. Time intensity profile for particles of various energies. $\lambda_0 = 0.4$ AU.

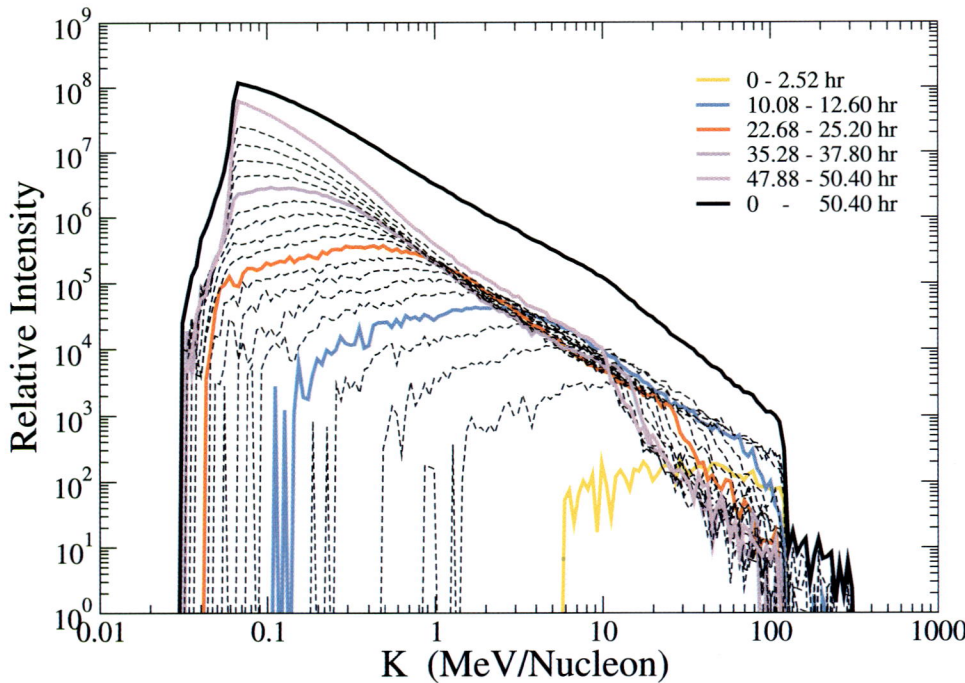

Plate 2. The particle spectra as a function of energy at 1 AU for different time periods. The uppermost solid curve is the event-integrated spectrum (multiplied by a factor of 2). See text for details.

particles with kinetic energies per nucleon of 80 keV/amu, 160 keV/amu, 320 keV/amu, 500 keV/amu and 1.23 MeV/amu. The corresponding observational data are also shown. The x-axis is the time, and the y-axis is the relative number density. The arrival of the shock is represented by the vertical line near $t = 113.15$ days. As shown in the illustration, we find that the model provides a reasonable intensity profile fit. For example, the predicted curve for particles of $K = 1.23$ MeV yields almost identical characteristics to those observed. The early rise before the shock arrival signals the earlier acceleration when the shock was still strong. The intensity reaches its maximum and begins to decrease before the shock arrival, suggesting the shock becomes too weak to accelerate particles to this energy. The predicted curves for particles of lower energies ($K = 80$ KeV/amu, 160 KeV/amu, 320 KeV/amu) seem to have earlier rises than the observations. This might due to the assumption of the mean free path in equation (8). In equation (8), we have assumed a momentum dependence of the mean free path as p^α, with $\alpha = 1/3$. A stronger momentum dependence would affect the low energy particles more, and thus squeeze the intensity profile preferentially for low-energy particles. In the paper of Li et al. [2003], the momentum dependence of the mean free path was investigated. Note, changing λ_0 will affect particles of all energies and thus cannot explain the predicted early rise for low-energy particles. It is worth noting that the April 21, 2002, SEP event has a rather peculiar velocity dispersion compared with the usual "well connected" event in the western hemisphere. One finds, for example, in the inverse velocity spectrogram for particles of mass 10 amu and up, a much slower rise of the intensity compared with a normal 'well connected' event. This could be because the turbulence near the shock in this event is particularly strong and particles (especially of lower energies) are thus trapped for a longer period of time.

We now discuss the evolution of particle spectra as a function of time. Plate 2 shows the simulation result of particle spectra at different times for an observer at 1 AU. At earlier times, more high-energy particles are observed at 1 AU due to their greater velocities, shown by the curves in the lower right corner of the illustration. At later times, more low-energy particles cross 1 AU and the number of high-energy particles begins to decrease as the shock weakens. The curves then evolve to the left, i.e. towards smaller K. The solid curves, from the lower right corner to the upper left corner, correspond to 0–2.52 hours, 10.08–12.60 hours, 22.68–25.20 hours, 35.28–37.80 hours and 47.88–50.40 hours. The uppermost solid curve is the time-integrated spectrum (times 2 for comparison) for the event, i.e. from the time the shock leaves the Sun until it reaches 1 AU. It is an approximate power law with a break (bump) around 10 MeV. The spectrum index is roughly ~1.6. Note a $f(p) \sim p^\alpha$ corresponds to $f(T) \sim T^\beta$ with $\beta = (\alpha + 1)/2$. Using $\alpha = 3s/(s-1)$, we find $\beta = 1.5$ for $s = 4$ and $\beta = 1.555$ for $s = 3.7$. This is in good agreement with our simulation. Such spectral indices are also seen observationally [Tylka, 2002]. The spectra at different times are also approximate power laws with a "broken-power-law-feature". For example, at time 10.08–12.60 hours, the spectrum for particles above $K = 70$ MeV/amu begins to deviate from a power law, suggesting that the shock is no longer accelerating particles to that energy. Similarly, at 10.08–12.60 hours, at energies above $K = 20$ MeV/amu, a deviation from a power law becomes noticeable.

4. CONCLUSION

We have investigated the acceleration and transport of energetic particles at CME-driven shocks. The acceleration is modeled numerically using a shell model. Particle transport is studied using a Monte Carlo code, where single-particle trajectories are followed. Particle spectra at 1 AU are then calculated. The time intensity profile at various energies is shown in Plate 1, and compared to CNO observation made during the 21 April 2002 event. The agreement is encouraging. The time evolution of the spectra is also shown in Plate 2, exhibiting a clear power law feature. As time evolves and the shock expands and weakens, the power-law spectra become "broken" at higher energies. The time-integrated spectrum is also approximately a power law with a spectral index in agreement with observations. To summarize, we have developed a dynamical model which enables us to study particle acceleration at an expanding CME-driven shock wave, together with the subsequent transport in the interstellar medium. We find promising agreement between our model simulations and observations, suggesting that this approach may provide an important step towards understanding and tracking the influence of large SEP events in the interplanetary and geospace environments.

Acknowledgments. This work has been supported in part by a NASA grant NAG5-10932 and an NSF grant ATM-0296113. GL acknowledges a visit to the University of Maryland in August 2002, during which period this work was mostly done.

REFERENCES

Axford, W.I., E. Leer, and G. Skadron, *Proc. 15th Int. Cosmic Ray Conf. (Plovdiv)*, *11*, 132, 1977.

Bell, A.R., The acceleration of cosmic rays in shock fronts. I, *Mon. Not. Roy. Astron. Soc.*, 182, 147-156, 1978a.

Bell, A.R., The acceleration of cosmic rays in shock fronts. II, *Mon. Not. Roy. Astron. Soc.*, 182, 443-455, 1978b.

Cane, H.V., The structure and evolution of interplanetary shocks and the relevance for particle-acceleration, *Nucl.Phys. B., 39A*, 35-44, 1995.

Cliver, E.W., Cane, H.V., Gradual and Impulsive Solar Energetic Particle Events, *EOS 83*(7), 2002.

Drury, L.O'C., An introduction to the theory of diffusive shock acceleration of energetic particles in tenuous *Rep. Prog. Phys., 46*, 973-1027, 1983.

Gosling, J.T., The solar-flare myth, *J. Geophys. Res., 98*, 18937-18949, 1993.

Gopalswamy, N., et al., Interacting coronal mass ejections and solar energetic particles, *Astrophys. J., 572*, L103, 2002.

Gordon, B.E., M.A. Lee, E. Mobius, and K.J. Trattner, Coupled hydrodynamic wave excitation and ion acceleration at interplanetary traveling shocks and Earth's bow shock revisited, *J. Geophys. Res., 104*, 28,263-28,277, 1999.

Jackman, C.H., et al., Northern Hemisphere atmospheric effects due to the July 2000 solar proton event interstellar pickup ions, *Geophys. Res. Lett., 28*, 2883, 2001.

Lee, M.A., Coupled hydrodynamic wave excitation and ion- acceleration at interplanetary traveling shocks, *J. Geophys. Res., 88*, 6109-6119, 1983.

Li, G., Zank, G.P., and Rice, W.K.M., Energetic particle acceleration and transport at coronal mass ejection-driven shocks. *J. Geophys. Res. 108*, 1082 doi:10.1029/2002JA009666, 2003.

Mason, G.M., Mazur, J.E., Looper, M.D., and Mewaldt, R.A., Charge-state measurements of solar energetic particles observed with SAMPEX, *Astrophys. J., 452*, 901-911, 1995.

Mason, G. M., Mazur, J. E., Dwyer, J. R., ^{3}He Enhancements in Large Solar Energetic Particle Events. *Astrophys. J., 525*, L133-L136, 1999.

Ng, C.K., Reames, D.V., and Tylka, A.J., Modeling Shock-accelerated Solar Energetic Particles Coupled to Interplanetary Alfvén Waves, *Astrophys. J., 591*, 461-485, 2003.

Oetliker, M., et al., The ionic charge of solar energetic particles with energies of 0.3-70 MeV per nucleon, *Astrophys. J., 477*, 495-501, 1997.

Parker, E.N., Dynamics of the interplanetary gas and magnetic fields, *Astrophys. J., 123*, 664-676, 1958.

Reames, D.V., Coronal abundances determined from energetic particles, *Adv. Space Res., 15*, 41-51, 1995.

Reames, D.V., Particle acceleration at the Sun and in the heliosphere, *Space Sci. Rev., 90*, 413-491, 1999.

Rice, W.K.M., Zank, G.P., and Li, G., Particle acceleration at coronal mass ejection drive shocks: for arbitrary shock strength. *J. Geophys. Res. 108*, 1369 doi:10.1029/2002JA009756, 2003.

Tylka, A.J., et al., The mean ionic charge state of solar energetic Fe ions above 200 MeV per nucleon, *Astrophys. J., 444*, L109-L113, 1995.

Tylka, A.J., private communication, 2002.

Völk, H.J., Zank, L.A., and Zank, G.P., Cosmic-ray spectrum produced by supernova-remnants with an upper limit on wave dissipation, *Astron Astrophys., 198*, 274-282, 1988.

Zank, G.P., W.K.M. Rice, and Wu, C.C., Particle acceleration and coronal mass ejection drive shocks: A theoretical model, *J. Geophys. Res., 105*, 25079-25095, 2000.

Zank, G.P., W. H., Matthaeus, J. W., Beiber, H., Moraal, The radial and latitudinal dependence of the cosmic ray diffusion tensor in the heliosphere, *J. Geophys. Res. (Space), 103*, 2085-2097, 1998.

M. I. Desai and G. M. Mason, Department of Physics, University of Maryland, College Park, Maryland 20742.

Gang Li and G. P. Zank, Institute of Geophysics and Planetary Physics, University of California, Riverside, California 92026. (gang.li@ucr.edu)

W.K.M. Rice, University of St. Andrews, St. Andrews, Fife KY16 9SS, Scotland, UK.

Cosmic Ray Acceleration at Relativistic Shock Waves With a "Realistic" Magnetic Field Structure

Jacek Niemiec

Institute of Nuclear Physics, Polish Academy of Sciences, Kraków, Poland

Michał Ostrowski

Astronomical Observatory, Jagiellonian University, Kraków, Poland

First results of our Monte Carlo simulations of the cosmic ray first-order Fermi acceleration at relativistic shock waves are presented. The simulations are based on numerical integration of particle equations of motion in a turbulent magnetic field near the shock. The field consists of a mean field component inclined at some angle to the shock normal and finite-amplitude sinusoidal perturbations imposed upon it. The perturbations are assumed to be static in the local plasma rest frame. Their flat or Kolmogorov spectra are constructed upstream of the shock with randomly drawn wave vectors from the wide range (k_{min}, k_{max}). The downstream field structure is derived from the upstream one as compressed at the shock. We present and discuss the particle spectra and angular distributions obtained at mildly relativistic shocks. In particular, we discuss changes in the resulting spectra due to varying the mean field inclination and varying amplitude and spectrum of the field perturbations.

1. INTRODUCTION

At a relativistic shock wave the bulk velocity of the flow is comparable to particle velocity. This leads to anisotropy of the particle angular distribution which can substantially influence the process of particle acceleration. The first consistent method to tackle the problem for a parallel shock, where the mean magnetic field is perpendicular to the shock front, was proposed by Kirk and Schneider (1987a). They solved the pitch-angle diffusion equation on both sides of the shock by taking into account the higher-order terms in the anisotropy of the particle distribution. Then, from the condition of the distribution function continuity at the shock, they obtained a power-law index of the resulting spectrum and a particle angular distribution at the shock. The extension of the above approach was proposed by Heavens and Drury (1988), who investigated the problem of particle acceleration for more general conditions at the shock. They found that the particle spectral indices depend on the spectrum of magnetic field irregularities. A situation where particle anisotropy plays an essential role in forming the spectrum was discussed by Kirk and Heavens (1989). They considered the acceleration process in oblique subluminal shocks and showed that such shocks can lead to much flatter spectra than the parallel ones. The results relied on the assumption of adiabatic invariant conservation for particles interacting with the shock. This assumption is valid only in a case of a weakly perturbed magnetic field. However, if finite-amplitude MHD waves are present in the medium the above mentioned approaches are no longer valid [*Decker and Vlahos* 1985; *Ostrowski* 1991] and numerical methods have to be used.

A role of finite-amplitude perturbations of the magnetic field in forming the particle spectrum was investigated by means of the Monte Carlo particle simulations by a number of authors [e.g. *Ostrowski*, 1991, 1993; *Ballard and Heavens*, 1992; *Bednarz and Ostrowski*, 1996, 1998]. The main result of these considerations is a direct dependence of the spectra derived on the conditions at the shock. The power-law spectra can either be very steep or very flat for different mean magnetic field inclinations to the shock normal and varying amplitudes of perturbations [*Ostrowski*, 1991, 1993; *Bednarz and Ostrowski*, 1996].

The acceleration studies considered by the above authors applied very simple approaches for numerical modeling of the perturbed magnetic field structure. The purpose of the present work is to simulate the first-order Fermi acceleration process at mildly relativistic shock waves propagating in a more realistically modeled perturbed magnetic fields, taking into account a wide wave vector range turbulence with the power-law spectrum and continuity of the magnetic field across the shock, involving the respective matching conditions. The much extended version of this work will be published in ApJ.

Below, c is the speed of light. All calculations are performed in the respective local (upstream or downstream) plasma rest frames. The upstream (downstream) quantities are labeled with the index '1' ('2'). We consider ultrarelativistic particles with $p = E$. In the units we use in our simulations a particle of unit energy $E = 1$ moving in an uniform mean upstream magnetic field, $B_{0,1}$, has the unit maximum (for $p\perp = E$, where $\vec{p}\perp \cdot \vec{B}_{0,1} = 0$) gyroradius $r_g(E = 1) = 1$ and the respective resonance wave vector $k_{res}(E = 1) = 2\pi$.

2. SIMULATIONS

In the simulations, trajectories of test particles are derived by integrating their equations of motion in the perturbed magnetic field. We consider a relativistic planar shock wave propagating in a rarefied electron-proton plasma with the turbulent magnetic field frozen in it. Upstream of the shock the field consists of the uniform component, $B_{0,1}$, inclined at some angle ψ_1 to the shock normal and finite-amplitude perturbations imposed upon it. The irregular component has either a flat $(F(k) \sim k^{-1})$ or a Kolmogorov $(F(k) \sim k^{-5/3})$ wave power spectrum in the (wide) wave vector range (k_{min}, k_{max}). The perturbations are assumed to be static in a local plasma rest frame, both upstream and downstream of the shock. Thus the possibility of second-order Fermi acceleration is excluded from our considerations. The shock moves with velocity u_1 with respect to the upstream plasma. The downstream flow velocity u_2 and the magnetic field structure are obtained from the hydrodynamic jump conditions at the shock propagating in the electron-proton plasma. Derivation of the shock compression ratio, defined in the shock rest frame as $R = u_1/u_2$, was based on the approximate formulae derived by Heavens and Drury (1988)[1].

We consider the acceleration process in the particle energy range where radiative (or other) losses can be neglected. The particle spectra and angular distributions are measured for different magnetic field amplitudes δB (see below), field inclinations ψ_1, and shock velocities u_1.

2.1. The Magnetic Field Structure

The magnetic field perturbations upstream of the shock are modeled as a superposition of sinusoidal static waves of finite amplitude [cf. *Ostrowski*, 1993]. In the magnetic field related primed coordinate system[2] they take the form:

$$\delta B_{x'} = \sum_{l=1}^{294} \delta B_{x'l} \sin(k^l_{x'y'} y' + k^l_{x'z'} z'), \qquad (1)$$

and analogously for components $\delta B_{y'}$ and $\delta B_{z'}$. Such form of $\delta \vec{B}$ ensure that $\nabla \cdot \vec{B} = 0$. The index 'l' enumerates the unit logarithmic wave vector range from which the wave vectors k^l_i ($i = x',y',z'$) are randomly drawn. The components $k^l_{x'y'}$, $k^l_{x'z'}$ of the wave vectors $k^l_{x'}$ are selected by choosing a random phase angle $\phi^l_{x'}$ so that $k^l_{x'y'} = k^l_{x'} \cos \phi^l_{x'}$ and $k^l_{x'z'} = k^l_{x'} \sin \phi^l_{x'}$ ($k^{l\,2}_{x'y'} + k^{l\,2}_{x'z'} = k^{l\,2}_{x'}$) and analogously for other components. The wave vectors span the range (k_{min}, k_{max}), where $k_{min} = 0.0001$ and $k_{max} = 10$. It was checked by simulations that the number of wave vectors used in the simulations is well sufficient for the perturbed magnetic field to diffusively scatter particles. The selected form of the perturbations and the method of k^l_i vector components drawing both produce isotropic turbulence upstream of the shock. Using larger number of waves would result in respectively longer simulation times.

The respective wave amplitudes δB_{il} are selected at random in such a way to satisfy $\delta B_l^2 = \sum_i \delta B_{il}^2$ and the amplitudes δB_l are chosen to reproduce the turbulence power spectrum assumed. In the wave vector range considered, this spectrum can be written:

$$\delta B_l(k) = \delta B_l(k_{min}) \left(\frac{k}{k_{min}}\right)^{(1-q)/2}, \qquad (2)$$

---

[1] The work of Heavens and Drury (1988) assumes thermal equilibrium between electrons and protons downstream from the shock. This assumption may not properly account for a physical situation near astrophysical shocks.

It leads to the compression ratios greater than 4 for mildly relativistic shocks, as opposed to the more realistic estimates that give R between 3 and 4 (see, e.g., *Double et al.* [2004]). However, the usage of more exact formulae for R would not change qualitative results of the present paper.

[2] In the upstream plasma rest frame the x-axis of an (unprimed) Cartesian coordinate system (x, y, z) is perpendicular to the shock surface. The shock wave moves in the negative x direction and the regular magnetic field lies in the xy-plane. The primed system (x', y', z') is obtained from the unprimed one by its rotation about the z-axis by an angle ψ_1, so that the x'-axis is directed along $\vec{B}_{0,1}$.

where q is the wave spectral index ($q = 1$ for the flat spectrum and $q = 5/3$ for the Kolmogorov one). The constant $\delta B_l(k_{min})$ is scaled to match the model parameter $\delta B \equiv [\sum_l \delta B_l^2]^{1/2}$. Its ratio to the upstream mean magnetic field, $\delta B/B_{0,1}$, is our measure of the field perturbations amplitude.

2.2. Details of the Simulation Method

The particle equations of motion are integrated in the local (upstream or downstream) plasma rest frame, where the electric field vanishes. Particles are injected at the shock into the upstream region with initial energies E_0 such that their resonance wave vectors $k_{res}(E_0) \gg k_{max}$. The trajectory of each particle is followed until it crosses the shock surface or reaches a boundary located 200 gyroradii upstream of the shock. The boundary is required to treat particles whose resonance wave vectors are much smaller than k_{min}. Such particles are only weakly scattered by the turbulence and can propagate very far upstream of the shock. The upstream particles which cross the shock front are Lorentz-transformed into the downstream plasma rest frame. At this side of the shock a free escape boundary is located at $x_{max}(E) = 30\, r_g(E)$ behind it. In the conditions considered by us, very few particles are able to diffuse from such a distance back to the shock. For particles crossing the shock back to the upstream region, the momentum vectors are transformed to the upstream plasma rest frame and the whole procedure is repeated. The spectrum of accelerated particles and their angular distribution is measured at the shock as particles cross it. The spectra presented are averaged over a few (usually 10) different sets of particles and realizations of the turbulent magnetic field. The computations are finished when further calculations do not modify the spectrum formed.

We use a method of trajectory splitting to derive spectra in a wide energy range [cf. *Kirk and Schneider*, 1987b; *Ostrowski*, 1991, 1993]. At the beginning of the simulations the same initial weight, w_0, is ascribed to each particle. During calculations some particles are lost by escaping through the boundaries. To keep the total number of particles active in the simulations constant, for each particle lost downstream or upstream we duplicate (or multiplicate) one of those that crossed the shock upstream, with the weight of the original particle equally divided between the 'daughter' particles [see e.g. *Ostrowski*, 1993]. Then, we weakly perturb the motion of these particles to create different trajectories out of the splitted one. With this method, the effective number of particles in the simulation is much higher than the injected number, and the particle spectrum is determined with approximately the same accuracy at all energies.

2.3. Derivation of Particle Trajectories: A Hybrid Approach

The particle orbits are calculated using the Runge-Kutta 5th order method with the adaptive stepsize control [*Press et al.*, 1992]. However, in the turbulent magnetic field the procedure is very time consuming for higher energy particles because exact derivation of a trajectory with short wave perturbations requires extremely short time-steps to be used. In the same time, the short waves considered, with $\lambda \ll r_g$, only weakly influence trajectories of such particles. Therefore, in our modeling the motion of a higher energy particle is derived within the proposed hybrid approach. It involves exact integration of particle trajectories in the turbulent field including long and resonance waves with $\lambda > 0.05\, r_g(B = B_{0,1})$ and the short wave influence at the trajectory is included as a respective small amplitude pitch-angle scattering term. The scattering probability distribution is determined by additional simulations involving the full set of the short waves considered.

3. RESULTS

In Figure 1 we present particle spectra for the oblique subluminal ($u_{B,1} \equiv u_1/\cos \psi_1 < c$) shock wave propagating with velocity $u_1 = 0.5c$ and the mean upstream magnetic field inclination to the shock normal $\psi_1 = 45°$. The shock velocity along the mean magnetic field is then $u_{B,1} = 0.71c$ and the shock compression ratio is $R = 5.11$. The particle spectra are measured at the shock for three different magnetic field perturbation amplitudes and the flat (Figure 1a) and the Kolmogorov (Figure 1b) wave power spectra. One may note the following features in the spectra:

- the particle spectra diverge from a power-law in the full energy range;
- before the spectrum cut-off a harder spectral component can appear;
- the exact shape of the spectrum depends on both the amplitude of the magnetic field perturbations and the wave power spectrum.

The power-law part of the particle spectrum steepens with increasing amplitude of the field perturbations. For the flat turbulence spectrum the (phase space distribution function) spectral index is $\alpha \approx 3.08$ for a weakly perturbed field, $\delta B/B_{0,1} = 0.3$. At larger amplitudes, $\alpha \approx 3.24$ for $\delta B/B_{0,1} = 1.0$ and $\alpha \approx 3.53$ for $\delta B/B_{0,1} = 3.0$. The spectral indices obtained are consistent with previous numerical calculations of Ostrowski (1991, 1993) and the analytic results obtained in the limit of small perturbations by Kirk and Heavens (1989).

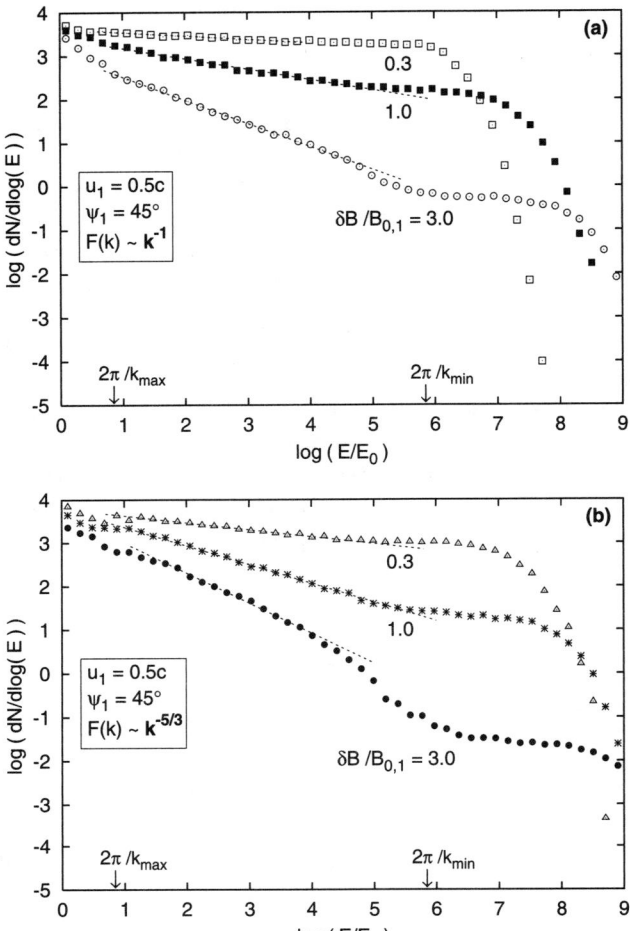

by the magnetic field inhomogeneities. The character of the spectrum changes at highest particle energies where $k_{res} \leq k_{min}$. These particles are only weakly scattered. Then the anisotropically distributed upstream particles can effectively reflect from the region of compressed magnetic field downstream of the shock leading to the spectrum flattening [cf. *Ostrowski*, 1991]. The effectiveness of reflections and the resulting flattening of the spectrum depend on the amplitude and spectrum of the field perturbations, as can be seen in the figures. The cut-off in the spectrum is formed mainly due to very weakly scattered particles escaping through the upstream boundary.

Examples of the particle angular distributions derived at the shock, in the shock normal rest frame, are presented in Figure 2.

The spectra obtained for superluminal shocks with $u_{B,1} \approx 2c$ are presented in Figure 3. For the low amplitude turbulence ($\delta B / B = 0.3$) we approximately reproduce results of Begelman and Kirk (1990), with a 'super-adiabatic' compression of injected particles, but no power-law spectral tail. At larger turbulence amplitudes power-law sections in the spectra are produced again, but the steepening and the cut-off occur at lower energies in comparison to the subluminal shocks (Figure 1).

Figure 1. Accelerated particle spectra at the subluminal shock wave ($u_1 = 0.5c$, $\psi_1 = 45°$ and $u_{B,1} = 0.71c$) in the shock rest frame for (a) the flat ($F(k) \propto k^{-1}$) and (b) the Kolmogorov ($F(k) \propto k^{-5/3}$) wave spectrum of the magnetic field perturbations. Any individual point in the spectrum represents a particle weight recorded per a logarithmic energy bin. The upstream perturbation amplitude $\delta B / B_{0,1}$ is given near the respective results. Linear fits to the power-law parts of the spectra are also presented. Particles of energies in the range indicated by arrows can effectively interact with the magnetic field inhomogeneities ($k_{min} < k_{res} < k_{max}$).

The differences in particle spectra obtained for the Kolmogorov and the flat wave power spectrum are visible from comparison of Figure 1a and 1b. In particular, the power-law part of the spectrum is steeper for the Kolmogorov case for a given perturbation amplitude ($\alpha \approx 3.16$ ($\delta B / B_{0,1} = 0.3$), $\alpha \approx 3.43$ ($\delta B / B_{0,1} = 1.0$), $\alpha \approx 3.69$ ($\delta B / B_{0,1} = 3.0$)).

The non-power-law character of the particle spectra results from the limited dynamic range of the magnetic field perturbations. In the energy range where the approximate power-law spectrum forms, particles can be effectively scattered

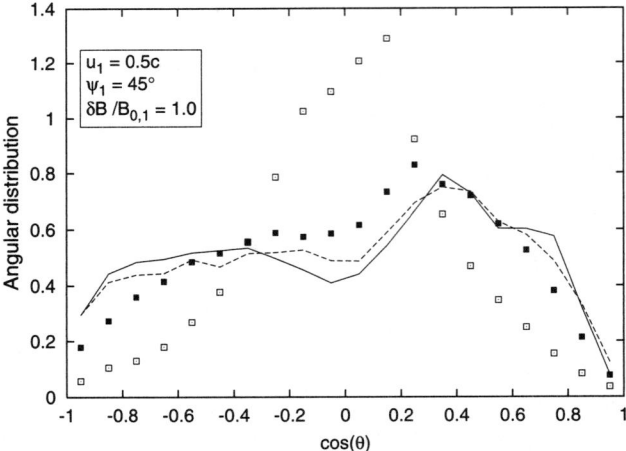

Figure 2. Particle angular distributions at the shock front with the parameters indicated. θ is the angle between the particle momentum and the shock normal. The angular distributions for particles forming the power-law part of the energy spectrum, $1 \leq \log (E/E_0) \leq 5$, are shown by solid and open squares for the flat and the Kolmogorov wave power spectrum, respectively. Angular distributions of particles forming the harder spectral component at $5 \leq \log (E/E_0) \leq 7$, are presented by solid (flat spectrum) and dashed (Kolmogorov one) lines. The distributions presented are normalized to unity.

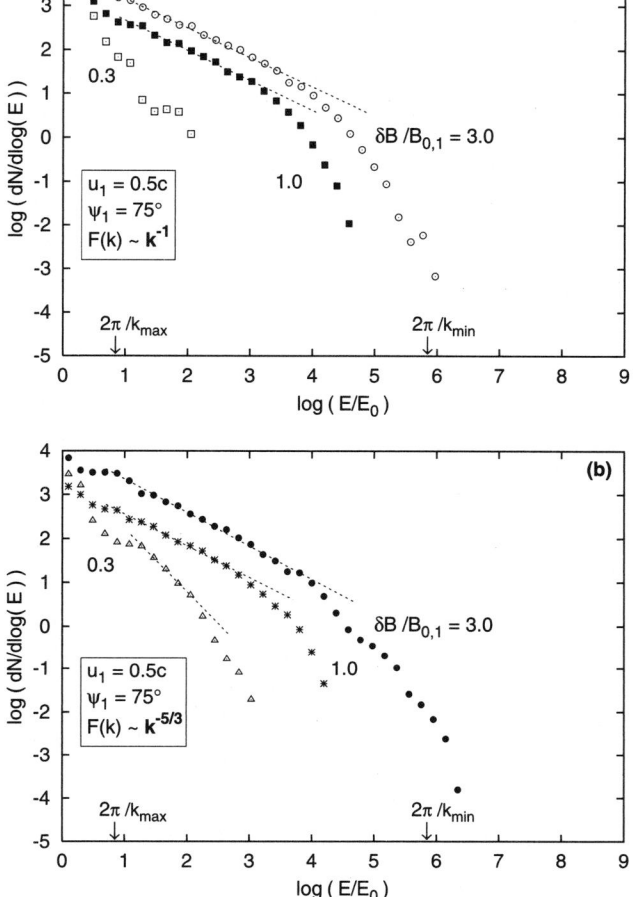

Figure 3. Accelerated particle spectra at the superluminal shock and (a) the flat and (b) the Kolmogorov spectra of magnetic field perturbations. The considered shock parameters are: $u_1 = 0.5c$, $\psi_1 = 75°$ and $u_{B,1} = 1.93c$.

4. SUMMARY

The present work is intended to study some aspects of the first-order Fermi acceleration process acting at relativistic shocks. We apply the test particle approach and we neglect possibility of radiative losses or the second order acceleration process acting in the turbulent medium near the shock. The results are presented for the mildly relativistic shock with velocity $0.5c$ and the mean magnetic field inclinations resulting in the subluminal ($\psi_1 = 45°$) and the superluminal ($\psi_1 = 75°$) shock configurations. In comparison to the previous work we include a few "realistic" features of the considered turbulence:

- an analytic model for the turbulence, with the power-law spectrum within a finite (wide) wave vector range

- conditions with different mean magnetic field inclinations to the shock normal and a range of turbulence amplitudes

- continuity of the turbulent magnetic field structure across the shock (the downstream field is derived from the upstream one with the respective shock jump conditions)

The modeling shows how the resulting spectra of accelerated particles depend on the perturbed magnetic field structure considered. In particular, we demonstrate

- variation of the accelerated particle spectral index with the turbulence amplitude for a given mean field inclination

- variation of this spectral index for varying the mean field inclination

- effects of the finite wave vector range of the turbulence leading to deviations of the derived spectra from the usually considered power-law form.

Acknowledgments. The work was supported by the Polish State Committee for Scientific Research in 2002-2004 as research project 2 P03D 008 23 (J.N.) and in 2002-2005 as research project PBZ-KBN-054/P03/2001 (M.O.)

REFERENCES

Ballard, K.R., and A. Heavens, Shock acceleration and steep- spectrum synchrotron sources, *MNRAS*, 259, 89-94, 1992.

Bednarz, J., and M. Ostrowski, The acceleration time scale for first-order Fermi acceleration in relativistic shock waves, *MNRAS*, 283, 447-456, 1996.

Bednarz, J., and M. Ostrowski, Energy spectra of cosmic rays accelerated at ultrarelativistic shock waves, *Phys. Rev. Lett.*, 80, 3911, 1998.

Begelman, M.C., and J.G. Kirk, Shock-drift particle acceleration in superluminal shocks: A model for hot spots in extra-galactic radio sources, *Astrophys. J.*, 353, 66-80, 1990.

Decker, R.B., and L. Vlahos, Shock-drift acceleration in the presence of waves, *J. Geophys. Res.*, 90, 47-56, 1985.

Double, G.P., M.G. Baring, F.C. Jones, and D.C. Ellison, Magneto-hydrodynamic jump conditions for oblique relativistic shocks with gyrotropic pressure, *Astrophys. J.*, 600, 485-500, 2004.

Heavens, A., and L'O.C. Drury, Relativistic shocks and particle acceleration, *MNRAS*, 235, 997-1009, 1988.

Kirk, J.G., and A. Heavens, Particle acceleration at oblique shock fronts, *MNRAS*, 239, 995-1011, 1989.

Kirk, J.G., and P. Schneider, On the acceleration of charged particles at relativistic shock fronts, *Astrophys. J.*, 315, 425-433, 1987a.

Kirk, J.G., and P. Schneider, Particle acceleration at shocks: A Monte Carlo method, *Astrophys. J.*, 322, 256-265, 1987b.

Ostrowski, M., Monte Carlo simulations of energetic particle transport in weakly inhomogeneous magnetic fields-I. Particle acceleration in relativistic shock waves with oblique magnetic fields, *MNRAS*, 249, 551-559, 1991.

Ostrowski, M., Cosmic ray acceleration at relativistic shock waves in the presence of oblique magnetic fields with finite-amplitude perturbations, *MNRAS*, 264, 248-256, 1993.

Press, W., S.A. Teukolsky, W.T. Vetterling, and B.P. Flannery, Numerical recipes in FORTRAN 77, *Cambridge University Press, Cambridge, 1992.*

Jacek Niemiec, Institute of Nuclear Physics, Polish Academy of Sciences, ul. Radzikowskiego 152, 31-342 Kraków, Poland. (Jacek.Niemiec@ifj.edu.pl)

Michał Ostrowski, Astronomical Observatory, Jagiellonian University, ul. Orla 171, 30-244 Kraków, Poland.

Kinetics of Particles in Relativistic Collisionless Shocks

Mikhail V. Medvedev[1], Luis O. Silva[2], Ricardo A. Fonseca[2], J. W. Tonge[3], and Warren B. Mori[3]

Charged plasma particles form an anisotropic counter-streaming distribution at the front of a shock. In the near- and ultra-relativistic regimes, Weibel (or two-stream) instability produces near-equipartition, chaotic, small-scale magnetic fields. The fields introduce effective collisions and "thermalize" the plasma particles via pitch-angle scattering. The properties of jitter radiation emitted by accelerated electrons from these small-scale fields are markedly different from synchrotron spectra. Here we give a theoretical summary of the results of recent numerical 3D PIC plasma kinetic simulations. The relation of the obtained results to the theory of gamma-ray bursts is outlined.

INTRODUCTION

It is well known that hydrodynamic shocks may occur in neutral media, in which collisions between gas particles mediate the pressure. However, the interplanetary space gas is very dilute and highly ionized. Hence Coulomb collisions are very rare and electric and magnetic fields, instead, mediate particle-particle interactions. Explosions in such rarefied plasma also drive shock waves, but these shocks are very different from their hydrodynamic counterparts and are referred to as "collisionless." The structure and properties of collisionless shocks vary dramatically and depend on many parameters, e.g., the value and orientation of magnetic fields in a plasma, the temperatures and masses of different plasma species, the shock speed compared to the speed of light, and others. Recently we made a great progress in theoretical understanding of ultra-relativistic collisionless shock moving in unmagnetized plasma and showed that non-linear Weibel (two-stream) instability plays a major role. A novel plasma radiation mechanism, which we discovered and called the "jitter" mechanism, is an unambiguous spectral benchmark of strong Weibel turbulence present in the shock. First three-dimensional particle-in-cell plasma kinetic simulations, performed with the OSIRIS code by the USA–Portugal collaboration, fully confirm our predictions. Moreover, this model receives an independent and very strong observational support from astrophysics, namely spectral characteristics of gamma-ray bursts [*Frontera et al.*, 2000, *Preece et al.*, 2000], in particular the existence of too steep spectra inconsistent with the simple synchrotron origin, but naturally explained by the jitter theory [*Medvedev*, 2000].

In this paper we review the theory of Weibel instability in application to ultra-relativistic shocks, which likely form in gamma-ray bursters, and present some 3D PIC plasma kinetic simulations, which confirm the theoretical predictions. We also discuss observational predictions of the theory based on the jitter radiation mechanism, which replaces the conventional synchrotron mechanism in the case of very small-scale, random magnetic fields.

2. RELATIVISTIC SHOCK THEORY

In general, relativistic shocks, as well as other shocks with the Mach number greater than three, must be highly turbulent. The source and the mechanism of the turbulence is thought to be kinetic in order to prevent multi-stream

---

[1]Department of Physics and Astronomy, University of Kansas, Lawrence, Kansas.
[2]GoLP/Instituto Superior Tecnico, Lisbon, Portugal.
[3]Department of Physics and Astronomy, University of California at Los Angeles, Los Angeles, California.

motion of plasma particles. It has been shown that the relativistic Weibel instability operates on the shock front [*Medvedev and Loeb*, 1999]. This instability is driven by the anisotropy of the particle distribution function (PDF) associated with a large number of particles reflected from the shock potential.

The instability under consideration was first predicted by [*Weibel*, 1959] for a non-relativistic plasma with an anisotropic distribution function. The simple physical interpretation provided later by *Fried* [1959] treated the PDF anisotropy more generally as a two-stream configuration of cold plasma. Below we give a brief, qualitative description of this two-stream magnetic instability.

Let us consider, for simplicity, the dynamics of the electrons only, and assume that the protons are at rest and provide global charge neutrality. The electrons are assumed to move along the x-axis (as illustrated in Figure 1) with velocities $\mathbf{v} = +\mathbf{x}v_x$ and $\mathbf{v} = -\mathbf{x}v_x$, and equal particle fluxes in opposite directions along the x-axis (so that the net current is zero). Next, we add an infinitesimal magnetic field fluctuation, $\mathbf{B} = \mathbf{z}\,B_z\cos(ky)$. The Lorentz force, $-e(\mathbf{v} \times \mathbf{B})/c$, deflects the electron trajectories. As a result, the electrons moving to the right and those moving to the left will concentrate in spatially separated current filaments. The magnetic field of these filaments appears to increase the initial magnetic field fluctuation. The growth rate is $\Gamma = \omega_p(v_x/c)$. Similar considerations imply that perpendicular electron motions along y-axis, result in oppositely directed currents, which suppress the instability. The particle motions along $\mathbf{z}$ are insignificant as they are unaffected by the magnetic field. Thus, the instability is indeed driven by the PDF anisotropy and should quench for the isotropic case.

The Lorentz force deflection of particle orbits increases as the magnetic field perturbation grows in amplitude. The amplified magnetic field is random in the plane perpendicular to the particle motion, since it is generated from a random seed field. Thus, the Lorentz deflections result in a pitch angle scattering, which makes the PDF isotropic. If one starts from a strong anisotropy, so that the thermal spread is much smaller than the particle bulk velocity, the particles will eventually isotropize and the thermal energy associated with their random motions will be equal to their initial directed kinetic energy. This final state will bring the instability to saturation.

Below we note the following points about the nature of the instability:

1. The free energy source is the anisotropy of the PDF at and near the shock associated with a two-stream motion of plasma particles: the inflowing (in the shock frame) particles of the external medium and the group of outgoing particles reflected from the shock front.
2. Despite its intrinsically kinetic nature, the instability is non-resonant, i.e., it is impossible to single out a group of particles that is responsible for the instability. Since the bulk of the plasma participates in the process, the energy transferred to the magnetic field could be comparable to the total kinetic energy of the plasma.
3. The characteristic growth time (inverse growth rate) for each species (electron, proton, positron, if any) is $\sim\omega_p^{-1}$ [$\omega_p = (4\pi e^2 n/m)^{1/2}$ is the plasma frequency] in the shock frame, which is very short compared to a shock evolution time scale.
4. The instability is *aperiodic*, i.e., Re $\omega = 0$. Thus, it can be saturated by nonlinear effects only, and not by kinetic effects such as collisionless damping or resonance broadening. Hence, the magnetic field strength may reach a large amplitude.
5. The instability generates randomly oriented magnetic fields on a spatial scale of order the plasma skin-depth $\sim c/\omega_p$. These fields predominantly lie in the plane of the shock front.
6. The instability introduces an effective scattering process into the otherwise collisionless system. This validates the use of the MHD approximation in the

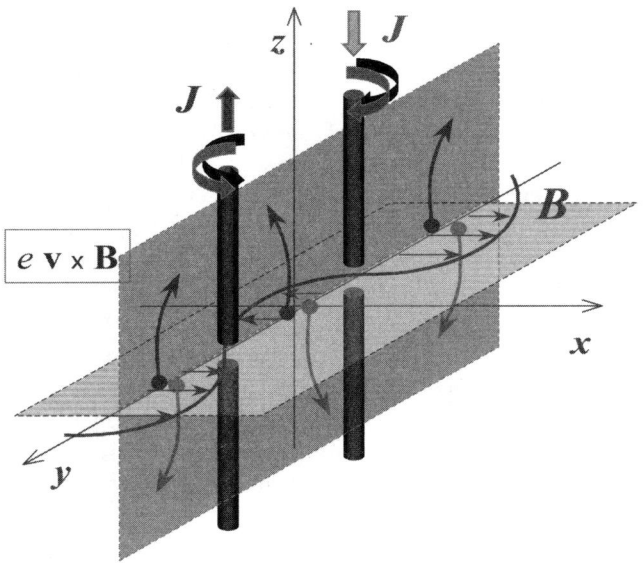

Figure 1. A diagram representing the mechanism of the Weibel instability. Counter-propagating charges are deflected by infinitesimally small magnetic fields to form current filaments. The magnetic fields of the filaments add up to (amplify) the initial magnetic field and create a positive feedback loop.

study of the large-scale dynamics of collisionless shocks. It makes the PDF isotropic via pitch-angle scattering and, thus, effectively heats the electrons and protons.

3. NUMERICAL SIMULATIONS OF THE WEIBEL INSTABILITY

The dynamics of the Weibel instability has recently been simulated by several research groups using 3D plasma kinetic codes [*Nishikawa et al.*, 2003; *Frederiksen et al.*, 2004]. We examined the instability, which occurs in a collision of two inter-penetrating unmagnetized plasma blobs with zero net charge [*Silva et al.*, 2003]. This is the simplest model for the formation region of a shock front, as well as a classic scenario unstable to electromagnetic and/or electrostatic plasma instabilities. To probe the full nonlinear dynamics and the saturated state of this system it is necessary to employ kinetic numerical simulations.

The fully electromagnetic relativistic PIC code OSIRIS [*Fonseca et al.*, 2002] was used to perform the first three-dimensional kinetic simulations of the collision of two plasma shells, and to observe the three-dimensional features of the electromagnetic filamentation instability, or Weibel instability.

The simulations were performed on a $256 \times 256 \times 100$ grid [the box size is $25.6 \times 25.6 \times 10.0 \ (c/\omega_{pe})^3$] with the total of 105 million particles for 2900 time steps [corresponding to $150.0 \ \omega_{pe}^{-1}$], with periodic boundary conditions. In all runs, energy is conserved down to 0.025%. In our simulations, at $t = 0$ there are two groups of particles moving along the vertical ($x3$) axis in opposite directions in the center of mass frame and occupying the entire simulation volume. The particles in both groups have a small thermal spread. The system has no net charge and no net current, and initially the electric and magnetic fields are set to zero. We performed both sub-relativistic ($\gamma_0 \sim 1.17$, γ_0 is the initial Lorentz factor) and ultra-relativistic ($\gamma_0 \sim 10.05$) simulations. Note that in the dimensionless units used, the results of the simulations are equally applicable to the collision of electron-proton plasmas as well.

The 3D structure of current filaments is shown in Figure 2. The temporal evolution of the total energy in the produced magnetic and electric fields is shown in Figure 3. During the linear stage of the instability there is rapid generation of a strong magnetic field, which predominantly lies in the plane of the shock ($x1x2$ plane), i.e., perpendicular to the direction of motion of the plasma shells. The magnetic field energy density reaches $\sim 5...20\%$ for $\gamma_0 \sim 1.17...10.05$. In all cases, the produced electric field is significantly weaker

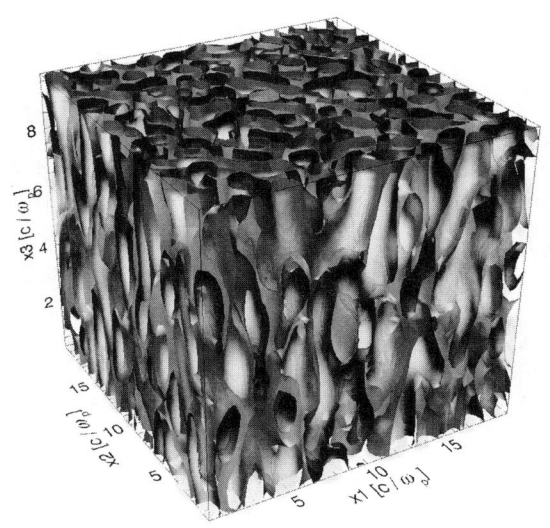

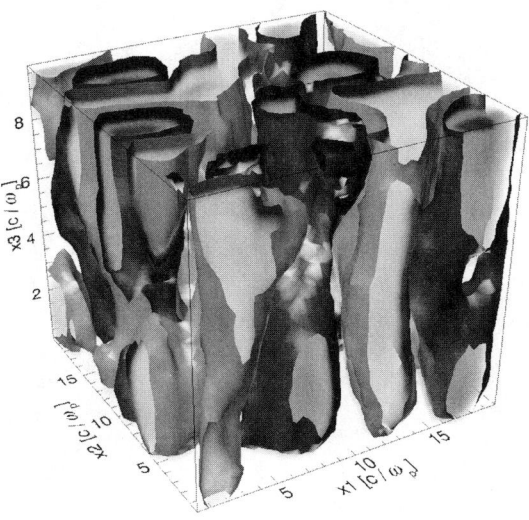

Figure 2. Three-dimensional structure of current filaments (depicted as iso-density contours; no gray scale coding is used) obtained from plasma kinetic 3D PIC simulations, at two different times ($t = 10$ and $t = 50$ electron plasma times; $\gamma\beta = 0.6$, where $\beta = v/c$ is the dimensionless velocity and γ is the corresponding Lorentz factor).

than the magnetic field. The linear growth rate agrees well with the theoretical estimates for the Weibel instability.

After a short linear stage, the instability enters the nonlinear regime (at $t \sim 10\omega_{pe}^{-1}$) in which current filaments begin to interact with each other, forcing like currents to approach each other and merge. During this phase, initially randomly

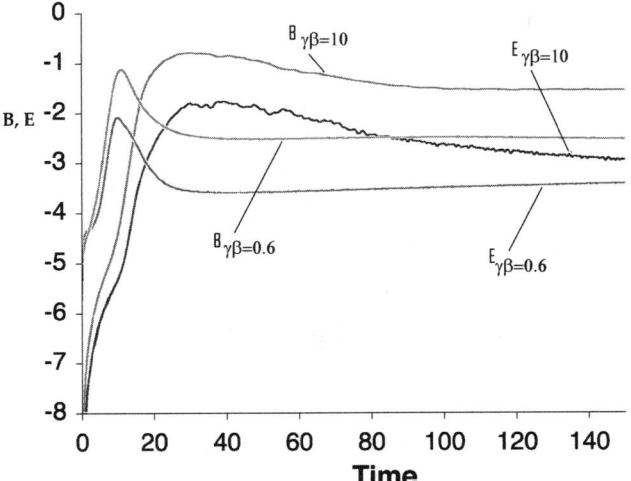

Figure 3. Temporal evolution of the magnetic and electric fields normalized to the total initial kinetic energy of plasma streams for sub- and ultra-relativistic shock conditions.

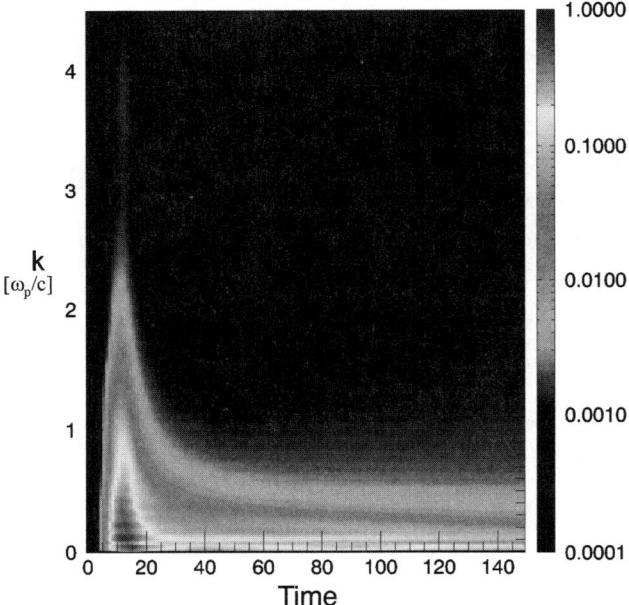

Figure 4. Temporal evolution of the magnetic field spectral density distribution, normalized to the peak spectral density for the sub-relativistic shock.

away. The total magnetic field energy is ~0.25% and does not change any more. Note that the residual magnetic field is highly inhomogeneous, seen as a collection of magnetic field filaments or "bubbles." The amplitude of the field in the bubbles is close to equipartition. Therefore, the overall decrease of the B-field energy is mostly associated with decreasing *filling factor* of the field. Note also that the magnetic domains separate current filaments of opposite polarity.

The topological evolution of the magnetic field is accompanied by heating and non-thermal particle acceleration, as illustrated in Figure 5. The particle energization is due to the pitch angle scattering in the produced B-field. The generation of non-thermal fast particles is more pronounced in the highest magnetic field scenarios. The presence of such high-energy particles is fundamental to provide the mildly relativistic particles to be injected in the accelerating structure of a shock that forms in the collision region.

We should comment here two things. First, the shock thickness is, in fact, greater that just a single skin depth. The thickness depends on the efficiency of pitch-angle scattering of the particles. The thickness would be on the order of skin depth if the generated fields were nearly equipartitioned. The simulations indicate that the fields saturate at the lower level, so it takes a bit longer to completely isotropize the particles. Second, the particles reflected from the shock may have a different bulk velocity and thermal spread. Our simulations with non-identical colliding electron-positron clouds indicate that it is the relative bulk velocity that matters, whereas the thermal spread usually diminished the growth

oriented filaments cross each other to form a more organized, large-scale quasi-regular pattern, hence much current and B-field is annihilated. At later times ($t \sim 30\omega_{pe}^{-1}$ and later), the filament coalescence continues, as indicated by the increase of the correlation scale, $\sim k^{-1}$, of the B-field in Figure 4. However, the spatial distribution of currents is now quite regular, so that filaments with opposite polarity no longer cross each other but simply interchange, staying always far

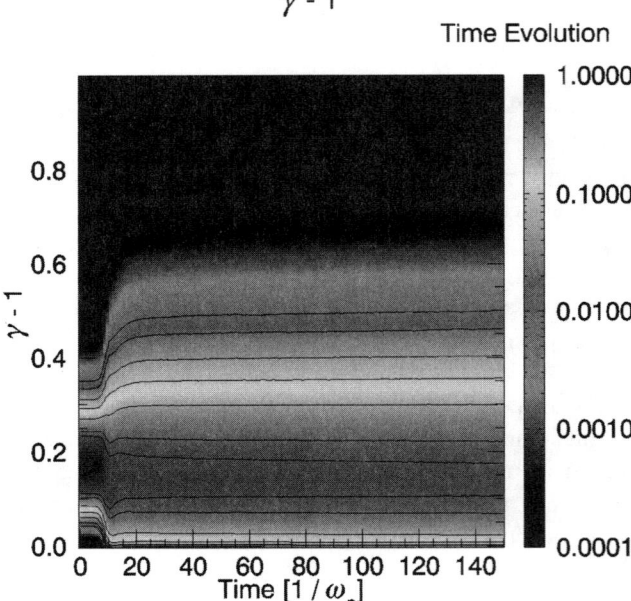

Figure 5. The evolution of the particle distribution function.

rate of the instability and the saturated value of the field, in accordance with the theory [*Medvedev and Loeb*, 1999].

4. JITTER RADIATION FROM SHOCKS

There is lore that radiation observed from astrophysical shocks is the synchrotron radiation emitted by accelerated electrons in nearly homogeneous (at least on a Larmor scale) magnetic fields. We have shown that, instead, the magnetic field in Weibel shocks is randomly tangled and its correlation length is less then the Larmor radius of an emitting electron. Hence, such an electron experiences random deflections as it moves through the field and its trajectory is, in general, stochastic. The synchrotron theory fails here.

In collisionless shocks dominated by Weibel turbulence, the particle moves almost straight, along the line of sight, and experiences high-frequency jittering in the perpendicular direction due to the random Lorentz force. We therefore refer the emerging radiation to as "jitter" radiation. Its spectrum is determined by random accelerations of the particle; it will be peaked at some frequency ω_j, which is estimated as follows. In the rest frame of the electron, the magnetic field inhomogeneity with wavenumber k_B is transformed into a transverse pulse of electromagnetic radiation with frequency $k_B c$. This radiation is then Compton scattered by the electron to produce observed radiation with frequency $\omega_j \sim \gamma^2 k_B c$ in the lab frame. It has been shown [*Medvedev*, 2000] that this frequency, referred to as the jitter frequency, is higher than the synchrotron frequency in the uniform magnetic field of the same strength. The calculated spectrum of jitter radiation is shown in Figure 6. The crucial difference of its spectrum from synchrotron is (i) harder spectrum: the photon spectra flux goes at low energies as $F \sim \nu^\alpha$ with $\alpha = 0$ in contrast to synchrotron $\alpha = -2/3$, and (ii) much sharper spectral breat at the jitter frequency. These properties can naturally explain the violation of the "synchrotron line of death" in time-resolved GRB spectra and the nature of the broken power-law spectra of GRBs.

In nature, magnetic fields at shocks may occupy a range of scales, from the smallest scales set by the electron skin depth through the proton skin depth, and perhaps, up to much larger scales set by the quasi-regular progenitor fields. In this case, we expect that the small-scale (electron skin depth scale) component will produce jitter spectrum, whereas all the larger-scale fields will produce synchrotron

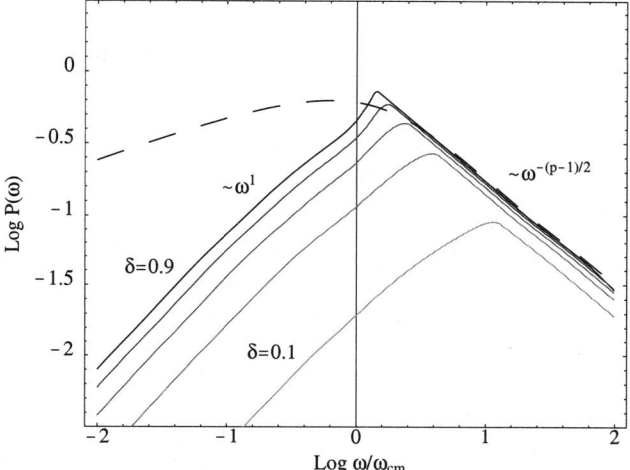

Figure 6. Spectra of jitter radiation; the synchrotron spectrum (dashed curve) is shown for comparison.

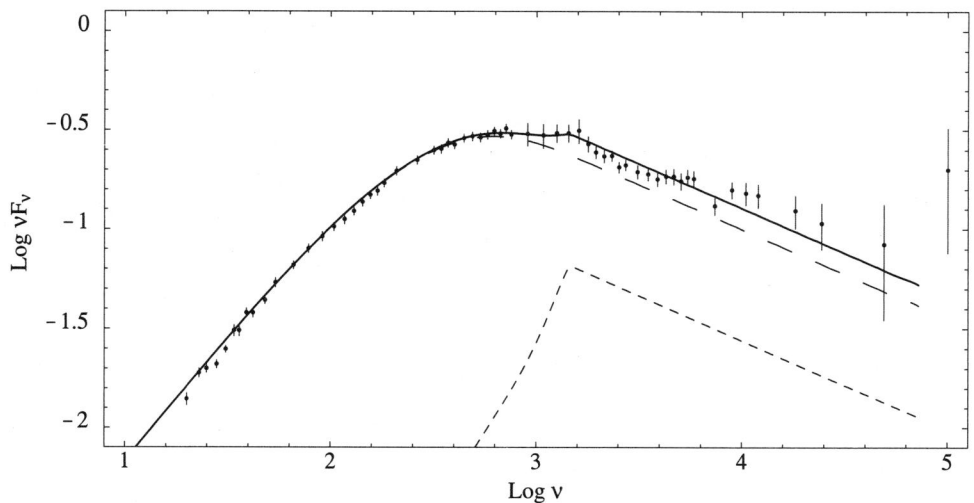

Figure 7. The spectrum GRB910503 fitted with the jitter + synchrotron composite spectral model.

spectrum. The natural spectral model for the GRBs involves both components. This composite jitter + synchrotron spectrum is illustrated in Figure 7. A more detailed discussion goes beyond of the scope of the present paper (but see *Medvedev*, 2000).

REFERENCES

Fonseca, R. A., et al., Lecture notes in computer science, 2329, III-342, Springer-Verlag, Heidelberg, 2002.

Frederiksen, J. T., Hededal, C. B., Haugboelle, T., Nordlund, A., Collisionless shocks – Magnetic field generation and particle acceleration, *Astrophys. J.* accepted, 2004 (astro-ph/0303360)

Frontera, et al., Prompt and delayed emission properties of gamma-ray bursts observed with BeppoSAX, *Astrophys. J. Suppl.*, 127, 2000.

Fried, B. D., Mechanism for instability of transverse plasma waves, *Phys. Fluids*, 2, 337, 1959.

Medvedev, M. V., Theory of "jitter" radiation from small-scale random magnetic fields and prompt emission from gamma-ray burst shocks, *Astrophys. J.*, 540, 704, 2000.

Medvedev, M. V., and Loeb, A., Generation of magnetic fields in the relativistic shock of gamma-ray burst sources, *Astrophys. J.*, 526, 697, 1999.

Nishikawa, K.-I., Hardee, P., Richardson, G., Preece, R., Sol, H., Fishman, G. J., Particle acceleration in relativistic jets due to Weibel instability, *Astrophys. J.*, 595, 555, 2003.

Preece, R. D., Briggs, M. S., Malozzi, R. S., Pendleton, G. N., Paciesas, W. S., and Band, D. L., The BATSE gamma-ray burst spectral catalog. I. High time resolution spectroscopy of bright bursts using high energy resolution data, *Astrophys. J. Suppl.*, 126, 2000.

Silva, L. O., Fonseca, R. A., Tonge, J. W., Dawson, J. M., Mori, W. B., and Medvedev, M. V., Interpenetrating plasma shells: Near equipartition magnetic field generation and nonthermal particle acceleration, *Astrophys. J. Lett.*, 596, L121, 2003.

Weibel, E. S., Spontaneously growing transverse waves in a plasma due to anisotropic velocity distribution, *Phys. Rev. Lett.*, 2, 83, 1959.

R. A. Fonseca and L. O. Silva, Centro de Fisica de Plasmas, Instituto Superior Tecnico, Av. Rovisco Pais, 1049-001 Lisboa, Portugal.

M. V. Medvedev, Department of Physics and Astronomy, University of Kansas, 1251 Wescoe Hall Drive, #1082, Lawrence, Kansas, 66045-7582. (medvedev@ku.edu)

W. B. Mori and J. W. Tonge, Department of Physics and Astronomy, University of California at Los Angeles, 405 Hilgard Avenue, Box 951361, Los Angeles, California, 90095-1361.

Studies of Relativistic Shock Acceleration

A. Meli

Max Planck Institut für Radioastronomie, Bonn, Germany

J. J. Quenby

Astrophysics Group, Imperial College of Science, Technology and Medicine, London, UK

We present Monte Carlo simulations of diffusive shock acceleration at highly relativistic parallel and oblique (sub-luminal and super-luminal) shock waves with high upstream flow Lorentz gamma factors (Γ), which could be relevant to models of highly relativistic particle shock acceleration. We investigate numerically the acceleration properties in the relativistic and ultra relativistic flow regime ($\Gamma \sim 10 - 10^3$), such as angular distribution, acceleration time constant, particle energy gain versus number of crossings and spectral shapes. Especially for the case of oblique shocks, the dependence on whether or not the scattering is pitch angle diffusion or large angle scattering is studied. The large angle model exhibits a distinctive structure in the basic power-law spectrum which is not nearly so obvious for small angle scattering. However, both models yield significant "speed-up" or faster acceleration rates when compared with the conventional, non-relativistic expression for the time constant, or alternatively with the time scale r_g/c, where r_g is Larmor radius. The Γ^2 energization for the first crossing cycle and the significantly large energy gain for subsequent crossings as well as the high speed-up factors found, are important in supporting the Vietri and Waxman work on Gamma Ray Bursts (GRBs), ultra-high energy cosmic ray output, and consequently to neutrino and gamma-ray production. For super-luminal shocks, the energy gain and the spectral shapes of the accelerated particles are given. For the last investigation only large angle scattering is considered, partly because of computational time limitations and partly because this model provides the most favourable situation for acceleration and high gamma flows are used with Lorentz factors in the range 10–40. The particle's trajectory is followed along the magnetic field lines during shock crossings where the equivalent of a guiding centre approximation is inappropriate, constantly measuring its phase space co-ordinates in the fluid frames where $\mathbf{E} = 0$. We find that a super-luminal "shock drift" mechanism—using large angle scatters—is less efficient in accelerating particles to the highest energies observed, allowing the speculation that the pitch angle scattering could be the key for very high energies to be attained (results to be published soon) in such a condition of high obliquity.

1. INTRODUCTION

The concept of the mechanism of diffusive Fermi acceleration at collisionless plasma shock waves has been around for quite a while. Fermi acceleration at shocks occurs in many physical and astrophysical systems such as in the solar system, in stellar winds, in supernovae, in accreting X-ray binaries, gamma ray bursts (GRBs), pulsar wind termination shocks, in our galaxy and throughout the universe. At the present it is generally believed that three distinct astrophysical situations, where the bulk plasma flow is extremely relativistic with Lorentz factors $\Gamma = (1 - V^2/c^2)^{-1/2} \geq 10$, could stand as a plausible source of Ultra High Energy Cosmic Rays (UHECRs) production. These are some active galactic nuclei (AGN) jets and "hot spot" sites, GRB fireballs and pulsar polar winds. Within these flows, relativistic shocks appear and diffusive particle acceleration takes place. In each case, there is evidence for energetic particle acceleration to high Γ factors but the upper limit to the possible energy attained becomes less certain with increasing bulk flow velocity.

At oblique shock fronts the motion of the particles is more complicated, compared to a parallel shock configuration, as they may either be transmitted or reflected by the shock surface. Apart from the diffusive acceleration mechanism, an oblique shock could also accelerate particles in the absence of scattering (*Armstrong and Decker*, 1979), via the namely "shock drift" mechanism. Provided that the motion of the particles is diffusive in the oblique shock configurations, it has been shown for an analytical non-relativistic flow approach, that the same result as in parallel shocks, connects the power-law index of accelerated particles with the compression ratio and the acceleration rate with the diffusion coefficients in the shock normal direction (e.g. *Axford et al.*, 1978; *Bell*, 1978a; *Drury*, 1983).

Also, following energy gain in the shock frame on shock crossing via the drift approach is equivalent to following the gain via Lorentz transformations from the shock frame into an $\mathbf{E} = 0$ frame (defined by *de Hoffmann and Teller*, 1950) at the shock, and then back into the shock frame. On the other hand, the theory of oblique shocks considers features that are not involved in the parallel shock case, in particular the ability of the shock to reflect particles meeting a weaker to stronger field transition and the dependence of the acceleration time on the field-shock normal angle (e.g. *Jokipii*, 1982). Oblique shocks can be classified into sub-luminal and super-luminal. The difference between these two categories is that, in the sub-luminal case, it is possible to find a relativistic transformation to the frame of reference (de Hoffmann-Teller frame), in which the shock front is stationary and the electric field is zero ($\mathbf{E} = 0$) in both upstream and downstream regions. On the other hand, super-luminal shock fronts do not admit a transformation to such a frame, as they correspond to shock fronts in which the point of intersection of the front with a magnetic field line moves with a speed greater than c. Acceleration time under non-relativistic theory clearly depends on shock obliquity through variation of K_n, the diffusion coefficient along the shock normal, with ψ, where $K_n = K_\parallel \cos^2 \psi + K_\perp \sin^2 \psi$. $K_\parallel, K_\perp$ are, respectively, the diffusion coefficients parallel and perpendicular to the field. The acceleration time is given by

$$\tau = \frac{3}{V_1 - V_2} \left(\frac{K_{n,1}}{V_1} + \frac{K_{n,2}}{V_2} \right) \quad (1)$$

Here the suffixes 1 and 2 refer to upstream and downstream, and V is plasma flow velocity. Because this equation (1) is clearly an inadequate measure of acceleration time in the highly relativistic case, we note it may also be written as τ lying in the range $4 \rightarrow (8/3)(Nr_g/c)\cos^2 \psi$ for a parallel mean free path $\lambda_\parallel = Nr_g$ (ψ is the angle between the field and the shock normal), a shock frame compression ratio of 4 and a downstream reduction in $\lambda_\parallel$ by a factor of 4. Hence while it will be convenient to measure any "speed-up" of acceleration relative to equation (1), we may equally regard the time to move over the relativistic Larmor radius as the unit of acceleration time for comparison purposes. To recover the Bohm diffusion limit, it is simply necessary to put $N = 1$ in the above expression for γ in terms of r_g. *Quenby and Lieu* (1989) and *Ellison et al.* (1990) employed Monte Carlo guiding centre approximation calculations to show that relativistic parallel shock flows with $\Gamma \sim 10$ produced a speed-up of acceleration by factors $\sim 3 \rightarrow 10$ when compared with equation (1) for τ. The former authors argued for large angle scattering while the latter found essentially the same result with either large or small angle scattering. *Lieu et al.* (1994) showed the speed-up as a function of flow velocity was similar for inclined, sub-luminal shocks, provided the results are plotted against flow seen in the $\mathbf{E} = 0$ frame. *Quenby and Drolias* (1995), using an inclined shock model which included the expected shock structure due to cosmic ray pressure back reaction, found some speed-up still occurred. *Bednarz and Ostrowski* (1996) employ a Monte Carlo computation scheme to investigate acceleration times for shock speeds up to $0.9c$ for the inclined case, with both small angle and large angle scattering and allowing for cross field motion of guiding centres. Speed-up is noticed especially at high shock speed and turbulence, although cross-field diffusion eventually reverses the correlation with turbulence level.

Both parallel and oblique shock acceleration numerical investigations are presented for both small and large angle scattering models, using the guiding centre Monte Carlo approach. A Γ^2 energy enhancement is found for the first shock crossing cycle, first emphasized by *Quenby and Lieu* (1989), and significant energy gains (a large factor times Γ) subsequently. The structure in the accelerated particle spectrum at the shock, indicated in previous lower Γ simulations, became enhanced for large angle scattering, although not in the small angle case, rendering the idea of a simple power-law output spectrum for photon production uncertain; these results confirm the existence of the speed-up effect. Truncated power-law spectra occurring in the super-luminal case for isotropic scattering are also presented, and acceleration time scales shorter than the upstream gyroperiod have been obtained.

2. NUMERICAL METHOD

Both large angle and pitch angle scattering, assuming elastic scattering in the fluid frames, are considered as the important aim is to investigate the role of different scattering models in reference to spectral shape, acceleration time scales and angular distribution in different frames of reference with varying shock obliquity at very high gamma plasma flows, ranging from $\Gamma = 10–10^3$. The relativistic frames of references used are the local fluid frame (1 and 2: upstream and downstream, respectively), the normal shock frame (s), and the de Hoffman-Teller (HT) frame. In general, flow into and out of the shock discontinuity is not along the shock normal, but a transformation is possible into the normal shock frame to render the flows along the normal (e.g. *Begelman and Kirk*, 1990), and for simplicity we assume such a transformation has already been made.

2.1. Parallel and Sub-Luminal Oblique Shocks

The purpose of these simulations is to find a solution to the particle transport equation for highly relativistic parallel flow velocities. The appropriate time-independent Boltzmann equation is given by the following as we assume steady state,

$$\Gamma(V + v\mu)\frac{df}{dx} = \frac{df}{dt}\bigg|_c \quad (2)$$

where V is the fluid velocity, v the velocity of the particle, Γ the Lorentz factor of the fluid frame, μ the cosine of the particle's pitch angle, and $(df/dt)|_c$ the collision operator. The reference frames used during the simulations are the upstream and downstream fluid frames and the shock frame. Pitch angle is measured in the local fluid frame, while x is in the shock rest frame and the model assumes variability in only one spatial dimension. In order for the above equation to be solved by a simple Monte Carlo technique, we assume that the collisions represent scattering in pitch angle, the scattering is elastic in the fluid frame and a phase-averaged distribution function may be employed. The collision operator may be characterized by two types of scattering that the particles suffer, large angle and small angle scattering (see overview by *Quenby and Meli*, this volume).

To establish the conditions for large angle scattering, for parallel shocks, the scatter centre Larmor radius of the particle is $r_g < \lambda_\parallel/2\Gamma^2$, assuming motion along the shock normal after upstream entry (above overview). In the overview article, we suggest that for oblique shocks, the condition becomes $r_g < (\pi/N)(\lambda_\parallel/2\Gamma^2)$ where the ambient field Larmor radius is $r_\alpha = \lambda/N$. Moreover, since this is derived for initial propagation perpendicular to the shock, it turns out that the scatter must occur within one half Larmor period (see overview by *Quenby and Meli*, this volume). This means $\lambda < \pi c/\omega_b = \pi r_\alpha$. Hence the criteria for large angle scatter are similar in the two cases if $N = 3$, a strong turbulence limit. In section 2.4 of the Quenby and Meli overview we discuss a GRB field "blob scenario" where sufficiently high r_g may be realized, where we note that more unfavourable N values were assumed.

A suitable scattering centre would be a rotational discontinuity or a much enhanced field region. Quenby and Meli (overview, this volume) discuss the parameter space where the guiding centre approximation may be used. It is found that for a reasonable fluctuation spectrum fitting the dimensions of the accelerating region is $K_\perp/K_\parallel \sim 0.05$ at Larmor radii where field line wandering is not dominant, i.e. high r_α. Here it is therefore resonable to neglect perpendicular diffusion (cross-field diffusion).

At low r_g where the field line angle, ψ, with the normal is crucial, the combination of field line wander and cross-field scattering jumps limits the model to $\psi < 66°$ shocks and also lambda (λ) needs to be greater than 10 r_g, which *Moussas et al.* (1992) make very likely. Cross-field jumps on large angle scatter change distances to the shock by $\sim r_\alpha \sin\psi$, which in the worst case for $N = 3$ would be $\lambda_\parallel/3$, rendering the super-luminal large angle scatter model a rather poor approximation, which needs improving in future work. This discussion of the validity of the guiding centre approximation is valid in the plasma frame. If the field in this frame is very oblique, close to the shock interface, the regime investigated by *Newman et al.* (1992) becomes applicable and it is likely that energy gains per crossing are underestimated, although time constants upstream and downstream should also shorten.

A "test particle" approximation is used for simplicity to allow a step by step approach to understanding the full problem, although eventually non-linear effects must be included (currently under investigation). Relativistic particles of initial $\gamma \sim (\Gamma+10)$ are injected upstream towards the shock and are allowed to scatter in the respective fluid frames with their basic motion described by the guiding centre approximation. It is noted here that *Bell* (1987) and *Jones and Ellison* (1991) have shown that "thin" sub-shocks appear even in the non-linear regime, so at some energy above the plasma Γ value, the accelerated particles may be dynamically unimportant while they recross the discontinuity. Another way of arriving at the test particle regime is if particles are injected well above the plasma particle energy, remaining dynamically unimportant and thus requiring the seed particles to have been already pre-accelerated. In AGN jets, traveling shocks superimposed on the relativistic flows accelerated by pressure at the jet base could provide the seed for the terminal hot spot acceleration. Such seed particles appear in the neutron star binary merger scenario for GRBs (*Narayan et al.*, 1992), which includes the presence of pre-accelerated particles before any terminal shock acceleration phase, possibly due to explosive reconnection. For isolated pulsars, we note that there is strong observational evidence (e.g. PSR 1913+16, PSR 2127+11C) for the continuous injection of relativistic particles into the surrounding medium over their lifetime. While this plasma will probably be predominantly in the form of e^+e^- pairs, created in the pulsar magnetosphere, it has been argued that pulsar winds must also contain ions in order to account for the electric currents in the Crab Nebula (*Hoshino et al.*, 1992; *Gallant and Arons*, 1994). These authors employ a wave acceleration process at the shocked termination of the flow, to provide a non-thermal pair spectrum to yield the X-ray and gamma-ray output. However their input distribution function is also not mono-energetic and therefore containing "seed particles" above the bulk flow energy. The supposed nature of the slot/gap electric field acceleration of the pair plasma is unlikely to give a mono-energetic output.

During the simulation a relativistic transformation is performed from the local plasma frames to the shock frame to check for shock crossings. We change units so that $m \approx c = 1$. Particles leave the system if they "escape" far downstream at the spatial boundary or reach a defined maximum energy E_{max} for computational convenience, even though other physics describing particle escape or energy loss would probably need to be taken into account in realistic situations (under investigation). The downstream spatial boundary required can be estimated from the solution of the convection-diffusion equation in a non-relativistic, large angle scattering approximation in the downstream plasma which gives the chance of return to the shock, $exp(-V_2 r_b/D)$, yielding a probability of return of $2 \cdot 10^{-3}$ if $r_b = 8\lambda_\parallel$. In fact, runs are performed with different spatial boundaries to investigate the effect of the size of the acceleration region on the spectrum, as well as to find a region where the spectrum is size independent. In the small angle scatter case, the inherent anisotropy due to the high downstream sweeping effect may greatly modify this analytical estimate.

The particles ($\sim 10^5$), weight (w_p) equal to 1.0, are injected far upstream and are allowed to move towards the shock, along the way colliding with the scattering centers. Provided multiple scattering between the upstream and downstream regions of the shock can occur and energy is gained in each crossing, diffusive shock acceleration is simulated.

A splitting technique is applied, similar to that of *Bednarz and Ostrowski* (1998), in order to obtain statistical accuracy over a wide range of particle Lorentz factors. The splitting technique helps to avoid the consequence that in highly relativistic flow environments, only a few high energy accelerated particles, which remain in the acceleration process, dominate the recorded distribution function, thereby limiting the statistical accuracy to an energy range of only two or three decades above injection energy E_o. By introducing more particles, but of corresponding lower weight when some energy well above injection is attained, a greater sampling of the possible regions within the accessible phase space of the model is achieved.

The mean free path is calculated in the respective fluid frames by the formula: $\lambda = \lambda_0 p_{1,2}$, assuming a momentum dependence to this mean free path for scattering along the field related to the spatial diffusion coefficient, K, in the shock normal or x direction by $K_\parallel = \lambda v/3$ and $K = K_\parallel cos^2 \psi$ (for the oblique shocks).

At the scattering centers the energy (momentum) of the particle is kept constant and only the direction of the velocity vector v is randomized, using a computational random number generator. The downstream mean free path is taken as a factor 4 less than upstream while the absolute value is arbitrary since the model boundaries are specified in units of the mean free path.

Gallant and Achterberg (1999) have demonstrated that small angle scattering (pitch angle diffusion) with $\delta\theta < 1/\Gamma$ applies with θ measured in the upstream fluid frame for scattering in a uniform field or a randomly orientated set of uniform field cells. This arises because particles attempting to penetrate upstream from the shock are swept back into the shock before they can move far in pitch. For this case, we allow particles' pitch angle diffusion to angles chosen at random, up to an angle $\sim 1/\Gamma$ (see also the overview by *Quenby and Meli*), where Γ is the upstream gamma, measured in the shock frame. This allows a direct investigation and prediction of the claimed limited energy gain on all crossings subsequent to the first cycle (*Gallant and Achterberg*, 1999), as it is important to establish the dependence of the model results on whether or not large angle or pitch angle diffusion operate in highly relativistic flows.

In the shock rest frame, the flow velocity (V_1, V_2 for upstream and downstream respectively) is parallel to the shock normal and the magnetic fields $\mathbf{B}_1$ and $\mathbf{B}_2$ are at an angle ψ_1 and ψ_2 to the shock normal respectively. We adopt a geometry with x in the flow direction, positive downstream, $\mathbf{B}_1, \mathbf{B}_2$ in the x-y plane and directed in the negative x and y directions and only $\mathbf{E}_z$ finite and in the positive z direction. A relativistic transformation is performed to the local fluid frame each time the particle scatters across the shock. Now, for the transformations between the shock frame and the de Hoffmann-Teller frame, we need to boost by a V_{HT} speed along the shock frame, where V_{HT} is equal to $V_1 \tan\psi_1$. The transformation to the de Hoffmann-Teller frame, though, is only possible if V_{HT} is less or equal to the speed of light, c. This means that $\tan\psi_1 \leq 1$ (sub-luminal case).

For the oblique shock cases, provided the field directions encountered are reasonably isotropic in the shock frame, we are about to show that $\tan\psi_1 = \Gamma_1^{-1}\tan\psi_s \sim \Gamma_1^{-1} \sim \psi_1$ where 1 and s refer to the upstream and shock frames respectively. Given the current interest in following shock acceleration in the relativistic flow in test particle regime, we outline a typical approach to such computations. Using the guiding centre approximation, a test particle moving a distance, d, along a field line at ψ to the shock normal, in the plasma frame has a probability of collision within d given by $P(d) = 1 - exp(-d/\lambda) = R$, where the random number R is $0 \leq R \leq 1$. Weighting the probability by the current in the field direction, μ, then $d = -\lambda\mu \ln R$. Between the ith and $(i+1)$th scattering shock crossing is checked by finding the position and time in the shock frame with $c = 1$ unit,

$$\Delta x_{i+1} = \Gamma \left(\Delta x_i + V \left| \frac{\Delta x_i sec\psi}{v\mu} \right| \right) \quad (3)$$

$$\Delta t_{i=1} = \Gamma \left(\left| \frac{\Delta x_i sec\psi}{v\mu} \right| + V\Delta x \right)$$

Let us comment here that whereas in the oblique superluminal case (next section), proper account of phase angle is taken via helical trajectory integration, conservation of the first adiabatic invariant in the de Hoffmann-Teller is assumed to determine reflection or transmission in the oblique subluminal case. Since in this frame, the allowed and forbidden angles for transmission depend only on the input pitch and phase, not on rigidity, the results of *Hudson* (1965) apply. Also for an isotropic input flux, the reflection coefficient ζ is simply given by particle flux conservation between upstream magnetic flux tube area FB_1/FB_2 ratio where F is a constant (e.g. *Parker*, 1965). *Hudson* (1965) shows the percentage of phase angles with reflection against incident pitch for a $\psi_1 = 15°$, $\psi_2 = 75°$ shock, very close to the compression ratio = 4 shock to be considered in the following. Weighting the phase angle space with parallel current and using his results gives excellent agreement with the predicted 0.732 reflection coefficient obtained from flux conservation. *Hudson* (1965) shows that the reflection percent against pitch never varies more than 20% from the mean value. Therefore although we are aware that the anisotropy in the relativistic shock situation renders the input to the shock from upstream, very anisotropic in pitch angle, we believe it to be a reasonable first approximation to randomize phase before transforming to the de Hoffmann-Teller frame and then using the adiabatic invariant to decide on reflection/transmission because of what we deduce from the Hudson result. Our deduction is that taking approximate constancy of reflection probability, averaged over phase, against pitch yields the same reflection coefficient as adiabatic invariance. A next stage in our work will be to apply the helical crossing algorithm to oblique, sub-luminal shocks (under preparation).

We would like to add here that it may be argued that current cannon-ball models for GRB ejecta imply large amplitude turbulence at the ejecta interface. Previous work on the effect of large amplitude turbulence at a near perpendicular shock in the non-relativistic regime has been done by *Newman et al.* (1992), using Monte Carlo 3D integrations in a realistic turbulent plus highly inclined shock model. They concluded that changes in the first adiabatic invariant were limited to 20%, while the average energy gain per shock crossing exceeded the non-relativistic, unperturbed field value by 18% or more. Their work encourages us both to use adiabatic theory as an approximate, lower limit to the relativistic shock acceleration in turbulent, inclined situations and in the future to introduce scattering in the shock region integrations. We note the work of *Niemiec and Ostrowski* (this volume) but point out our objectives are more to do with qualitative trends in acceleration rate and spectral structure as a function of Γ than of scattering regime details and believe our work is a pointer to the Γ dependence, coexisting with the Niemiec and Ostrowki approach. Physically, as we mentioned, for the occurrence of a shock crossing the conservation of the adiabatic invariant in the de Hoffman-Teller frame is used. However, to get into this frame we need all components of momentum and so need to assign a particle phase angle ϕ at random. Then, in $m = 1$ units, the momentum components for particle energy γ are

$$p_{1,x} = \gamma_1 v_1 cos\theta_1 cos\psi_1 - \gamma_1 v_1 sin\theta_1 cos\phi_1 sin\psi_1 \quad (4)$$

$$p_{1,y} = \gamma_1 v_1 cos\theta_1 sin\psi_1 + \gamma_1 v_1 sin\theta_1 cos\phi_1 cos\psi_1 \quad (5)$$

$$p_{1,z} = -\gamma_1 v_1 sin\theta_1 sin\phi_1 \quad (6)$$

So, for all sub-luminal shocks, shock crossing may be followed by conserving the first adiabatic invariant in the de Hoffmann-Teller frame.

Defining the fields $\mathbf{B}_s$ and $\mathbf{E}_s$ in the shock normal frame with compression ratio r, we arrive at upstream frame quantities: $B_{1,x} = B_{s,x}$, and $B_{1,y} = \Gamma_1(B_{s,y} + V_1 E_{s,z})$, and $E_{1,z} = 0$, where $E_{s,z} = V_1 B_s \sin\psi_{1,s}$. Also $\tan\psi_1 = \Gamma_1^{-1} \tan\psi_s$. Note that the high relative velocity of the frames in the x direction induces an electric field that via a back-induction suppresses the y component of B and swings the field lines towards the x axis in the plasma frame.

A two-stage transform is then initiated, first to the shock frame,

$$\gamma_s = \Gamma_1(\gamma_1 + V_1 p_{x,1}) \quad (7)$$

$$p_{s,x} = \Gamma_1(p_{x,1} + V_1 \gamma_1) \quad (8)$$

$$p_{s,y} = p_{y,1} \quad (9)$$

$$p_{s,z} = p_{z,1} \quad (10)$$

and secondly into the de Hoffmann-Teller frame with a boost along the negative y axis with $V_{HT} = V_1 \tan\psi_1$. Then in this new frame, the B components become $-B_{HT,x} = \Gamma_{HT}(-B_{s,x} - V_{HT} E_{s,z})$, $B_{HT,y} = B_{s,y}$, and hence $\tan\psi_{HT,1} = \Gamma_{HT,1} \tan\psi_1$. Here the field lines are swept towards the transformation velocity in the y direction under frame transform. Energy and the y component of momentum are then transformed into the de Hoffmann-Teller frames according to

$$\gamma_{HT} = \Gamma(\gamma_s + V_{HT} p_{s,y}) \quad (11)$$

and

$$p_{HT,y} = \Gamma_{HT}(p_{s,y} + V_{HT} \gamma_s) \quad (12)$$

allowing the particle pitch angle to be obtained from $\cos\theta_{HT} = \mathbf{p} \cdot \mathbf{B}/pB$. Then the particle that crosses the shock from upstream is transmitted only if its pitch angle is less than the critical pitch angle:

$$\theta_c = \arcsin\sqrt{\frac{\mathbf{B}_{HT,1}}{\mathbf{B}_{HT,2}}} \quad (13)$$

From the conservation of the first adiabatic invariant we can find the new pitch angle in the downstream frame and similar transformations allow the particle scattering to be followed in this frame.

2.2. Super-Luminal Oblique Shocks

A Monte Carlo scheme considers the motion of a particle of momentum p in the magnetic field $\mathbf{B}$. In the super-luminal situation it is not possible to transform to a single frame where $\mathbf{E} = 0$ (de Hoffmann-Teller frame). Hence, for our investigation the most convenient frames of reference to use are the fluid frames (where still the electric field is zero) and the shock frame, which we will use only as a frame to check whether upstream or downstream conditions apply. $\sim 8 \times 10^5$ particles are injected in order to keep reasonable statistics throughout the simulations. Large angle scattering is considered, which is calculated in the respective fluid rest frames, based on the possible situation previously discussed. Also *Bednarz and Ostrowski* (1996) find that only in the large angle scattering case does a power law develop and thus this is the most favourable case to further investigate at present. It will at least yield a limit to the acceleration a super-luminal shock may provide. However, there is further work under way (to be published soon) considering the pitch angle scattering implemented in our scheme of super-luminal shocks, which will provide further insights.

Initially we inject the particles 100λ from the shock and we follow their guiding center in the upstream frame as in the sub-luminal case until after the appropriate transformation to the shock frame the particle reaches the shock at $x_{sh} = 0$. At this juncture, there is no easy approximation to determine the probability of shock crossing or reflection. We change the model to following a helical trajectory, in the fluid frames upstream ("1") or downstream ("2"), where the velocity coordinates of the particle are calculated in a three-dimensional space as follows:

$$v_{x1} = v_1 \cos\theta_1 \cos\psi_1 - v_1 \sin\theta_1 \cos\phi_1 \sin\psi_1 \quad (14)$$

$$v_{y1} = v_1 \cos\theta_1 \sin\psi_1 + v_1 \sin\theta_1 \cos\phi_1 \cos\psi_1 \quad (15)$$

and

$$v_{z1} = -v_1 \sin\theta_1 \sin\phi_1 \quad (16)$$

where θ_1 is the pitch angle and ψ_1 is the angle between the magnetic field and the shock normal.

We follow the trajectory in time, using $\phi_1 = \phi_0 + \omega t$, where $\phi_1 \in (0, 2\pi)$ and t is the time from detecting shock presence at x_{sh}, y_{sh}, z_{sh} by using

$$dx = x_{sh} + v_{x1} dt \quad (17)$$

$$dy = y_{sh} + v_{y1} dt \quad (18)$$

and

$$dz = z_{sh} + v_{z1}dt \qquad (19)$$

assuming that $dt = r_g/Hc$, where r_g is the Larmor radius, and $H \geq 100$. The particle's gyrofrequency ω is given by the relation:

$$\omega_1 = \frac{e|\mathbf{B}_1|}{\gamma_1} \qquad (20)$$

where $\mathbf{B}_1$ is the magnetic field, γ_1 is the particle's gamma and e is its charge in gaussian units. For a matter of convenience though, the last relation is transformed in units of c/sec. All the above relations apply to the downstream case by changing only the signs from 1 to 2 respectively, and all the calculations are performed in the upstream or downstream frames. Because of the peculiar properties of the helix we need to establish where a particle—in the upstream frame—of a particular θ, ϕ first encounters the shock. To establish when this happens, we choose to go back a whole period,

$$T_1 = \frac{2\pi}{\omega_1} \qquad (21)$$

by reversing signs of the helix velocity coordinates and keep checking throughout the simulation to see if the particle's trajectory encounters the shock again. The trajectory integrations automatically select the allowed phase-pitch angles for transmission or reflection from an effectively random input. The starting point for transforming to the downstream frame is the *furthest* upstream shock crossing.

After making the suitable relativistic transformations to the downstream fluid frame by calculating the (x_2, y_2, z_2, t_2) and (v_{x2}, v_{y2}, v_{z2}) coordinates, and the momentum P_2 and the gamma γ_2 of the particle, we follow the trajectory of the particle for a whole downstream period

$$T_2 = \frac{2\pi}{\omega_2} \qquad (22)$$

checking to see whether the particle meets the shock again, by transforming to the shock frame. If the particle meets the shock then the suitable transformations to the upstream frame are made again and we follow the particle's trajectory as described above. If the particle never meets the shock, then its guiding center is followed, the same way as mentioned earlier for the upstream side after the injection and it is left to leave the system if it reaches a well defined E_{max} momentum boundary or a spatial boundary of 100λ.

3. RESULTS

3.1. Parallel Shocks

Our preliminary code runs show that the qualitative trend of the results is insensitive to the exact compression ratio (r) value (3 or 4), due to the mildly-relativistic nature of the downstream plasma. Initial Lorentz factors of the flow investigated are 5, 50, 200, 500 and 990. Test runs to verify the validity of the codes, which have been performed in the mildly relativistic limit (e.g. $V_1 = 0.1c - 0.6c$), tend to excellent agreement with the non-relativistic theory, giving smooth spectra shapes and spectral indexes close to -2.2. In addition, similar runs show that as the flow becomes more relativistic, the spectrum flattens, which is in agreement with the analytical and numerical work of *Kirk and Schneider* (1987a,b). We also compare our results with similar studies for $\Gamma \sim 5$ (*Baring*, 1999) and find agreement concerning the spectral index and the spectral shape for both large angle scattering and pitch angle diffusion mechanisms.

All results are given at the shock, in the shock frame. The particle distribution function is obtained in a particular energy and μ space cell by recording the passage of each simulation particle crossing the cell, weighted by the time taken in cell crossing. In Figure 1 we show the ratio of the computational time constant to the non-relativistic analytical acceleration time constant, as defined earlier in the Introduction, calculated as a function of the upstream Lorentz factor Γ flow for large and small angle scattering. Note we are effectively measuring speed-up of the acceleration time in units of a few times r_g/c. We obtain the time constant t_{acc} from the computation by recording the energy increment for each complete cycle, up-down-up and the time taken for the cycle at a particular energy level, Γ, and finding the mean increments, $\Delta \Gamma$ and Δt. Then $t_{acc}/\Delta t = \Gamma/\Delta\Gamma$.

We find a speed-up of about a factor of 5 for large angle scattering and a considerable speed-up of a factor ~ 20 for pitch angle diffusion. In the latter case, the parallel mean free path is estimated for the analytical formula from the time to multiple scatter in pitch angle through $\pi/2$. That is $\lambda_\parallel = (\pi^2/4)(c\lambda/\delta\theta^2)$ where $\lambda \propto \gamma$ is the step length between collisions yielding $\delta\theta$. Particle energy is specified in units of γ and hence applies both to electron and ion acceleration. Figure 2 shows the logarithm of the particles' mean energy against the number of shock crossings for the two scattering models, at $\Gamma = 50$. Here crossings 1 to 3 (downstream $\rightarrow$ upstream $\rightarrow$ downstream) represent one cycle. The energy is measured immediately after the particle has crossed the shock front in the shock frame. In all cases, the energy gain on the first complete cycle, specified by crossing numbers 1 and 3, is $\sim \Gamma^2$ but subsequent crossings show a reduced gain, passing through a region where the increase is $\sim \Gamma$ and tending towards the factor ~ 2 predicted by *Gallant and Achterberg* (1999).

In Figure 3 the angular distribution is measured in the shock frame just downstream for the upstream to downstream transition and just upstream for the down to up transition, for pitch angle diffusion and upstream flow $\Gamma = 200$. We only show the contribution to the distribution function

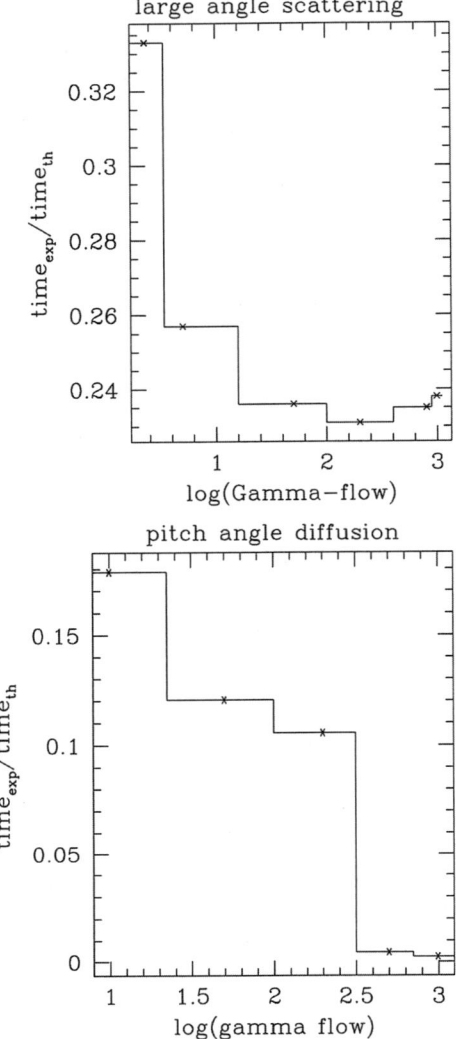

Figure 1. The ratio of the computational time to the non-relativistic analytical constant (measured in the shock frame) versus the logarithm of the Lorentz Γ flow, for large angle scattering (top) and for pitch angle diffusion (bottom).

of particles which have just crossed, not the complete time-averaged distribution. Similar results were obtained for Γ = 10 – 990. It is evident that there is extreme peaking in the angular distribution towards smaller pitch angles in all but one case, continuing the trend noticed by *Quenby and Lieu* (1989) and *Ellison et al.* (1990) at lower flow Γ, who found that as the velocities become more relativistic, the anisotropies in pitch angle become greater. If the distribution function at the shock is isotropic, the number of particles between μ and μ+Δμ is proportional to μ, taking into account solid angle weighting. Both scattering cases reveal highly peaked distributions pointing downstream on up to down transition,

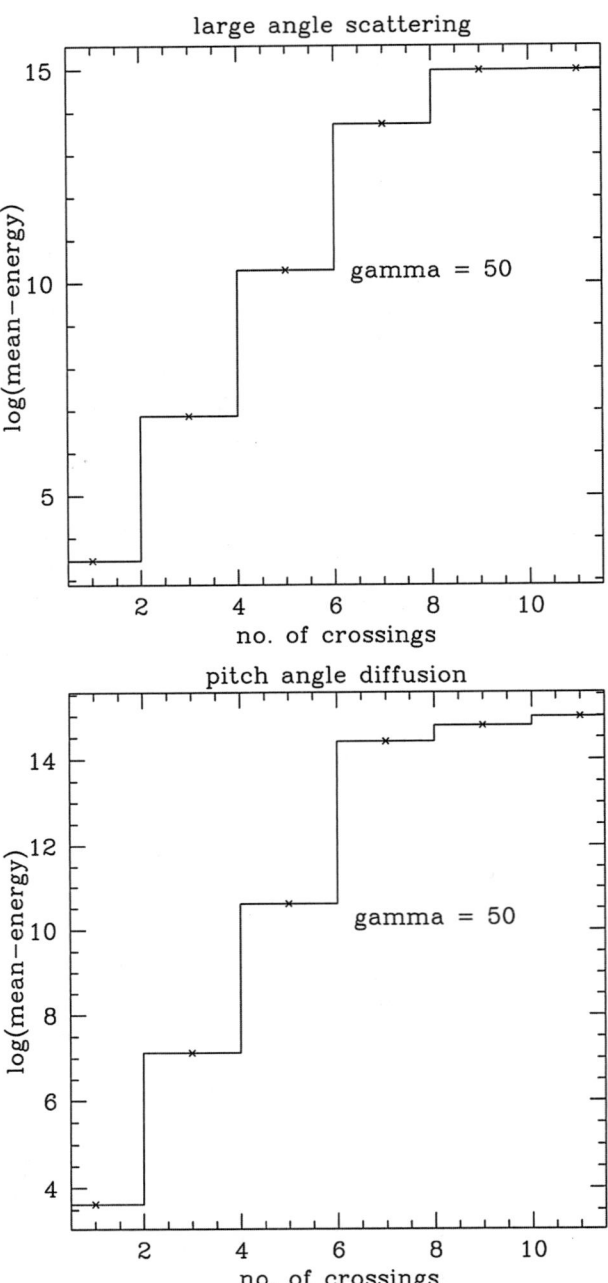

Figure 2. The logarithm of the mean-energy gain of particles versus the no. of shock crossings (1-3-5-7-9-11), where upstream Γ equals 50. The energy is measured immediately after the particle has crossed the shock front. Large angle scattering (top), pitch angle diffusion (bottom).

due to the beaming of the upstream distribution function as seen by the shock.

Downstream to upstream for large angle scattering produces near triangular distributions in the upstream directed half space, due to the very efficient scattering in the model.

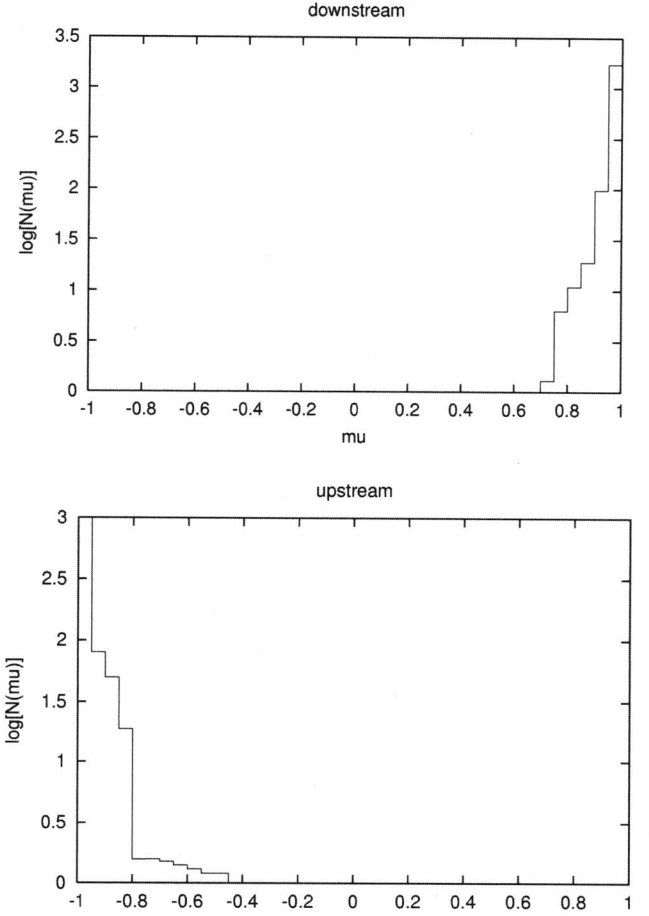

Figure 3. Angular distribution (mu = μ = $\cos\theta$) in the shock frame for $\Gamma = 200$ and pitch angle diffusion, after the particles have crossed the shock from upstream $\to$ downstream (top) and downstream $\to$ upstream (bottom).

For small angle scattering, highly peaked distributions are found due to the lack of time for significant deflection from nearly zero pitch angle before downstream re-entry and presumably for those particles which can re-enter upstream. We investigated the sensitivity of these spectral shapes for small angle scattering to downstream boundary condition, and found that the shapes did not change significantly for the boundary distance, $r_b \geq 2 \cdot 10^3 \lambda$.

Probably the small chance of return to the shock except for a relatively small subset of downstream pitch angle particle histories produced this result and the anisotropy in Figure 3 is seen for these returning particles. Note the relatively smooth spectra for the relativistic flow with $\Gamma = 5$ (Figure 4) which becomes plateau-like at the more extreme value of $\Gamma = 990$ (Figure 5) where the effects of individual acceleration cycles are clearly evident in the large angle case, but not

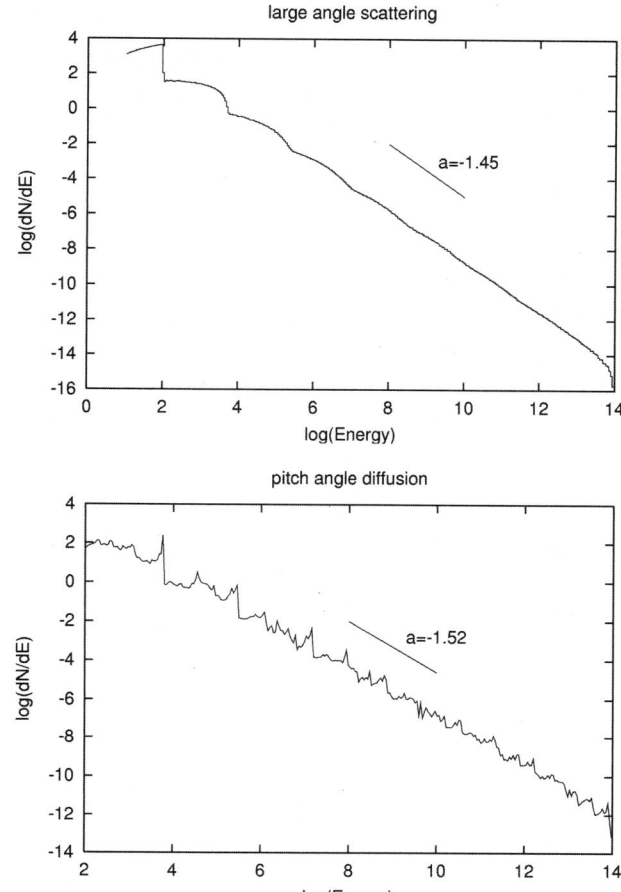

Figure 4. Spectral shape for upstream $\Gamma = 5$. Large angle scattering (top) and pitch angle diffusion (bottom). The line shows the mean slope of the spectrum and the number is the spectral index found. The smoothness (compared to higher gamma flows used) and the spectral index (especially for the large angle scattering) are consistent with similar work presented by *Baring* (1999).

for small angle scattering, where the spectral shape remains smooth.

Features can be seen also developing in the lower Γ simulations of *Quenby and Lieu* (1989) and *Ellison et al.* (1990), while *Protheroe* (2001) shows a similar contrast in behaviour between large and small angle scattering models up to $\Gamma = 20$. On any individual cycle, greater energy gain is available for the large angle case because of the greater upstream scattering, as demonstrated in the simple analytical approach mentioned above. Hence, the plateaus develop for the large angle case with flatter segments than the smoother pitch angle case spectrum. An opposing effect, however, is that because the accelerated particles re-entering downstream have pitch angles more favourable to eventual loss downstream, the drop between plateaus is greater and so the overall spectral slopes in the two cases are rather similar.

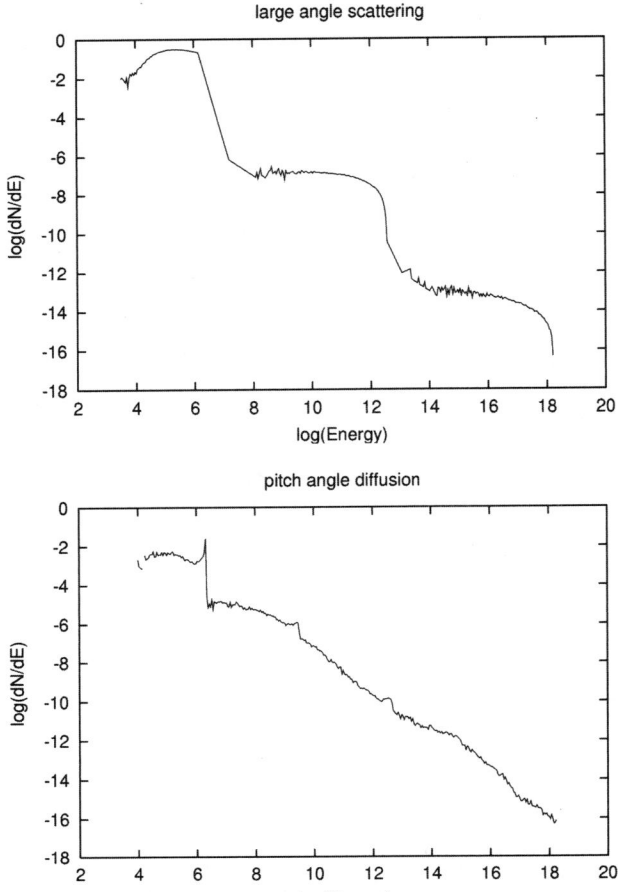

Figure 5. Spectral shape for an upstream $\Gamma = 990$, for large angle scattering (top) and for pitch angle diffusion (bottom) for the same gamma.

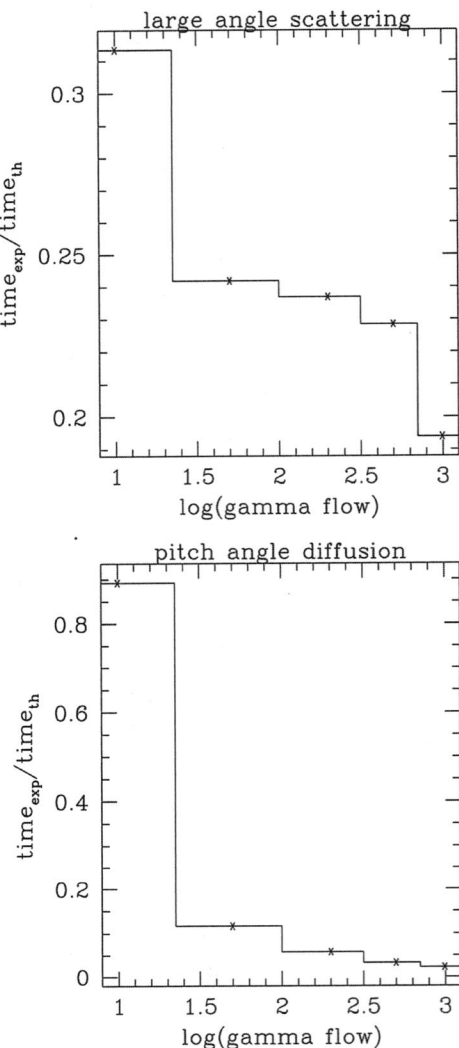

Figure 6. The ratio of the computational time to the theoretical acceleration time constant versus the logarithm of the upstream gamma flow. Measurements are made at the de Hoffmann-Teller frame. Shown is the acceleration time decrease in the case of the large angle scattering (top) and by pitch angle diffusion (bottom), both for $\psi_1 = 35°$.

The actual mean slope of the differential number spectrum is well above -2, as illustrated by the lines of slope -1.45 and -1.52 shown for $\Gamma = 5$. This relativistic flattening effect is consistent with previous studies by *Baring* (1999) while the mechanism of plateau development as an acceleration cycle effect is implicit in Figure 6 of *Protheroe* (2001). It is perhaps not surprising to observe a step-like behaviour in the accelerated spectrum, clearly related to groups of particles undergoing different numbers of acceleration cycles, because the simulations also reveal that about 90% of all particles in the simulation box are lost downstream each cycle. One might also expect that small angle scattering has the greater chance of smoothing the inherent structure in the acceleration process. Clearly, smoothing is much more efficient at lower flow Γ where the gain per cycle is less. One can infer that at extreme relativistic flow velocities and different scatter models, it is difficult to assume that the spectral shape of the accelerated particles follows a *smooth* power-law shape, as in the classic non-relativistic first-order Fermi acceleration mechanism, under all circumstances.

3.2. Oblique Sub-Luminal and Super-Luminal Shocks

We chose two sub-luminal inclination cases (following $tan\psi_1 \leq 1$) to be studied, equal to 15° and 35° in the shock frame. Particles are injected isotropically at 100λ upstream in the shock frame and when they reach the spatial boundary placed at $100\lambda_{\parallel}$ downstream they exit the simulation for the large angle case. We apply large angle and pitch angle scatters as well. We add here that preliminary code runs show that the qualitative trend of the results is insensitive to the exact compression ratio (*r*) value (3 or 4), due to the mildly-relativistic nature of the downstream plasma, but we will show some result plots (e.g. Figure 8) for the reader's consideration.

For pitch angle scatter, the stability of the results against boundary position enabled a suitable downstream exit to be chosen (1000λ) where here, λ is the distance between scatterings. The mean free path (λ) is chosen to have the value of $\sim 20 r_g$, close to the value that *Quenby and Lieu* (1989) used, based on a summary of interplanetary transport simulations referring to the only astrophysical plasma where we have any detailed knowledge of the magnetic field properties and performed by *Moussas et al.*, 1992 (and references therein), but in most investigations it is only the ratio of up/downstream mean free paths and their ratio to the spatial boundary which matter.

We observe a significant acceleration time decrease of a factor $\sim$ 5–10, which is comparable to that of the parallel shock speed-up found for example in *Ellison et al.* (1990). Basically, this speed-up effect observed is due to the Γ_1 energy increase each shock crossing, Γ_1^2 per cycle, although it can be reduced by the very small pitch angles the particle may have in the small angle case as it crosses the shock upstream to downstream. Our computations also show a similar, to the parallel shock acceleration, mean energy gain per crossing for large angle scattering and pitch angle diffusion, where the relative independence on *r* and the similarity in the trend from Γ^2 to Γ in energy dependence and finally saturation are seen. The gain per crossing falls off noticeably faster as Γ_1 increases.

It must be emphasized that a full model would include loss processes, which would limit the maximum Γ obtain-

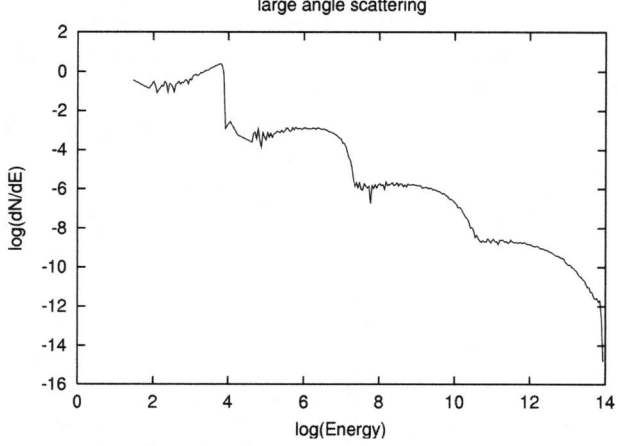

Figure 7. Spectral shape for upstream gamma flow equal to 50 and large angle scattering, $r = 4$ and $\psi = 35°$.

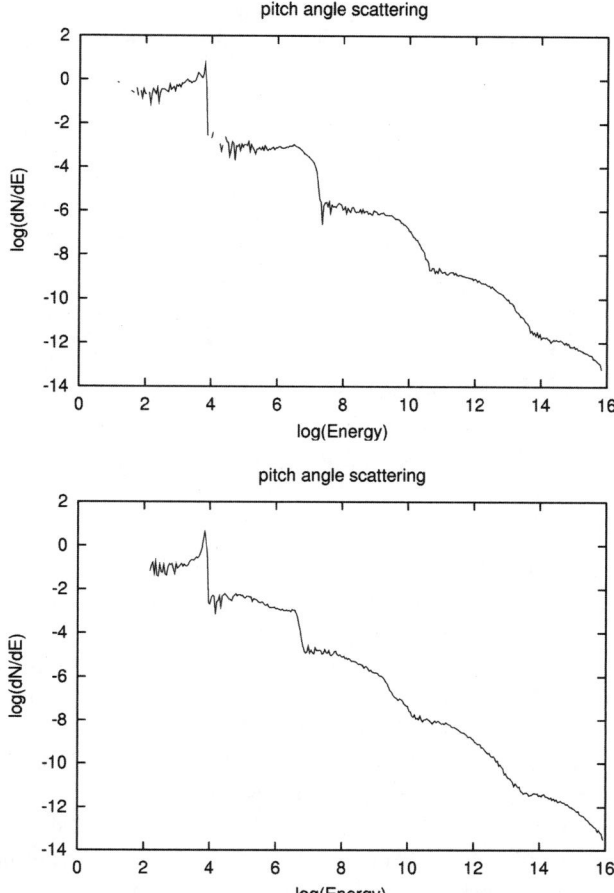

Figure 8. Spectral shapes for pitch angle diffusion, for upstream $\Gamma = 50$. Top plot: $\psi = 15°$ and $r = 3$. Bottom: $\psi = 35°$ and $r = 3$.

able. Note that proton energies $\sim 10^{20}$ eV can be reached in all models computed before the saturation sets in, provided faster loss mechanisms are not present. This is despite the relatively small amount of upstream deflection allowed before particles are swept downstream on all cycles except the first, which limits the first-order Fermi gain available (see *Gallant and Achterberg*, 1999, where the full Γ_1^2 factor requires isotropization in each frame). Rather different spectral shapes are seen in Figures 7–9 for $\Gamma = 10$, 50, 500 and 990, depending on whether it is pitch angle or large angle scattering, at $\psi_1 = 15°$ and 35°. The sensitivity to downstream boundary position has been checked up to a distance $r_b = 2 \cdot 10^5 \lambda$, where λ is distance between small angle scattering for this model.

For oblique sub-luminal shocks, we show in Figure 6 the ratio of computational acceleration time to the analytic non-relativistic time constant, versus the logarithm of the upstream flow gammas used. The measurements are all made in the de Hoffmann-Teller frame as in *Lieu et al.* (1994)

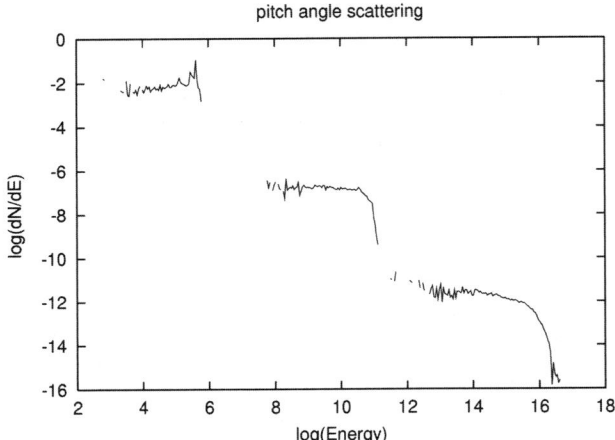

Figure 9. Spectral shapes for $\Gamma = 500$, $\psi = 15°$, $r = 4$, and pitch angle diffusion.

where, it was found that this was a relevant frame to observe acceleration speed-up. Then the non-relativistic acceleration time is given by equation (1).

The plateau-like shapes are again prominent and more disrupted, compared to parallel shock cases for large angle scatter. In the case of pitch angle diffusion the shapes are much smoother, particularly at lower Γ_1. The disrupted shapes are probably due to high anisotropy allowed in some situations. We qualitatively understand the spectral shapes in terms of each plateau exhibiting the acceleration gain in one shock cycle, down to up to down, with a Γ_1^2 factor boosting energy, followed by a high probability of loss downstream due to the beaming of the particles away from the shock. Examples of the beaming are presented in Figure 10, where we see the angular distributions of the transmitted particles at the shock front in the de Hoffmann-Teller frame. We only show particles which have just crossed the shock, not the complete, time-averaged, distribution function. The observed effect is due to the anisotropies in pitch angle space expected for highly relativistic flows, which are in accordance with lower Γ_1 results of *Ellison et al.* (1990) and *Lieu et al.* (1994). These results show that, due to very relativistic velocities, the distribution functions are very anisotropic and so the spatial diffusion approximation cannot apply.

In Figure 11, we present results for the super-luminal case, where $\psi_1 = 75°$ and Lorentz factors range from 30 to 40, expecting to apply in the relativistic flows of AGN environments. The compression ratio is equal to 4 and large angle scattering is used.

To obtain the spectral shapes for the accelerated particles as the trajectories are followed during the simulations, we add the momentum of each particle in the corresponding

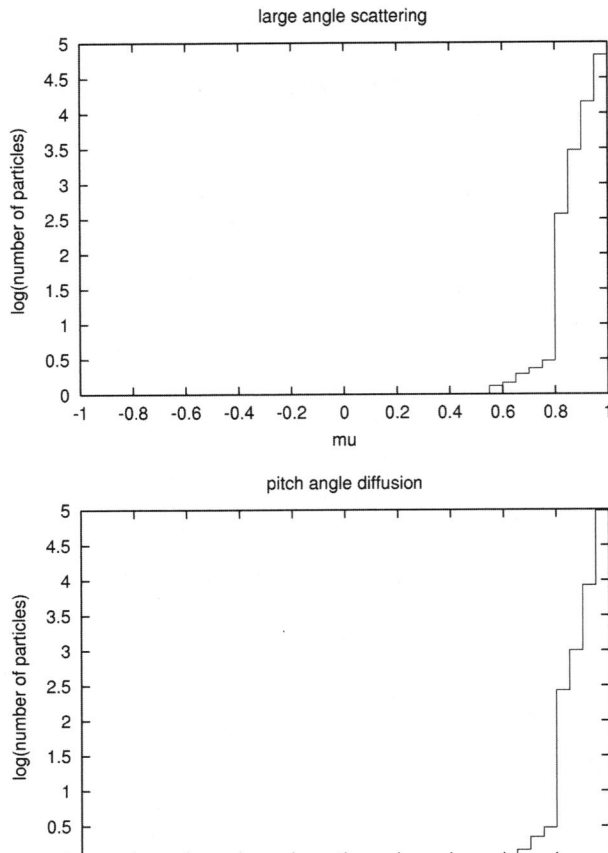

Figure 10. The angular distribution of the logarithm of the number of transmitted (just crossed the shock) particles versus μ at $\mu = cos\theta$ at the de Hoffmann-Teller frame, using upstream gamma flow equal to 200, $\psi = 15°$, and $r = 4$ for large angle scattering (top) and pitch angle diffusion (bottom). In both cases we observe the strong beaming of the particles' distribution.

bin every time the particle crosses the shock as before to obtain the mean energy gain and distribution function. We observe that the spectral shape falls away from a power law and descends rapidly to an upper cutoff which is expected, because the particles are swept away by the flow after one cycle and have a very limited chance of return upstream, especially due to the high field inclination to the shock normal.

In Figure 12, the energy gain of the particle is shown, versus the number of the helix-trajectory shock crossings for Lorentz factor 10 and $\psi_1 = 75°$ between the magnetic field and the shock normal. It is seen that after the 3rd crossing (after one cycle) the particle gains a considerable amount of energy ($\sim \Gamma^2$), but further gain is limited due to the fact that the particles are swept away by the flow very

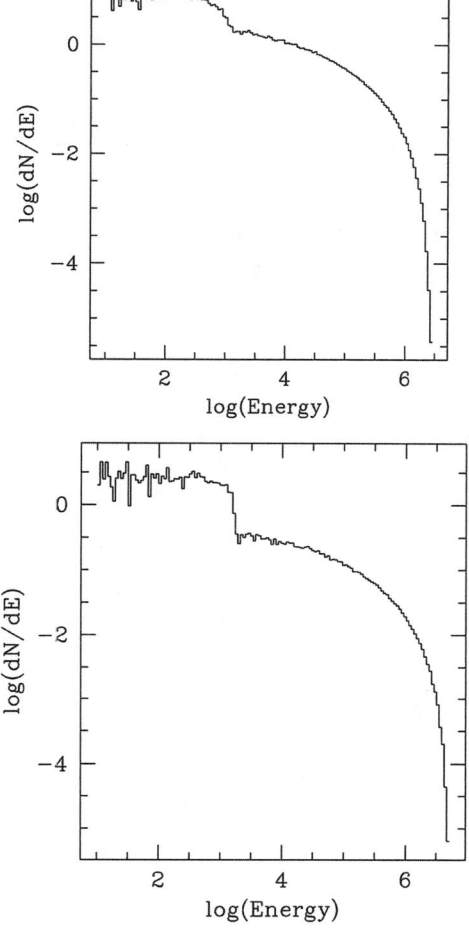

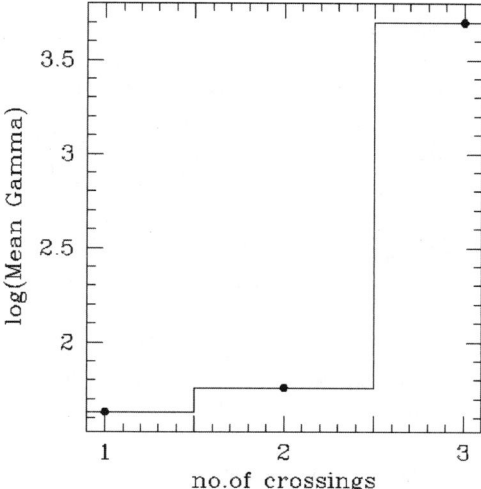

Figure 12. Energy gain versus number of shock crossings (1-2-3) of the particles' helix trajectory for $\Gamma = 10$ and $\psi = 75°$.

Figure 11. Spectral shape for the super-luminal case in the shock frame at the downstream side for $\Gamma = 30$ and $\psi = 75°$ (top) and $\Gamma = 40$ and $\psi = 75°$ (bottom).

rapidly with no chance of recycling the shock again. Our calculations show that ~60% of the particles cross the shock three times (upstream → downstream → upstream → downstream).

It is possible to conclude that "one-shot" drift acceleration has taken place, that is, on one up to down crossing, and there is energy increase as expected from drift in the shock frame electric field under the magnetic field gradient. Some particle return upstream has been possible under the action of large angle, downstream scattering. However this return flux is so limited that a power law spectrum cannot develop and a cutoff is soon reached. Computational time limitations have currently prevented an extension of the modeling to higher Γ flows and pitch angle scattering models for the super-luminal case.

4. DISCUSSION AND CONCLUSIONS

The presented results emphasize the critical question of to what extent is first-order Fermi acceleration universally applicable in shocked, turbulent, magnetohydrodynamic flows.

For parallel shocks it is found that the spectral shape in the very relativistic flow regime depends on the scatter model (large angle or pitch angle diffusion/small angle scattering), with the former model exhibiting a structure in the spectrum at the shock related to the cycle of acceleration, and the latter one showing a smoother shape.

Our results for the most interesting case of oblique particle shock acceleration are subject to the applicability of appropriate parameter regimes, as specified above, allowing a guiding centre approximation in the plasma frame and adiabatic invariant conservation in the sub- but not super-luminal case. A major finding within the limitations of our model is: In contrast to parallel shocks with small angle scattering, all other types of shocks and scattering models do not yield smooth, power law spectra at the shock in the $\Gamma \gg 1$ regime, even though subsequent smoothing during particle escape may arise. Reasons for this will be reiterated later. However, sub-luminal, inclined relativistic shocks provide spectra more extended in energy and therefore more approximating to a power law distribution than super-luminal shocks, the latter exhibiting a pronounced high energy cutoff feature. The result that up to 3 crossings of the shock can often occur in the adopted large angle scattering model implies that the super-luminal case is not simply one-shot shock drift. However, the large gain crossing 2 → 3 followed by loss to the system can well be described as the energy gain experienced on the last shock drift pass through the discontinuity standing out from the smaller, average diffusive gain.

In contrasting the inclined, sub-luminal and super-luminal shock results, we would add that the former exhibit large spectral, downward jumps with energy increase, qualitatively explained as the combined effect of a Γ^2 energy increase per crossing, together with a severe downstream particle loss. This loss probability clearly increases with shock inclination as the particles tend to follow magnetic field lines and it is therefore not surprising that negligibly few particles survive more than 2 to 3 crossings in the superluminal situation and a spectral cutoff develops. There is probably no critical transition for this behaviour, however. Work under preparation suggests that use of strong pitch angle scattering, including close to the shock transition, allows the recovery of a power law spectral shape, even at very large mean field inclination angles. Here, we have been constrained to a large angle scattering model. The reader is referred to *Quenby and Meli* (this volume) for a discussion of possible astrophysical sites of "cannon ball", large angle scattering.

Concerning the important speed-up effect in the sub-luminal case, this appears to be due to a combination of Γ^2 or Γ energy gain and the anisotropic angular distributions at the shock front (*Ellison et al.*, 1990) which as we showed, is confined to a very small angle transmission cone. Thus, only a narrow range of pitch angle particles return to the shock and are counted for the acceleration time calculation, as the ones with larger pitch angles are lost downstream. If the particle distribution is made isotropic each side of the shock before re-crossing, the energy gain per cycle is $\sim\Gamma_1^2$ (*Quenby and Lieu*, 1989), but a distribution remaining anisotropic will experience less energy gain (*Gallant and Achterberg*, 1999). Hence some part of the speed-up compared with the non-relativistic situation is expected to be connected with the unallowed for energy enhancement. The anisotropy is of course linked to the beaming effect found also in mild relativistic cases, in accordance with *Peacock* (1981) and *Kirk and Schneider* (1988). *Lieu and Quenby* (1990) and *Lieu et al.* (1994) note that the distribution of the particles' pitch angle is peaked towards the direction of the relativistic transformation to the de Hoffmann-Teller frame. In other words this beaming, which enhances the number of particles close to the field line, may provide an additional reason for the decrease in acceleration time as the flow becomes more relativistic for both large angle and pitch angle diffusion. This speed-up resembles a trend found in the parallel shock particle acceleration case. Furthermore, the last arguments could explain the prominent spectral "plateau" shapes found in many oblique, sub-luminal and parallel shock cases where each shock cycle crossing seems to stand out *separately*. Enhanced single cycle acceleration is followed by large flux loss downstream, preventing statistical smoothing into a single power law accelerated spectrum. The speed-up effect, which has also been observed in the mildly relativistic regime by *Quenby and Lieu* (1989) and *Ellison et al.* (1990), is important for a complete understanding of particle acceleration to the highest energies. An understanding of this crucial effect is necessary in any explanation for the maximum energy that particles (protons, electrons, iron nuclei) can achieve in GRB fireballs, AGN hotspots, and pulsar ultrarelativistic polar winds. For example results found in the previous sections dealing with the sub-luminal case for $\Gamma \sim 10^3$ show only few crossings are needed for the particle to achieve energies up to 10^{19}–10^{20} eV, although for $\Gamma \sim 10$ more than 10 crossings may be needed to gain such high energies. It is the competition between loss time and energy gain time which determines the maximum energy in many situations. Generally speaking the percentage increase of the maximum energy would depend on many factors. One of those is the velocity of the upstream flow. Also the dimensions of the shock could play a vital role in order to define the escape losses in a specific shock acceleration site. In addition another factor in the physical conditions which could alter the cosmic ray spectrum and decrease the spectral index is due to different energy losses occurring in the regime; for example the presence of high or low energy photons (Greisen-Zatsepin-Kuz'min cutoff effect) or γ–γ flux interactions in GRB will decrease the maximum energy of the particles. In the context of AGN it is well known from a number of interesting papers such as *Protheroe and Kazanas* (1983) and *Chakrabarti and Moltoni* (1993) that a shock in the central engine of the accreting flow ($\Gamma \simeq 5 - 10$) is likely to appear and particle acceleration can consequently take place. Also there is observational evidence that a common feature of many AGN is the formation of one or two jets where shock formations should appear in conjunction with $\Gamma \sim 5 - 10$ flows. Thus, the computed flattening of the spectra—they are all flatter than a −2 power law—as well as the acceleration speed-up should have important consequences when compared with predictions and observational data, regarding for example, the photons produced by emission from electrons which can be accelerated at the shocks or produced from accelerated protons. GRB seem to be another potential candidate for the acceleration of UHECRs, but the issue is debated by many workers. As we have mentioned *Vietri* (1995) and *Waxman* (1995), who have exploited the Γ^2 factor energy gain of the particle, were first to predict theoretically that indeed UHECRs can be produced in GRB where a pre-acceleration of the particles has been implied. Briefly, we note that observationally there is very strong evidence that pulsars (e.g. PSR 1913+16, PSR 2127+11C) are capable of injecting continuously relativistic particles into the surrounding medium over their lifetime. While this plasma will probably be predominantly in the form of e^+e^- pairs created in the pulsar magnetosphere, it has been argued that pulsar winds must also contain ions in order to account for the electrical current required in the Crab Nebula (*Hoshino et al.*, 1992; *Gallant and Arons*,

1994). The termination shock as the $\Gamma \sim 10^5$ flow meets the nebula is likely to be quasi-perpendicular and hence the super-luminal situation applies for the additional e^+e^- acceleration. The enhanced production of neutrinos in astrophysical sites such as in AGN and GRB is another consequence of the simulation findings of the speed-up effect and Γ^2 factor. *Mastichiadis and Kirk* (1992), *Protheroe and Szabo* (1992) and *Vietri* (1998) showed that the neutrino flux depends on the maximum energy attained from the primary protons; thus a decrease of the acceleration time constant could produce higher maximum energy limits for the accelerated particles and consequently an increase in the cutoff energy of the neutrino flux will be noted.

REFERENCES

Achterberg, A. A., Gallant, Y. A., Kirk, J. G. and Guthmann, A. W., 2001, MNRAS, 328, 393.

Armstrong, T. P., and Decker, R. B., 1979, Particle Acceleration Mechanisms in Astrophysics AIP Proc. No 55, p. 101.

Axford, W. I., Leer, E. and Skadron, G., 1978, 15th ICRC Plovdiv, Bulgaria, 11, 132.

Axford, W. I. 1981, 17th ICRC, Paris, 12, 155.

Baring, M. G., Ellison, D. C. and Jones, F. C., 1994, ApJS, 90, 547.

Baring, M. G., Ellison, D. C. and Jones, F. C., 1995, Adv. Sp. Res., 15, No 8/9, 397, COSPAR.

Bell, A. R., 1978a, MNRAS, 182, 147.

Bell, A. R., 1978b, MNRAS, 182, 443.

Bednarz, J. and Ostrowski, M., 1996, MNRAS, 283, 447.

Bednarz, J. and Ostrowski, M., 1998, Ph. Rev. Let., 80, 3911.

Begelman, M. C. and Kirk, J. G., 1990, ApJ., 353, 66.

Blandford, R. D., and Ostriker, J. P., 1978, ApJ. Lett., 221, L29.

Chakrabarti, S. K. and Moltoni, D., 1993, ApJ., 417, 671.

Drury, L. O'C, 1983, Rep. Prog. Phys., 1, 973.

de Hoffmann, F. and Teller, E., 1950, Phys. Rev., 80, 692.

Ellison, D. C., Jones, F. C., and Reynolds, S.P., 1990, ApJ., 360, 702.

Gallant, Y. A. and Arons, J., 1994, ApJ., 435, 230.

Gallant, Y. A. and Achterberg, A., 1999, MNRAS, 305, 6.

Greisen, K., 1966, Ph. Rev. Let., 16, 748.

Hoshino, M., Arons, J., Gallant, Y. A., and Langdon, A. B., 1992, ApJ., 390, 454.

Hudson, P. D., 1965, MNRAS, 131, 23.

Jokipii, J. R., 1966, ApJ., 146, 480.

Jokipii, J. R., 1982, ApJ., 255, 716.

Jokipii, J. R. and Parker, E. N., 1969, ApJ., 155, 177.

Kirk, J. G. and Schneider, P., 1987a, ApJ., 315, 425.

Kirk, J. G. and Schneider, P., 1987b, ApJ., 322, 256.

Kirk, J. G. and Schneider, P., 1988, A&A, 201, 177.

Kirk, J. G. and Heavens, A. F.,1989, MNRAS, 239, 995.

Krymsky, G. F., 1977, Dokl. Acad. Nauk., SSSR, 234, 1306.

Lieu, R. and Quenby, J. J., 1990, ApJ., 350, 692.

Lieu, R., Quenby, J. J., Drolias, B. and Naidu, K., 1994, ApJ., 421, 211.

Lucek, S. G. and Bell, A. R., 1994, MNRAS, 268, 581.

Mastichiadis, A. and Kirk, J. G., 1992, High Energy Neutrino Astrophysics, p.63, World Scientific.

Meli, A. and Quenby, J. J., 2003a, Ast.Part. Phys., 19..649

Meli, A. and Quenby, J. J., 2003b, Ast.Part. Phys., 19, 637

Moussas, X., Quenby, J. J., Theodossiou-Ekaterinidi, Z., Valdes-Galicia, J. F., Drillia, A. G., Roulias, D. and Smith, E. J., 1992, Sol. Phys., 140, 161.

Newman et al., Astron Astro, 225, 443, 1992

Ostrowski, M., 1991, MNRAS, 249, 551-559.

Peacock, J. A., 1981, MNRAS, 196, 135.

Protheroe, R. J. and Kazanas, D., 1983, ApJ., 265, 620.

Protheroe, R. J. and Szabo, A. P., 1992, Phys. Rev. Lett., 69, 2885.

Protheroe, R. J., Meli, A. and A.-C. Donea, 2002, Sp. Sc. Rev., 107, 369.

Quenby, J. J. and Lieu, R., 1989, Nature, 342, 654.

Quenby, J. J. and Drolias, B., 1995, 24th ICRC Rome, 3, 261.

Vietri, M., 1995, ApJ., 453, 883.

Vietri, M., 1998, ApJ. Lett, 448, L105.

Waxman, E., 1995, Phys. Rev. Let, 75, 386.

Zatsepin, G. T. and Kuz'min, V. A., 1966, JETP Let., 4, 78.

A. Meli, Max Planck Institut für Radioastronomie, Auf dem Hügel 69, 53121, Bonn, Germany. (ameli@mpifr-bonn.mpg.de)

J. J. Quenby, Astrophysics Group, Blackett Laboratory, Imperial College of Science, Technology and Medicine, Prince Consort Rd., SW7 2BW, London, UK.

Energy Spectra of Energetic Ions Around Quasi-Parallel Shocks

T. Sugiyama

The Earth Simulator Center, Japan Agency for Marine-Earth Science and Technology (JAMSTEC), Yokohama, Japan

M. Fujimoto

Department of Earth and Planetary Sciences, Tokyo Institute of Technology, Tokyo, Japan

H. Matsumoto

Research Institute for Sustainable Humanosphere (RISH), Kyoto University, Kyoto, Japan

We have performed a number of simulations of quasi-parallel shocks to investigate how the energy spectra of non-thermal components are controlled by the presence of alpha (He^{2+}) particles. The simulations are done for the Earth's bow shock-like situation by an one-dimensional hybrid code that treats ions as particles but electrons as a massless fluid. The upstream conditions are modified by varying the He^{2+} ions content while keeping the proton content unchanged. The range of the He^{2+} ion density ratio (R) variation relative to proton is from 0.1% to 30% including the typical ratio of 4 ~ 5% for the solar wind. As observed in the upstream of the bow shock, the differential flux spectra of the two ion species obtained in the simulations from the downstream region are found to be well represented by exponential shapes. When the energy scale is presented in the energy-per-charge (E/Q) unit, the two spectra have the same characteristic energy (E_C) that increases in time. While E_C is a function of the density ratio R and increases with increasing R, the equality between the two species holds throughout. Detailed analyses show that larger magnetic fluctuations brought about by the presence of He^{2+} particles enable more efficient acceleration at the shock front. Thus it is via exciting, stronger turbulence that the additional upstream ram energy carried by He^{2+} ions is poured into the process of hardening the energy spectrum of the non-thermal particles of both species.

1. INTRODUCTION

Non-thermal populations are produced at shocks in collisionless space plasmas, since the time it takes for the particle energy distribution to relax to the Maxwellian state is much longer than one's interest. One of the most well documented examples of these populations is the upstream particles of the Earth's bow shock where the supersonic flow of the solar wind dominated by protons are decelerated and deflected. In the region upstream of a quasi-parallel part of the bow shock, non-thermal ions are frequently detected by spacecrafts making in-situ measurements. One of the most characteristic feature of these ions is in their spectral shape. *Ipavich et al.* [1981] have shown that the upstream energy spectrum is well described by an exponential shape. Furthermore it is found that the characteristic energy E_C of the exponential spectrum is independent of ion species (protons and He^{2+} and CNO ions) if the energy scale is presented in energy-per-charge (E/Q) unit.

Scholer et al. [1999] have performed hybrid simulations (ions as particles, electrons as a massless charge neutralizing fluid) and have shown that He^{2+} ions are also accelerated at quasi-parallel shocks. In their simulations, upstream fluctuations are initially imposed and non-thermal particles of both species are produced by interactions with the given fluctuations and the shock. The energy spectra of non-thermal components of the two species in the simulation turned out to have a nearly identical characteristic energy when spectra are presented by energy per charge, as in the observations. The He^{2+} to proton flux ratio in the non-thermal energies was found to be enhanced by about a factor of 5 relative to the same ratio in the thermal range.

The upstream fluctuation, which was not treated entirely self-consistently in *Scholer et al.* [1999], is essential in producing the non-thermal populations. Indeed, in the traditional framework, the acceleration process at quasi-parallel shocks is described by a diffusive process and the fluctuations are supposed to be responsible for this diffusion. In this framework, the ions are described to diffuse in the real space and in the velocity space through successive cyclotron resonances with ensemble of waves having different frequencies [*Blandford and Eichler*, 1987]. Acceleration results as an ion scattered back from the downstream to the upstream region is scattered again to be redirected into the shock. This is because the momentum gain in the upstream redirection is larger than the loss in the downstream. It is the flow deceleration at the shock front that causes this unbalance: the scattering agent in the upstream has faster downstream directed velocity. The ion energy is multiplied in this one cycle and the energy spectrum resulting from this process in its simplest form has a power-law shape.

Eichler [1981] noted that some ions escape to the flanks of the curved bow shock during the course of acceleration and proposed that this explains the discrepancy in the form of the energy spectrum between the theory (power-law) and the observations (exponential). As some of the accelerated particles start to escape to the magnetic-field lines which are not connected to the shock surface, the cut-off effect at higher energies makes the energy spectrum assume an exponential shape. *Scholer et al.* [1999] showed in their hybrid simulations that the energy spectrum gets harder with time and interpreted this time-dependence in terms of the growing of the cut-off energy in time (*Jokipii*, [1987]).

Regarding the wave-particle interaction in the shock-upstream region, a new perspective has been recently proposed by *Sugiyama et al.* [2001a]. This new approach is based on the recognition that the upstream/downstream waves are rather monochromatic which are large in amplitude so that their interaction with the particles may not necessarily involve ensemble of the waves and may not necessarily have resonance as the basic mechanism. Another new point is that the wave-particle interaction at the shock front itself makes substantial contribution to the ion energization, at least during the very initial phase of the whole extended acceleration process. Indeed it has been shown that the interaction with large amplitude upstream waves affects ions over a wide energy range, from thermal to non-thermal, and a quick injection of non-thermal ions from the thermal population can be attained even if the wave is dominated by the single mode that has the largest growth rate in the upstream environment. This should be contrasted with the slow resonance picture where the injection is attained after multiple transfers from one wave to another. Again, because resonance is not required, He^{2+} ions can be accelerated even in an off-resonant wave field excited by protons, which has been confirmed by a test particle study [*Sugiyama et al.*, 2001b]. *Scholer* [1998] has also discussed the particle acceleration process at the shock surface by the wave electric field. However, as discussed in Sugiyama et al. [2001a], the wave electric field mainly contributes not to the acceleration but to the bounce motion of particles around the shock surface.

According to this new framework, protons and He^{2+} ions are accelerated by the same wave field, with substantial contribution from the interaction at the shock front, as long as its amplitude is large. Then the exponential shape and the equality of the characteristic energies between the two species may be the nature of this acceleration process. That is, it is not the discrepancy between the traditional framework and the observations but may be due to the dominance of this other mechanism that makes the energy spectrum shape to be exponential but not power-law. In this study, by performing hybrid simulations for typical Earth's bow shock-like situations, with ion dynamics and the wave excitation treated fully self-consistently, we first confirm that both species are indeed accelerated and have the same characteristic energy (E_C) in the non-thermal energy range. We then vary the He^{2+} to proton density ratio (R) over the range expected for the solar wind [*Ogilvie and Wilkerson*, 1969] and investigate how the ratio affects the characteristic energy. We show that the equality of E_C between the two species stay for various R while E_C increases with increasing R. The equal E_C values between protons and He^{2+} ions are predicted under various density ratios expected in the solar wind, from the lowest, when He^{2+} ions would behave more or less like test particles in a given proton-excited wave field, to the highest, when the He^{2+} upstream particles also excite upstream waves of significant intensity.

It should be noted that the present runs were stopped while the characteristic energies of the exponential spectra were increasing. As such, what we study here is the injection process of non-thermal ions at the very beginning of the whole acceleration process. What we show is that the injection process itself is a wave-particle interaction process dominated by a small number of largest amplitude modes that produces the exponential energy spectra which is hardening in time. What happens at later times as the non-thermal ions further hardens? This is something we cannot answer definitively now, but one can reasonably think that the interaction with the ensemble of lower-frequency waves, which may eventually become more efficient at higher energies and in an environment where extended time interval is allowed for the acceleration, to show up. This may transform the spectrum shape to a power-law at higher energies. What we see at the Earth's bow shock, however, is the very initial injection process. The time allowed for the ion acceleration seems well within the range where the effects of the large amplitude monochromatic waves dominate, and this would be why the spectrum shapes observed are exponential but not power-law.

2. SIMULATION MODEL

The hybrid model used in this study treats the ions as macro-particles while the electrons are treated as a charge-neutralizing massless fluid. The simulations are one-dimensional in space (shock normal direction x: positive x directed downstream) but fully three-dimensional in velocity. Hereafter, magnetic field is normalized by the upstream value B_0. Density is normalized by the upstream proton density N_{p0}. Velocity, time, and length are, by the Alfvén velocity V_{A0}, the inverse of proton gyro-frequency, Ω_i^{-1}, and the ion inertia length $\lambda_i \equiv V_{A0}/\Omega_i$, respectively, based on the upstream parameters. The Alfvén velocity V_{A0} is defined by $B_0/(\mu_0 M_p N_{p0})^{1/2}$, where M_p is the proton mass. With the Earth's bow shock situation in mind, we have the following fundamental values for normalizing parameters of $V_{A0} \sim 90$ km/s, $\Omega_i^{-1} \sim 0.75$ s, and $\lambda_i \sim 120$ km by using $N_{p0} = 4$/cc and $B_0 = 8$ nT. We have run the simulations up to $T = 500$. By using the typical solar wind speed of 400 ~ 500 km/s, the convection distance of the interplanetary magnetic field during the time interval $T = 500$ is estimated to be 23 ~ 29 Re, where Re is the Earth radius. The stand-off distance of the earth's bow shock is about 15 Re, which is shorter than the above convection distance. The IMF has passed behind the terminator position in the time interval of 500.

A shock wave is set up by the conventional way (conductive wall method) and the simulation frame is in the downstream rest frame. The upstream particles are injected at $V_{inj} = 3.4$ from the left boundary and the shock propagates at ~ 1.2 in the upstream direction. As a result, the upstream bulk flow speed of proton U_{p1} is about 4.6 V_{A0} in the shock frame ($M_{A0} = U_{p1}/V_{A0} \sim 4.6$). The upstream proton beta is $\beta = 1$. For simplicity, electron beta is $\beta = 0$. The upstream He^{2+} density is given by the parameter R. Note that we add the He^{2+} component while keeping the proton content unchanged. The upstream temperature of He^{2+} ions is 4 times that of the upstream protons. This characteristics of He^{2+} ions is observed in the solar wind [Ogilvie et al., 1980]. A quasi-parallel case with the shock angle (the angle between the magnetic field and the shock normal) of $\theta = 20°$ is studied. The initial magnetic field is quiet and the fluctuating component is generated in a self-consistent way.

The grid-cell size Δx and the time step size are 0.5 and 0.01, respectively, in the normalized unit. The simulation system length is 5000. 200 particles per grid-cell represent the initial upstream distribution. Since only a few percent of thermal ions are accelerated, we have applied the particle-splitting method in order to obtain statistically meaningful distributions in the higher energy range. When a particle crosses an energy threshold, it is split into two daughter particles, whose contributions to the bulk flow and the density are each one-half of the mother's. The daughters are displaced by $\pm 0.05\Delta x$ from their mother's position. We have chosen 30 energy levels which are linearly spaced in 10-445 E_0 (E_0 is the shock ram energy equal to $1/2 M_p(U_{p1})^2$, that is, $1/2 M_{A0}^2$ in the normalized unit). This allows us to follow the distribution function over ten orders of magnitude.

The shock position, which is denoted by $x = 0$ in the following, is defined as the position where the proton density exceeds 3.0 times that of the far upstream value. As time elapses, a shock front propagates into the upstream region and the upstream portion of the simulation domain shrinks. Although a part of the energetic upstream particles crosses the upstream boundary of the simulation system, the total energy of such escaping particles E_{esc} is very small relative to the total energy E_{tot} at $T = 500$, and the escaping particles should not affect the results in this paper (at least up to the end of the simulation time, $T = 500$). Here, E_{esc} is summed up for all escaping particles when it crosses the upstream boundary until $T = 500$, and $E_{esc} \sim 0.0014 \, E_{tot}$.

3. SIMULATION RESULTS

Hereafter we discuss the energy spectrum obtained in the downstream region. This is because the waves are compressed and have shorter wavelength in the downstream so that energy spectra that are well averaged over the wave phase is obtained even if the short sampling extent (= 50)

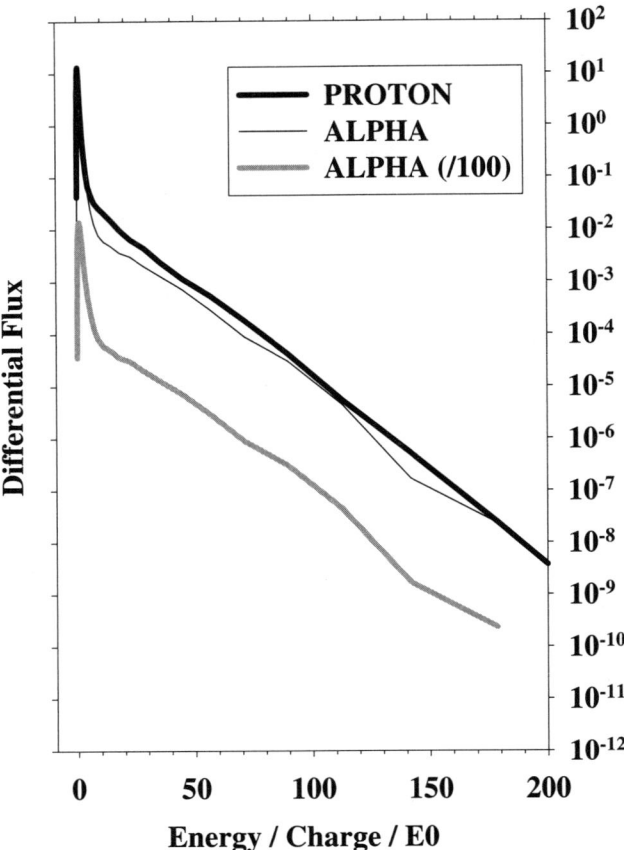

Figure 1. Energy spectra of protons and He^{2+} ions for the case of $R = 20\%$. Plotted are the differential number flux. Both black lines, thick and fine, represent the results for protons and He^{2+} ions, respectively. The thick gray line shows the intensity of He^{2+} ion shifted by two-decades. The nearly parallel lines in the non-thermal energy range show that the characteristic energies for both species are similar.

is adopted. The short sampling extent is necessary to avoid aliasing from the spatial dependence.

Figure 1 shows the proton and He^{2+} ion spectra sampled in the downstream region $100 < x < 150$ at the end of simulation run ($T = 500$). The spectra consist of 13 data points produced by averaging over the duration of $2\pi\,\Omega_i^{-1}$. The density ratio of He^{2+} ions to protons in the upstream thermal population is assumed to be $R = 20\%$, which is higher than the normal density ratio in the solar wind of $R = 4 \sim 5\%$ but in the range that is realized sometimes [*Ogilvie and Wilkerson*, 1969]. Plotted are the differential number fluxes, with thick and thin black curves representing the results for protons and He^{2+} ions, respectively. The energy scale is in E/Q units. Since the two curves almost overlap, the shape of the He^{2+} ions spectrum can be more clearly inspected by the plot in which the flux of He^{2+} ion is shifted by two-decades (gray curve). For both species, the spectra in the log-linear scales are characterized by straight lines above the non-thermal energy range of >10 E/Q, which indicates that the spectra are exponential-like over the wide non-thermal energy range. Moreover, since both curves are parallel to each other, the characteristics (e-folding) energy E_C is almost identical. For this large density ratio $R = 20\%$ case as well as for a more typical solar wind situation of $R = 5\%$, we find the characteristic energies for the two species to be mostly identical. As can be seen from the fact that the two spectra almost overlap, the He^{2+} to proton flux ratio in the non-thermal energy range is enhanced by about a factor of ~ 5 relative to the same ratio for thermal (solar wind) ions. This is also found in the case of $R = 5\%$. Here we note that while our results agree with the previous study [*Scholer et al.* 1999], the wave excitation is now treated fully consistently.

By performing the simulation runs with the particle-splitting, we managed to investigate the time evolution of the energy spectra for both species. The spectrum shapes turn out to be well modeled by exponential shapes throughout and we show in Figure 2 the time history of E_C for the case of $R = 20\%$. Black and gray curves represent the results for proton and He^{2+} ions, respectively. The spectra are sampled in the downstream region of $100 < x < 150$ over the sampling duration of $2\pi\,\Omega_i^{-1}$ and the function $exp(-E/E_C)$ is fitted by the least-squares method to the data in the energy range

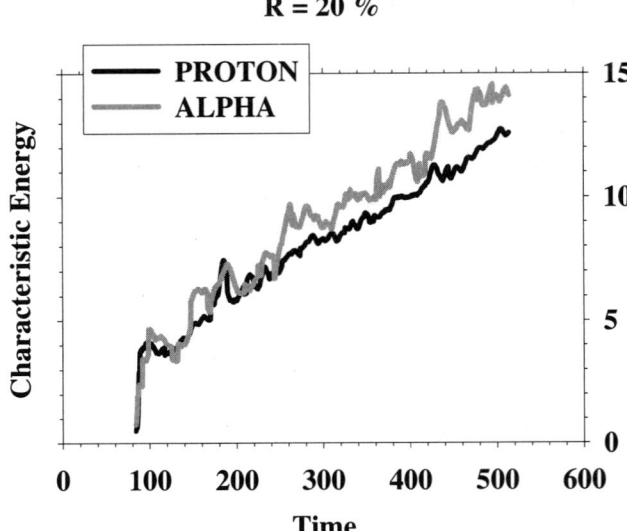

Figure 2. Time history of the characteristic energy E_C. The spectra continue to grow in time, which suggests that the acceleration process has not reached a steady state. The growth rate is similar for both species.

0.1 to ~ 0.9 E_m, where the E_m is the maximum energy end of each spectrum. The E_m is defined by the highest energy bin which has more than 10 particles contained in it. To reduce the statistical error, we omitted the portion close to the high-energy end of the spectrum by setting the upper limit for fitting at 0.9 E_m. The lower limit 0.1 E_m is necessary to cut-off the thermal population. While E_C continues to grow in time, its values stay almost identical between the two species. The continuous increase of E_C in time implies that the acceleration process has not reached a steady state by the end of the run. This does not allow us to apply any time-independent model to analyze the acceleration process quantitatively.

In order to investigate how the characteristic energy E_C depends on the value of R, we have performed simulations with R = 0.1, 1.0, 5.0, 10, 20, and 30%, while keeping other conditions the same. Since the upstream injection velocity V_{inj} is fixed and the simulation is done in the downstream-rest frame, the jump of the velocity across the shock $\Delta U = U_{p1} - U_{p2}$, where U_{p2} is the downstream bulk flow speed of proton in the shock frame, stays the same at V_{inj}. Moreover the shock propagation speed turns out to depend little on the density ratio R. Indeed, the shock Mach number M_{A0} does not change more than 2%. Figure 3a shows the dependence of E_C on R. Characteristic energy values E_C are seen to grow in time in all the runs, and the values sampled at the ends of the runs (T = 500) are plotted. Black and gray lines with solid and open circles represent the results for protons and He^{2+} ions, respectively. Over the wide range of the density ratio R, the characteristic energy is almost independent of ion species or slightly if any larger for He^{2+} ions. This means that the equality of E_C holds not only in the cases where the system is predominantly determined by protons (small R) but also in the cases where the dynamics are significantly affected by He^{2+} ions. We also find that the spectrum becomes harder with increasing R. Increasing R does increase the upstream mass density of the ions so that the average ram energy of upstream ions increases. In this sense, the hardening of the non-thermal spectra with increasing R implies that some of the increased ram energy is used for producing more non-thermal components.

According to the traditional framework, particles are ac-

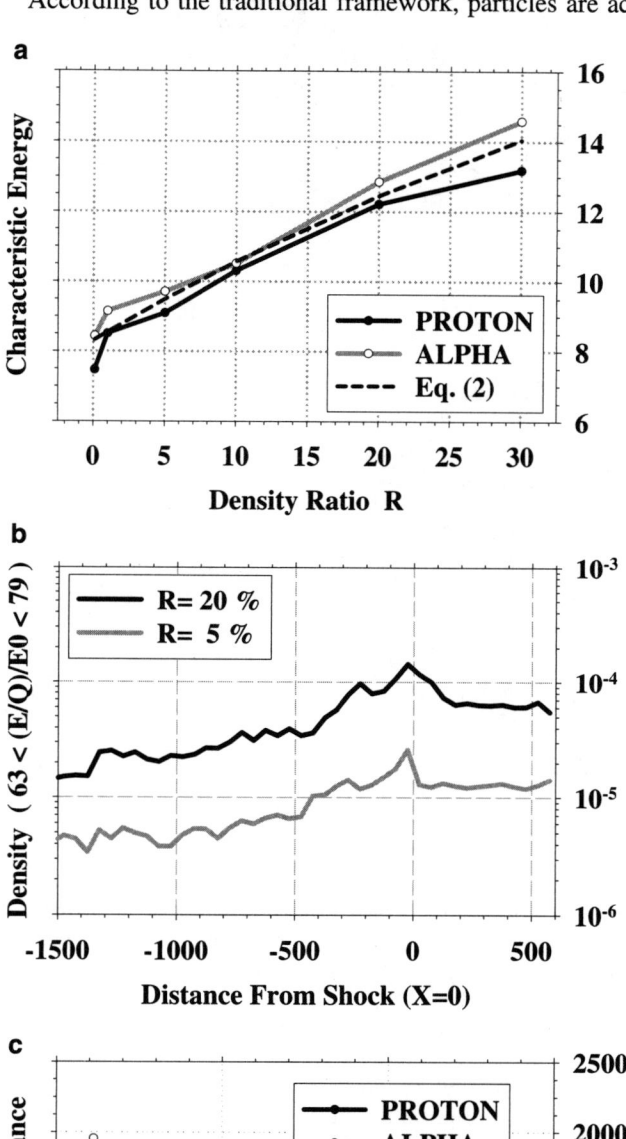

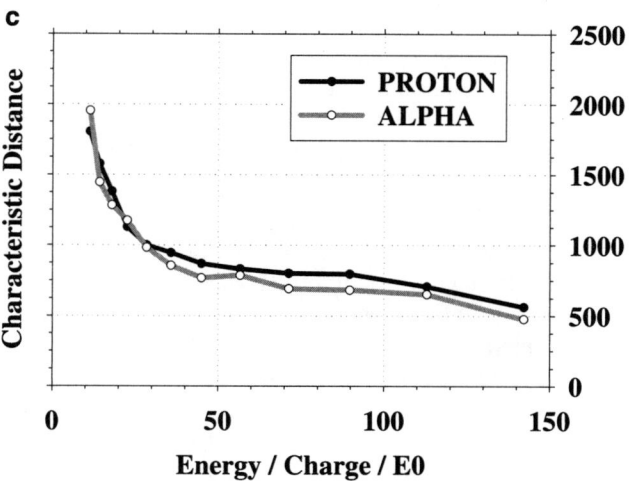

Figure 3. (a) Dependence of the characteristic energy E_C on the He^{2+} density ratio R. In the wide range of R, E_C is almost identical for both species. Spectrum becomes harder with increasing R, even though the jump velocity across the shock (ΔU) is almost constant. (b) Spatial density profile of the energetic protons of 63 < (E/Q)/E_0 < 79 for the cases of R = 5% and 20%. The spatial profiles are almost identical, which suggests that the spatial diffusion coefficient is similar in both cases even though the upstream wave amplitude of R = 20% is larger than that of R = 5% (shown in Figure 4a). (c) Dependence of the characteristic distance X_C on the normalized particle energy. In the wide range of energy, X_C is almost identical for both species. (d) See next page.

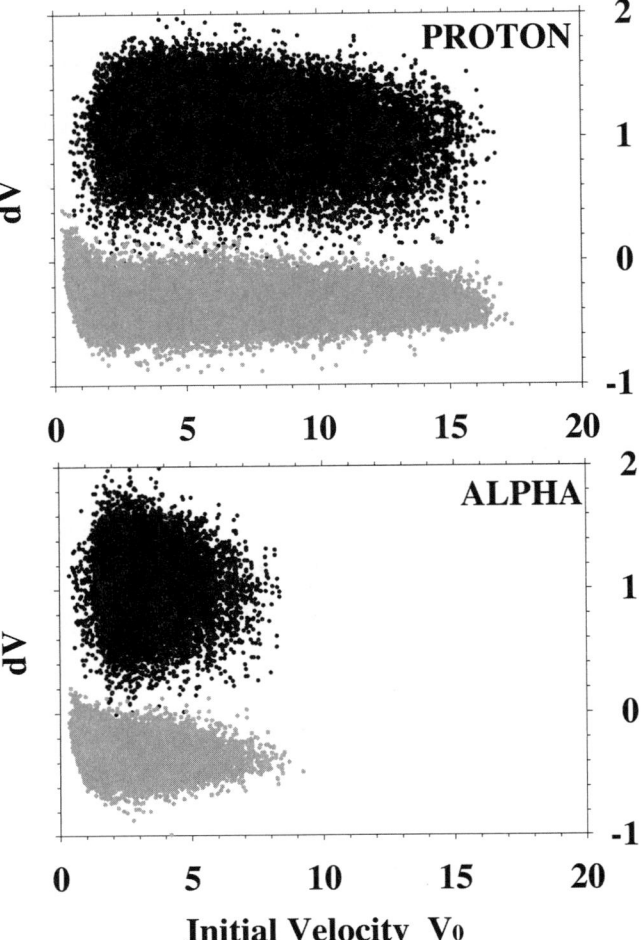

Figure 3d. Velocity increase of particles (ΔV) versus their initial velocity V_0 for protons and He^{2+} ions. The values of ΔV and V_0 are normalized by U_{p1} and $Q^{1/2}$; that is, the energy is normalized by initial energy and charge. The black and gray dots represent the results for the upstream and downstream regions, respectively. The increase of the velocity is almost constant in the wide range of particle initial velocity. For both species, the increase in velocity is almost same. This means that the He^{2+} ions have larger acceleration per charge, due to their larger mass in the wave scattering process in the upstream region.

celerated as they repeatedly meander around the shock. (1) The upstream/downstream spatial profiles of non-thermal ions would give information on how the accelerated ions meander. Since the spatial profiles are determined by the balance between the spatial diffusion and the convection by the flow, it would show the scattering efficiency of the wave-particle interaction that causes the ions to meander. (2) The velocity difference (ΔU) between the upstream and the downstream bulk flow will be the basic parameter that determines the energy gain per cycle. Since E_C is seen to increase with increasing R while the velocity jump ΔU is unchanged, both ions should undergo more cycles of acceleration with increasing R during the same amount of time. The independence of E_C on species in the E/Q unit for a given R implies that protons with smaller M/Q should experience more acceleration cycles. Here we focus on these two items and analyze the simulation data according to the traditional framework.

As seen in the previous simulations (e.g. *Scholer et al.*, [1999]), the density of energetic ions is found to fall off exponentially from the shock into the upstream region in our runs. The density profiles from the $R = 20\%$ (black lines) and the $R = 5\%$ (gray lines) runs are compared in Figure 3b. Plotted are the number densities of the energetic protons in the energy range $63 < (E/Q)/E_0 < 79$ at the end of runs ($T = 500$). The shock location is denoted by $x = 0$. Since these curves are almost parallel, the scattering efficiency from the two different R cases are almost the same. It is unlikely that the increased density ratio R makes the nonthermal ions wander more often around the shock front.

We have calculated the fractional densities for various particle energies in the upstream region and have fitted their spatial profiles by exponential functions. The region $-2000 < x < -100$ is considered to obtain the e-folding distance X_C. The upstream e-folding distances at the end of runs are plotted versus E/Q, normalized by E_0, in Figure 3c. The black (gray) line with solid (open) circles represents the result for proton (He^{2+}) ions. Over the wide range of the particle energies, the characteristic distance X_C is independent of the ion species. This implies that the diffusion coefficient along the magnetic field is also independent of the ion species. These results imply that both non-thermal protons and He^{2+} ions wander between the upstream and the downstream of the shock in a similar manner. It is unlikely that protons with the smaller M/Q experience more cycles of acceleration.

In the traditional framework only the acceleration (deceleration) that takes place upon scattering in the upstream (downstream) region is taken into account. Figure 3d shows more direct information on this sort of particle acceleration/deceleration. When a particle crosses X = -150 (X = 150) from the shock-side to upstream (downstream), its velocity V_0 is recorded. When it comes back to the location from upstream (downstream), its velocity is recorded and compared with its previously recorded value. Shown are the amounts of the velocity increases for these particles (ΔV) versus their initial velocity V_0, that is, a plot showing how much velocity gain (loss) they have experienced in the upstream (downstream) region. The black and gray dots represent the results for the upstream and downstream regions, respectively. ΔV and V_0 are normalized by $Q^{1/2} U_{p1}$, which is in concert with the E/Q energy unit. As is well known, the head-on collision in the upstream region makes positive contribution $\Delta V > 0$

and vice versa. This can be seen in the panels of Figure 3d. Moreover, we can see clearly that (1) the increase/decrease of the velocity is almost constant over the wide range of particle initial velocity, and (2) for both species, the amount of the increase is almost the same. If the upstream and downstream scattering were the only agents for the particle acceleration, He^{2+} ions would be expected to have harder spectrum (in the E/Q energy scale) due to their larger mass, which is not the case.

The arguments given so far indicate that the traditional framework seemingly fails to explain the results. Something else would be dominating for acceleration. Meanwhile, the new perspective by *Sugiyama et al.* [2001b] proposes that the interaction at the shock front with the large amplitude wave can make substantial contribution to the ion acceleration in its initial phase. A test particle study shows that (1) both species are injected into the non-thermal energies and the resulting spectrum shapes are exponential of the same E_C values (if plotted in E/Q scale) (see Figure 14 in *Sugiyama et al.*, [2001a]), and (2) E_C increases with increasing upstream wave amplitude (see Figure 6 in *Sugiyama et al.*, 2001a]. It is interesting to see how the interaction at the shock front is varied by the substantial presence of He^{2+} ions in the self-consistent system.

Now let us inspect what happens to the upstream waves when R is varied. Figure 4a shows the dependence of the wave amplitude on R. For the calculation of the amplitude, we Fourier-decomposed the magnetic field in the upstream region of $-1074 < x < -50$ in the frame comoving with the shock during the interval $120 < T < 247.75$. Then the wave power from the wavenumber range of $[-0.5, 0.5]$ in and $[-1,2]$ in frequency is integrated. The fluctuation level increases with increasing R because the wave excited by He^{2+} ions is superposed on the proton wave in larger R cases. To show this clearly, we show the $\omega - k$ diagrams in Figure 4b. The results from the two cases of $R = 5\%$ and 20% are compared. R/L and +/-, respectively, denote polarization and propagation direction: For instance, R+ stands for right-hand polarized waves propagating into the shock surface.

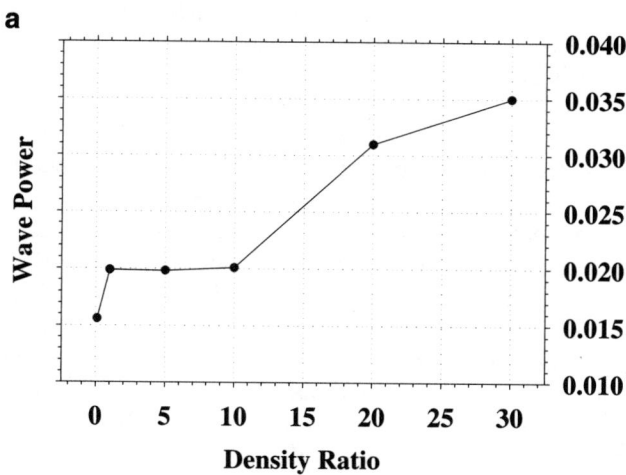

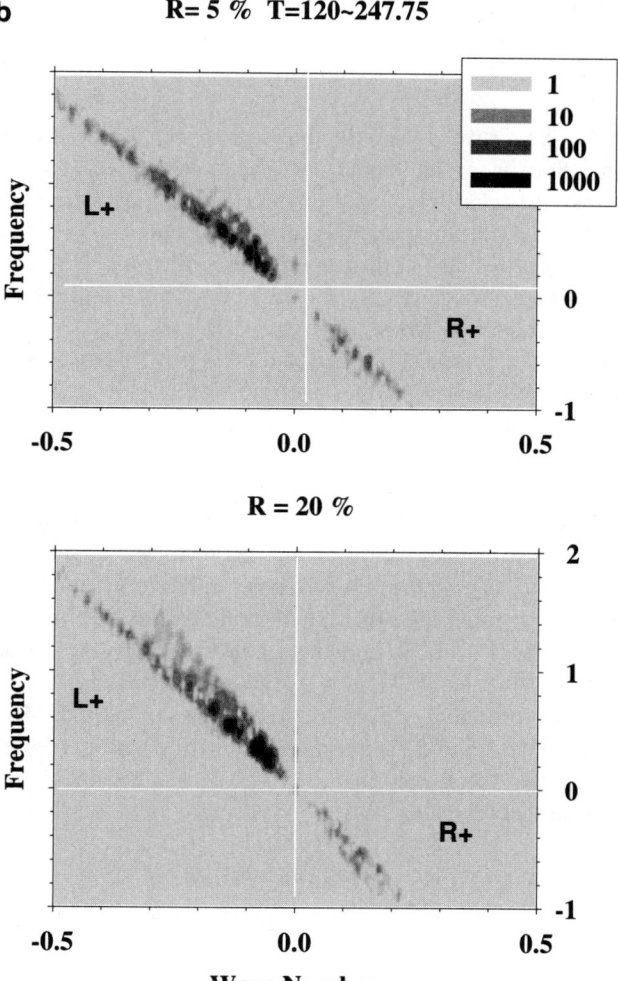

Figure 4. (a) Relation between wave amplitude and R. The wave amplitude becomes larger for higher R. (b) $\omega - k$ diagrams for the cases of $R = 5\%$ and 20%. The diagrams are calculated in the upstream region and in the shock frame. R+ and L+ represent the wave polarity and propagation direction, respectively. For instance, L+ denotes the left-hand polarized waves propagating into shock surface. In the case of $R = 5\%$, the L+ wave with $k \sim -0.1$ is mainly excited by proton. In the case of $R = 20\%$, there is a significant additional wave with $k \sim -0.05$ which is excited by He^{2+} ions.

The L+ wave at $k \sim -0.1$ is excited by the upstream escaping proton beam produced at the shock, while the L+ wave at $k \sim -0.05$ is excited by the upstream He^{2+} beam [*Winske and Leroy*, 1984]. The wave excited by the backstreaming He^{2+} ions ($k \sim -0.05$) is clearly observed in the case of $R = 20\%$ due to the higher density of the He^{2+} beam, resulting in more fluctuating field around the shock.

To summarize, a higher density of He^{2+} ions adds a larger amplitude He^{2+} wave to the proton wave and thus higher magnetic fluctuations around the shock. Higher efficiency in the ion acceleration process at the shock-front with larger wave amplitude is expected, which causes E_C to be higher with increasing R. It is the nature of this acceleration process that makes the two E_C values stay the same. The fact that they do stay the same implies that the shock-front acceleration does dominate during the injection interval, which is the very initial phase of the whole acceleration, and that is what the simulation runs are dealing with.

4. DISCUSSION

We have investigated the initial phase (up to $T = 500$, the injection phase) of the ion acceleration process at quasi-parallel shocks including the effects of He^{2+} ions. The energy spectra of non-thermal ions produced by the parallel shocks have exponential-like shapes for both protons and He^{2+} ions. The characteristic energy E_C for both protons and He^{2+} is almost identical when the energy scale is represented in energy per charge (E/Q) unit and they grow in time. When the relative density of the upstream He^{2+} ions is increased, the spectrum shapes retain exponential form and the equality of E_C stays, while the value itself increases with R.

We have also shown that the additional wave is excited in the upstream region as R is increased. Curiously, the associated intensification of the upstream waves does not contribute to enhance ion scattering in the upstream region (Figure 3b). On the other hand, the acceleration efficiency at the shock front is known to increase [*Sugiyama et al.*, 2001a]. This would be how E_C increases with R.

In the present model, increasing R means that more He^{2+} ions are added to the unchanged proton population. Accordingly, the shock ram energy (E_{ram}, normalized by E_0) per particle increases as

$$E_{ram} = \frac{1 + 4(R/100)}{1 + (R/100)} \quad (1)$$

The hardening of the non-thermal spectra with increasing R implies that some of the increased ram energy is used for producing energetic components. From the present simulations, an empirical relation shown by the dotted line in Figure 3c is obtained.

$$E_C = 8.3 \, E_{ram} \quad (2)$$

While this relation seems so simple, it is reminded that the redistribution of energy is attained via complicated chains of physical processes around the shock region [*Sugiyama et al.*, 2001a].

We have performed simulations with another type of setting to see if E_{ram} is indeed the key determining factor. In these runs, proton and He^{2+} contents and the density ratio are varied while E_{ram} is kept constant. Figure 5 shows the results in the same format as Figure 3a. The values of E_C are almost constant over the wide range of R. The almost identical E_C values persist while they grow with time, as shown in Figure 2. The results strongly support the idea that E_{ram} is the basic determining parameter of E_C.

E_{ram} increases with R because upstream mass density increases. Increasing R also increases the true Alfvén Mach number of the shock

$$M_A = \sqrt{1 + 4(R/100)} \, M_{A0}. \quad (3)$$

Then, the dependence of E_C on R can be alternatively written as

$$E_C = 8.3 \, \frac{1}{(1 + (R/100))} \frac{M_A^2}{M_{A0}^2} \quad (4)$$

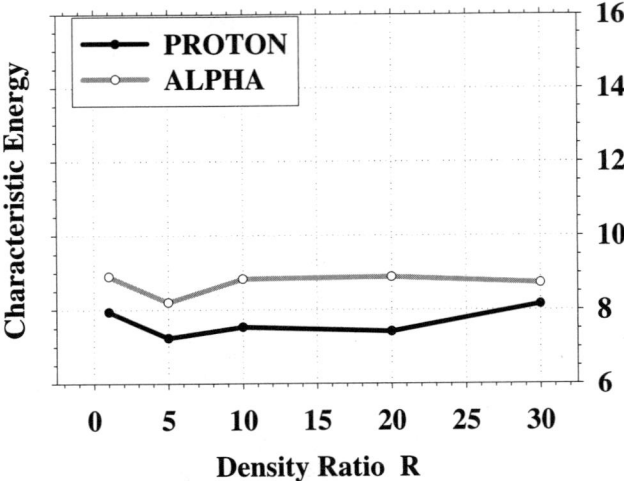

Figure 5. The results for the case of constant E_{ram} in the same format as Figure 3a. E_C shows almost constant value in the wide range of R, which strongly supports the idea that E_{ram} is the basic determining parameter of E_C.

The interpretation would be that R increases the true Mach number of the shock and thus brings about more turbulence around the shock front. The enhanced wave activity boosts up the acceleration efficiency. Note, however, that E_0 (upstream proton kinetic energy) works as the basic energy unit in the above argument based on eq.(3). Since this is not necessarily true in the present cases, where He^{2+} ions carry negligible kinetic energy into the shock, we would think that eq.(2) gives a more transparent picture than eq.(3).

The exponential shape is what appears in the initial phase of shock acceleration when the wave-particle interaction at the shock front dominates and it is the limited acceleration time at the Earth's bow shock that causes the non-thermal ions to assume this spectrum shape. As in the traditional framework, the scattering process in the upstream (downstream) region makes the particle velocity increase (decrease) by a constant amount over the wide range of wandering particle energy, as shown in Figure 3d. If nothing else comes into play while the particles bounce back and forth about the shock front, the energy spectrum would have the power-law shape. The acceleration at the shock front is this something else that is not only present but seems to be dominate in making the exponential spectrum. The simplest way to produce an exponential spectrum is to repeat the cycle that consists of (1) adding a constant energy to particles, and (2) abandoning a certain fraction of the particles. The acceleration at the shock front seems to have the aspect of adding energy to particles and would naturally dominate over the traditional framework, which multiplies the particle energy, when the particle energy is still not very large. The effects of the limited simulation box size is negligible because the number of the particles that have hit the upstream boundary of the simulation box before the end of simulation time is quite small. Thus the deviation of the spectrum from the power-law to exponential shape is not due to an escape of the more energetic ions.

The time allowed for an ion to interact with the Earth's bow shock would not be much longer than this simulation as mentioned in Section 2. Therefore, noting that E_0 is ~1 keV for the solar wind, it would be reasonable to conclude that the soft, exponential-like form in the energy range from 10 keV/Q up to ~ 200 keV/Q is what would be produced by the quasi-parallel part of the Earth's bow shock assigned with limited acceleration time. It is quite conceivable that the traditional framework, which is indeed operative but is masked by the other agent in the present runs, becomes more efficient and dominates as the energy spectrum is further hardened. Indeed, the characteristic energy of the exponential spectrum is seen to increase in time. The turn over to the power-law spectrum is well expected in an accelerating object that has larger spatial extent than the Earth's bow shock.

Acknowledgments. We thank J. Quenby for his constructive comments. A part of the simulations have been performed on the KDK system of Research Institute for Sustainable Humanosphere (RISH) at Kyoto University as a collaborative research project and on VPP 5000 at Nagoya University Computer Center via Collaborative Research Program of STE Laboratory.

REFERENCES

Blandford, R. D., and J.P.Ostriker, Particle acceleration by astrophysical shocks, *Astrophys. J.*, *221*, L29, 1978.

Eichler, D., Energetic particle spectra in finite shocks: the earth's bow shock, *Astrophys. J.*, *244*, 711-716, 1981.

Ipavich, F. M., A. B. Galvin, G. Gloeckler, M. Scholer, and D. Hovestadt, A Statistical Survey of Ions Observed Upstream of the Earth's Bow Shock: Energy Spectra, Composition, and Spatial Variation, *J. Geophys. Res.* 86, 4337-4342, 1981.

Jokipii, J. R., Rate of energy gain and maximum energy in diffusive shock acceleration, *Astrophys. J.*, *313*, 842-846, 1987.

Ogilvie, K. W., T. D. Wilkerson, Helium abundance in the solar wind, *Solar Phys.*, *8*, 435-449, 1969.

Ogilvie, K. W., P. Bochsler, J. Geiss, and M. A. Coplan, Observations of the velocity distribution of solar wind ions, *J. Geophys. Res.* 85, 6069-6074, 1980.

Scholer, M., H. Kucharek, K. -H. Trattner, Injection and acceleration of *PROT* and He^{2+} at Earth's bow shock, *Ann. Geophys.*, *17*, 583-594, 1999.

Scholer, M., Injection and Acceleration of Energetic Particles at Collisionless Shocks, *Adv. Space Res.*, *21* (4) 533-542, 1998.

Sugiyama, T., M. Fujimoto, and T. Mukai, Quick ion injection and acceleration at quasi-parallel shocks, *J. Geophys. Res. 106*, 21,657-21,673, 2001a.

Sugiyama, T., M. Fujimoto, and M. Scholer, Injection of He^{2+} at a Parallel Shock, *Adv. Space Res.*, *27* (3) 637-642, 2001b.

Winske, D., and M. M. Leroy, Diffuse Ions Produced by Electromagnetic Ion Beam Instabilities, *J. Geophys. Res. 89*, 2673-2688, 1984.

M. Fujimoto, Department of Earth and Planetary Sciences, Tokyo Institute of Technology, Tokyo, Japan.

H. Matsumoto, Research Institute for Sustainable Humanosphere (RISH), Kyoto University, Kyoto, Japan.

T. Sugiyama, The Earth Simulator Center, Japan Agency for Marine-Earth Science and Technology (JAMSTEC), 3173-25, Showa-machi, Kanazawa-ku, Yokohama, Kanagawa 236-0001, Japan. (tsugi@jamstec.go.jp)

Particle Acceleration in Shell Supernova Remnants: Observational Evidence

Stephen P. Reynolds

Harvard-Smithsonian Center for Astrophysics and North Carolina State University, Raleigh, North Carolina

The origin of Galactic cosmic-ray electrons and ions up to energies of order a few thousand TeV (the "knee" in the cosmic-ray spectrum) is generally attributed to shell supernova remnants (SNRs). Until recently, the only direct observational evidence for this conclusion was the presence in SNRs of synchrotron radio emission, indicating electrons with a power-law energy distribution between roughly 0.1 and 100 GeV (in the best cases). Further arguments are necessary to rule out the possibility that these electrons are just "borrowed" from the ambient Galactic pool and compressed in the remnant shock wave. In the last few years, however, striking evidence at much higher photon energies has emerged: X-ray synchrotron emission in several objects, indicating electrons with energies up to 100 TeV, and TeV photon emission in a few, most likely due to inverse-Compton upscattering of cosmic microwave background photons by these same electrons. Evidence for cosmic-ray ions is so far inconclusive; the most unmistakable form would be detecting the expected "bump" at a few hundred MeV due to photons from the decay of neutral pions produced in inelastic collisions between cosmic-ray ions and interstellar gas. I shall review these various lines of evidence, and the constraints they impose on theoretical models.

INFERENCES FROM RADIO DATA

The most obvious evidence for the presence of energetic ($E \gg mc^2$) particles in shell supernova remnants is synchrotron radio emission, associated with supernova remnants ever since Shklovsky's proposal in 1953. Standard synchrotron theory gives the energy of an electron radiating its peak power at frequency ν_9 GHz in a magnetic field $B_{\mu G}$ microGauss (sky-plane component $B_{\perp \mu G}$) as $E = 14.7 \, (\nu_9/B_{\perp \mu G})^{1/2}$ GeV, so for typical estimates of $B_{\perp \mu G} \sim 10$ for shell SNRs, observed between 100 MHz and 10 GHz in many cases, we infer electron energies of 1.5–15 GeV. The largest plausible range takes $B_{\perp}$ in the range 1–100 μG and observing frequency in the range 10^7–10^{11} Hz, implying electron energies of about 0.1–100 GeV.

These electrons could in principle simply be borrowed from the interstellar medium, compressed in the remnant shock wave. Two arguments suggest that this mechanism cannot explain radiation from young remnants with adiabatic shocks and compression ratios r of order 4. First, the spectrum is incorrect. Since an electron's power peaks at a frequency $\nu_m \propto E^2 B_{\perp}$, the post-shock peak will be boosted by a factor between r^2 and r^3 (since $E \propto r$, and the magnetic field will increase by a power of r between 0 and 1 depending on orientation). So electrons radiating between 1 and 5 GHz, where most remnant spectral indices are produced, would have radiated at frequencies below ~ 100 MHz in front of the shock. We know that the diffuse Galactic radio synchrotron radiation has a spectrum $S_\nu \propto \nu^\alpha$ with $\alpha \sim -0.4 \pm 0.05$ below 100 MHz [*Lawson et al.*, 1987], whereas shell remnants have

Particle Acceleration in Astrophysical Plasmas
Geophysical Monograph Series 156
Copyright 2005 by the American Geophysical Union
10.1029/156GM12

$\alpha < -0.5 \pm 0.05$ typically, ≤ -0.6 for historical shells like Cas A, Kepler, Tycho, and SN 1006. Furthermore, the 220 or so Galactic radio supernova remnants have spectral indices with a dispersion of order 0.2 about -0.5 [*Green*, 2001]—hard to explain if all remnants borrow their electrons from the same fairly homogeneous pool of Galactic cosmic-ray electrons.

Second, many shells have too high a brightness to be explainable simply by compression of interstellar electrons (and magnetic field). For an electron distribution $N(E) = KE^{-s}$ electrons cm^{-3} erg^{-1} between E_ℓ and $E_h \gg E_\ell$, the synchrotron emissivity $j_\nu \propto K B^{1-\alpha}$ with $\alpha = (1-s)/2$. Compression by a factor r raises the electron energy density u_e by $r^{4/3}$ and B by r (if the field is tangential to the shock), so $K \propto E_\ell^{s-2} u_e \propto r^{4/3}$. So $j_\nu \propto r^{-\alpha + 7/3} = r^{(3s+11)/6}$. For $s \sim 2$, this gives $j_\nu \propto r^3$. However, the mean surface brightness of a young shell remnant at 1 GHz is about 10^{-19} W m^{-2} Hz^{-1} (see compilation by *Green* [2001]), larger than that of the mean Galactic-plane diffuse synchrotron emission (e.g., *Beuermann, Kanbach, and Berkhuijsen*, 1985) by a factor of order 20, in spite of a typical path length through a remnant smaller than that through the Galaxy by factors of order 1,000–10,000. Thus we infer mean synchrotron emissivities in young remnants of order 10^5 times higher than the diffuse Galactic ambient medium. So unless young remnants have shocks with compression ratios far larger than the expected 4 for adiabatic shocks into monatomic gas, compression alone cannot provide the requisite surface brightness; newly accelerated electrons are required.

While various acceleration mechanisms have been proposed to explain the presence of these energetic electrons in SNRs, the general consensus for about 20 years has been that the dominant process is first-order Fermi (diffusive) shock acceleration (DSA), whose basic characteristics are reviewed by *Blandford and Eichler*, 1987; BE87). Evidence includes very sharp edges seen in radio images of many SNRs (several examples shown in *Achterberg, Blandford, and Reynolds*, 1994), and the rough agreement between the simplest predicted test-particle power-law index, $s = 2$ for a compression ratio of 4, and the mean SNR inferred electron index. The upstream MHD turbulence required by DSA to scatter particles has been inferred to be present based on the sharp rims [*Achterberg et al.*, 1994]. The electrons are assumed to be drawn from the thermal population by some poorly understood "injection" mechanism to an energy high enough to see the shock as a discontinuity (e.g., *Levinson*, 1992). Nonlinear shock modification, in which a more gradual velocity transition brought about by non-negligible pressure from accelerated particles means effective compression ratios varying with distance ahead of the shock, predicts that if particle diffusion lengths increase with energy, the spectrum should flatten at higher energies, as more energetic particles sample larger compression ratios. This means both that mean spectra at relatively low energies (radio-emitting) should be steeper than the ultimate asymptotic spectral slope, and that the spectra should show positive (concave-up) curvature. Both effects have been seen in the integrated spectra of young remnants [*Reynolds and Ellison*, 1992].

However, the considerable numbers of Galactic shell SNRs (about 40%; *Green*, 2001) with $s < 2$ ($\alpha > -0.5$) require some modifications of the simple theory. It is possible that second-order Fermi (stochastic) acceleration plays a role (e.g., *Ostrowski*, 1999). The acceleration time to energy E for stochastic acceleration (once particles are relativistic) scales as [*Melrose*, 1974]

$$t_{acc}(\text{Fermi II}) \sim \left(\frac{\lambda_{mfp}}{c}\right)\left(\frac{v_A}{c}\right)^{-2}, \quad (1)$$

where v_A is the Alfvén speed (in the downstream medium), while the rate for DSA scales as (e.g., BE87)

$$t_{acc}(\text{Fermi I}) \sim \left(\frac{\lambda_{mfp}}{c}\right)\left(\frac{u_{sh}}{c}\right)^{-2}, \quad (2)$$

where for the DSA time we have assumed a diffusion coefficient proportional to particle energy, which is proportional to mean free path λ_{mfp}. We have also assumed a parallel shock [*Jokipii*, 1987]. The ratio of acceleration rates of the two processes then scales as

$$\frac{t_{acc}(\text{Fermi I})}{t_{acc}(\text{Fermi II})} \propto \mathfrak{M}_{A2}^{-2} \quad (3)$$

where $\mathfrak{M}_{A2} \equiv u_{sh}/v_A$ is the Alfvén Mach number of the downstream flow (since we expect the dominant turbulence to be in the downstream region). Since we expect the bulk post-shock flow to be super-Alfvénic (equivalent to requiring that the post-shock magnetic energy density not exceed the thermal energy density ρu_{sh}^2), under normal conditions we expect Fermi I acceleration to be more rapid.

There is certainly plenty of energy available in supernovae to explain the required populations of fast particles. Simple equipartition (minimum-energy) arguments (e.g., *Pacholczyk*, 1970) give for a typical remnant

$$E_{min} \sim (10^{48} - 10^{49})(1+k)^{4/7} \text{ erg} \quad (4)$$

where $k \equiv E(\text{ions})/E(\text{electrons})$ (in cosmic rays ~ 5 GeV, $k \sim 50$; BE87). Of course, there is no guarantee that equipartition holds between magnetic field and particles, so the energy required per supernova may be greater than this minimum.

STANDARD SHOCK-ACCELERATION THEORY AND ITS PREDICTIONS

For the simplest test-particle case, with a plane shock and compression ratio r, DSA predicts (BE87)

$$f(p) \propto p^{-\sigma} \text{ with } \sigma = 3r/(r-1) \Rightarrow \\ N(E) \propto E^{-s}, s = (r+2)/(r-1) \quad (5)$$

for $E \gg mc^2$, where $f(p)$ is the momentum-space distribution function, and $N(E)\,dE = f(p)\,(4\pi p^2)\,dp$. For strong shocks in fluids with adiabatic index $\gamma = 5/3$, $s = 2 \Rightarrow S_\nu \propto \nu^{-0.5}$ for synchrotron radiation. This result is *independent* of the value or energy-dependence of the diffusion coefficient κ. Acceleration can continue until it is limited by geometric effects (finite shock size; Larmor radius comparable to shock size), finite remnant age (acceleration time comparable to age), or radiative losses (on electrons; acceleration time comparable to radiative lifetime). (The acceleration *rate* does depend on $\kappa(E)$). This simple benchmark result also ignores the possibility of particle escape, in any case expected to be important only for the highest energy particles.

However, the test-particle limit is not likely to be relevant, since to explain the Galactic cosmic-ray pool, SNRs need to put of order (1–10)% or more of their energy into fast particles. Various nonlinear effects are expected when accelerated-particle energies can no longer be neglected. The most significant is the broadening of the shock transition due to the pressure of upstream-diffusing accelerated particles ("dynamical precursor"), but others include particles scattering from turbulence which they themselves generate, especially upstream ("turbulence precursor"); possible self-regulation of particle "injection" from thermal energies; the escape of particles either due to finite remnant size or absence of scattering waves at the highest energies; and a drop in the mean adiabatic index below 5/3 as a significant fraction of accelerated particles become relativistic. The latter two effects can cause substantial increases in the shock compression ratio above the nonrelativistic, monatomic-gas value of 4 [*Berezhko, Elshin, and Ksenofontov*, 1996], but the increase is less for spherical shocks than for plane shocks. Sphericity also has observational implications, as the postshock electron distribution evolves due both to radiative losses and adiabatic deceleration and decompression. These various effects are reviewed in BE87 as well as *Drury* [1983], *Jones and Ellison* [1991], *Malkov and Drury* [2001], and others.

RADIATIVE SIGNATURES OF ACCELERATED PARTICLES

We expect four basic nonthermal radiative processes to be important for supernova remnants: three leptonic processes and one hadronic [*Drury, Aharonian, and Völk*, 1994; *Gaisser, Protheroe, and Stanev*, 1998; *Sturner et al.*, 1997; *Baring et al.*, 1999]. The leptonic processes are synchrotron radiation (radio to X-ray bands), nonthermal bremsstrahlung (X-rays and gamma-rays), and inverse-Compton emission from any of various possible populations of seed photons (infrared and higher energies). The hadronic process is inelastic scattering of cosmic-ray protons from thermal nuclei, producing π^0 mesons which decay to gamma rays.

While nonlinear effects are expected to produce some concavity in accelerated-particle spectra, as described above, a power-law is still a fairly good gross approximation, with a high-energy cutoff due to one of the mechanisms listed above that should be roughly exponential [*Reynolds*, 1998]. So we shall assume an input spectrum $N(E) \propto E^{-s} \exp(-E/E_{max})$ for illustrative purposes. For π^0-decay emission, kinematic constraints cause the spectrum to cut off below $E_\gamma \sim 100$ MeV, and extend up to a broad cutoff above $E_\gamma \sim E_{max}$(protons). The spectrum $F(E_\gamma)$(ph cm^{-2} s^{-1} GeV^{-1}) $\propto E_\gamma^{-s}$.

In synchrotron radiation, the above input electron spectrum produces a power-law photon spectrum $S_\nu \propto \nu^\alpha$ until a slow rolloff begins near a frequency corresponding to E_{max}: $h\nu_c \sim 2\,(E_{max}/100\text{ TeV})^2\,(B/10\,\mu\text{G})$ keV. Inverse-Compton emission produces photons at characteristic energies $h\nu$ (IC) $\sim (E/m_e c^2)^2 h\nu$(seed). When Klein-Nishina effects are unimportant (that is, when incoming photons have $h\nu' \ll m_e c^2$ in the electron rest-frame, or equivalently $E\,h\nu \ll (m_e c^2)^2$ for electrons of energy E), the output flux is proportional to the seed-photon energy density, with a spectrum independent of the photon spectrum, having the same slope as the radio synchrotron spectrum, $F(E_\gamma) \propto E_\gamma^{(-1-s)/2}$, until a cutoff near $E_\gamma(\text{max}) \sim 36\,(E_{max}/100\text{ TeV})^2$ TeV. For TeV photon emission, Klein-Nishina cutoffs reduce contributions from more energetic seed photons, so cosmic microwave background (CMB) seed photons dominate. Electrons lose energy in inverse-Compton losses, compared to synchrotron losses, in the ratio of seed-photon energy density to magnetic energy density $B^2/8\pi$. The minimum energy-loss rate for relativistic electrons is then to inverse-Compton upscattering of CMB photons, corresponding to an equivalent magnetic-field strength of 3.27 μG.

Bremsstrahlung processes are dominantly conventional electron-proton bremsstrahlung; other flavors (e.g., relativistic electron-electron bremsstrahlung) are less important. Since an electron of energy E typically radiates photons with $E_\gamma \sim E/3$, the photon spectrum basically mirrors the electron spectrum: $F(E_\gamma) \propto E_\gamma^{-s-1} \log(E_\gamma)$, cutting off near $E_\gamma \sim E_{max}$. Detailed calculations of these processes in the case of a self-consistent nonlinear shock-acceleration model are contained in *Baring et al.* [1999].

OPEN QUESTIONS

We believe we have a basic idea of the process of particle acceleration in SNR shocks, based on the above considerations. Many important questions remain unanswered, however: To how high energies are SNRs observed to accelerate particles? Is there any evidence for accelerated *ions?* What is the obliquity-dependence of the process? Injection efficiency (percentage of thermal particles that become cosmic rays)? Acceleration rate? (The obliquity is the angle θ_{Bn} between the upstream magnetic field direction and the shock normal). What is the nature of the turbulence doing the scattering? What is its spectrum?

Major advances have occurred in the last few years in bringing observational evidence to bear on these questions. While radio synchrotron emission has confirmed the presence of newly accelerated electrons in SNRs, the approximately power-law portion of the spectrum contains relatively little information on the detailed physics of shock acceleration. Since the integrated synchrotron flux $S_\nu \propto K B^{1-\alpha}$, in general one cannot even determine the energy density of the radio-emitting electrons, since there are no generally applicable methods of estimating mean magnetic-field strengths. However, the departures from power-laws, chiefly at high energies where cutoffs set in, contain much more information, since cutoff photon energies depend on the physical processes limiting particle acceleration: radiative losses, particle escape, or finite remnant size or age, competing with acceleration rates. Various microphysical parameters can thus be constrained by observing spectral cutoffs, due to any of the four processes mentioned above.

X-RAY OBSERVATIONS

Historically, the first suggestion of nonthermal shell-SNR emission above radio energies was made by *Reynolds and Chevalier* [1981], who attempted to explain the featureless X-ray spectrum of the remnant of SN 1006 AD as the loss-steepened tail of the radio spectrum. However, further observations showed the presence of some thermal X-ray line emission, and later models were proposed that could produce a lineless thermal spectrum, under certain conditions [*Hamilton, Sarazin, and Szymkowiak*, 1986]. The issue was finally settled when the *ASCA* satellite showed conclusively that emission from the bright limbs showed almost no line emission, while that from the center was dominantly thermal [*Koyama et al.*, 1995]. The synchrotron model was then resuscitated [*Reynolds*, 1996], where it was shown that cutoff energies of order 50 TeV could explain the observations, and were naturally to be expected from an object with the physical parameters of SN 1006. Subsequent work by various authors has established that synchrotron emission is almost certainly the dominant process in producing the bright limb emission in SN 1006 [e.g., *Dyer et al.*, 2001; *Long et al.*, 2003], and a fairly elaborate spectral model incorporating electron escape has been shown to describe the X-ray spectrum from *ASCA* and *Chandra* quite well [*Dyer et al.*, 2001, 2004; *Long et al.*, 2003]. In such models, the X-ray emission arises from a slow rolloff of the spectrum; but a crude characterization in terms of a simple power-law gives photon indices Γ of about 3 ($F_\nu \propto (h\nu)^{-\Gamma}$ photons cm^{-2} s^{-1} keV^{-1}).

There are two other Galactic remnants with essentially lineless X-ray spectra. G347.3–0.5 (RX J1713.7–3946) was found to have a featureless spectrum in *ASCA* observations [*Koyama et al.*, 1997] and has been examined in more detail with *ASCA* [*Slane et al.*, 1999] and *Chandra* [*Uchiyama et al.*, 2003; *Lazendić et al.*, 2004]; simultaneous spectral fitting of *ROSAT*, *ASCA*, and *RXTE* spectra was reported by *Pannuti et al.* [2003]. G347.3–0.5 is a large (~ 1°) remnant, perhaps interacting with molecular gas seen nearby in CO emission. Radio images [*Lazendić et al.*, 2004] show emission chiefly from the NW region. At high angular resolution, the X-ray emission breaks up into small-scale filaments and knots. Little or no evidence for any thermal emission is seen. Fitting with a simple model describing synchrotron emission from a power-law electron distribution with an exponential cutoff (XSPEC model *srcut*; *Reynolds and Keohane*, 1999) indicates a characteristic cutoff energy of about 15 TeV [*Lazendić et al.*, 2004], a similar result to one obtained with a nonlinear shock model [*Ellison, Slane*, and *Gaensler*, 1999]. Again, a rough power-law fit gives $\Gamma \sim 3$.

A third remnant, G266.2–1.2 (also known as Vela Jr.), was discovered, again by *ROSAT*, in an image of the Vela SNR restricted to relatively high photon energies (> 2 keV) [*Aschenbach*, 1998]. More extensive studies with *ASCA* [*Slane et al.*, 2001] show a large shell (~ 2°) projected against the even larger shell of the Vela remnant, with a featureless X-ray spectrum with a power-law index of $\Gamma \sim 2.6$.

Essentially lineless X-ray spectra are almost certainly due to synchrotron radiation, because inverse-Compton emission would have far too hard a spectrum (the same slope as the radio spectrum, which would give $\Gamma \sim 1.5$). (A drastically steeper spectrum, $s \sim 5$, setting in just above the highest measured radio frequencies, could in principle produce the observed X-ray spectra, but no particle spectrum this steep has ever been observed in any synchrotron source, and the energy density in requisite seed photons, which would need $\nu \lesssim 10^8$ Hz, fails by many orders of magnitude in any SNR.) Bremsstrahlung from the beginning of the power-law tail to the electron distribution that will continue on to radio-emitting electrons could produce an appropriately shaped continuum, but since photons from this process emitted at a few keV originate from electrons also of a few keV energy, these electrons would collisionally excite line transitions quite as effectively as electrons drawn from a thermal distribution. The almost complete absence of line emission essentially rules out bremsstrahlung (though contrived models of shocks moving into pure ejecta, with appropriately chosen ionization timescales, have been produced to try to explain featureless spectra; see *Hamilton, Sarazin, and Szymkowiak* [1986] for one example). Synchrotron emission, on the other hand, arises from electrons of higher energy by many orders of magnitude, hopelessly inefficient at exciting line transitions.

The decoupling of the continuum from the thermal emission processes means that one might expect to see objects with a thermal spectrum partially but not totally swamped by a synchrotron continuum. The observational clue to such a situation would be a remnant whose X-ray spectrum showed anomalously weak lines, unexplainable by conventional ideas such as depletion onto grains. A clear example is RCW 86 (G315.4–2.3), where the designation refers to bright optical emission in the SW quadrant. *ASCA* observations revealed weak lines whose analysis with standard thermal models gave extremely peculiar results [*Vink et al.*, 1997].

However, a combination of a thermal shock model (describing line and continuum emission from a solar-abundance shocked plasma) with a model for a synchrotron component described the data very well (Figure 1; *Borkowski et al.*, 2001), a result confirmed with higher spatial and spectral resolution by *Chandra* [*Rho et al.*, 2002]. It has been argued that the continuum could still be nonthermal bremsstrahlung [*Vink et al.*, 2002], but such a model requires a large pool of slightly suprathermal electrons, which lose energy rapidly by collisions with thermal electrons, leading to a high energy requirement and a prediction that the emission would be confined very closely to the shocks in which the suprathermal electrons are produced.

Other examples of possible synchrotron-dominated remnants have been found in the *ASCA* Galactic Plane Survey [*Bamba et al.*, 2003]; another candidate is reported by *Ueno et al.* [2003]. It seems clear that the phenomenon is widespread. Finally, evidence at slightly higher X-ray energies (5 – 20 keV) was produced with the *Rossi X-ray Timing Explorer* (*RXTE*) of nonthermal "hard tails" in the spatially integrated spectra of SN 1006, Tycho, Kepler, RCW 86, and Cas A by *Allen, Gotthelf, and Petre* [1999]. It is not completely clear what fraction of these "hard tails" might be synchrotron, though evidence for small regions dominated by nonthermal, perhaps synchrotron, emission has been reported for Cas A [*Gotthelf et al.*, 2001; *Bleeker et al.*, 2001], Tycho [*Hwang et al.*, 2002], and Kepler [*Cassam-Chenaï et al.*, 2003]. Cas A has been detected to 100 keV with the OSSE instrument aboard *CGRO* [*The et al.*, 1996], but this is almost certainly bremsstrahlung. (Above 10 keV, no common elements produce lines, so the above arguments against nonthermal bremsstrahlung no longer apply.)

Although several remnants now definitively show evidence of electrons with energies of hundreds of TeV, it is important to note that in no case so far examined is the X-ray flux on the straight power-law extrapolation of the radio. In fact, out of 14 remnants in the Galaxy [*Reynolds and Keohane*, 1999] and 11 in the Large Magellanic Cloud [*Hendrick and Reynolds*, 2001], not one shows X-ray emission as bright as the extrapolation of its radio spectrum. Whether the observed X-rays are thermal or not, the electron spectrum must therefore begin to steepen at energies of order 10–100 TeV, or X-ray synchrotron emission would be produced in excess of the observations. Now standard DSA should not differentiate between electrons and protons once both are highly relativistic (as long as MHD waves of both helicities are present to scatter charges of both signs). If in all these cases, radiative losses steepen the electron spectrum, then it is still possible that these SNRs may harbor proton spectra that extend unbroken as far as the "knee" energy of about 3000 TeV, to which it is generally assumed that shell SNRs accelerate particles. However, in five Galactic remnants with independently known ages and shock velocities, the observed upper limit for E_{max} is lower than losses would produce [*Reynolds and Keohane*, 1999]. Any other mechanism for limiting acceleration (finite size, age, sphericity, lack of scattering) should act as well on protons as electrons, suggesting a substantial problem in the standard picture of the acceleration of Galactic cosmic rays.

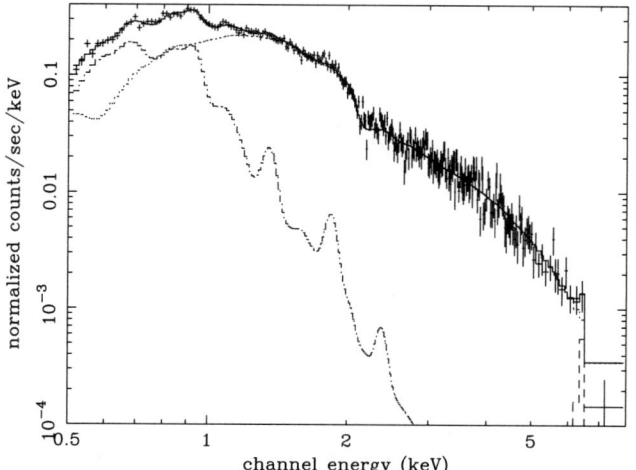

Figure 1. Chandra X-ray spectrum of a synchrotron-dominated filament in RCW 86 [*Rho et al.*, 2002]. A model for thermal emission from a shocked plasma is shown, accounting for faint line emission below 1 keV, as well as a synchrotron model (*SRCUT*).

GAMMA-RAY OBSERVATIONS

Between 100 keV and 1 TeV, observations of SNRs are essentially confined to the *EGRET* instrument on *CGRO*, sensitive to energies between 20 MeV and 30 GeV. At energies of order 1 TeV and above, ground-based air Čerenkov detectors such as the Whipple Observatory, CANGAROO, HEGRA, and now HESS analyze optical observations of air showers and separate photon-initiated from cosmic-ray-initiated events on the basis of imaging data. The angular resolution of *EGRET* observations varies widely, but is normally the better part of a degree, so associations of observed sources with particular SNRs tend to be statistical (see [*Torres et al.* [2003] for an extensive discussion). Some of the 74 unidentified *EGRET* sources near the Galactic plane ($|b| < 10°$) are probably SNRs, but observations with instruments with higher angular resolution will be necessary to make convincing cases for particular associations. However, the spectra observed for some possible associations (e.g., G78.2+2.1 [the γ Cyg SNR]) can be reasonably well-described by emission models invoking a superposition of π^0-decay emission and bremsstrahlung [*Gaisser, Protheroe, and Stanev*, 1998; *Baring et al.*, 1999].

Only a few SNRs have been detected in the TeV range. First to be reported was SN 1006 [*Tanimori et al.*, 1998] with the CANGAROO instrument, where the NE limb, but not the SW, was found to produce emission of a few TeV energy. That emission could be well described as inverse-Compton upscattering of CMB photons (ICCMB) [*Tanimori et al.*, 1998; *Dyer et al.*, 2001], giving a direct measurement of the relativistic-electron population and allowing an inference of the magnetic-field strength from observed synchrotron fluxes. The resulting value, ~ 5 – 10 μG, was reasonable for a relatively weak ambient magnetic field, to be expected at SN 1006's galactic latitude of 15°, compressed in the shock but not otherwise amplified. However, an alternative model with a much higher magnetic-field strength [*Berezhko, Ksenofontov, and Völk*, 2002] accounts for the TeV emission with a relatively high upstream density and π^0 decay, thus claiming evidence for accelerated ions. TeV emission has also been reported from G347.3–0.5 [*Enomoto et al.*, 2002], again with CANGAROO, and has been explained by the ICCMB mechanism [*Ellison et al.*, 2001], though fitting both the detailed TeV spectrum, as well as GeV upper limits from *EGRET*, demanded a very small filling factor for magnetic field, of order 1% [*Lazendić et al.*, 2004]. Cas A has been detected at the 5σ level with the HEGRA instrument [*Aharonian et al.*, 2001a], while only upper limits were reported for Tycho [*Aharonian et al.*, 2001b]. All three SNR detections in TeV gamma-rays have relatively steep spectra between 1 and 10 TeV: photon indices –2.5 ± 0.5 (Cas A), –2.3 ± 0.2 (SN 1006), and –2.8 ± 0.2 (G347.3–0.5). The Cas A detection may point most plausibly to π^0-decay [*Berezhko, Pühlhofer, and Völk*, 2003], providing at last some direct confirmation for the presence of relativistic protons in SNRs. It is important that all these results be confirmed by other instruments; extensions to lower energies are particularly useful for modeling.

BROADBAND MODELING OF NONTHERMAL SNR EMISSION

Now that a few SNRs are detected at wavelengths from radio to TeV gamma-ray (a range of about 18 orders of magnitude in photon energy!), it is possible to create simple spectral models for the spatially integrated emission (since the TeV observations have very poor angular resolution, this approximation is necessary). Of particular interest is whether leptonic models can explain the entire spectrum, or whether energetic protons are required. If TeV emission is inverse-Compton, it is produced by the same electrons producing X-ray synchrotron emission, allowing the direct inference of the magnetic-field strength—very valuable information deducible in no other way.

First, the steepness of the observed photon spectra implies that if the radiative process is inverse-Compton emission, it must result from the steepening high-energy tail of the electron distribution. The observed spectrum of G347 is also too steep for bremsstrahlung or π^0-decay emission unless it also results from particles in the cutoff region of the distribution.

Simple relations can be obtained between amplitudes and peaks (in νS_ν) of synchrotron and ICCMB emission as functions of $\langle B \rangle$ and the filling factor f_B of the magnetic field [*Lazendić et al.*, 2004; *Aharonian and Atoyan*, 1999]. If the synchrotron and inverse-Compton components have peaks in νS_ν at $\nu_m(SR)$ and $\nu_m(IC)$ respectively, then we can infer

$$B \cong 9 \times 10^4 \left(\frac{\nu_m(IC)}{\nu_m(SR)} \right)^{-1} \text{G}$$

$$f_B \sim 2 \times 10^{-14} \left(\frac{S_{\text{synchr}}(\nu_m(SR))}{S_{\text{IC}}(\nu_m(SR))} \right) B^{-(s+1)/2}$$

Applied to G347.3–0.5 [*Lazendić et al.*, 2004], these estimates yield values similar to those obtained from a model fit (Figure 2): $B \sim 15$ μG and $f_B \sim 0.01$. This is a small, but not impossible, filling factor, consistent with the observed filamentary X-ray (synchrotron) emission. The low filling factor is required by the comparable peaks (in νS_ν) of synchrotron and inverse-Compton emission, and is very roughly

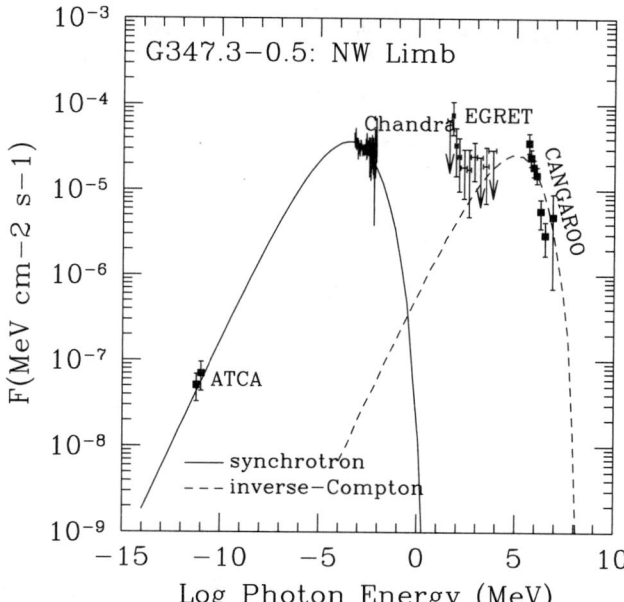

Figure 2. Broadband fit to radio through gamma-ray integrated spectrum of G347.3–0.5 (see text for details) [*Lazendić et al., 2004*].

the inverse of the factor by which the energy density in the 15 μG field exceeds that in the CMB. The total energy in magnetic field was of order 5×10^{46} erg, and that in electrons about 3×10^{46} erg—remarkably close to equipartition. (Somewhat different conclusions were reached by *Pannuti et al.* [2003], who found such small values of f_B implausible). *Enomoto et al.* [2002] made a case that the TeV emission had to result from π°-decay, but based on the assumption that $f_B \equiv 1$. Our estimate shows that an inverse-Compton model cannot be ruled out, at the price of the small filling factor. In any case, stringent upper limits on emission at *EGRET* energies constrain the modeling severely [*Butt et al., 2002*].

SUMMARY AND CONCLUSIONS

Observations of radio, X-ray, and gamma-ray emission from SNRs have provided powerful new constraints on the physics of particle acceleration in strong shocks. They confirm the operation of electron acceleration to energies of order 100 TeV, but show that the electron spectrum has always begun to steepen below that energy. In 25 observed cases, unless radiative losses are limiting electron acceleration in every one, proton acceleration to the "knee" at 2000 TeV is problematic [*Reynolds and Keohane*, 1999; *Hendrick and Reynolds*, 2001]. X-ray synchrotron emission is confirmed in several remnants, and may be a significant contributor in many more. The study of this emission gives information on the energies at which electron acceleration begins to roll off, allowing constraints on both global SNR parameters such as age, shock velocity, and dynamical history, and microphysical parameters such as the diffusion coefficient or shock obliquity. Sharp edges in emission at radio and X-ray wavelengths, and the absence of an obvious X-ray synchrotron halo [*Long et al.*, 2003], demand short diffusion lengths and strong scattering, perhaps from self-generated turbulence [*Achterberg et al.*, 1994], but also suggest that magnetic-field amplification above simple flux-freezing is required [*Lucek and Bell*, 2000]. At this time, evidence suggests, but does not demand, the presence of relativistic protons in SNRs. The unambiguous detection of π^0-decay emission, perhaps from the bump predicted near 100 MeV, should be a major target of upcoming gamma-ray missions such as AGILE and GLAST.

Acknowledgments. This work has been supported by NASA through the Astrophysics Theory Program (grant NAG5-10940).

REFERENCES

Achterberg, A., R. D. Blandford, and S. P. Reynolds, Evidence for enhanced mhd turbulence outside sharp-rimmed supernova remnants, *Astron. Ap.*, 281, 220–230, 1994

Aharonian, F. A., and A. M. Atoyan, On the origin of TeV radiation of SN 1006, *Astron. Astrophys.*, 351, 330–340 (1999).

Aharonian, F., et al. Evidence for TeV gamma ray emission from Cassiopeia A, *Astron. Astrophys.*, 370, 112–120, 2001a.

Aharonian, F., et al. A study of Tycho's SNR at TeV energies with the HEGRA CT-System, *Astron. Astrophys.*, 373, 292–300, 2001b.

Allen, G. E., E. V. Gotthelf, and R. Petre, Evidence of 10-100 TeV electrons in supernova remnants, Proc. 26th ICRC, Salt Lake City, 3, 480–483 (astro-ph/9908209), 1999.

Aschenbach, B., Discovery of a young nearby supernova remnant, *Nature*, 396, 141–142, 1998.

Bamba, A., M. Ueno, K. Koyama, and S. Yamauchi, Diffuse hard X-ray sources discovered with the ASCA Galactic Plane Survey, *Astrophys. J.*, 589, 253–260, 2003.

Baring, M. G., D. C. Ellison, S. P. Reynolds, I. A. Grenier, and P. Goret, P. Radio to gamma-ray emission from shell-type supernova remnants: predictions from nonlinear shock acceleration models. *Astrophys. J.*, 513, 311–338, 1999

Berezhko, E. G., V. K. Elshin, & L. T. Ksenofontov, Numerical studies of the acceleration of cosmic rays in supernova remnants, *Astr. Repts.*, 40, 155–166, 1996.

Berezhko, E. G., L. T. Ksenofontov, and H. J. Völk, Emission of SN 1006 produced by accelerated cosmic rays, *Astron. Astrophys.*, 395, 943–953, 2002.

Berezhko, E. G., G. Pühlhofer, and H. J. Völk, Gamma-ray emission from Cassiopeia A produced by accelerated cosmic rays, *Astron. Astrophys.*, 400, 971–980, 2003.

Beuermann, K., G. Kanbach, and E. M. Berkhuijsen, Radio structure of the Galaxy: thick disk and thin disk at 408 MHz, *Astron. Astrophys.*, **153**, 17–34, 1985.

Blandford, R. D., & D. Eichler, Particle acceleration at astrophysical shocks: a theory for cosmic ray origin, *Phys. Reports*, **154**, 1–75, 1987 (BE87).

Bleeker, J. A. M., R. Willingale, K. van der Heyden, K. Dennerl, J. S. Kaastra, B. Aschenbach, and J. Vink, Cassiopeia A: On the origin of the hard X-ray continuum and the implication of the observed O VIII Ly-α/Ly-β distribution, *Astron. Astrophys.*, **365**, L225–L230, 2001.

Borkowski, K. J., J. Rho, S. P. Reynolds, & K. K. Dyer, Thermal and Nonthermal X-Ray Emission in Supernova Remnant RCW 86, *Astrophys. J.*, **550**, 334–345, 2001.

Butt, Y. M., D. F. Torres, G. E. Romero, T. M. Dame, and J. A. Combi, Supernova-Remnant Origin of Cosmic Rays? *Nature*, **418**, 499, 2002.

Cassam-Chenaï, G., A. Decourchelle, J. Ballet, U. Hwang, J. P. Hughes, and R. Petre, XMM-Newton observation of Kepler's supernova remnant, *Astron. Astrophys.*, **414**, 545–558 (2004).

Drury, L. O.'C., An introduction to the theory of diffusive shock acceleration of energetic particles in tenuous plasmas, *Rep. Progr. Physics*, **46**, 973-1027, 1983

Drury, L. O.'C., F. Aharonian, and H. Völk, GeV/TeV gamma-ray emission from dense molecular clouds overtaken by supernova shells, *Astron. Astrophys.*, **285**, 645–647, 1994.

Dyer, K. K., S. P. Reynolds, Borkowski, K.J., et al., Separating Thermal and Nonthermal X-Rays in Supernova Remnants. I. Total Fits to SN 1006 AD, *Astrophys. J.*, **551**, 439–453, 2001.

Dyer, K. K., S. P. Reynolds, and Borkowski, K. J., Separating thermal and non-thermal X-rays in supernova remnants II: Spatially resolved fits to SN 1006 AD, *Astrophys. J.*, **600**, 752–768 (2004).

Ellison, D. C., & S. P. Reynolds, Electron Acceleration in a Nonlinear Shock Model with Applications to Supernova Remnants, *Astrophys. J.*, **382**, 242–254, 1991.

Ellison, D. C., P. O. Slane, and B. M. Gaensler, Broadband observations and modeling of the shell-type supernova remnant G347.3-0.5, *Astrophys. J.*, **563**, 191–201, 2001.

Enomoto, R., et al., The acceleration of cosmic-ray protons in the supernova remnant RX J1713.7-3946, *Nature*, **416**, 823–826, 2002.

Gaisser, T. K., R. J. Protheroe, and T. Stanev, Gamma-Ray production in supernova remnants, *Astrophys. J.*, **492**, 219–227, 1998

Ghavamian, P., P. F. Winkler, J. C. Raymond, & K. S. Long, Balmer-dominated Spectra of Nonradiative Shocks in the Cygnus Loop, RCW 86, and Tycho Supernova Remnants, *Astrophys. J.*, **572**, 888–896, 2002.

Gotthelf, E. V., B. Koralesky, L. Rudnick, T. W. Jones, U. Hwang, and R. Petre, Chandra detection of the forward and reverse shocks in Cassiopeia A, *Astrophys. J.*, **552**, L39–L43, 2001.

Green D. A., A catalogue of Galactic supernova remnants, Mullard Radio Astronomy Observatory, Cavendish Laboratory, Cambridge, United Kingdom (available on the World-Wide-Web at http://www.mrao.cam.ac.uk/surveys/snrs/), 2001.

Hamilton, A. J. S., C. L. Sarazin, and A. E. Szymkowiak, The X-ray spectrum of SN 1006, *Astrophys. J.*, **300**, 698–712, 1986.

Hendrick, S. P., & S. P. Reynolds, Maximum Energies of Shock-accelerated Electrons in Large Magellanic Cloud Supernova Remnants, *Astrophys. J.*, **559**, 903–908, 2001.

Hwang, U., A. Decourchelle, S. S. Holt, and R. Petre, Thermal and nonthermal X-ray emission from the forward shock in Tycho's supernova remnant, *Astrophys. J.*, **581**, 1101–1115, 2002.

Jokipii, J. R., Rate of energy gain and maximum energy in diffusive shock acceleration, *Astrophys. J.*, **313**, 842–846, 1987.

Jones, F. C., and D. C. Ellison, The plasma physics of shock acceleration, *Sp. Sci. Rev.*, **58**, 259–346 (1991)

Koyama, K., et al., Evidence for shock acceleration of high-energy electrons in the supernova remnant SN 1006, *Nature*, **378**, 255–258, 1995.

Koyama, K., K. Kinugasa, K. Matsuzaki, M. Nishiuchi, M. Sugizaki, K. Torii, Y. Yamauchi, B. Aschenbach, Discovery of nonthermal X-rays from the northwest shell of the new SNR RX J1713.7-3946: The second SN 1006? *Pub. Astr. Soc. Japan*, **49**, L7–L11, 1997.

Lawson, K. D., C. J. Mayer, J. L. Osborne, and M. L. Parkinson, Variations in the spectral index of the galactic radio continuum emission in the northern hemisphere, *Mon. Not. Royal Astr. Soc.*, **225**, 307–327, 1987.

Lazendić, J. S., P. O. Slane, B. M. Gaensler, S. P. Reynolds, P. P. Plucinsky, J. P. Hughes, A high-resolution study of nonthermal radio and X-ray emission from SNR G347.3-0.5, *Astrophys. J.*, **602**, 271–285 (2004).

Levinson, A. Electron injection in collisionless shocks, *Astrophys. J.*, **401**, 73–80, 1992.

Long, K. S., S. P. Reynolds, J. C. Raymond, P. F. Winkler, K. K. Dyer, and R. Petre, *Chandra* CCD imagery of the NW and NE limbs of SN 1006, *Astrophys. J.*, **586**, 1162–1178, 2003.

Lucek, S. G., & A. R. Bell, Non-linear Amplification of a Magnetic Field Driven by Cosmic Ray Streaming, *MNRAS*, **314**, 65–74, 2000.

Malkov, M. A., and L. O.'C. Drury, Nonlinear theory of diffusive acceleration of particles by shock waves, *Rep. Progr. Physics*, **64**, 429–481. 2001.

Melrose, D. B., Resonant Scattering of Particles and Second Phase Acceleration in the Solar Corona, *Sol. Phys.*, **37**, 353–365, 1974.

Ostrowski, M., Supernova remnants in molecular clouds: on cosmic ray electron spectra, *Astron. Astrophys.*, **345**, 256–258, 1999.

Pacholczyk, A.G., *Radio Astrophysics*, Freeman, San Francisco, 1970.

Pannuti, T. G., G. E. Allen, J. C. Houck, and S. J. Sturner, RXTE, ROSAT, and ASCA observations of G347.3-0.5 (RX J1713.7-3946): Probing cosmic-ray acceleration by a galactic shell-type supernova remnant, *Astrophys. J.*, **593**, 377–392, 2003.

Reynolds, S. P., Synchrotron models for X-rays from the supernova remnant SN 1006, *Astrophys. J.*, **459**, L13–L16, 1996.

Reynolds, S. P., Models of synchrotron X-rays from shell supernova remnants, *Astrophys. J.*, **493**, 375–396, 1998.

Reynolds, S. P., and R. A. Chevalier, nonthermal radiation from supernova remnants in the adiabatic phase of evolution, *Astrophys. J.*, **245**, 912–919, 1981

Reynolds, S. P., & D. C. Ellison, Electron acceleration in Tycho's and Kepler's supernova remnants—Spectral evidence of Fermi shock acceleration, *Astrophys. J.*, **399**, L75–L78, 1992.

Reynolds, S. P., & J. W. Keohane, Maximum Energies of Shock-accelerated Electrons in Young Shell Supernova Remnants, *Astrophys. J.*, **525**, 368–374, 1999.

Rho, J., K. K. Dyer, K. J. Borkowski, and S. P. Reynolds, X-Ray Synchrotron-emitting Fe-rich Ejecta in Supernova Remnant RCW 86. *Astrophys. J.*, **581**, 1116–1131, 2002.

Slane, P., B. M. Gaensler, T. M. Dame, et al., Nonthermal X-Ray Emission from the Shell-Type Supernova Remnant G347.3-0.5, *Astrophys. J.*, **525**, 357–367, 1999.

Slane, P., J. P. Hughes, R. J. Edgar, P. P. Plucinsky, E. Miyata, H. Tsunemi, B. Aschenbach, RX J0852.0-4622: Another nonthermal shell-type supernova remnant (G266.2-1.2), *Astrophys. J.*, **548**, 814–819, 2001.

Sturner, S. J., J. G. Skibo, C. D. Dermer, J. R. Mattox, Temporal Evolution of Nonthermal Spectra from Supernova Remnants, *Astrophys. J.*, **490**, 619–632, 1997.

Tanimori, T., Y. Hayami, S. Kamei, et al., Discovery of TeV Gamma Rays from SN 1006: Further Evidence for the Supernova Remnant Origin of Cosmic Rays, *Astrophys. J.*, **497**, L25–L28, 1998.

The, L.-S., et al., CGRO/OSSE observations of the Cassiopeia A SNR, *Astron. Astrophys. Supp.*, **120**, 357–360, 1996

Torres, D. F., G. E. Romero, T. M. Dame, J. A. Combi, and Y. M. Butt, Supernova remnants and γ-ray sources, *Phys. Reports*, **382**, 303–380, 2003.

Uchiyama, Y., F. A. Aharonian, T. Takahashi, Fine-structure in the nonthermal X-ray emission of SNR RX J1713.7-3946 revealed by Chandra, *Astron. Astrophys.*, **400**, 567-574, 2003.

Ueno, M., A. Bamba, K. Koyama, and K. Ebisawa, Chandra observations of a nonthermal supernova remnant candidate AX J1843.8-0352 and its surroundings, *Astrophys. J.*, **588**, 338–343, 2003.

Vink, J., J. S. Kaastra, & J. A. M. Bleeker, X-ray spectroscopy of the supernova remnant RCW 86. A new challenge for modeling the emission from supernova remnants, *A&A*, **328**, 628–633, 1997

Vink, J., J. A. M. Bleeker, J. S. Kaastra, K. van der Heyden, & J. Dicke, XMM-Newton observations of MSH 14-63 (RCW 86), in Neutron Stars in Supernova Remnants, ASP Conference Series, Vol. 271, ed. P.O. Slane & B.M. Gaensler (San Francisco: ASP), 423–426, 2002

Stephen P. Reynolds, Physics Department, North Carolina State University, 110 Lox Hall, 2700 Stinson Drive, Raleigh, NC 27695-8202. (steve_reynolds@ncsu.edu)

Overview: Particle Acceleration by Waves and Turbulence

Robert L. Lysak

School of Physics and Astronomy, University of Minnesota, Minneapolis, Minnesota

Waves and turbulence are frequently invoked as a mechanism to accelerate particles. This overview will describe the importance of this form of acceleration and introduce the papers on this subject that are included in this volume.

In addition to well-ordered structures such as shocks and double layers, waves and turbulence can play a significant role in the astrophysical acceleration of particles. Waves can accelerate particles due to the resonant interactions between waves and particles at the Landau resonance (the point at which the particle velocity matches the wave phase velocity) and the cyclotron resonances (where the motion of the particle Doppler shifts the wave frequency to a multiple of the particle's cyclotron frequency). At such resonances, particles see the wave at a nearly constant phase and can be accelerated. Such resonant heating of particles usually manifests itself as a broadened distribution of particles, as opposed to the more coherent, often mono-energetic distributions produced by parallel electric fields or shocks. Thus, wave acceleration and heating of particles can be distinguished from more coherent acceleration by observation of the particle distributions.

Since most of the waves that can resonate with particles are low phase velocity waves that do not propagate long distances, observations of wave accelerated particles are mostly confined to the Solar System where in situ measurements are possible. Indeed, even close to the Earth, such as in the auroral zones, it can be difficult to establish a one-to-one correspondence between accelerated particles and the wave modes that provide the acceleration. In addition, there is the additional problem that if the waves are giving energy to the particles, the waves themselves must be excited and grow to large amplitudes. Therefore, a description of the free energy sources for these waves is necessary. Wave acceleration therefore requires a knowledge of the large-scale behavior of the system, since it is usually these large-scale structures that ultimately provide energy for the waves.

Even though the study of wave acceleration has largely been confined to solar system plasmas, it seems reasonable to believe that such processes are ubiquitous in astrophysical plasmas. Since wave processes can often produce superthermal tails from the bulk cold plasma, it is tempting to speculate that such processes may provide the means to create warm plasmas that can be further accelerated, for example by Fermi acceleration, to produce high energy cosmic rays. This so-called "injection problem" is crucial to the production of cosmic rays since Fermi acceleration produces energy gains that scale with the original energy of the particles; thus a mechanism to accelerate particles to energies at which the Fermi mechanism can operate efficiently is necessary.

The majority of work on wave acceleration mechanisms focuses on the magnetosphere of the Earth, where satellite observations of both waves and particles are available. Wave-accelerated particles have been observed in the Earth's auroral zones, where the so-called "ion conic" distributions are formed. These distributions are caused by wave heating of particles perpendicular to the magnetic field, which are then folded into a cone shape by the action of the magnetic mirror force (for a recent review, see *André et al.* [1998]). Such a distribution of ions cannot be produced by quasi-static fields alone. Electron distributions that are broad in energy but narrow in pitch angle have also been recently observed in the auroral zone [e.g., *Chaston et al.*, 2002]. Similar acceleration in the presence of kinetic Alfvén waves at the plasma sheet boundary layer has been observed by Polar [e.g., *Wygant et al.*, 2002]. Wave acceleration processes can contribute to the energization of radiation belt and ring current particles [e.g., *Horne and Thorne*, 1997]. Lower hybrid heating of electrons in the magnetotail has also been observed [*Shinohara et al.*, 1998]. This is only a brief and incomplete list of the variety of wave modes and

Particle Acceleration in Astrophysical Plasmas
Geophysical Monograph Series 156
Copyright 2005 by the American Geophysical Union
10.1029/156GM13

regions in which wave acceleration is thought to be important.

The papers in this section present a variety of approaches to problems of wave heating in astrophysical plasmas. Lund considers the well-known problem of wave heating of ions in the Earth's auroral zone, which is treated through a test particle approach in an assumed wave spectrum, including the effects of parallel electric fields and the magnetic mirror force. In the spirit of extending well-measured magnetospheric models to other astrophysical situations, he proposes that a similar model could apply to heating of the solar corona.

The other two papers in this section consider the wave acceleration problems through computer simulations. Wang and Lin perform hybrid simulations of ion-beam generated Alfvén waves to study their effect on the heating of ions. In such hybrid models, the ions are treated using particle-in-cell (PIC) techniques while the electrons are treated as a neutralizing fluid. They show that ions can be heated to 100 times their initial temperature. This model is motivated by instabilities caused by ions reflected at the Earth's bow shock, but the model is more general in its validity. Finally, Lui considers a model in which plasma instabilities disrupt the magnetotail current, using a PIC code to describe these instabilities and the resulting ion heating. The rapid development of ion heating and the resulting disruption of the current are proposed as means by which magnetospheric substorms may be initiated.

These papers provide a variety of different mechanisms and models by which wave acceleration processes can be studied. It is clear that wave processes can be very efficient at heating thermal particles to higher energies. While it can be difficult to study these waves without the benefit of in situ observations such as those that can be made in near-Earth space, there can be little doubt that wave acceleration plays an important role in a variety of astrophysical contexts.

Acknowledgments. This work was supported in part by NSF grant ATM-0201703.

REFERENCES

André, M., P. Norqvist, L. Andersson, L. Eliasson, A. I. Eriksson, L. Blomberg, R. E. Erlandson, and J. Waldemark, Ion energization mechanisms at 1700 km in the auroral region, *J. Geophys. Res., 103,* 4199, 1998.

Chaston, C. C., J. W. Bonnell, L. M. Peticolas, C. W. Carlson, J. P. McFadden, and R. E. Ergun, Driven Alfven waves and electron acceleration: A FAST case study, *Geophys. Res. Lett., 29*(11), 10.1029/2001GL013842, 2002.

Horne, R. B., and R. M. Thorne, Wave heating of He+ by electromagnetic ion cyclotron waves in the magnetosphere: Heating near the H+-He+ bi-ion resonance frequency, *J. Geophys. Res., 102,* 11,457, 1997.

Shinohara, I., T. Nagai, M. Fujimoto, T. Terasawa, T. Mukai, K. Tsuruda, and T. Yamamoto, Low-frequency electromagnetic turbulence observed near the substorm onset site, *J. Geophys. Res., 103,* 20,365, 1998.

Wygant, J. R., A. Keiling, C. A. Cattell, R. L. Lysak, M. Temerin, F. S. Mozer, C. A. Kletzing, J. D. Scudder, V. Streltsov, W. Lotko, and C. T. Russell, Evidence for kinetic Alfvén waves and parallel electron energization at 4–6 RE altitudes in the plasma sheet boundary layer, *J. Geophys. Res., 107*(A8), 1201, doi:10.1029/2001JA900113, 2002.

Robert L. Lysak, School of Physics and Astronomy, University of Minnesota, 116 Church Street SE, Minneapolis, Minnesota 55455. (bob@belka.space.umn.edu)

The Connection Between Parallel Electric Fields and Ion Acceleration in Astrophysical Plasmas

E. J. Lund

Space Science Center, University of New Hampshire, Durham, New Hampshire

It is now well-established that magnetic field-aligned electric fields play a significant role in particle dynamics in the auroral region, and strong evidence for similar electric fields exists in other astrophysical plasmas. However, ionospheric ion outflow in the auroral region is primarily associated with downward or oscillating parallel electric fields rather than upward electric fields. The reason for the discrepancy is that waves which produce transverse ion heating are found in regions of parallel electric fields, and downward electric fields enhance the heating by trapping ions. We discuss a model for the distribution of parallel electric fields along auroral flux tubes, as well as implications for ion heating in other astrophysical plasmas including the solar corona.

1. INTRODUCTION

Heating and acceleration of ions produces a wide variety of important mass transport effects in magnetospheric and astrophysical plasmas. The most prominent of these effects are solar and stellar winds in astrophysics and ion outflow from planetary ionospheres. Thus understanding how these ions are accelerated is important to understanding the physics of space plasmas.

In the auroral region, this ion acceleration has been shown to be closely associated with electric fields parallel to the background geomagnetic field. The most obvious association is between upward flowing ion beams [*Shelley et al.*, 1976] and upward electric fields located above visible auroral arcs. At altitudes below the auroral acceleration region, however, most ionospheric ion outflow is observed on flux tubes lacking the "inverted-V" electrons which signify the presence of upward parallel electric fields [*André et al.*, 1998; *Norqvist et al.*, 1998]. Instead, outflowing ionospheric ions are accelerated transversely to the magnetic field, producing ion conics [*Sharp et al.*, 1977]. Most of these ion conics occur on flux tubes with either downward or fluctuating parallel current; such flux tubes have downward or oscillating parallel electric fields, respectively.

It is reasonable to assume that ion heating in the corona occurs through processes similar to the aurora. This paper will briefly review what is known about the relationship between auroral ion outflow and parallel electric fields in the aurora and discuss how some of these physical processes relate to the solar corona.

2. AURORAL ION OUTFLOW

Most heavy ion outflow, and much of the proton outflow, from the Earth's ionosphere occurs in the auroral region. On large scales, O^+ outflow in particular is correlated with auroral emissions in the Lyman-Birge-Hopfield long band [*Wilson et al.*, 2001]. On shorter scales, however, ion conics are usually anticorrelated with the electron precipitation which produces visible aurora [*André et al.*, 1998; *Norqvist et al.*, 1998]. *André et al.* [1998] divided ion conics into four types: types 1 and 2, which are correlated with suprathermal electron bursts (type 2 is also correlated with precipitating ions) and broadband waves observed in the ion cyclotron frequency regime (also called broad-band extremely low frequency or BBELF waves); type 3, which is correlated with auroral electrons and either lower hybrid or electromagnetic ion cyclotron (EMIC) waves; and type 4, which is correlated with precipitating ions and lower hybrid waves but not with any electron signatures. Types 1 and 2 were found to predominate on the nightside and dayside, respectively. Type 3 is a minor contributor except under active conditions *Norqvist et al.*, 1998], while type 4 conics are rare. Note that only type 3 conics contribute to ion beams; all other types occur in regions of either downward [*André et al.*, 1998] or oscillating [*Tung et al.*, 2001] current. The overall distribution of conics associated with BBELF and EMIC waves found in the study of *Lund et al.* [2000] is shown in Figure 1. Note that the BBELF events occur at all local times and dominate all but the pre-midnight sector.

The nature of BBELF remains controversial. At first glance the spectra support the idea of ion cyclotron resonance heating [*Chang et al.*, 1986]; only a small fraction of the measured power at the O^+ gyrofrequency Ω_{O+} suffices for this mechanism. However, in many cases the spectrum is dominated by Doppler-shifted ULF turbulence with $k_\perp \gg k_\parallel$ [*Temerin*, 1978], which has been interpreted as Alvfénic fluctuations with $k_\perp c/\omega_{pe} \sim 1$ [*Stasiewicz et al.*, 2000]. Nevertheless, at least one published observation has been shown to be inconsistent with the Alvfénic interpretation [*Kintner et al.*, 2000]; the latter authors believe that BBELF is not associated with any known normal mode. None of these observations have ruled out the existence of real temporal fluctuations in combination with ULF turbulence, so the ion cyclotron resonant mechanism remains a viable explanation.

3. PARALLEL ELECTRIC FIELDS IN THE AURORAL REGION

The Freja and FAST ion outflow studies [*André et al.*, 1998; *Norqvist et al.*, 1998; *Tung et al.*, 2001] show that most auroral region outflow is associated with regions of downward or ambiguous current. The parallel electric fields which occur in these regions would tend to inhibit ion outflow. *Gorney et al.* [1985] suggested a resolution to this paradox: downward parallel electric fields can trap ion conics, thereby causing their energy to be enhanced by the need to overcome this parallel electric field. This conceptual model was subsequently verified with FAST data [*Carlson et al.*, 1998]. This model also explains why the energy of upflowing ions is independent of ion mass, contrary to what would be expected from any heating mechanism with finite altitude extent or explicit mass dependence [*Lund et al.*, 1999, 2001].

An example of how Gorney's "pressure cooker" scenario works is shown in Figure 2, which traces H^+, He^+, and O^+ ions through a triangular electric field profile with maximum $E_\parallel = 0.2$ mV/m at 4000 km altitude and a half-width of 1000 km [*Lund et al.*, 2001]. The heating rate of the ions is $m^{1/2}$ eV/s where m is given in AMU, corresponding to a spectral index of -1.5 in the ion cyclotron resonant heating paradigm [*Chang et al.*, 1986]. Below the $E_\parallel$ region, the heavier O^+ ions are heated significantly faster than He^+, which in turn are heated significantly faster than H^+. The O^+ ions acquire enough parallel momentum to punch through the $E_\parallel$ region on the first try. Not so the He^+, which takes a second pass, or H^+, which is pushed back nine times before gaining enough energy to clear the $E_\parallel$. By choosing parameters appropriately, any desired ordering of energy of these three species within and above the parallel electric field region can be obtained.

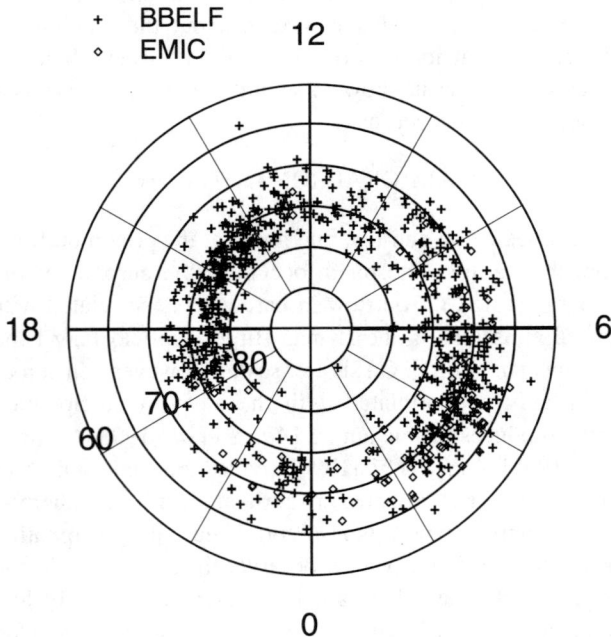

Figure 1. Occurrence distribution of BBELF (plusses) and EMIC (diamonds) ion conic events from the study of *Lund et al.* [2000].

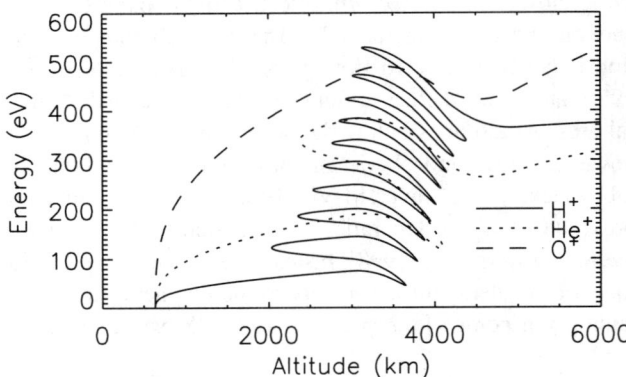

Figure 2. Energies of H^+ (solid line), He^+ (dotted line), and O^+ (dashed line) as a function of altitude for a triangular electric field profile [*Lund et al.*, 2001]. The electric field is centered at 4000 km with half-width 1000 km and maximum magnitude 0.2 mV/m. The ions are heated at $m^{1/2}$ eV/s, where m is in AMU.

The physical need for parallel electric fields in the aurora is to compensate for a lack of charge carriers. In the upward current region the lack of charge carriers is due to the mirroring of plasma sheet electrons; the magnitude of the potential difference needed to maintain current continuity can be calculated from the well-known *Knight* [1973] relation, but Knight's formula gives no indication of how this potential difference is distributed along the field line. In the downward current region, even though ionospheric electrons provide a readily available reservoir of charge carriers, the need to maintain quasineutrality along most of the field line limits the mobility of the electrons away from ionospheric ions [*Temerin and Carlson*, 1998]. No direct analog to the Knight relation has been found for the downward current region.

Parallel electric fields in the auroral region fall into two categories: strong double layers [*Mozer and Kletzing*, 1998; *Ergun et al.*, 2001] and long-range fields due to differential anisotropy [*Alfvén and Fälthammar*, 1963; *Jasperse and Grossbard*, 2000]. Although weak double layers have been observed in the auroral region, their effect is limited to short perpendicular scales [*Mälkki et al.*, 1993]. Gravity-induced ambipolar electric fields also exist [*Pannekoek*, 1922; *Rosseland*, 1924], but their magnitude is typically ~ 0.1 μV/m}, three orders of magnitude smaller than Alfvéen-Fälthammar fields. In the upward current region, the double layer at the bottom of the auroral density cavity typically accounts for 10 to 50 percent of the total potential difference along the field line. There may be more than one double layer present on the bottom side [*Pottelette et al.*, 2001]; in addition, under certain circumstances a second double layer has been predicted to form at the top of the density cavity [*Ergun et al.*, 2000], but the latter double layer has not been definitively observed. In the downward current region the double layers move with the ion acoustic speed [*Ergun et al.*, 2001; *Newman et al.*, 2001; *Andersson et al.*, 2002]. The long-range parallel electric fields are more difficult to measure unambiguously; however, *Jasperse and Grossbard* [2000] give a formula for the parallel electric field self-consistent with moments of the ion distribution. Using this formula, *Lynch et al.* [2002] inferred parallel electric fields of a few hundred μV/m and showed that the Ohmic dissipation $E_\parallel j_\parallel$ is well-correlated with observed wave power at Ω_{O+}.

Incorporating this assumption into a multimoment fluid model, *Jasperse et al.* [2001] showed that the resulting parallel electric field asymptotically approaches 0, as it must in the actual acceleration region. The basic approach of the multimoment fluid model, the mathematical basis of which is described in more detail in *Jasperse and Grossbard* [2000], is a one-dimensional time-independent Fokker-Planck-Poisson system of equations for the particle distributions, moments of which are solved for self-consistently.

Even though most of the BBELF emissions appear to be Doppler-shifted ULF turbulence [*Temerin*, 1978; *Stasiewicz et al.*, 2000], numerical experiments indicate that the cyclotron resonant term still dominates the ion heating. Observed particle distributions and wave spectra are used as boundary conditions. Solutions consistent with Alfvéen-Fälthammar fields are usually found for some distance below the satellite, but the ion anisotropy vanishes before all of the potential difference inferred from the electron distributions is accounted for. Thus we infer the existence of a double layer similar to those reported in FAST data [*Andersson et al.*, 2002]. Further details will be reported in a future paper.

4. APPLICATION TO THE SOLAR CORONA

There are many similarities between ion heating in the auroral region and ion heating in the solar corona. Coronal ions, like auroral ion conics, have significant temperature anisotropies; O^{5+}, for example typically has $10 < T_\perp/T_\parallel < 100$ [*Kohl et al.*, 1998]. The straightforward mass (or mass per charge) ordering expected from straightforward cyclotron resonant heating is not observed in the corona, where O^{5+} is typically more energetic than Mg^{9+} [*Esser et al.*, 1999]. The corona is optically thick to waves with $0.1 < \omega/\Omega_{H+} < 0.5$ [*Cranmer*, 2000], implying that a local generation mechanism for the waves must exist. Recently, *Ofman et al.* [2002] have investigated O^{5+} and proton heating in the presence of a power law spectrum. It has also been pointed out that the need to drive field-aligned currents along coronal loops can, according to the *Knight* [1973] relation, produce parallel potential drops of up to ~ 100 kV [*Kan and Lyu*, 1990]. It is therefore reasonable to ask whether these parallel electric fields can account for the departure from what is expected of ion cyclotron resonant heating of species-dependent effects in coronal heating.

For this purpose we will adapt the cartoon model developed for the auroral region by *Gorney et al.* [1985] and later extended to demonstrate mass-dependent effects [*Lund et al.*, 1999] to the solar corona. This model treats the ions as test particles in an imposed configuration of ion heating and parallel electric fields. The relevant equations are

$$\frac{d}{dt}(v_\perp^2) = -\frac{2fv_\parallel v_\perp^2}{z} + \frac{2}{m}\left(\frac{d\epsilon_\perp}{dt}\right)_{WPI} \quad (1)$$

$$\frac{dv_\parallel}{dt} = \frac{q}{m}\left(E_\parallel - \frac{GM_s m}{(q+e)z^2}\right) + \frac{fv_\perp^2}{z} - k_B T_e \frac{dn}{dz} \quad (2)$$

$$\frac{dz}{dt} = v_\parallel \quad (3)$$

which includes a pressure gradient term due to the density profile given by *Feldman et al.* [1997] with a constant $T_e = 10^6$ K. Here f is an expansion factor ranging from 1 to 3, representing the strength of the mirror force; 1.5 corresponds to a dipole. Note that the effect of gravity in (2) is modified by charge separation effects [*Pannekoek*, 1922; *Rosseland*, 1924]. The heating rate $(d\epsilon_\perp/dt)_{WPI}$ due to wave-particle interactions is 1 eV/s for $1.5 \leq z/R_s < 2.5$ (R_s is the solar radius) and 0 elsewhere for protons; for heavier ions it scales as $(m/q)^{\alpha-1}$ where α is the power law index of the electric field fluctuations in the ion cyclotron resonant heating paradigm. We have examined four cases: with and without a constant parallel electric field of 0.1 µV/m downward over the region $1.5 \leq z/R_s < 2.5$, and with and without gravity. The ions used are protons, alpha particles, O with charge states 5–7, Mg with charge states 6–9, and Fe with charge states 8–10. These charge states are the most common for their respective elements, according to the model of *Cranmer* [2000].

Representative results are shown in Figure 3, which displays the perpendicular and parallel velocities and energies of H^+, O^{5+}, Mg^{9+}, and Fe^{9+} for $\alpha = 1.67$ and $f = 2.0$ as a function of z with both the parallel electric field and gravity turned on. Initial behavior is similar to results with similar electric field profiles in the aurora [*Gorney et al.*, 1985; *Lund et al.*, 1999] in that all ions must acquire a certain amount of energy before they can move into the coronal heating region; the more significant influence of gravity in the corona causes this energy to depend on mass. In this run the protons, because of their light weight, emerge at a significantly higher speed than the heavier ions, but their energy at the top of the heating region is only 0.4 fJ (1 fJ = 6.2 keV). O^{5+} and Mg^{9+} acquire similar energies (≈ 1.5 fJ) on the first pass; this is enough energy for the former, but not the latter, to escape. The iron ions also return, as does Mg^{8+} (not shown), and are further heated on the second pass. Higher values of f and α promote faster escape of all ions, especially the heavy ions.

The effect of ionic charge is isolated in Figure 4, which compares the trajectories for different charge states of oxygen for the same parameters as for Figure 3. There is only a slight effect for $v_\perp$, while $v_\parallel$ and energy are higher for lower charge states (hence higher m/q). For these parameters the energy scales as $q^{-0.54}$. Here the higher heating rate associated with the higher m/q overcomes the higher effective gravity for low charge state ions [*Pannekoek*, 1922].

Figure 5 compares the results for different masses for the same charge state (O^{6+} and Mg^{6+}) and same charge-to-mass ratio (O^{6+} and Mg^{9+}). Velocities for the $q = +6$ ions are similar (the oxygen is slightly faster), producing final energies of 1.36 and 1.85 fJ for O^{6+} and Mg^{6+}, respectively. The Mg^{9+} acquires a slightly slower perpendicular speed and a significantly slower parallel speed. The result is that Mg^{9+} has similar energies to Mg^{6+} at low altitudes but drops to only a slightly higher energy than O^{6+} at the top of the heating region (it subsequently returns to be heated further).

The energy at the top of the heating region depends exponentially on α, as shown for O^{5+} in Figure 6, and as a power of f, as shown in Figure 7. This behavior is similar to the weak field limit of the auroral case [*Chang et al.*, 1986; *Lund et al.*, 1999]. The coefficient of α in the best-fit expo-

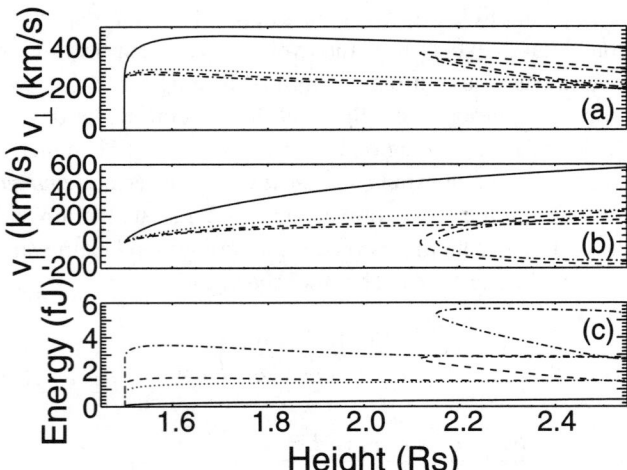

Figure 3. (*a*) Perpendicular velocities, (*b*) parallel velocities, and (*c*) energies for H^+ (solid curves), O^{5+} (dotted curves), Mg^{9+} (dashed curves), and Fe^{9+} (dot-dash curves) as a function of altitude in the $\alpha = 1.67, f = 2.0$ model with both gravity and parallel electric field.

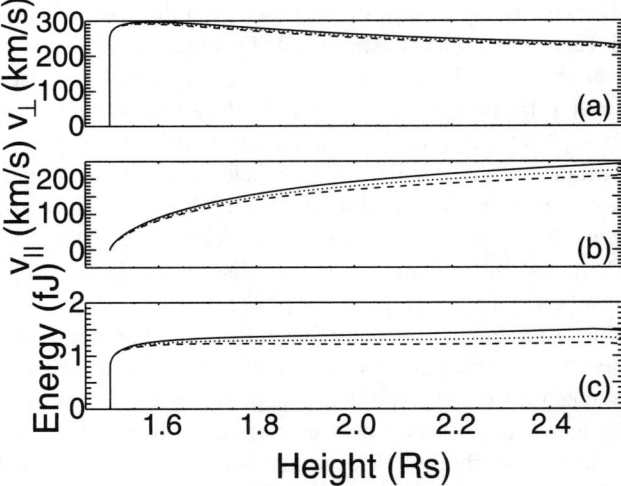

Figure 4. (*a*) Perpendicular velocities, (*b*) parallel velocities, and (*c*) energies for O^{5+} (solid curves), O^{6+} (dotted curves), and O^{7+} (dashed curves) as a function of altitude in the $\alpha = 1.67, f = 2.0$ model with both gravity and parallel electric field.

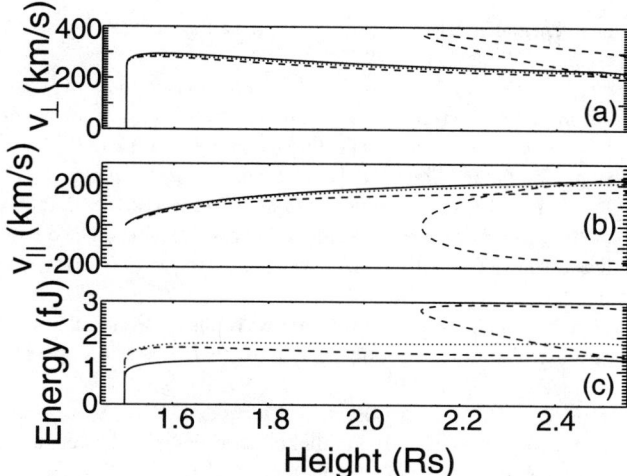

Figure 5. (a) Perpendicular velocities, (b) parallel velocities, and (c) energies for O^{6+} (solid curves), Mg^{6+} (dotted curves), and Mg^{9+} (dashed curves) as a function of altitude in the $\alpha = 1.67$, $f = 2.0$ model with both gravity and parallel electric field.

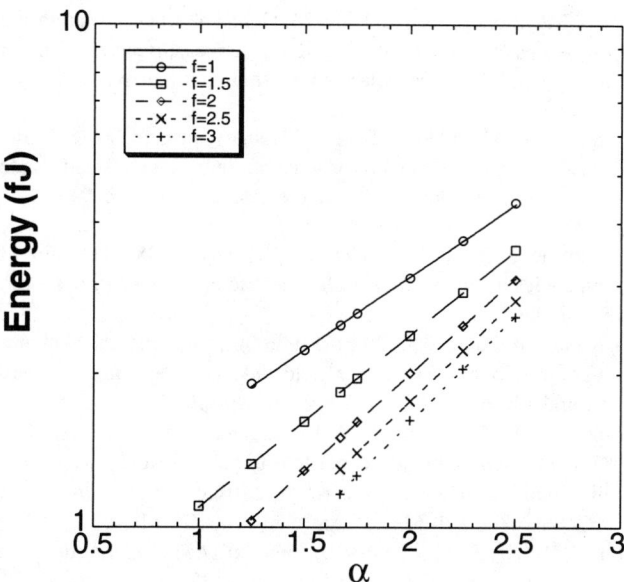

Figure 6. Variation of O^{5+} energy with α for various values of f at the top of the heating region. Both gravity and the parallel electric field are included.

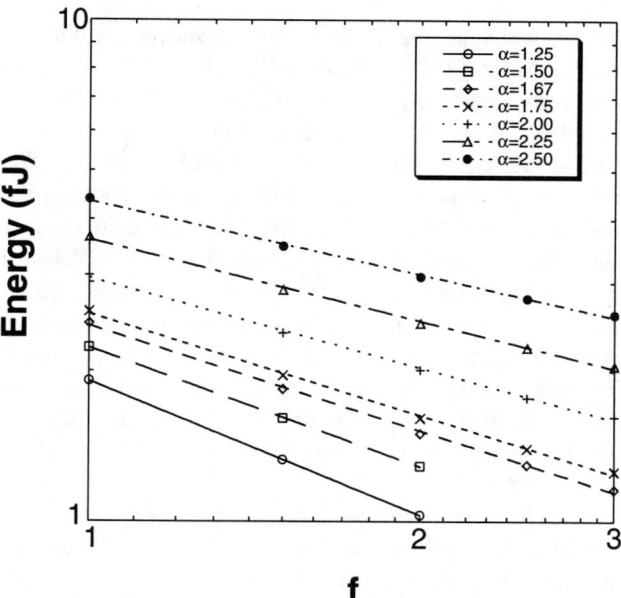

Figure 7. Variation of O^{5+} energy with f for various values of α at the top of the heating region. Both gravity and the parallel electric field are included.

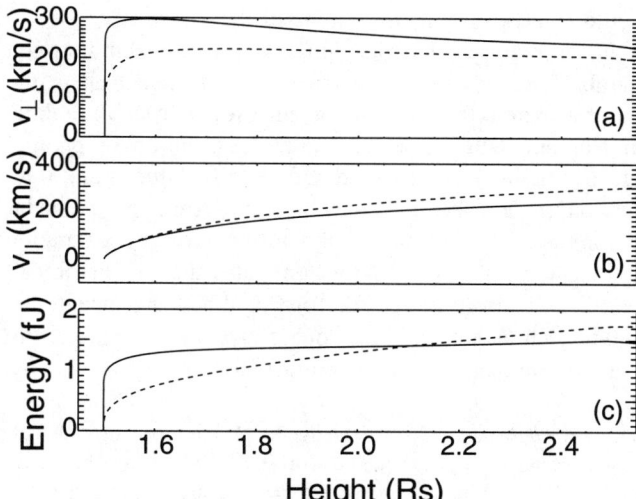

Figure 8. (a) Perpendicular velocities, (b) parallel velocities, and (c) energies for O^{5+} as a function of altitude in the $\alpha = 1.67$, $f = 2.0$ model with parallel electric field, both with (solid curves) and without (dashed curves) gravity.

nential varies logarithmically with f, while the power f varies exponentially with α.

Finally, Figure 8 shows what happens if gravity is neglected. Gravity produces higher perpendicular speeds and consequently higher ion energies at low altitudes, but the parallel velocity lost in overcoming gravity eventually outweighs the higher perpendicular speed to produce a lower overall energy when gravity is turned on. Surprisingly, the constant parallel electric field has little effect: the curves with the parallel electric field turned off, which are not shown in the figure, would lie nearly on top of the curves shown.

5. DISCUSSION AND SUMMARY

Clearly there are some differences between the auroral and coronal heating. The most important of these is that gravity, which is negligible in the aurora for all but cold heavy ions, plays a significant role in the solar corona as a species-dependent effective parallel electric field. Thus we should expect that, unlike in the aurora [*Lund et al.*, 1999], ion energies in the corona should depend on species. This species dependence is in fact observed in the corona [e.g., *Kohl et al.*, 1998; *Esser et al.*, 1999] as well as in the solar wind. The presence of gravity also provides an explanation for why heavy ions have such pronounced temperature anisotropies when protons do not: gravity keeps the heavy ions trapped longer, so they must acquire more energy to escape. The assumed power law wave spectrum heats the ions which are most strongly trapped by gravity more rapidly than ions with higher charge states. More work is needed to sort out these effects.

It should be noted also that the parallel electric field profile used in the coronal simulations is a cartoon sketch, and that the actual profile, as in the auroral region, will be significantly more complicated. It is not known whether these electric fields have a smooth profile as predicted by the *Alfvéen and Fälthammar* [1963] formula, or whether double layers develop as they do in the aurora. The latter scenario could significantly modify some of these results.

In summary, we have examined the relationship between parallel electric fields and ion heating in plasmas. In the auroral region, the two phenomena are inextricably linked, and a model that can predict the distribution of parallel electric fields from observed particle distributions and fluctuation spectra is under development. Although the solar corona is similar in many ways to the auroral acceleration region, the effect of gravity dominates that of distributed parallel electric fields in the corona. Whether concentrated fields such as are found in double layers are significant to coronal ion heating is an open question.

Acknowledgments. I acknowledge helpful discussions with J. R. Jasperse and K. A. Lynch. The work on FAST at the University of New Hampshire was supported by National Aeronautics and Space Administration grant number NAG5-12590 via a subcontract from the University of California, Berkeley (C. W. Carlson, principal investigator).

REFERENCES

Alfvéen, H., and C. G. Fälthammar (1963), *Cosmical Electrodynamics*, Oxford University Press, Oxford.

Andersson, L., R. E. Ergun, D. L. Newman, J. P. McFadden, C.W. Carlson, and Y.J. Su (2002), Characteristics of parallel electric fields in the downward current region of the aurora, *Phys. Plasmas*, *9*, 3600.

André, M., P. Norqvist, L. Andersson, L. Eliasson, A. I. Eriksson, L. Blomberg, R. E. Erlandson, and J. Waldemark (1998), Ion energization mechanisms at 1700km in the auroral region, *J. Geophys. Res.*, *103*, 4199.

Carlson, C. W., *et al.* (1998), FAST observations in the downward auroral current region: Energetic upgoing electron beams, parallel potential drops, and ion heating, *Geophys. Res. Lett.*, *25*, 2017.

Chang., T., G. B. Crew, N. Hershkowitz, J. R. Jasperse, J. M. Retterer, and J. D. Winningham (1986), Transverse acceleration of oxygen ions by electromagnetic ion cyclotron resonance with broadband left-hand-polarized waves, *Geophys. Res. Lett.*, *13*, 636.

Cranmer, S. R. (2000), Ion cyclotron wave dissipation in the solar corona: The summed effect of more than 2000 ion species, *Astrophys. J.*, *532*, 1197.

Ergun, R. E., C. W. Carlson, J. P. McFadden, F. S. Mozer, and R. J. Strangeway (2000), Parallel electric fields in discrete arcs, *Geophys. Res. Lett.*, *27*, 4053.

Ergun, R. E., Y. J. Su, L. Andersson, C. W. Carlson, J. P. McFadden, F. S. Mozer, D. L. Newman, M. V. Goldman, and R. J. Strangeway (2001), Direct observation of localized parallel electric fields in a space plasma, *Phys. Rev. Lett.*, *87*, 045003.

Esser, R., S. Fineschi, D. Dobrzycka, S. R. Habbal, R. J. Edgar, J. C. Raymond, J. L. Kohl, and M. Guhathakurta (1999), Plasma properties in coronal holes derived from measurements of minor ion spectral lines and polarized white light intensity, *Astrophys. J.*, *510*, L63.

Feldman, W. C., S. R. Habbal, G. Hoogeveen, and Y.-M. Wang (1997), Experimental constraints on pulsed and steady-state models of the solar wind near the Sun, *J. Geophys. Res.*, *102*, 26,905.

Gorney, D. J., Y. T. Chiu, and D. R. Croley (1985), Trapping of ion conics by downward parallel electric fields, *J. Geophys. Res.*, *90*, 4205.

Jasperse, J. R., and N. J. Grossbard (2000), The Alfvéen-Fälthammar formula for the parallel *E* field and its analogue in downward auroral-current regions, *IEEE Trans. Plasma Sci.*, *28*, 1874.

Jasperse, J. R., E. J. Lund, K. A. Lynch, and C. W. Carlson (2001), New results for the self-consistent parallel E-field and particle distributions in the auroral return current region, *Eos Trans. AGU*, *82*(47), Fall Meet. Suppl., Abstract SM11C-06.

Kan, J. R., and L. H. Lyu (1990), Field-aligned potential drops in the solar corona, *J. Geophys. Res.*, *95*, 4239.

Kintner, P. M., J. Franz, P. Schuck, and E. Klatt (2000), Interferometric coherency determination of wavelength or what are broadband ELF waves?, *J. Geophys. Res.*, *105*, 21,237.

Knight, S. (1973), Parallel electric fields, *Planet. Space Sci.*, *21*, 741.

Kohl, J. L., *et al.* (1998), UVCS/SOHO empirical determinations of anisotropic velocity distributions in the solar corona, *Astrophys. J.*, *501*, L127.

Lund, E. J., E. Möbius, R. E. Ergun, and C. W. Carlson (1999), Mass-dependent effects in ion conic production: The role of parallel electric fields, *Geophys. Res. Lett.*, *26*, 3593.

Lund, E. J., E. Möbius, C. W. Carlson, R. E. Ergun, L. M. Kistler, B. Klecker, D. M. Klumpar, J. P. McFadden, M. A. Popecki, R. J. Strangeway, and Y. K. Tung (2000), Transverse ion acceleration mechanisms in the aurora at solar minimum: Occurrence distributions, *J. Atmos. Solar Terr. Phys.*, *62*, 467.

Lund, E. J., E. Möbius, K. A. Lynch, D. M. Klumpar, W. K. Peterson, R. E. Ergun, and C. W. Carlson (2001), On the mass dependence of transverse ion acceleration by broad-band extremely low frequency waves, *Phys. Chem. Earth*, *26*, 161.

Lynch, K. A., J. W. Bonnell, C. W. Carlson, and W. J. Peria (2002), Return current region aurora: $E_\parallel$, j_z, particle energization and BBELF wave activity, *J. Geophys. Res.*, *107*(A7), 1115, doi:10.1029/2001JA900134.

Mälkki, A., A. I. Eriksson, P. O. Dovner, R. Boström, B. Holback, G. Holmgren, and H. E. J. Koskinen (1993), A statistical survey of auroral solitary waves and weak double layers 1. Occurrence and net voltage, *J. Geophys. Res.*, *98*, 15,521.

Mozer, F. S., and C. A. Kletzing (1998), Direct observation of large, quasi-static, parallel electric fields in the auroral acceleration region, *Geophys. Res. Lett.*, *25*, 1629.

Newman, D. L., M. V. Goldman, R. E. Ergun, and A. Mangeney (2001), Formation of double layers and electron holes in a current-driven space plasma, *Phys. Rev. Lett.*, *87*, 255001.

Norqvist, P., M. André, and M. Tyrland (1998), A statistical study of ion energization mechanisms in the auroral region, *J. Geophys. Res.*, *103*, 23459.

Ofman, L., S. P. Gary, and A. Viñas (2002), Resonant heating and acceleration of ions in coronal holes driven by cyclotron resonant spectra, *J. Geophys. Res.*, *107*(A12), 1461, doi:10.1029/2002JA009432.

Pannekoek, A. (1922), Ionization in stellar atmospheres, *Bull. Astron. Soc. Netherlands*, *19*, 107.

Pottelette, R., R. A. Treumann, and M. Berthomier (2001), Auroral plasma turbulence and the cause of auroral kilometric radiation fine structure, *J. Geophys. Res.*, *106*, 8465.

Rosseland, S. (1924), The electrical state of a star, *Mon. Not. Royal Astron. Soc.*, *84*, 720.

Sharp, R. D., R. G. Johnson, and E. G. Shelley (1977), Observation of an ionospheric acceleration mechanism producing energetic (keV) ions primarily normal to the geomagnetic field direction, *J. Geophys. Res.*, *82*, 3324.

Shelley, E. G., R. D. Sharp, and R. G. Johnson (1976), Satellite observations of an ionospheric acceleration mechanism, *Geophys. Res. Lett.*, *3*, 654.

Stasiewicz, K., Y. Khotyaintsev, M. Berthomier, and J.-E. Wahlund (2000), Identification of widespread turbulence of dispersive Alfvéen waves, *Geophys. Res. Lett.*, *27*, 173.

Temerin, M. (1978), The polarization, frequency, and wavelengths of high-latitude turbulence, *J. Geophys. Res.*, *83*, 2609.

Temerin, M., and C. W. Carlson (1998), Current-voltage relationship in the downward auroral current region, *Geophys. Res. Lett.*, *25*, 2365.

Tung, Y.-K., C. W. Carlson, J. P. McFadden, D. M. Klumpar, G. K. Parks, W. J. Peria, and K. Liou (2001), Auroral polar cap boundary ion conic outflow observed on FAST, *J. Geophys. Res.*, *106*, 3603.

Wilson, G., D. M. Ober, G. Germany, and E. J. Lund (2001), The relationship between suprathermal ion outflow and auroral electron energy deposition: Polar/UVI and FAST/TEAMS observations, *J. Geophys. Res.*, *106*, 18,981.

E. J. Lund, Space Science Center, Morse Hall, University of New Hampshire, Durham, New Hampshire 03824. (eric.lund@unh.edu)

Simulation Study of Beam–Plasma Interaction and Associated Acceleration of Background Ions

X. Y. Wang and Y. Lin

Physics Department, Auburn University, Auburn, Alabama

Wave evolution and ion heating in ion beam-plasma interaction are studied by one-dimensional (1-D) and two-dimensional (2-D) hybrid simulations for cases with a relatively strong field-aligned ion beam (e.g., the ratio of beam ion density to background ion density >0.06 for beam velocity = $10V_A$, where V_A is the Alfvén speed). It is found that the wave evolution possesses four different phases. In the final phase, nonlinear shear Alfvén waves with right-hand polarization in magnetic field are generated. These Alfvén waves propagate mainly with $\mathbf{k}\cdot\mathbf{B}_0 > 0$, and the dispersion relation $\omega = kV_A\cos\alpha$ is satisfied, where α is the angle between $\mathbf{k}$ and $\mathbf{B}_0$. The shear Alfvén waves are excited through a parametric process via beam-ion cyclotron instability. In addition, the background plasma is accelerated and heated significantly. The ratio of the final to initial temperature, varying almost linearly with beam energy density $n_b V_b^2$, can reach 80–100 for typical plasmas in space. It is found that the ion heating in obliquely propagating waves ($\alpha \neq 0$) is more effective than that in the case with ($\alpha = 0$).

1. INTRODUCTION

Ion beam-plasma is a fundamental plasma process in space plasma physics [*Wu and Davidson*, 1972; *Gary*, 1991; *Akimoto et al.*, 1993]. Research on this topic was motivated in space physics by the study of reflected ions in upstream of the earth's bow shock [*Dubouloz and Scholer*, 1995], solar wind [*Feldman et al.*, 1974; *Daughton et al.*, 1999] and the pickup of cometary ions [*Yoon and Wu*, 1986]. The most significant conclusion found in these works is that an ion beam can excite low-frequency electromagnetic waves, and subsequently the enhanced waves can pitch-angle scatter and diffuse the beam ions. Alfvénic and whistler fluctuations associated with ion beams have also been observed in situ by spacecraft around cometary environments [*Tsurutani and Smith*, 1986], terrestrial as well as interplanetary shock upstream [*Hoppe and Russell*, 1983] and the solar wind [*Belcher and Davis*, 1971].

Most of the early studies focused on the ion electromagnetic instabilities and the time evolution of beam ions, but the detailed properties of waves and modification of background ions were somehow ignored. For example, in the work of *Winske and Quest* [1986], they demonstrated, in one-dimensional (1-D) and two-dimensional (2-D) hybrid simulations, the presence of the right-hand resonant and nonresonant instabilities predicted by the linear theory. The 1-D simulation of *Akimoto et al.* [1993] illustrated in detail the nonlinear evolution of ion beam instabilities and the associated pulsations in a transient stage. *Hellinger and Mangeney* [1999] showed in a 2-D simulation oblique pulsations generated by the beam–plasma interaction. On the other hand, the rapid pickup of newborn ions in cometary exosphere was also extensively studied [*Yoon and Wu*, 1986, and therein].

The purpose of this paper is to report a hybrid simulation of an ion beam-plasma interaction. In this study we pay special attention to the analysis of properties and evolution of nonlinear waves and the acceleration and heating of background ions. Unlike previous simulations, our calculation is made for the evolution of the beam-plasma system on a long (~400 ion gyroperiods) time scale beyond the pulsation stage.

2. SIMULATION RESULTS

Since the typical electromagnetic instabilities driven by ion beams streaming in plasma are associated with frequen-

Particle Acceleration in Astrophysical Plasmas
Geophysical Monograph Series 156
Copyright 2005 by the American Geophysical Union
10.1029/156GM15

cies ω smaller or on the order of Ω_i, where Ω_i is the ion cyclotron frequency, hybrid simulation is most suitable for the study of the nonlinear system of beam-plasma interaction. In the hybrid simulation, the ions (protons) are treated as particles and electrons are treated as a massless fluid. Quasi-charge neutrality is assumed. Both 1-D and 2-D hybrid simulations are performed, with periodic boundary conditions in all dimensions.

In our simulation, the length is expressed in units of ion inertial length of the background plasma c/ω_{pi}, and the time t is expressed in units of $1/\Omega_i$, where the proton gyrofrequency $\Omega_i = eB_0/m_i$, and m_i is the proton mass. Thus the velocity is normalized to $\overline{V} \equiv V_{A0}$. The temperature is normalized in unit of $\overline{T} \equiv m_0 V_{A0}^2$. The magnetic field is expressed in units of $\overline{B} \equiv B_0$, density n is normalized to $\overline{N} \equiv n_0$, and the electric field is in units of $\overline{E} \equiv V_{A0}B_0$. The simulation domain is along x for 1-D simulations with system length $L_x = 4000c/\omega_{pi}$ and in xz plane for 2-D simulations with $L_x \times L_z = 800 \times 400 c/\omega_{pi}$. The grid size $\Delta = 0.5$–$2c/\omega_{pi}$. Zero resistivity is used in our simulation. In this study, we only show the results in the cases with relatively strong ion beam (e.g., the ratio of beam ion density to background ion density >0.06 for beam velocity $= 10V_A$, where V_A is the Alfvén speed). The beam is assumed to be field-aligned. The results for cases with a weak beam can be found in a recent paper by *Wang and Lin* [2003].

In the following, we first show a case with the initial beam velocity relative to the background plasma $V_b = 10$, and the beam density (normalized to n_0) $n_b = 0.1$. The magnetic field $\mathbf{B}_0$ is assumed to be along $\mathbf{x}$, and thus the resulting waves propagate along the unperturbed magnetic field, corresponding to $\alpha = 0°$, where α is the propagation angle between $\mathbf{k}$ and $\mathbf{B}_0$. The β values of both the beam and the background ions are 0.01, where $\beta = 8\pi nkT/B_0^2$. The electron beta is also equal to 0.01. Note that the normalized ion initial temperature is equal to β. We have simulated the cases with $\beta = 0.01$–0.5 and the results are found insensitive to the initial β in those cases.

The formation of low-frequency electromagnetic waves in the beam plasma interaction, which was not demonstrated in previous self-consistent simulation, is one of the central focuses of this paper. In order to illustrate the formation of such waves, in the following we present a detailed analysis of simulation results for the properties and evolution of waves.

2.1. Properties and Evolution of Waves

The 1-D simulation results at, from the top, $t = 14, 20, 48$, and 200, are shown in Figure 1. Column (a) of Figure 1 shows the spatial profiles (in partial domain) of B_y component of magnetic field, V_y component of bulk flow velocity. Column (b) shows the spatial profiles of magnetic field B, total (plasma plus beam) density n, and the electric field component E_x. Column (c) of presents the hodogram of B_y and B_z.

It is found that the wave evolution can be divided into four phases, with typical results corresponding to the four times shown in Figure 1. In phase 1 (from $t = 0$ to $t = 16$), the wave amplitudes are small, wavy structures are present in the B_y and V_y profile. The fluctuations of magnetic field are in the transverse components, while the total field B, density n, and the electric field E_x have nearly no fluctuation. The right-hand circular polarization of magnetic fields can also be seen in the magnetic field (B_y and B_z) hodogram shown in the top right plot of Figure 1. The simulation results are found to be consistent with the linear theory [*Gary et al.*, 1984; *Wang and Lin*, 2003].

In phase 2 (from $t = 16$ to $t = 23$), the fluctuation in magnetic field is large enough that linear theory cannot be satisfied. The wave energy still grows exponentially, reaching its maximum at $t = 23$ (not shown). It is seen from Figure 1 that the amplitude of the right-hand polarized waves have grown significantly at time $t = 20$, with $\Delta B_y \sim 2$. During this phase, the total density and magnetic field are not constant. At $t = 20$, large-amplitude fluctuations are also present in B, n, and E_x, with shorter wavelength waves lying above some longer wavelength components, as shown in Figure 1. The fluctuations in n is nearly positively correlated with those in B and E_x for waves with short wavelengths. The phase difference between V_y and B_y is nearly 90°.

During phase 3 (from $t = 23$ to $t \sim 100$), the wave structure becomes quite different. The wave amplitudes are greatly reduced and then saturated as $t \to 100$. The third row of Figure 1 shows the results at $t = 48$ in phase 3. The fluctuations in total n becomes very large, and they are nearly anti-correlated with the fluctuations in B, while the fluctuations in E_x are nearly positively correlated with those in B. The presence of similar density and magnetic field fluctuations due to beam plasma interaction has also been discussed by *Akimoto et al.* [1993] for pulsations. At this stage, the dominant wave number in the transverse magnetic field fluctuations is greatly reduced, while the fluctuations in B_y and B_z have no clear pattern of polarization. Moreover, the waves are found to evolve toward Alfvén waves in long wavelength limit through this phase. In $t > 100$, Alfvén waves can be identified.

It is found during phase 3 that the nonlinear wave evolution is mainly a parametric decay process [*Wang and Lin*, 2003]. A beat wave appears at a long wavelength with wave number $k_1 \sim 0.1$, acting on a pump wave with wave number $k_0 \sim -0.43$, which is a dominant mode in early times. As a result,

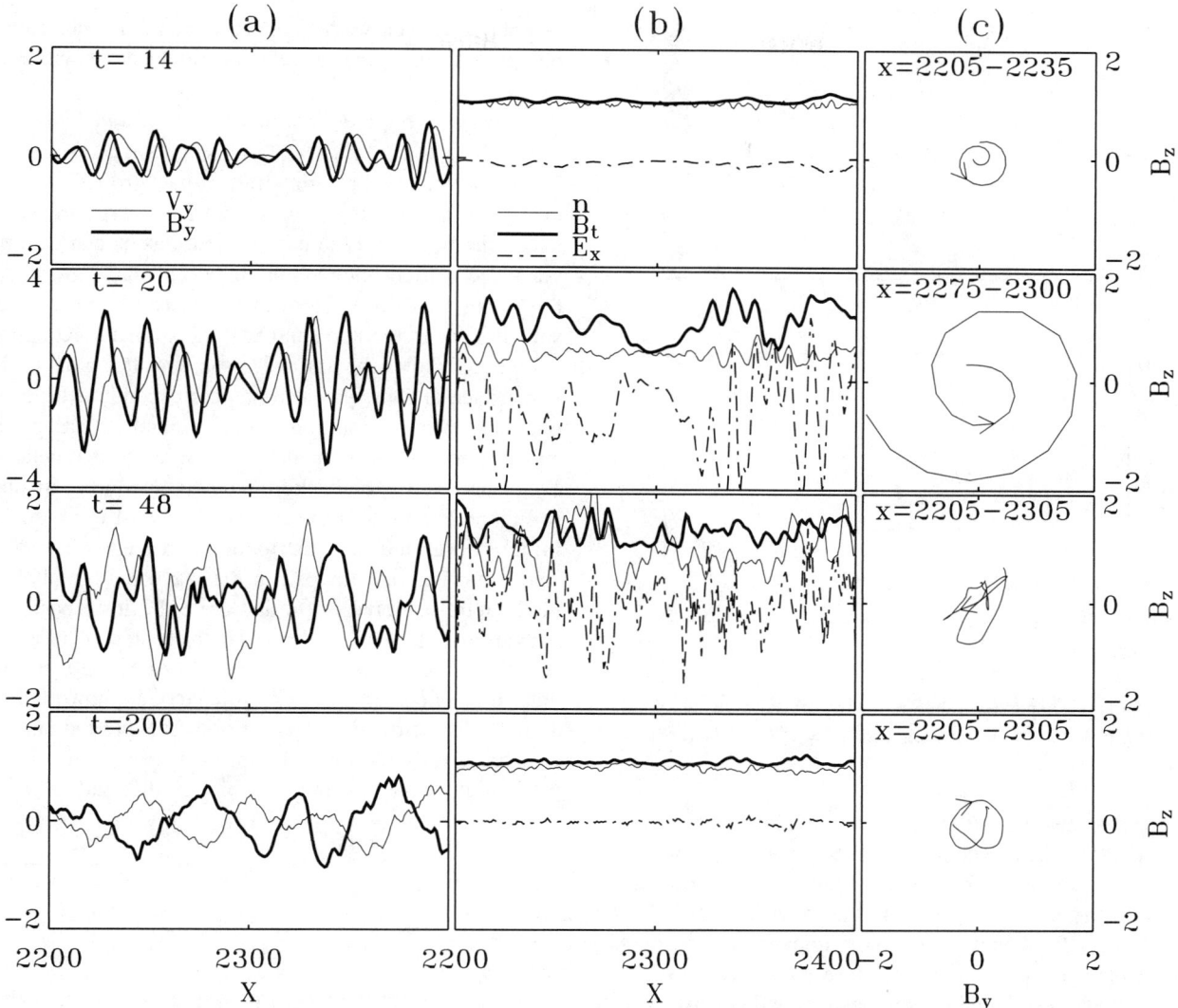

Figure 1. (Column a) Spatial profiles of B_y (thick solid line) and total velocity V_y (thin solid lines) at, from the top, $t = 14$, 20, 48, and 200. (Column b) Spatial profiles of total density n (thin solid line), total magnetic field B_t (thick solid line), and electric field E_x (dot-dashed line) at the corresponding times. (Column c) Hodograms of magnetic field (B_y - B_z) at the corresponding times. Only partial domain is shown.

the pump wave decays, and the beat wave grows. The latter eventually dominates the process. In addition to these two modes, a sound wave carrying fluctuation in B and n is found at $k_2 = k_0 - k_1 \sim -0.53$ during the decay of the pump wave. The appearance of these wave modes is associated the ion trapping, back scattering, and ion Landau damping.

In phase 4 (from $t \sim 100$ to the end of the run $t = 400$), the wave energy nearly remains the same. In this phase, shear Alfvén waves are found to dominate the evolution. The bottom row of Figure 1 depicts the resulting wavy structure at $t = 200$. It is seen that $B \sim$ const, $n \sim$ const, and the electric field component $E_x \sim 0$. The magnetic field fluctuation is found to be mainly in the transverse, B_y and B_z, directions, as shown in the field hodogram in the bottom right plot of Figure 1. The waves are found to be right-hand polarized, with a dominant wave vector $\mathbf{k}$ in the $+\mathbf{x}$ direction, different from the $\mathbf{k}$ in early phases. The flow component V_y and the field component B_y are in an anti-phase relation, consistent with the Walen relation of MHD Alfvén waves for $\mathbf{k} \parallel \mathbf{B}_0$.

2.2. Dispersion Relation of the Resulting Waves

The top plot of Figure 2 shows the dispersion relation of the resulting Alfvén waves in the final phase. The results in the (ω, k) space are obtained by performing Fourier trans-

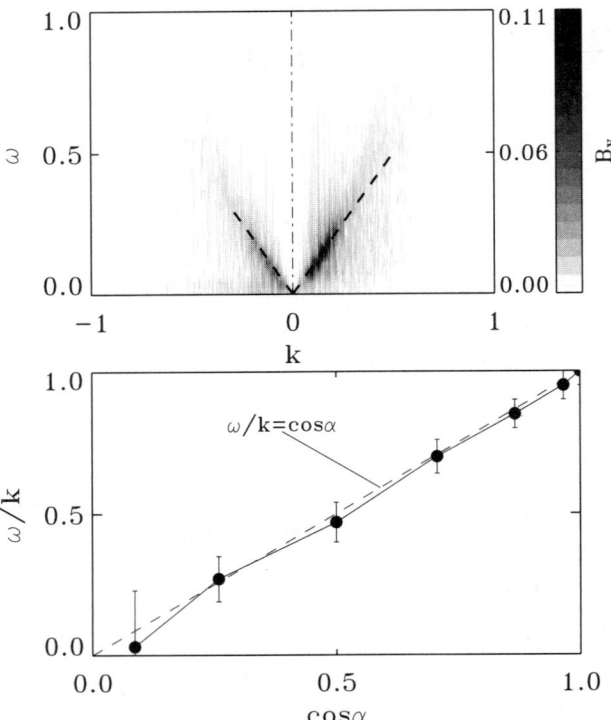

Figure 2. (Top) Dispersion relation of the Alfvén waves obtained during phase 4. (Bottom) Phase speed $V_p = \omega/k$ as a function of $\cos\alpha$.

form for data samples of B_y in the simulation. Here $k > 0$ (< 0) represents wave vectors parallel (anti-parallel) to $+\mathbf{x}$. The data are sampled in space at all grid points in the domain and in time from $t = 100$ to $t = 400$ at every $\Delta t = 0.4$. It is found that the frequency ω is almost linearly proportional to the wave number k, and the phase speed $\omega/|k| \sim V_A$. Although waves are found in both positive and negative k's, the dominant wave power is with $k > 0$. From the analysis in section 2.1, waves with both $k > 0$ and $k < 0$ are right-hand polarized in magnetic field. Therefore, the resulting waves are found to satisfy the properties of right-hand polarized shear Alfvén waves.

We have also carried out the simulation of beam-plasma interaction for cases with the background $\mathbf{B}_0$ not parallel to $\mathbf{x}$, which correspond to a wave vector $\mathbf{k}$ (along $\mathbf{x}$) oblique to $\mathbf{B}_0$, i.e., $\alpha \neq 0°$. In these cases, the initial beam velocity is still assumed to be parallel to $\mathbf{B}_0$. The simulation results are found to be similar to those in the case with $\alpha = 0°$, with four phases in the wave evolution process. And the Alfvén waves are also dominant in the final phase. The dots in the bottom plot of Figure 2 show the phase speed $V_p = \omega/k$ as a function of $\cos\alpha$ obtained from runs with various α. It is seen that the dispersion relation

$$\omega = kV_A\cos\alpha \qquad (1)$$

of oblique Alfvén waves is nearly satisfied. Therefore, the oblique Alfvén waves are generated in the cases with $\alpha \neq 0°$.

2.3. Acceleration and Heating of Background Ions

The modification of properties of background ions is another central focuses of this paper. The energy transfer between the beam and the background plasma can take place via wave-particle interactions between the beam ions and the background ions. The waves discussed above have resulted in the acceleration and heating of the ambient plasma.

Figure 3 shows the ion phase-space scatter plots in the $v_{\parallel} - v_{\perp}$ space at four typical times corresponding to the four phases of wave evolution, where $v_{\parallel}$ and $v_{\perp}$ are particles speeds parallel and perpendicular to the dc magnetic field $\mathbf{B}_0$, respectively. The background and beam ions are shown in Figure 2a and 2b, respectively. At $t = 14$ in phase 1, the initial background ion distributions are nearly unchanged. The beam ions are trapped in the wave field, and slightly pitch-angle scattered. Note that the bulk flow speed of the background plasma is $V_{0\parallel} = -0.9$ in the center-of-mass frame, and that of beam ions is $V_{b\parallel} = 9.1$. At $t = 20$, the wave amplitude in B_y fluctuation is quite large as shown in Figure 1, and thus the distributions for both background and beam ions are greatly modified. Beam ions are not only strongly pitch-angle scattered, but also energy diffused. The back-

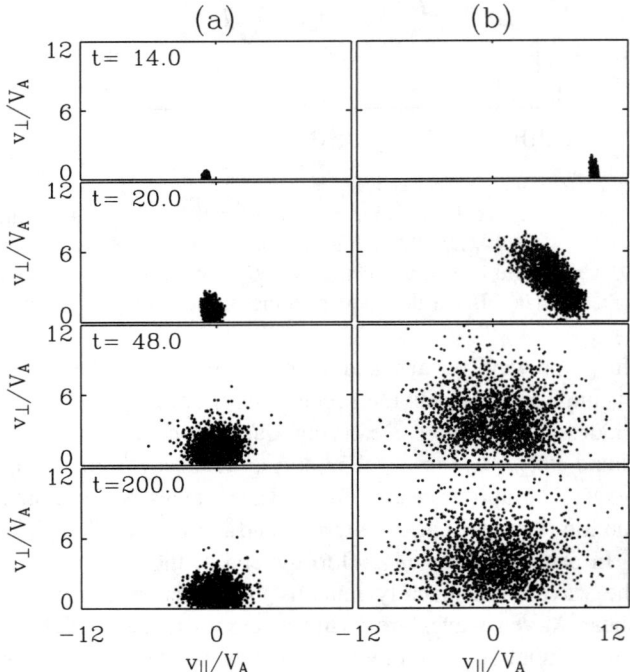

Figure 3. (Left) Phase space scatter plots at, from the top, $t = 14$, 20, 48, and 200 for background ions. (Right) Phase space scatter plots for beam ions at corresponding times.

ground population is somewhat heated. At $t > 40$, the background ions are greatly accelerated. Both the beam ions and background plasma have been strongly (and nearly isotropically) heated, and their temperatures increase relative to the initial values by a factor of 900 and 100, respectively. At $t = 48$ in phase 3, the ion heating of background plasma has almost reached its maximum. The central bulk flow speeds of beam and background ions have merged and become nearly the same, ~0 in the simulation frame. At $t > 100$, the structure of waves has nearly reached its final stage and the background ion distribution also remain the same.

As mentioned above, the background ions are accelerated and heated because of the wave particle interaction. The significantly heating and acceleration occur in a short time, from $t \sim t_1 = 14$ at the end of phase 1 to $t \sim 40$ when the amplitude of waves is quite large. It is found the ion heating is also associated with the electric field E_x; the increase in temperature is also correlated with E_x.

The top plot in Figure 4 shows the change of the ion kinetic energy density $\Delta W_{p\parallel}$ ($\Delta W_{p\perp}$) parallel (perpendicular) to the dc magnetic field $\mathbf{B}_0$ in the whole system at the end of the run, where the energy density is normalized to $n_0 V_{th0}^2$, and V_{th0} is the initial thermal speed of background ions. The energy density gain $\Delta W_{p\parallel}$ (stars) and $\Delta W_{p\perp}$ (diamonds) are obtained in cases with various α. It is found that both $\Delta W_{p\parallel}$ and $\Delta W_{p\perp}$ increase slightly with α for $\alpha \leq 60°$. Nevertheless, for $\alpha > 60°$, the kinetic energy density decreases with α. Note that the temperature gain of the background plasma is slightly lower (by 10–20%) than the energy gains shown in Figure 4. The results are qualitatively consistent with the theoretical analysis by *Dilday et al.* [2002].

The middle plot in Figure 4 shows $\Delta W_{p\parallel}$ and $\Delta W_{p\perp}$ as a function of $n_b V_b^2$, which corresponds to the free energy density provided by the ion beam. The solid line represents $\Delta W_{p\parallel}$, and the dashed line represents $\Delta W_{p\perp}$. The diamonds represent the cases with various V_b for a fixed $n_b = 0.1$, while the stars correspond to the cases with various n_b for a fixed $V_b = 10$. It appears that both $\Delta W_{p\parallel}$ and $\Delta W_{p\perp}$ increase with the beam energy, and the increase in energy density is much faster when $n_b V_b^2 \leq 5$. The final energy density gain of the background ions is linearly proportional to $n_b V_b^2$ with $\Delta W_p/n_0 V_{th0}^2 \sim 4 n_0 V_b^2$ when $n_b V_b^2 < 5$, and $\Delta W_p/n_0 V_{th0}^2 \sim 15 n_0 V_b^2 - 3.5$ when $n_b V_b^2 > 5$. When normalized to $n_b V_b^2$, the final energy gains $\Delta W_{p\parallel}$ and $\Delta W_{p\perp}$ of the background ions are seen to increase with $n_b V_b^2$, as shown in the bottom plot of Figure 4. The efficiency of the conversion of energy from beam to background ions is about 10% for $n_b V_b^2 > 5$.

2.4. 2-D Effects on Wave Properties and Ion Acceleration

The 2-D simulation results of the beam plasma interaction are found similar to those from the 1-D simulation, with

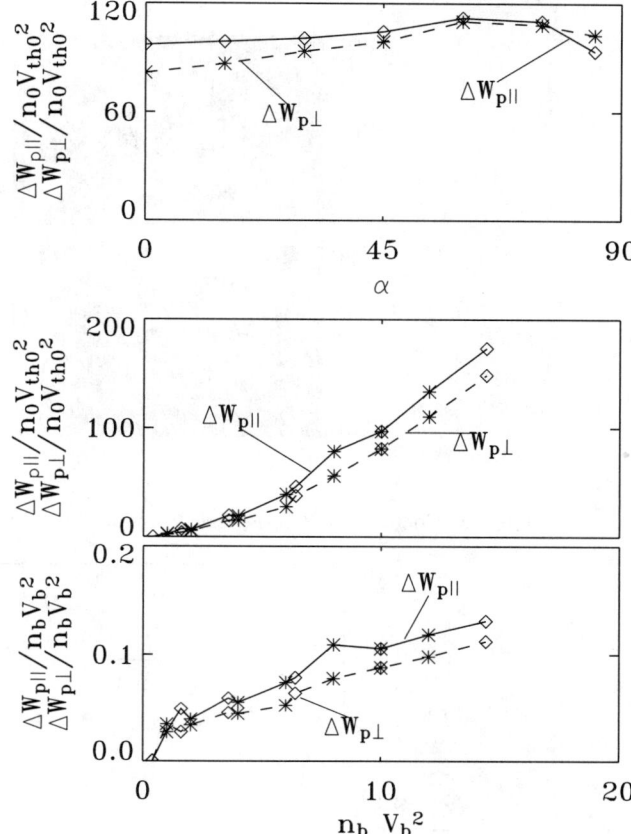

Figure 4. (Top) Energy density gains $\Delta W_{p\parallel}$ (stars) and $\Delta W_{p\perp}$ (diamonds) of background ions, normalized to $n_0 V_{th0}^2$, as a function of α, where V_{th0} is the initial ion thermal speed. (Middle) Energy density gains as a function of $n_b V_b^2$. The diamonds represent the cases with various V_b for a fixed $n_b = 0.1$, while the stars correspond to the cases with various n_b for a fixed $V_b = 10$. (Bottom) Energy density gains normalized to $n_b V_b^2$ as a function of $n_b V_b^2$.

four typical phases present in the wave evolution, as described above. Figure 5 shows the contours of B, n, B_y, and V_y at $t = 200$ corresponding to phase 4. It is found that a nearly anti-phase correlation is present between B_y and V_y, and the Walen relation of Alfvén waves is nearly satisfied. Therefore, similar to the 1-D results, Alfvén waves have formed in the nonlinear stage. In addition, nonlinear wavy structures with perpendicular wave vector $\mathbf{k}_\perp \neq 0$ also appears due to the 2-D effects, as seen in Figure 5. Field-aligned filaments are present in the fluctuations in n and B with wave power mainly along $\mathbf{k}_\perp$. It is found that these filaments are nearly phase-standing, with an anti-phase correlation between B and n. The presence of field-aligned filaments due to ion beam-plasma interaction has also been found in a recent 2-D global hybrid simulation of the quasi-parallel bow shock [*Lin*, 2003].

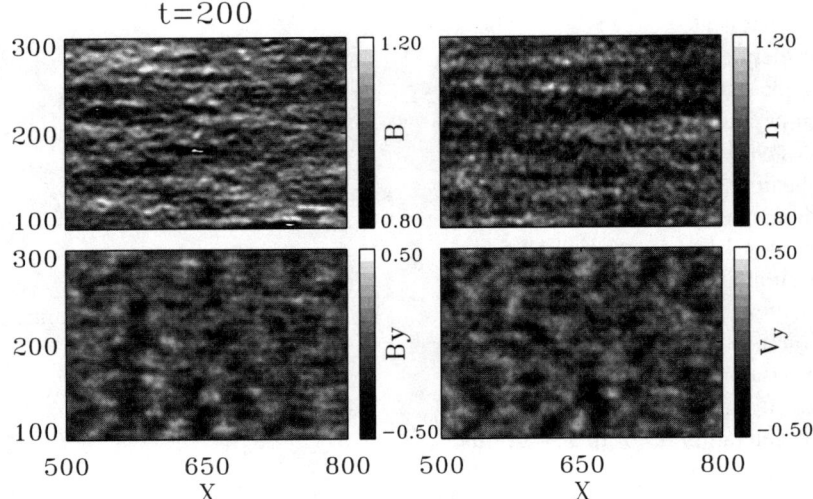

Figure 5. Contours of B, n, B_y, and V_y at $t = 200$ for the 2-D simulation. Only partial domain is shown.

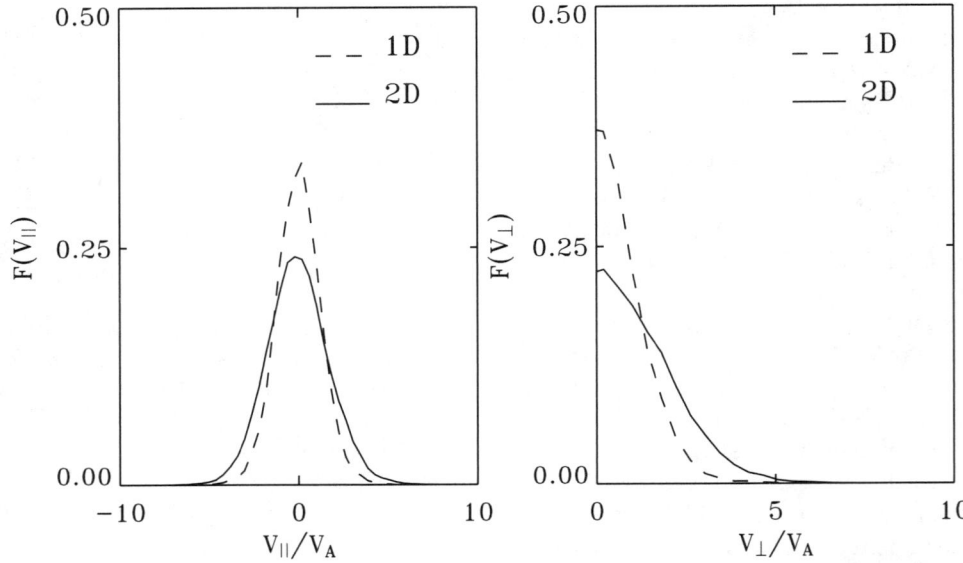

Figure 6. Background ion distribution functions $F(v_\parallel)$ (left) and $F(v_\perp)$ (right) obtained from 1-D simulations (dashed line) and 2-D simulations (solid lines).

Similar to the 1-D case, background ions in the 2-D case are also accelerated and heated. The temperature and bulk flow speed of both beam ions and plasma are saturated at the end of phase 3. The distribution function almost unchanged during the phase 4. The left plot of Figure 6 show the background ions distribution function $F(v_\parallel)$ at $t = 200$ for 2-D (solid line) and 1-D (dashed line) simulations. The perpendicular distribution function $F(v_\perp)$ of background ion is shown in the right plot of Figure 6 for 2-D (solid line) and 1-D (dashed line) simulations. It is found that the parallel thermal speed of ions in the 2-D case is $v_{th\parallel} = 1.52$, which is greater than that in the 1-D case ($v_{th\parallel} = 1.14$). Meanwhile, the perpendicular thermal speed of background ions in the 2-D case is about $v_{th\perp} = 1.57$, which is also larger than that in 1-D simulation ($v_{th\perp} = 1.06$). Therefore, ions in the 2-D case are heated more effectively than those in the 1-D case.

3. SUMMARY

1-D and 2-D hybrid simulations are conducted for the study of evolution of wave properties and heating of background ions in the beam-plasma interaction. It is found that the wave evolution can be divided into four phases. In phase 1, waves properties are found to be consistent with those predicted from linear theory. Phase 2, phase 3, and phase 4 represent the nonlinear evolution. In phase 3, the waves grow to form a non-linear structure such as pulsation, which was observed in previous work. The nonlinear evolution is mainly a parametric decay process.

In the final phase, phase 4, large-amplitude Alfvén waves with right-hand polarization in magnetic field are found to form. The magnetic field strength and the total plasma density are nearly constant, and the Walen relation of Alfvén waves is nearly satisfied. The dispersion relation of Alfvén waves is also satisfied. Note that in the cases with a weak ion beam (e.g. initially $n_b = 0.01$ and $V_b = 10$), fast magnetosonic/whistler waves and slow waves are found to form [*Wang and Lin*, 2003].

The properties of background ions are found to be significantly modified during ion beam-plasma interaction. Background ions are accelerated and heated, mainly through phase 2 and phase 3, due to the wave-particle interaction. The energy gain of background ions increases with the free energy $n_b V_b^2$ of the ion beam. The efficiency of the conversion of energy from beam to background ions is about 10% for $n_b V_b^2 > 5$. The ion heating is also found to be more efficient in the oblique cases ($\alpha \neq 0$) than in the case with $\alpha = 0°$ for $\alpha < 60°$.

In addition to Alfvén waves, 2-D simulation shows that field-aligned filaments (with $k \sim k_\perp$) in B and n can form in the final stage of the evolution due to the 2-D effects. The ion heating is also found more effective in the 2-D cases.

Acknowledgments. This work was supported by NASA grant NAG5-12899 and NSF grant ATM-0213931 to Auburn University and NNSFC grant 40074043 of China to X.Y.W. The authors thank L. Chen and C. S. Wu for helpful discussions and the instruction on nonlinear wave instabilities. X.Y.W. thanks X. X. Zhang for discussions. Computer resources were provided by the National Partnership for Advanced Computational Infrastructure and the Arctic Region Supercomputer Center.

REFERENCES

Akimoto, K., D. Winske, S. P. Gary, and M. F. Thomsen, Nolinear evolution of electromagnetic ion beam instabilities, *J. of Geophys. Res.*, 98, 1419-1433, 1993

Belcher, J. W., and L. Davis, Large-amplitude Alfvén waves in the interplanetary medium, 2, *J. of Geophys. Res.*, 76, 3534-3539, 1971.

Daughton, W., S. P. Gary, and D. Winske, Electromagnetic proton/proton instabilities in the solar wind: Simulations, *J. of Geophys. Res.*, 104, 4657-4668, 1999.

Dilday, B., Chen, L., and White, R. B., Plasma Heating Due to Spectrum of Obliquely Propagating Alfvén Waves, *American Geophysical Union*, Fall Meeting 2002.

Dubouloz, N., and M. Scholer, Two-dimensional simulations of magnetic pulsations upstream of the Earth's bow shock, *J. of Geophys. Res.*, 100, 9461-9474,1995.

Feldman, W. C., J. R. Asbridge, S. J. Bame, and M. D. Montgomery, Interpenetrating solar wind streams, *Reviews of Geophysics and Space Physics*, 12, 715-723, 1974.

Gary, S. P., Electromagnetic ion/ion instabilities and their conesquences in space plasmas: a review, *Space Sci.Rev.*, 56, 373-415, 1991.

Gary, S. P., D. W. Foosland, C. W. Smith, M. A. Lee, and M. L. Goldstein, Electromagnetic ion beam instabilities, *Phys. Fluids*, 27, 1852-1862, 1984.

Hellinger, P. and Mangeney, A., Electromagnetic ion beam instabilities: Oblique pulsations, *J. of Geophys. Res.*, 104, 4669-4680, 1999.

Hoppe, M. M., and C. T. Russell, Plasma rest frame frequencies and polarizations of the low-frequency upstream waves - ISEE 1 and 2 observations, *J. of Geophys. Res.*, 88, 2021-2027, 1983.

Lin, Y., Global-scale simulation of foreshock structures at the quasi-parallel bow shock, *J. of Geophys. Res.*, 108, SMP 3-1, 2003.

Tsurutani, B. T., and E. J. Smith, Hydromagnetic waves and instabilities associated with cometary ion pickup - ICE observations, *Geophys. Res. Lett.*, 11, 263-266, 1986.

Wang, X. Y., and Y. Lin, Generation of nonlinear Alfvén and magnetosonic waves by beam-plasma interaction, *Physics of Plasmas*, 10, 3528-3538, 2003.

Winske, D., and M. M. Leroy, Diffuse ions produced by electromagnetic ion beam instabilities, *J. of Geophys. Res.*, 89, 2673-2688, 1984.

Winske, D., and K. B. Quest, Electromagnetic ion beam instabilities - Comparison of one and two-dimensional simulations, *J. of Geophys. Res.*, 91, 8789-8797, 1986.

Wu, C. S., and R. C. Davidson, Electronmagnetic instabilities produced by neutral-particle ionization in planetary space. *J. of Geophys. Res.*, 11, 5399-5405, 1972.

Yoon, P. H., and C. S. Wu, Ion pickup by the solar wind via wave-particle interactions, *in Cometary Plasma Process*, edited by A. D. Johnstone, pp 241-258, American Geophysical Union, 1986.

Y. Lin and X. Y. Wang, Physics Department, Auburn University, Auburn, Alabama 36834. (xywang@physics.auburn.edu)

Particle Acceleration and Current Disruption From the Cross-Field Current Instability

Anthony T. Y. Lui

The Johns Hopkins University Applied Physics Laboratory, Laurel, Maryland

An ongoing topic prominent in space plasma research is thin current sheet dynamics. In recent years, particle acceleration associated with a current-driven instability known as the cross-field current instability (CCI) has been studied from the approaches of observational data analysis, analytical theory, and computer simulation. The CCI is a potential candidate for substorm onset and for the current disruption phenomenon in the magnetotail of the Earth's magnetosphere. One form of CCI is ion Weibel mode with waves excited mainly along the magnetic field while another form is the modified two-stream mode with waves excited nearly perpendicular to the magnetic field. In this paper, we use particle-in-cell simulation to elaborate the mechanisms by which particles are accelerated by these two modes and to determine the amount of current reduction at the nonlinear saturation stage of these modes.

1. INTRODUCTION

Evidence of immense particle acceleration is quite commonly found in many astrophysical plasma phenomena. An outstanding question in the space plasma discipline is how nature can so efficiently impart energy to charged particles, often accomplishing the energization process within rather short time scales. Recent in situ satellite observations of geospace plasmas have provided some clues to this quest. Thin current sheets, which often have thickness comparable to the ion gyroradius, seem to be the incubator for the accelerated particles. In addition, the dynamics associated with thin current sheets have been considered by many researchers to hold the key to the unsolved problem on the cause of magnetospheric substorms.

A generic mechanism commonly considered for particle acceleration in the astrophysics community is Fermi acceleration. This mechanism is based on motion of an individual charged particle to a different magnetic field configuration. The change in the magnetic field felt by the particle produces an induced electric field that leads to particle acceleration. No collective plasma effects are required for this consideration. A different and broader class of interaction exists in terms of wave-particle interaction. Collective effects of space plasmas can lead to localized but strong wave electric fields that can preferentially accelerate a portion of the particle population to extremely high energies. In this article, we investigate particle acceleration due to a plasma instability that may occur in thin current sheets prevalent in space plasmas. This instability, known as the cross-field current instability (CCI), is a potential candidate for the sought-after substorm onset process. Section 2 provides a brief background for the subject, with an overview on disturbances associated with magnetospheric substorms. The instability analysis behind the CCI theory is elaborated in Section 3. Some preliminary results from numerical simulation of the instability are presented in Section 4 to illustrate the expected nonlinear consequences of the instability.

2. MAGNETOSPHERIC SUBSTORM

There are episodic disturbances from the interaction between the solar wind and the Earth's magnetosphere. A class of these disturbances is magnetospheric substorms. These disturbances were originally recognized from studies of ground-based magnetograms and all-sky-camera pictures of auroral displays in the polar regions.

Today, substorm disturbances are seen over many regions in the near-Earth space. Three different phases of a substorm have been defined—growth, expansion, and recovery [*Akasofu*, 1964; *McPherron*, 1970]. Although substorms may occur so frequently that they overlap in time and space, repeatable substorm features in each of these substorm phases have been identified for isolated substorms by observations from multiple ground stations and satellites distributed within the magnetosphere.

The very quiet magnetospheric condition corresponds typically to northward interplanetary magnetic field (IMF) and a small polar cap encircled by a belt-shaped region, called the auroral oval, consisting of auroral arcs and diffuse aurora. The polar cap is the region at Earth which contains magnetic field lines connected to the IMF. Growth phase usually begins with the start of southward turning of the IMF. In the ionosphere, the polar cap size increases as the ionospheric currents in the auroral oval intensify. In the magnetosphere, the cross-section of the magnetotail enlarges and plasma sheet thins in the near-Earth magnetotail ($\sim$5–15 R_E downstream), altering the dipolar field configuration there to become tail-like.

Figure 1 illustrates some key substorm expansion disturbances. At expansion onset, one of the nightside auroral arcs, typically the most equatorward one in the midnight sector, brightens suddenly, breaks up, and expands poleward (Figure 1a). This activity builds up to a bulge of highly structured auroral luminosity in which its westward end exhibits one or multiple surge forms. Large negative excursions are detected in the horizontal component at high-latitude ground magnetic stations while small positive excursions are found at mid-latitude and equatorial stations (Figure 1b). Micropulsations with periodicities in the 40-150s range, called Pi2, are often observed in these magnetograms. In the near-tail region (Figure 1c) and at the geostationary altitude (Figure 1d), the stretched tail-like configuration developed during the growth phase relaxes abruptly to a dipolar-like field geometry. This sudden change in magnetic field configuration is referred to as dipolarization, which is often preceded by a short interval (a few minutes) of large magnetic fluctuations suggestive of a turbulent state. Earthward injection of energetic particles and thickening of the plasma sheet are often seen accompanying dipolarization. These injected particles can interfere with satellite operations and may lead to failure of an entire satellite. The loss of the AT&T Telstar 401 satellite on January 10, 1997, is believed to be such a case. In the mid-tail region, thinning of the plasma sheet commences with substorm expansion onset. Transient fast plasma flows are detected in the central plasma sheet, along with impulsive and highly fluctuating electric and magnetic fields.

Recovery phase begins when the poleward expansion of the auroral bulge halts. In the mid-tail region ($\sim$15–80 R_E downstream), plasma sheet suddenly thickens with fast plasma flows, predominantly field-aligned, at the plasma sheet boundary. Magnetic field in the distant tail region (Figure 1f) shows a northward then southward swing. This field change is typically accompanied by tailward streaming of energetic electrons at the plasma sheet boundary. These features are interpreted as signatures of a plasmoid with closed loops of magnetic field lines.

One substorm scenario pertinent to this paper is the current disruption model, which invokes a physical process acting close to the Earth [e.g., *Lui*, 1991; *Erickson et al.*, 2000]. One idea of the current disruption process is a plasma instability, known as the cross-field current instability [*Lui et al.*, 1991, 1993], which limits the strength of the cross-tail current and shunts it to the ionosphere when the current density exceeds a certain threshold. In a simplified description, the process invokes waves, out of the thermal noise spectrum, being excited by the free energy associated with the strong current density. The current density reduces its strength as its energy is tapped by the excited waves. The current continuity condition promotes the generation of currents along the magnetic field, thus setting up a current system tantamount to short-circuiting the cross-tail current. Particles are transported earthward as the magnetic field in the current disruption region relaxes from the reduction of the local current density. Another consequence of the plasma instability is the acceleration of electrons along the magnetic field, allowing them to precipitate into the polar ionosphere and cause the visual auroras. The initial disturbance of current disruption propagates as a rarefaction wave down the magnetotail [*Chao et al.*, 1977] and instigates other regions downstream in the magnetotail to be activated, spreading the activity like an avalanche and giving rise to a large-scale disturbance for substorms.

3. THEORY OF THE CROSS-FIELD CURRENT INSTABILITY

This instability, CCI, is proposed with the viewpoint that a high current density from the relative drift between ions and electrons is a source of free energy available to disrupt current in the magnetotail/magnetosphere. Intuitively, CCI may be considered as consisting of the ion-Weibel instability (IWI; *Chang et al.*, 1990) and the modified two stream instability (MTSI; *McBride et al.*, 1972) modes. The former has a wavenumber nearly parallel to the magnetic field and the latter with a wavenumber nearly perpendicular to the

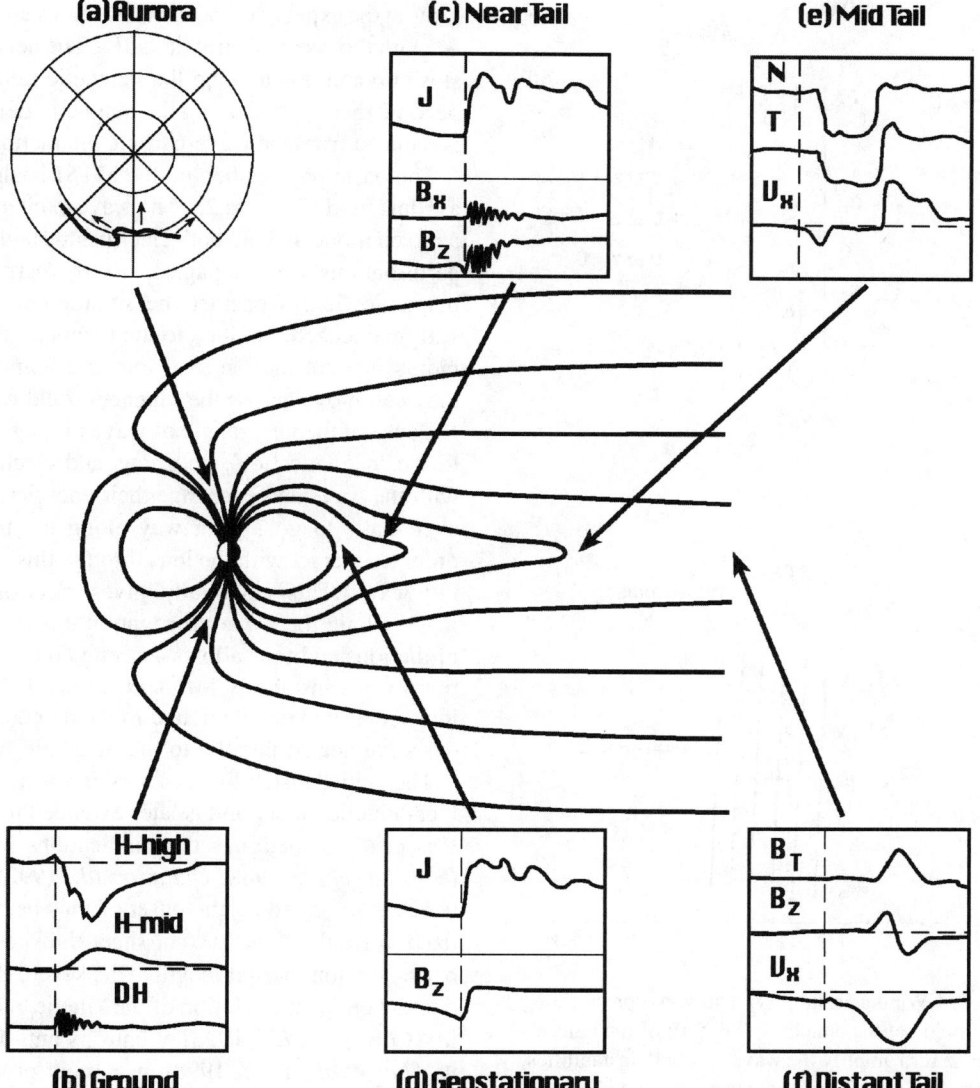

Figure 1. A diagram to illustrate several key substorm phenomenon. The substorm onset time is indicated by the vertical dashed lines in the panels. The ground activity is shown by the H-component of the magnetic field at high latitude stations (H-high) and at mid-latitude stations (H-mid). The micropulsation at onset is indicated by the ΔH trace. The increased fluxes of energetic particles at the geostationary altitude and in the near-tail region are indicated by the J traces. The magnetic field dipolarization in these regions is indicated by the B_z trace. Magnetic turbulence at the current sheet is often seen prior to dipolarization in the near-tail at substorm onset. Plasma sheet thinning in the mid-tail region is often seen by drops in number density (N trace) and temperature (T trace). Plasma flow (V_x trace) may occasionally be tailward before dropout and become earthward at plasma sheet recovery. Observation in the distant tail often show a transient increase in the total magnetic field magnitude (B_T trace) accompanied by a north-then-south excursion of the B_z component and tailward plasma flow (V_x trace). These are interpreted as signatures of a plasmoid being ejected down the tail (after *Lui*, 2000a).

magnetic field. These two modes are merely two special cases of CCI since both lie on the same dispersion surface (*Lui et al.*, 1991).

The basic physics of the IWI is rather simple and is illustrated in Figure 2a. In the electron rest frame, the free energy comes from the ion drift and electrons essentially provide the charge neutrality background. Let us consider an electromagnetic wave (out of noise) propagating in the z-direction along the magnetic field with the average ion drift in the y-direction while electrons are stationary. This wave is associ-

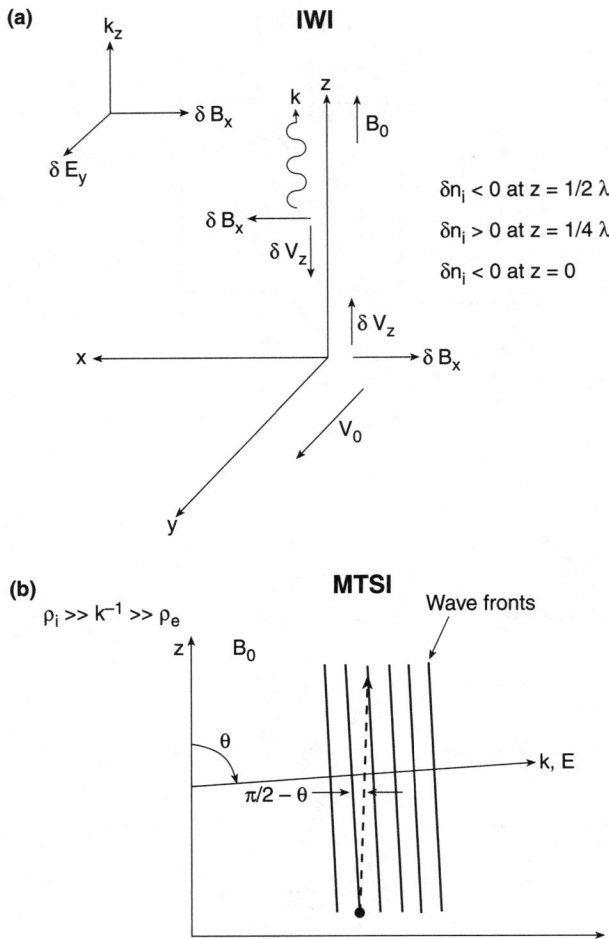

Figure 2. (a) The Ion Weibel instability. The wave propagating along the magnetic field causes bunching ($\delta n_i > 0$) of the current-carrying ions, which in turn amplify the wave. For both instabilities, the free energy from the relative drift between ions and electrons, i.e., electric current, is reduced in exciting the wave. (b) The modified two-stream instability. The excited wave propagates obliquely to the magnetic field as indicated by the wavenumber vector **k**. Ions have large gyroradius and can interact with the wave readily while electrons have small gyroradius and have to travel a long distance along the magnetic field line to interact with the wave (after *Lui et al.*, 1991).

ated with a magnetic perturbation δB_x in the x-direction and an electric perturbation δE_y in the y-direction. The ion drift (electric current) in the y-direction produces magnetic perturbation δB_x that has opposite signs at half a wavelength apart. This phase difference causes ion bunching in the z-direction due to the opposite signs of the Lorentz force $V_i \times \delta B_x$ at half a wavelength distance. Ions bunched or current filamentation in the z-direction amplifies the magnetic perturbations associated with the wave due to Ampere's law, allowing waves to grow at the expense of the ion drift energy since ions can interact with the wave electric field δE_y. The net effect is that ions slosh up and down along the magnetic field and eventually become thermalized along the magnetic field as the ion drift is reduced from the wave-particle interaction.

The basic physics behind the MTSI is also rather simple, as illustrated in Figure 2b. The wave excited by this MTSI is a mixed mode, having both electrostatic and electromagnetic perturbations and propagates nearly perpendicular to the magnetic field. Consider the situation when electrons are still magnetized, i.e., tied to the magnetic field by the small radius of gyro-motion, and ions are unmagnetized so that they can move across the magnetic field rather unimpeded. Because of the large angle of wave propagation with respect to the magnetic field, both ions and electrons can interact with the wave and exchange their energies. Electrons, however, have to go a long way along the magnetic field in order to interact with the ions through this oblique propagating wave. This essentially gives electrons an "effective mass" along the wave propagation direction similar to that of the ions and thus allows a strong coupling. Early simulation of this instability [*McBride et al.*, 1972] indicates that ions are heated perpendicular to the magnetic field and electrons are heated parallel to the magnetic field.

The initial instability analysis is formulated in terms of local kinetic theory and is later extended to nonlocal theory. It is also assumed, based on indications from observations (e.g., *Lui et al.*, 1988; *Ohtani et al.*, 1992), that the disturbance is triggered by the onset of ions becoming unmagnetized, a result of the current sheet thickness reducing to or below the ion inertial length. This severe thinning could be due to nonlinear evolution of the kinetic ballooning instability [*Cheng and Lui*, 1998] or entropy anti-diffusion instability [*Lee et al.*, 1995, 1998] or another process.

For the linear stability analysis, the magnetic field is taken to be in the z-direction. If waves propagating in the yz-plane are considered with perturbed quantities having a spatiotemporal dependence of $e^{i(k \cdot r - \omega t)}$, where $k = (0, k_y, k_z)$ is the wavenumber, r is the distance vector, ω is the wave frequency, and t is the time, the dispersion tensor $\mathbf{D}(k, \omega)$ is given by

$$D_{xx} = 1 - \frac{k^2 c^2}{\omega^2} + \frac{\omega_{pi}^2}{\omega^2} \xi_i Z(\xi_i) + \frac{\omega_{pe}^2}{\omega^2} e^{-\mu} \xi_e \times$$
$$\sum_{n=-\infty}^{\infty} \left\{ \frac{n^2}{\mu} I_n(\mu) - 2\mu \left[I'_n(\mu) - I_n(\mu) \right] \right\} Z(\xi_n),$$

$$D_{yy} = 1 - \frac{k_z^2 c^2}{\omega^2} + \frac{\omega_{pi}^2}{\omega^2} \times$$
$$\left\{ \xi_i Z(\xi_i) \cos^2 \theta - \left[\xi_i \sin \theta + V_{io}/u_i \right]^2 Z'(\xi_i) \right\}$$
$$+ \frac{\omega_{pe}^2}{\omega^2} \frac{e^{-\mu}}{\mu} \xi_e \sum_{n=-\infty}^{\infty} n^2 I_n(\mu) Z(\xi_n),$$

$$D_{zz} = 1 - \frac{k_y^2 c^2}{\omega^2} + \frac{\omega_{pi}^2}{\omega^2} \xi_i \times$$
$$\{Z(\xi_i)\sin^2\theta - \xi_i Z'(\xi_i)\cos^2\theta\}$$
$$+ \frac{\omega_{pe}^2}{\omega^2} 2\xi_e \left[\xi_e + e^{-\mu} \sum_{n=-\infty}^{\infty} \xi_n^2 I_n(\mu) Z(\xi_n)\right],$$

$$D_{xy} = -D_{yx} = i \frac{\omega_{pe}^2}{\omega^2} e^{-\mu} \xi_e \times$$
$$\sum_{n=-\infty}^{\infty} \{n[I'_n(\mu) - I_n(\mu)]\} Z(\xi_n),$$

$$D_{xz} = -D_{zx} = i \frac{\omega_{pe}^2}{\omega^2} \frac{2\Omega_e}{k_y v_e} \mu e^{-\mu} \xi_e \times$$
$$\sum_{n=-\infty}^{\infty} [I'_n(\mu) - I_n(\mu)] \xi_n Z(\xi_n),$$

$$D_{yz} = D_{zy} = k_y k_z c^2/\omega^2 - \frac{\omega_{pi}^2}{\omega^2} \xi_i \cos\theta \times$$
$$\{Z(\xi_i)\sin\theta + [\xi_i \sin\theta + V_{i0}/u_i]\xi_i Z'(\xi_i)\}$$
$$+ \frac{\omega_{pe}^2}{\omega^2} \frac{2\Omega_e}{k_y v_e} e^{-\mu} \xi_e \sum_{n=-\infty}^{\infty} n I_n(\mu)\xi_n Z(\xi_n) \quad (1)$$

Here, $\xi_i = (\omega - k_y V_{i0})/k u_i$, $\xi_e = \omega/k_z u_e$, $\xi_n = (\omega - n\Omega_e)/k_z u_e$, $\mu = k_y^2 \rho_e^2/2$, $\theta = \cos^{-1}(k_z/k)$. The other parameters of particle species α are u_α for the thermal speed, V_α for the drift speed, ρ_α for the gyroradius, Ω_α for gyrofrequency, and $\omega_{p\alpha}$ for the plasma frequency. The functions Z and I_n are the *Fried and Conte*'s plasma dispersion function and the modified Bessel function of order n, respectively. Numerical solution of the dispersion equation, as shown in Figure 3, indicates oblique whistler waves to be excited with a substantial growth rate (a small fraction of the lower hybrid frequency) when the relative motion between ions and electrons constituting the cross-tail current becomes a significant fraction of the ion thermal speed.

Quasi-linear analyses for two simpler situations were carried out to determine the saturation levels and the amount of current reduction resulting from onsets of these instabilities [*Lui et al.*, 1993; *Yoon and Lui*, 1993]. The moment method was used, i.e., the particle velocity distribution is assumed to retain its initial functional form of a drifting Maxwellian as time evolves. Changes in temperature and drift velocity are included. Since temperature and drift velocity can be calculated by taking the moments of the distribution function, these moments can be fed back into the linear dispersion relation to provide a self-consistent wave growth as time evolves. With this approach, the temporal evolutions of the moment parameters and magnetic field energy are as follows:

$$\frac{dV_{i0}}{dt} = \frac{2e^2}{m_i} \frac{V_{i0}}{T_{i\parallel}} \int_0^\infty dk \frac{\gamma_k \delta B_k^2}{c^2 k^2} \frac{\operatorname{Re} Z'(i\gamma_k/ku_{i\parallel})}{1 + |D_{xy}/D_{xx}|^2}, \quad (2)$$

$$\frac{dV_{e0}}{dt} = \frac{2e^2}{m_e} \frac{V_{e0}}{T_{e\parallel}} \int_0^\infty dk \frac{\gamma_k \delta B_k^2}{c^2 k^2} \frac{\operatorname{Re} Z'(i\gamma_k/ku_{e\parallel})}{1 + |D_{xy}/D_{xx}|^2}, \quad (3)$$

$$\frac{dT_{i\perp}}{dt} = \frac{2e^2}{m_i} \frac{T_{i\perp}}{T_{i\parallel}} \int_0^\infty dk \frac{\gamma_k \delta B_k^2}{c^2 k^2} \times$$
$$(\operatorname{Re} Z'(i\gamma_k/ku_{i\parallel}) + T_{i\parallel}/T_{i\perp}), \quad (4)$$

$$\frac{dT_{i\parallel}}{dt} = -\frac{2e^2}{m_i} \frac{T_{i\perp}}{T_{i\parallel}} \int_0^\infty dk \frac{\gamma_k \delta B_k^2}{c^2 k^2} \times$$
$$\left(1 + \frac{m_i V_{i0}^2/T_{i\perp}}{1 + |D_{xy}/D_{xx}|^2}\right) \operatorname{Re} Z'(i\gamma_k/ku_{i\parallel}), \quad (5)$$

$$\frac{dT_{e\perp}}{dt} = -\frac{2e^2}{m_e} \int_0^\infty dk \frac{\delta B_k^2}{c^2 k^2} \left(\gamma_k \frac{T_{e\perp}}{T_{e\parallel}} \times\right.$$
$$\left\{1 - \frac{\Omega_e}{ku_{e\parallel}}\left[\frac{2\gamma_k}{\Omega_e} \operatorname{Im} Z\left(\frac{i\gamma_k + \Omega_e}{ku_{e\parallel}}\right) - \operatorname{Re} Z\left(\frac{i\gamma_k + \Omega_e}{ku_{e\parallel}}\right)\right]\right\}$$
$$- \left(1 - \frac{T_{e\perp}}{T_{e\parallel}}\right)\left\{\gamma_k + \frac{\Omega_e^2}{ku_{e\parallel}} \times\right.$$
$$\left.\left.\left[\frac{2\gamma_k}{\Omega_e} \operatorname{Re} Z\left(\frac{i\gamma_k + \Omega_e}{ku_{e\parallel}}\right) + \operatorname{Im} Z\left(\frac{i\gamma_k + \Omega_e}{ku_{e\parallel}}\right)\right]\right\}\right), \quad (6)$$

$$\frac{dT_{e\parallel}}{dt} = \frac{4e^2}{m_e} \int_0^\infty dk \frac{\delta B_k^2}{c^2 k^2} \left(\gamma_k \frac{T_{e\perp}}{T_{e\parallel}} \times\right.$$
$$\left\{1 - \frac{\Omega_e}{ku_{e\parallel}}\left[\frac{\gamma_k}{\Omega_e} \operatorname{Im} Z\left(\frac{i\gamma_k + \Omega_e}{ku_{e\parallel}}\right) - \operatorname{Re} Z\left(\frac{i\gamma_k + \Omega_e}{ku_{e\parallel}}\right)\right]\right\}$$
$$- \left(1 - \frac{T_{e\perp}}{T_{e\parallel}}\right)\frac{\Omega_e^2}{ku_{e\parallel}} \times$$
$$\left.\left[\frac{\gamma_k}{\Omega_e} \operatorname{Re} Z\left(\frac{i\gamma_k + \Omega_e}{ku_{e\parallel}}\right) + \operatorname{Im} Z\left(\frac{i\gamma_k + \Omega_e}{ku_{e\parallel}}\right)\right]\right)$$
$$- \frac{2e^2 V_{e0}^2}{T_{e\parallel}} \int_0^\infty dk \frac{\gamma_k \delta B_k^2}{c^2 k^2} \frac{\operatorname{Re} Z'(i\gamma_k/ku_{e\parallel})}{1 + |D_{xy}/D_{xx}|^2}, \quad (7)$$

$$\frac{\partial}{\partial t} \delta B_k^2 = 2\gamma_k \delta B_k^2. \quad (8)$$

Numerical solutions based on Eqs. (2)–(8) show that the current density can be reduced by ~15–28% of its initial

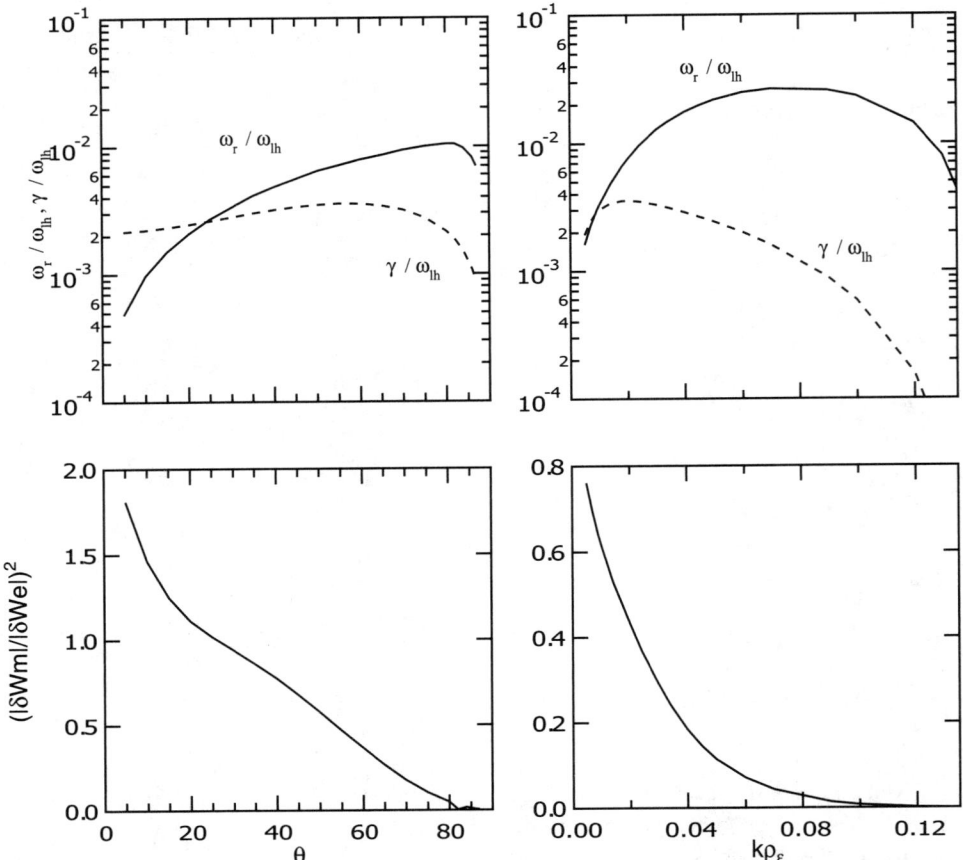

Figure 3. Local linear analysis of the cross-field current instability for a drift speed equal to half of the ion thermal speed. Comparable growth rates exist over a broad band of wave propagation angle with respect to the local magnetic field (after *Lui et al.*, 1991).

value. Figure 4 shows the quasi-linear calculations for a specific set of plasma parameters. The anomalous resistivity η_a is defined not by the wave amplitude at saturation but by the rate of decrease in the ion drift speed and is given by

$$\eta_a = 6.93 \times 10^{-7} \frac{B_0\,(\mathrm{nT})}{n_i\,(\mathrm{cm}^{-3})} \frac{v_c}{\Omega_i}\,\mathrm{s} \qquad (9)$$

Here v_c denotes the anomalous collision rate resulting from wave-particle interaction due to the instability activity. It has a value in the range of 10^{-7}–10^{-6} s, dependent on specific current sheet parameters such as ion and electron temperatures, relative drift speed between ions and electrons, number density, and magnetic field strength. These results are obtained from local analysis and may need to be revised for nonlocal analysis. However, a preliminary evaluation shows the local approximation to be valid for certain waves excited by this instability and thus applicable to certain thin current sheet configurations [*Lui*, 2000b].

4. NUMERICAL SIMULATIONS

Plasma simulations play an important role in understanding plasma instabilities. It serves to verify analytic theories, at least in the linear stage. It also provides valuable information on the nonlinear evolution of instabilities, often giving insights to the mechanism responsible for the saturation of the instability in the nonlinear regime. However, it should be reminded that numerical simulations have their intrinsic limitations. Artifacts due to the numerical scheme, compromised values of plasma parameters adopted in the simulation, the finite size of the simulation domain, and the boundary conditions imposed could all influence the outcome of the numerical experiment. Caution is required to interpret simulation results to avoid being misled by these potential artifacts. This section discusses numerical simulation results relating to the CCI discussed in the previous section. As a first step to probe the nonlinear consequence of CCI, only one-dimensional (1D) simulations are performed. Naturally,

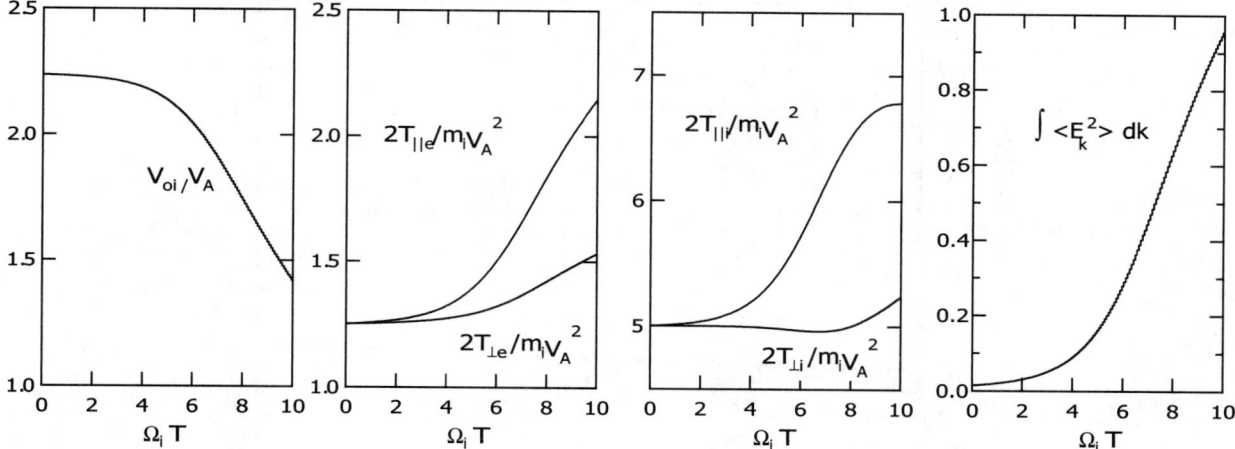

Figure 4. Local quasilinear calculation of the cross-field current instability showing the rapid increase of the parallel temperatures of both ions and electrons as the instability progresses. The perpendicular temperatures of ions and electrons are not affected significantly (after *Lui et al.*, 1993; *Yoon and Lui*, 1993).

the limitation of 1D raises the question whether or not the simulations portray adequately the realistic situation of a thin current sheet. However, if local approximation for this instability is satisfied in certain current sheet cases, then the 1D simulations may still provide some insight into the evolution of these current sheets. The appropriateness of local approximation has been examined by *Lui* [2000b] and is found to be valid for certain waves excited by CCI. The accuracy of these simulation results needs to be evaluated in the future with 3D simulations of a thin current sheet.

The simulation of CCI is based on the numerical scheme of cloud-in-cell (CIC) particle-in-cell (PIC) procedure, as described in *Birdsall and Langdon* [1985]. We employ a relativistic 1D fully electromagnetic CIC-PIC code. In this code, the trajectory of each particle is followed by solving the equation of motion in the presence of the self-consistent electric and magnetic fields, which are determined by solving Maxwell's equations with charge and current densities due to particles collected on a grid. Each simulated particle has a finite-size specified by a shaping function. The force exerting on each particle is calculated by interpolating the fields at its nearby grid points. The contribution of each particle to the charge and current densities is collected at the nearby grid points through interpolation also. Relativistic equation of motion is used in the code. The Maxwell equations are solved through Fast Fourier Transform (FFT) by decomposing the electric field and current into longitudinal and traverse components.

To proceed progressively on particle simulation of CCI, we first perform simulation for simple 1D situations. The first numerical experiment is on the ion Weibel instability (IWI) [*Chang et al.*, 1990]. The system length and the magnetic field are in the z-direction for this simulation. A uniform isotropic plasma is initialized and the cell size corresponds to one Debye length λ_d. The adopted plasma parameters for this run are $\Omega_e/\omega_{pe} = 0.078$, $m_i/m_e = 1836$, speed of light $c = 10u_e$, $u_i/u_e = 0.0467$, $V_i = 0.2u_e$. The system size is chosen to be $2048\lambda_d$. Periodic boundary condition is imposed in the simulation axis. A total of ~33,000 particles for each species is used. Solving the linear dispersion equation for IWI with these plasma parameters indicates a growth rate of ~$2.2 \times 10^4 \omega_{pe}$, corresponding to an e-folding growth time τ of 4.5×10^4 time steps with a step size of $0.1\omega_{pe}^{-1}$. Figure 5 shows the result from a simulation running for ~10τ. Shown in the figure are the magnetic and electrostatic field energies, averaged ion drift speed, and ion temperatures (parallel and perpendicular). Since IWI is a purely growing electromagnetic mode, only the magnetic energy is expected to increase exponentially while the electrostatic field energy is expected to be relatively unaffected. This is indeed what is seen from the simulation. The initial slight drop in magnetic energy is a result of not initializing the system with a quiet start procedure, which does not affect the later development. The growth rate of the magnetic energy, which is twice of the growth rate of the instability (see Eq. (8)), is determined to be ~$4.3 \times 10^{-4} \omega_{pe}$, corresponding to a growth rate for the instability of $2.1 \times 10^{-4} \omega_{pe}$. This value compares well with the theoretical expectation of ~$2.2 \times 10^{-4} \omega_{pe}$. Current density, as represented by the ion drift speed, is reduced on the average by ~28% at saturation in this case. This drift reduction occurs simultaneously with ion heating parallel to the magnetic field while the ion temperature perpendicular to the magnetic field remains rather unaffected. These characteristics are consistent with the expected evolution of the instability.

Figure 6 shows the ion phase space density in both the v_y and v_z phase space at four different time snapshots,

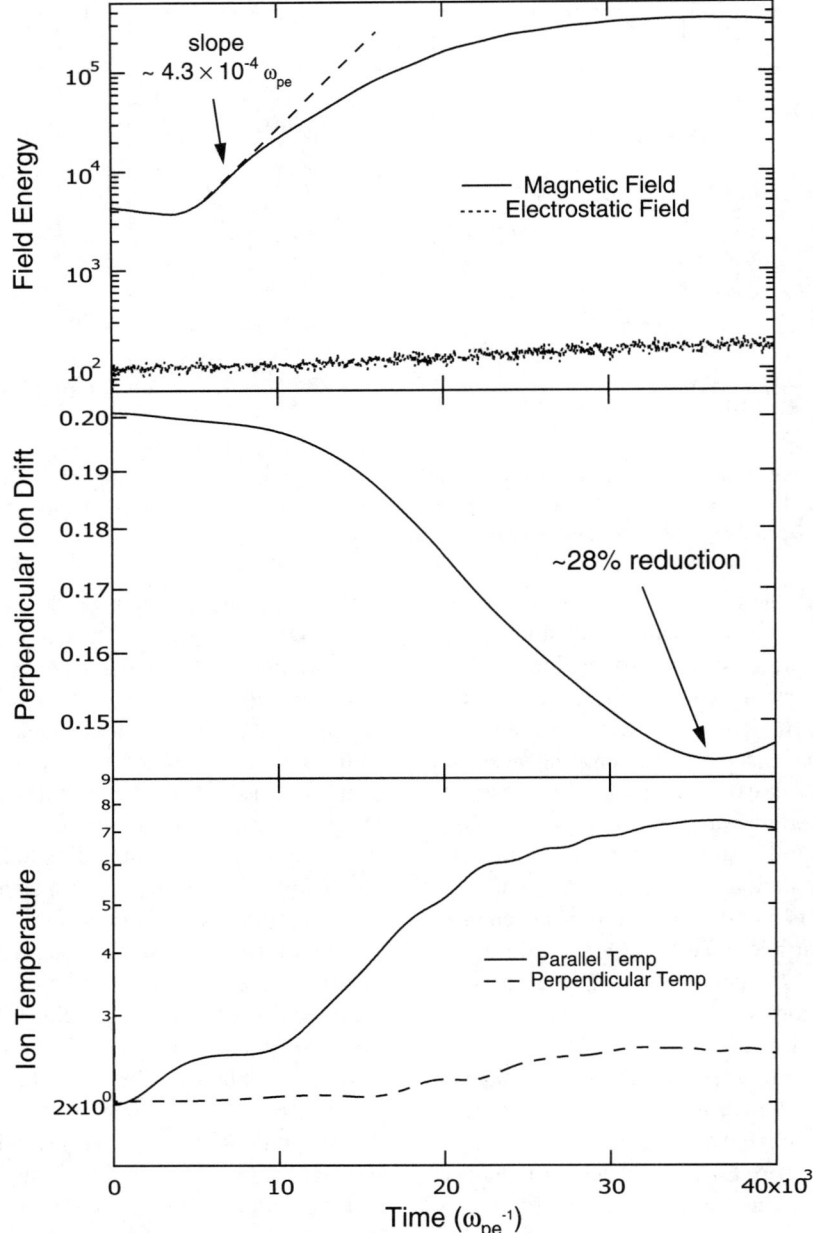

Figure 5. Temporal development of magnetic and electrostatic field energies, ion drift speed, and ion temperature (parallel and perpendicular) for the IWI simulation. Ions are heated in the direction parallel to the magnetic field as the instability grows with reduction of ion drift.

namely, $\omega_{pe}t = 0$, 10,000, 16,000, and 20,000. Ion bunching becomes noticeable at $\omega_{pe}t = 10,000$ and is clearly evident at $\omega_{pe}t = 20,000$. At the location of ion bunching, significant reduction of the ion drift is seen. Filamentation of the current layer is quite evident in both phase space plots.

The next step in the simulation is to investigate the modified two-stream instability (MTSI). In this simulation, the magnetic field is in the z-direction and the system length is tilted by 85° to the z-direction (i.e., in the wavenumber direction). The adopted plasma parameters for this run are $\Omega_e/\omega_{pe} = 0.5$, $m_i/m_e = 500$, speed of light $c = 10u_e$, $u_i/u_e = 0.0541$, $V_i = 0.5u_e$. The system size is chosen to be $256\lambda_d$. A total of ~16,000 particles for each species is used. Solving the linear dispersion equation for MTSI with these plasma parameters indicates a growth rate of $\sim 1.1 \times 10^2 \omega_{pe}$, corresponding to an e-folding growth time τ of 9.1×10^2 time steps.

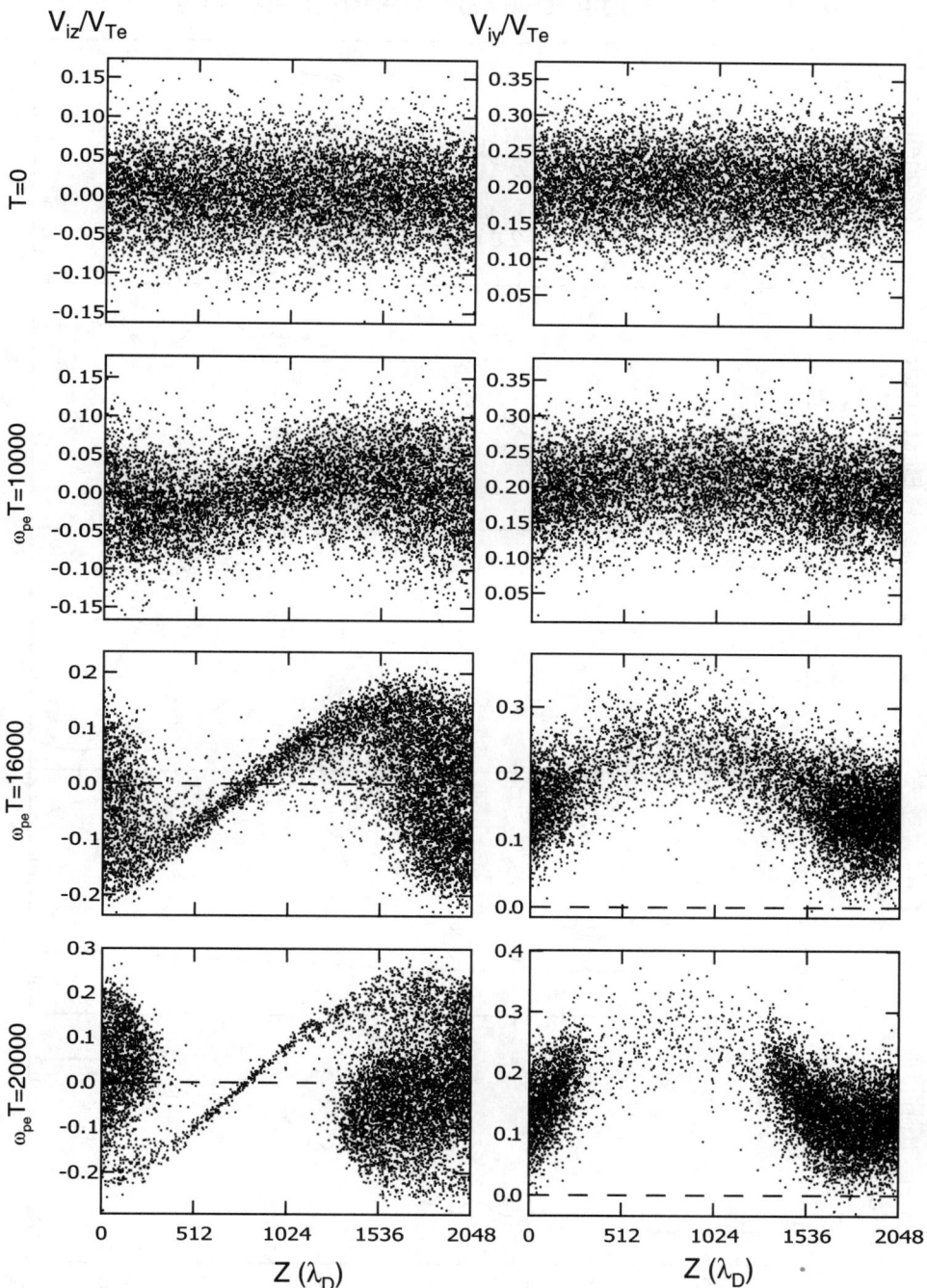

Figure 6. Ion phase space density plot for four time snapshots from the IWI simulation. Current filamentation (clumping of phase points at certain locations along the axis) and ion heating along the magnetic field are evident.

Figure 7 gives the temporal development of magnetic and electrostatic energies of waves, ion drift speed, ion and electron temperatures. The mixed mode of MTSI is reflected in the simultaneous increase in both the magnetic and electrostatic energies. The growth rate of these energies is found to be $\sim 2.6 \times 10^{-2}\omega_{pe}$, indicating an instability growth rate of $1.3 \times 10^{-2}\omega_{pe}$, in good agreement with the theoretical expectation of $\sim 1.1 \times 10^{-2}\omega_{pe}$. Current density at the nonlinear stage is reduced by $\sim 60\%$ in this case. This current reduction occurs simultaneously with ion heating perpendic-

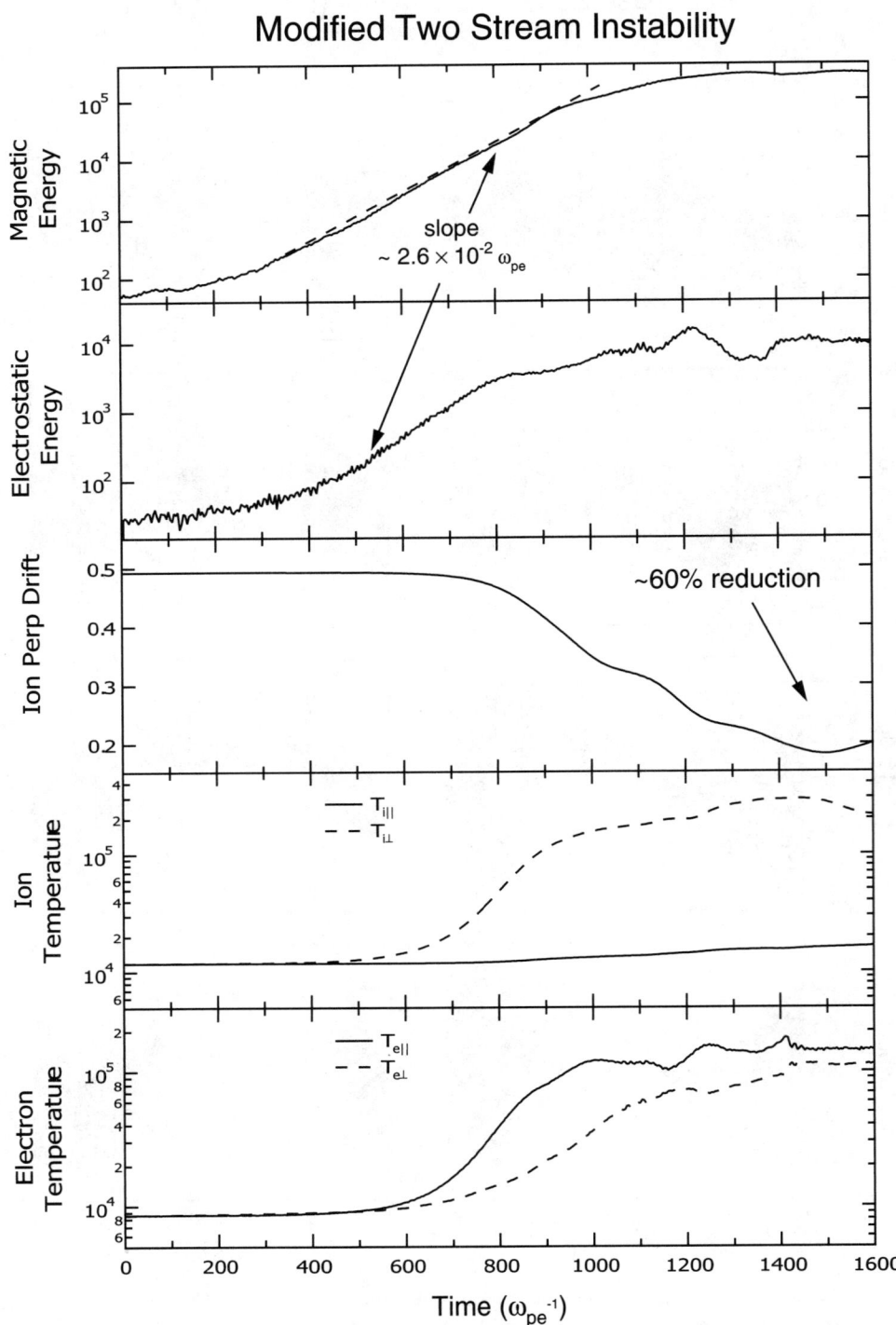

Figure 7. Temporal development of magnetic and electrostatic field energies, ion drift speed, and temperatures (parallel and perpendicular) of ions and electrons for the MTSI simulation. As the instability progresses, the ions are heated perpendicular to the magnetic field while electrons are heated in the parallel direction first but followed shortly by heating in the perpendicular direction.

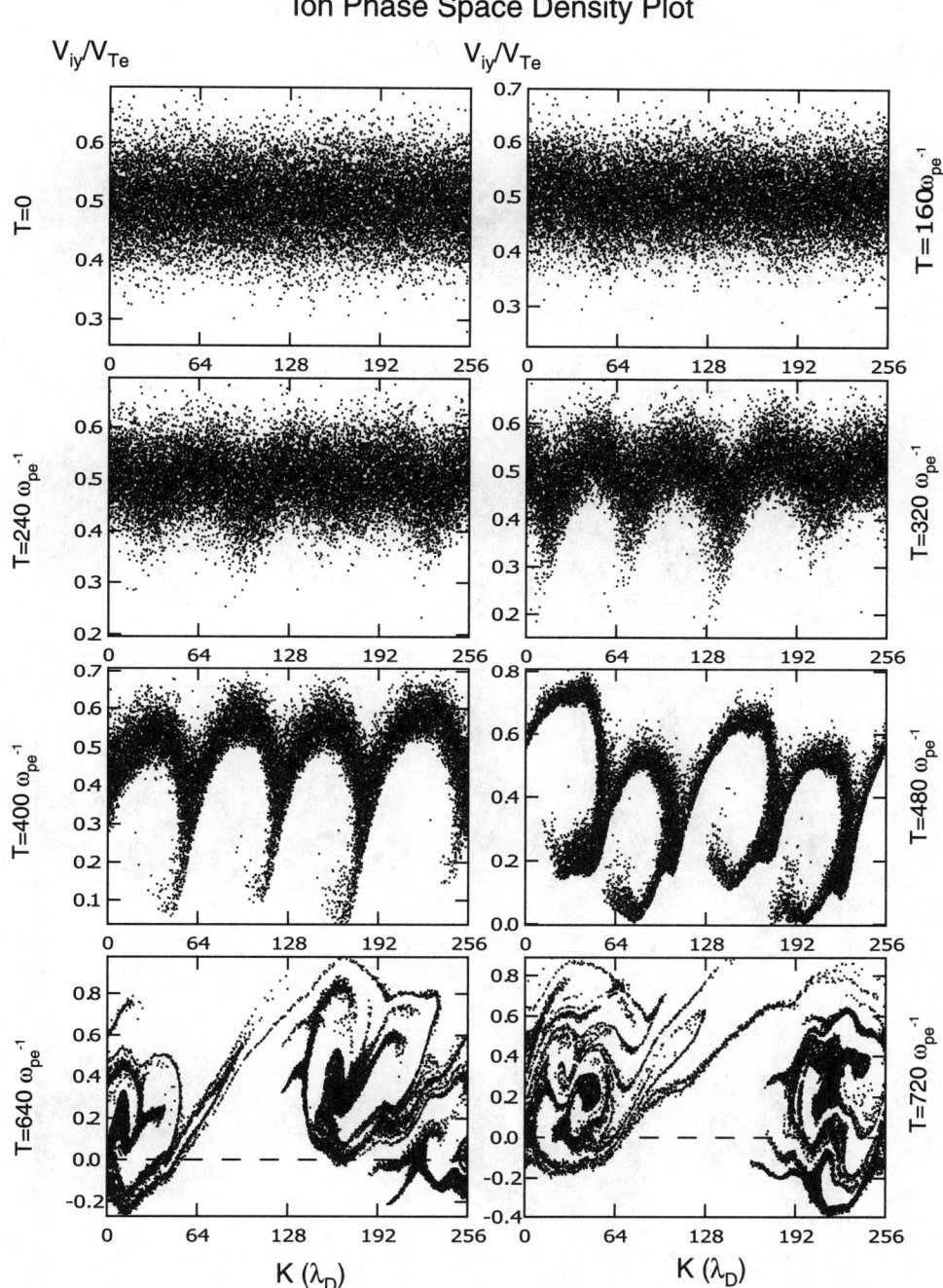

Figure 8. Ion phase space density plot for four time snapshots from the MTSI simulation. Note the gradual change of the velocity scale in the V_{iy} phase space plot showing the heating perpendicular to the magnetic field. Current filamentation is also evident at the nonlinear stage.

ular to the magnetic field while its parallel temperature remains rather unaffected. On the other hand, the electrons are heated first in the parallel direction but subsequently in the perpendicular direction as well.

At the end of the numerical run, the electron parallel temperature is only slightly above its perpendicular temperature. With the exception of perpendicular electron heating at the later stage of instability activity, the characteristics

136 PARTICLE ACCELERATION FROM PLASMA INSTABILITY

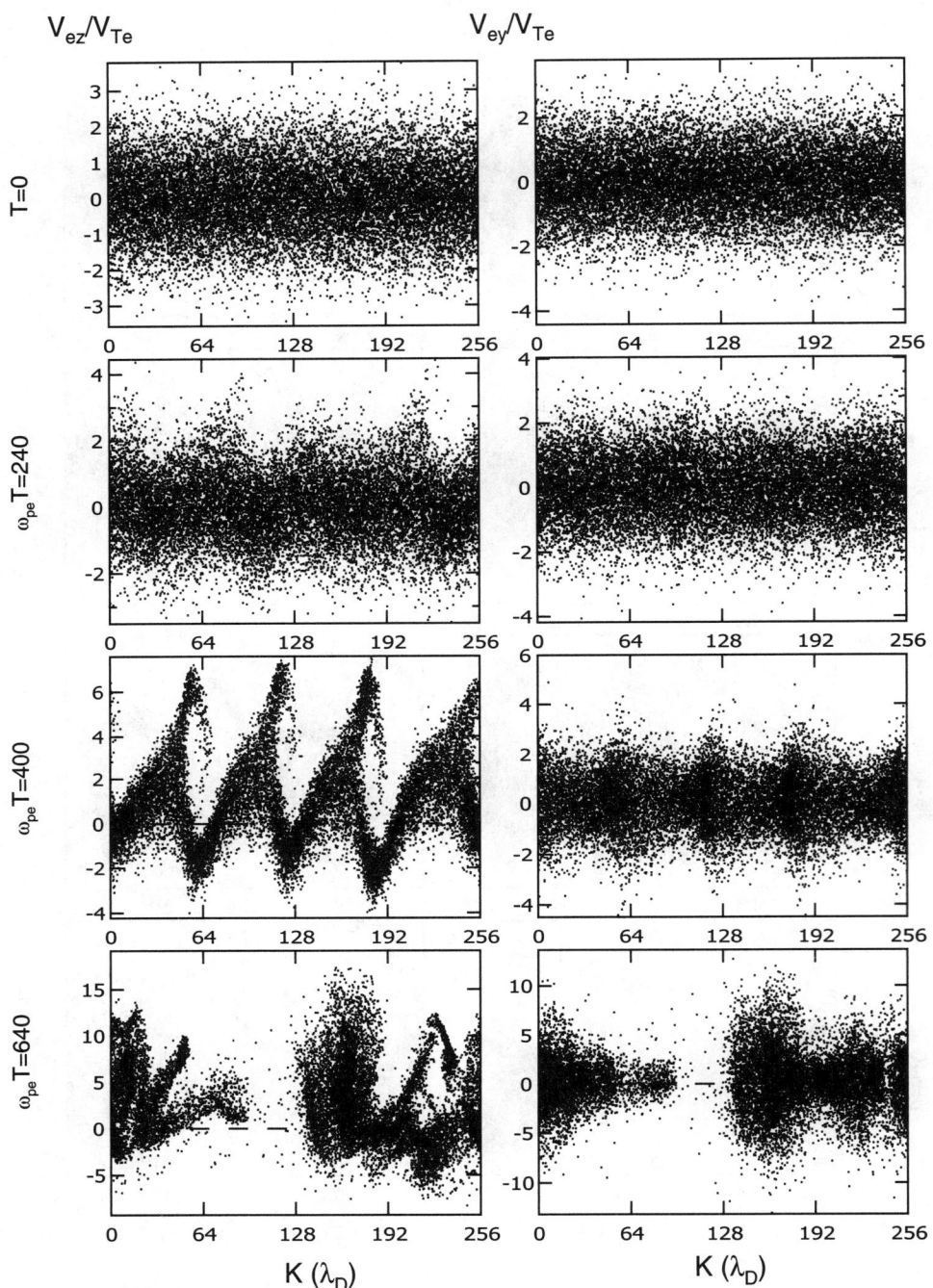

Figure 9. Electron phase space density plot for four time snapshots from the MTSI simulation. Note the gradual change of the velocity scale in these panels showing the electron energization by the instability growth. Vortices in the phase plot indicate electron trapping.

revealed from this simulation are in general agreement with the expected evolution of the instability.

Figure 8 shows the ion phase space plot in both the v_y and v_z phase space at four different time snapshots, namely, $\omega_{pe}t = 0$, 240, 400, and 640. The spatial location shown is along the simulation axis (the wavenumber direction), i.e., 85° from the magnetic field line in the z-direction. Ion bunching along the simulation axis is clearly evident at $\omega_{pe}t = 400$

and 640. At the location of ion bunching, large reduction of the ion drift and significant heating are seen. Filamentation of the current layer is again quite evident in these phase space plots. The corresponding electron phase space plot is given in Figure 9. At the location of ion bunching and ion drift reduction, electrons are heated along the magnetic field. Electron heating perpendicular to the magnetic field can be seen to develop later as well.

The preceding simulation results are encouraging in accounting for thin current sheet dynamics and particle acceleration through the cross-field current instability. However, it should be borne in mind that they should not be taken as confirmation of the relevance of this instability since only one-dimensional simulations are performed. Eventual assessment of this instability requires three-dimensional simulations in an appropriate current sheet configuration.

5. SUMMARY AND CONCLUDING REMARKS

This paper presents a brief review of substorm phenomena in the Earth's magnetosphere and local kinetic analysis of the cross-field current instability. This is followed by some recent numerical simulation results of particle acceleration due to this instability. The free energy of this instability comes from the current and can be excited in thin current sheets with strong current densities. It has a short development time-scale and thus can account for impulsive energetic phenomena. Its relevance to the initiation of substorm expansion and generation of substorm currents has been expounded. Since thin current sheets are abundant in astrophysical space plasmas, this instability may find application in other impulsive cosmic phenomena as well. Naturally, the instability considered here is only one of many plasma instabilities that can accelerate charged particles efficiently.

Acknowledgments. This work was supported by the Atmospheric Sciences Division of the NSF Grant ATM-0135667 and by the NASA Grant NAG5-10475 to The Johns Hopkins University Applied Physics Laboratory. The author appreciates the past collaborations with P. H. Yoon, C.-L. Chang, C. S. Wu, and K. Papadopoulos, from which much of the idea on plasma instability theory has emerged.

REFERENCES

Akasofu, S.-I., The development of the auroral substorm, *Planet. Space Sci., 12*, 273-282, 1964.

Birdsall, C. K. and A. B. Langdon, *Plasma physics via computer simulation*, McGraw-Hill, New York, 1985.

Chang, C. L., H. K. Wong, and C. S. Wu, Electromagnetic instabilities attributed to a cross-field ion drift, *Phys. Rev. Lett., 65*, 1104-1107, 1990.

Chao, J. K., J. R. Kan, A. T. Y. Lui, S.-I. Akasofu, A model of thinning of the plasma sheet, *Planet. Space Sci., 25*, 703-710, 1977.

Cheng, C. Z., and A. T. Y. Lui, Kinetic ballooning instability for substorm onset and current disruption observed by AMPTE/CCE, *Geophys. Res. Lett., 25*, 4091-4094, 1998.

Erickson, G. M., N. C. Maynard, W. J. Burke, G. R. Wilson, and M. A. Heinemann, Electromagnetics of substorm onsets in the near-geosynchronous plasma sheet, *J. Geophys. Res., 105*, 25265-25290, 2000.

Lee, L. C., L. Zhang, G. S. Choe, and H. J. Cai, Formation of a very thin current sheet in the near-earth magnetotail and the explosive growth phase of substorms, *Geophys. Res. Lett., 22*, 1137-1140, 1995.

Lee, L. C., L. Zhang, A. Otto, G. S. Choe, and H. J. Cai, Entropy antidiffusion instability and formation of a thin current sheet during geomagnetic substorms, *J. Geophys. Res., 103*, 29419-29428, 1998.

Lui, A. T. Y., A synthesis of magnetospheric substorm models, *J. Geophys. Res., 96*, 1849-1856, 1991.

Lui, A. T. Y., Tutorial on geomagnetic storms and substorms, *IEEE Trans. on Plasma Phys., 28 (6)*, 1854-1866, 2000a.

Lui, A. T. Y., Electric current approach to magnetospheric dynamics and the distinction between current disruption and magnetic reconnection, *Magnetospheric Current Systems*, AGU Monograph 118, AGU, Washington, DC, pp. 31-40, 2000b.

Lui, A. T. Y., R. E. Lopez, S. M. Krimigis, R. W. McEntire, L. J. Zanetti, and T. A. Potemra, A case study of magnetotail current sheet disruption and diversion, *Geophys. Res. Lett., 15*, 721, 1988.

Lui, A. T. Y., C.-L. Chang, A. Mankofsky, H.-K. Wong, and D. Winske, A cross-field current instability for substorm expansions, *J. Geophys. Res., 96*, 11389-11401, 1991.

Lui, A. T. Y., P. H. Yoon, and C.-L. Chang, Quasi-linear analysis of ion Weibel instability, *J. Geophys. Res., 98*, 153-163, 1993.

McBride, J. B., E. Ott, J. P. Boris, and J. H. Orens, Theory and simulation of turbulent heating by the modified two-stream instability, *Phys. Fluids, 15*, 2367, 1972.

McPherron, R. L., Growth phase of magnetospheric substorms *J. Geophys. Res., 75*, 5592-5599, 1970.

Ohtani, S., S. K. Takahashi, L. J. Zanetti, T. A. Potemra, R. W. McEntire, and T. Iijima, Initial signatures of magnetic field and energetic particle fluxes at tail reconfiguration: Explosive growth phase, *J. Geophys. Res., 97*, 19311-19324, 1992.

Yoon, P. H. and A. T. Y. Lui, Nonlinear analysis of generalized cross-field current instability, *Phys. Fluids B, 5 (3)*, 836-853, 1993.

A.T.Y. Lui, The Johns Hopkins University Applied Physics Laboratory, 11100 Johns Hopkins Road., Laurel, Maryland 20723-6099. (tony.lui@jhuapl.edu)

The Role of Electron Acceleration in Quick Reconnection Triggering

Masaki Fujimoto

Department of Earth and Planetary Sciences, Tokyo Institute of Technology, Tokyo, Japan

Iku Shinohara

Japan Aerospace Exploration Agency, Institute of Space and Astronautical Science, Kanagawa, Japan

We have recently found a quick spontaneous triggering mechanism of magnetic reconnection in an ion-scale current sheet, in which electron acceleration plays a crucial role. In the quick triggering, the lower hybrid drift waves are first excited at the outer-edges of the current sheet. The non-linear evolution of the lower hybrid drift instability modifies the current density distribution and thus the magnetic field profile. It is the inductive electric field associated with this change in the magnetic structure that efficiently accelerates the meandering electrons around the magnetic neutral sheet. The accelerated electrons sustain the imbedded thin current layer that is also formed in response to the re-arrangement of the current density. Such a current layer is highly unstable to the tearing mode and thus is crucial for making the quick triggering available.

INTRODUCTION

Magnetic reconnection plays crucial roles in various explosive phenomena in the plasma universe. Recent observational progresses, in-situ observations in the Earth's magnetosphere and X-ray imaging of the solar corona, have provided compelling evidence that magnetic reconnection truly occurs in the collisionless plasmas and that magnetic reconnection is an indispensable part of magnetospheric substorms and solar flares. The most impressive feature of these explosive phenomena is sudden breakup, however, the triggering process of magnetic reconnection has been only poorly understood (*Drake et al.* 2001).

In collisionless plasmas, an anomalous dissipation process is necessary for driving magnetic reconnection. So far, a number of theories for the anomalous dissipation process have been proposed, with the anomalous resistivity generated by the lower hybrid drift instability (LHDI) being the most popular candidate. However, using the Geotail wave data, *Shinohara et al.* (1998) have shown that the observed LHD waves cannot provide sufficient resistivity for quick triggering of magnetic reconnection in the magnetotail. Turbulence in an electron scale current sheet has been thought to be important for providing sufficient dissipation to trigger fast magnetic reconnection (e.g., *Drake et al.* 1994, *Biskamp et al.* 1996). While there is observational evidence that the magnetotail current sheet thickness becomes as thin as comparable to the relevant ion inertial length prior to a substorm onset (*Sergeev et al.* 1990), whether it thins further down to electron scale to initiate reconnection is an open question. Moreover, the solar corona situation is much more serious than the magnetotail, where even the ion inertial length is minuscule compared to the macroscopic current sheet scales (*Terasawa et al.* 2000). Recent results from the laboratory experiments of magnetic reconnection also suggest that the typical thickness of a reconnecting current layer is much broader than the electron inertial length (*Yamada et al.* 2000). It is therefore an interesting question to ask if some dissipation mechanisms set in already at an ion scale current sheet.

In a study by high mass-ratio three-dimensional (3-D) full particle simulations searching for a dissipation process that can lead to fast reconnection onset in an ion scale current

Particle Acceleration in Astrophysical Plasmas
Geophysical Monograph Series 156
Copyright 2005 by the American Geophysical Union
10.1029/156GM17

Hall current loop. *Nagai et al.* [2001] demonstrated that the counterstreaming electrons in the outermost layer are common characteristics of magnetic reconnection at substorm onsets and that outflowing currents of 6–13 nA m^{-2} exist in the outermost layer of the plasma sheet/tail lobe boundary, in the immediate vicinity of the magnetic reconnection site. These currents produce the quadrupole structure in the magnetic field. Hence, the Hall current system for magnetic reconnection has been established observationally [*Nagai et al.*, 2003].

The purpose of this paper is to provide an overview of the key findings of the *Nagai et al.* [2001] work and then further discuss the characteristics of the Hall current system in the magnetotail. Observations in the magnetotail are provided by the spacecraft Geotail [*Nishida*, 1994]. The magnetic field data were obtained from the magnetic field experiment MGF [*Kokubun et al.*, 1994], and the plasma data were obtained from the low-energy plasma experiment LEP [*Mukai et al.*, 1994].

2. DECEMBER 10, 1996, EVENT

2.1. Overview

Figure 1 presents Geotail magnetic field and plasma observations for the period 1500–2000 UT on December 10, 1996, with ground substorm activity. This event is one of the highly energetic ones in the Geotail observations of the magnetotail during 1996–2002. The magnetic field data (values averaged over 12 s) are Bx, By, Bz, and Bt (the total magnetic field) in the GSM coordinates. The plasma data (time resolution of 12 s) are the x component of ion flow velocity Vx derived from ions with energies <40 keV. Geotail is located near the midnight meridian at a radial distance of 25.5 R_E. The ground magnetic field data from Kakioka (magnetic latitude is 26.8° and local midnight is 1440 UT) indicate a substorm starting near 1700 UT. Pi2 pulsation activities show this is a multiple-onset substorm, with onsets at 1652, 1703, and 1719 UT, and that the substorm intensifies at 1740 UT. The Los Alamos National Laboratory geosynchronous spacecraft 1994-084 near the midnight meridian (local midnight of this spacecraft is 1700 UT) observes increases in the electron and proton fluxes near 1650, 1700, 1720, and 1740 UT. In the period 1500–1650 UT, there is no Pi2 pulsation activity and no evident change in particle fluxes at synchronous orbit. Furthermore, there is no substorm onset on the ground or at synchronous orbit in the period 1800–1900 UT.

2.2. Counterstreaming Electrons

First, we examine the Geotail data for 1740–1810 UT, which corresponds to the substorm intensification period. As seen in Figure 1, Geotail observes fast earthward ion flows (almost field-aligned) near the plasma sheet/tail lobe boundary in the Southern Hemisphere ($Bx < 0$) for the interval 1742–1746 UT. Then, Geotail observes fast tailward ion flows near the equatorial plane and near the plasma sheet/tail lobe boundary in the Northern Hemisphere for the interval 1747–1751 UT. Bz becomes negative during this interval. Finally, Geotail observes fast earthward ion flows, mostly in the Southern Hemisphere ($Bx < 0$), at 1752–1804 UT. Geotail is located briefly near the plasma sheet/tail lobe boundary of the Northern Hemisphere at about 1754 UT. Hence, Geotail surveyed three different boundary regions in this period: the Northern Hemisphere earthward of the magnetic reconnection site (EN), the Southern Hemisphere earthward of the magnetic reconnection site (ES), and the Northern Hemisphere tailward of the magnetic reconnection site (TN). The ion velocities in Figure 1 are calculated as plasma moment values from LEP. Ion and electron energy–time diagrams are presented in Plate 5 of *Nagai et al.* [2001]. Ion energies exceed the upper limit of the LEP instrument (40 keV). In addition, simultaneously present are cold, dense, downward-flowing ions. These ions are common in the boundary region [*Nagai et al.*, 1998a, 2001]. The speed of the field-aligned ion component occasionally exceeds 2800 km/s (the upper velocity limit of the LEP instrument). Furthermore, electrons show flat-top velocity distributions near the equatorial plane, indicating strong acceleration and heating (the electron temperature exceeds 5 keV).

Plate 1 presents electron velocity distribution functions (each distribution function is taken over 12 s) in the plane that includes the magnetic field vector and the ion bulk flow vector. The horizontal red line is parallel to the magnetic field direction; earthward is to the left and tailward to the right. To clarify the differences between the parallel component and the antiparallel component, cuts of distributions (in the red line of each distribution) are also presented.

Three representative electron distribution functions from the three boundary regions are presented in Plate 1. The distribution at 1754:18 UT (upper left panel, EN) is obtained in the Northern Hemisphere earthward of the magnetic reconnection (ion bulk flows are earthward). Low-energy (<5 keV) electrons show an excess in the tailward part, and high-energy (>5 keV) electrons show an excess in the earthward part. The distribution at 1742:23 UT (lower left panel, ES) is obtained in the Southern Hemisphere earthward of the magnetic reconnection (ion bulk flows are earthward). Low-energy (< 5 keV) electrons show an excess in the tailward

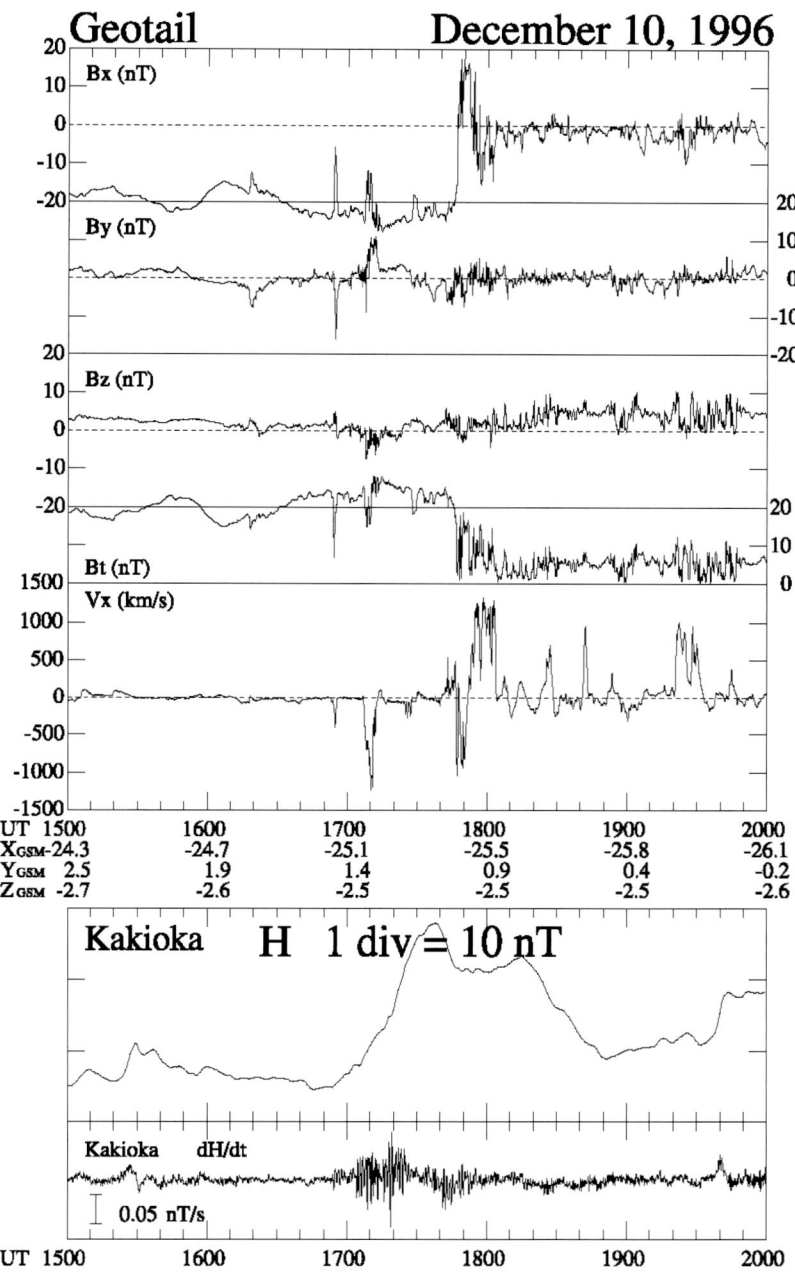

Figure 1. Magnetic field variations in the GSM coordinate system (Bx, By, Bz, and Bt), and ion flow velocity Vx for the period 1500–2000 UT on December 10, 1996. Ground magnetic field data (the H component and dH/dt for Pi2 pulsations) are also presented.

part, and high-energy (> 5 keV) electrons show an excess in the earthward part. Finally, the distribution at 1750:52 UT (upper right panel, TN) is obtained in the Northern Hemisphere tailward of the magnetic reconnection (ion bulk flows are tailward). Low-energy (<5 keV) electrons show an excess in the earthward part, and high-energy (>5 keV) electrons show an excess in the tailward part. Hence, the electrons show counterstreaming features, in which low-energy electrons flow into the magnetic reconnection site, whereas high-energy electrons flow out of the magnetic reconnection site. Three other electron distribution functions are presented in Plate 6 of *Nagai et al.* [2001].

Second, we examine the Geotail data for the period from 1700 to 1720 UT, which corresponds to the second Pi2 onset

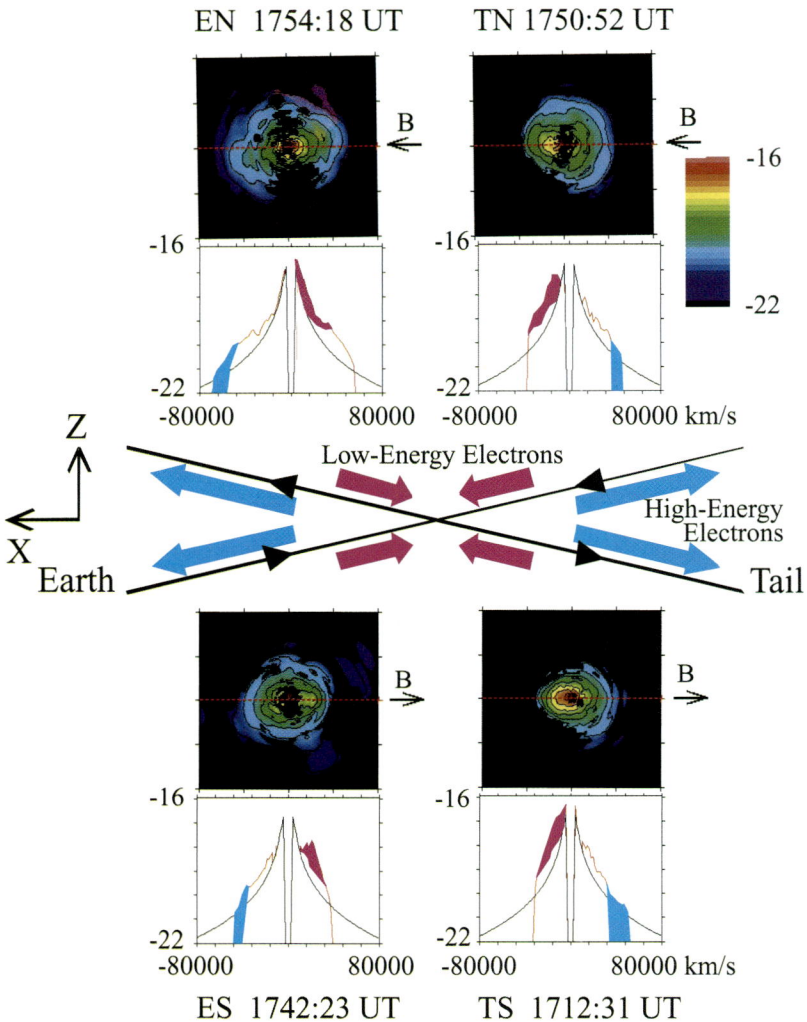

Plate 1. Electron distribution functions at 1754:18 (EN), 1742:23 (ES), 1750:52 (TN), and 1712:31 UT (TS) on December 10, 1996. The distribution functions are presented in the plane that includes the magnetic field vector and the ion bulk-flow vector. The x-axis is the magnetic field direction. Electron phase space densities from 10^{-22} to 10^{-16} s^3m^{-6} are color-coded according to the color bar. Cuts of distribution functions in the red line are also presented. In the cuts, asymmetric parts relative to zero are colored red for the low-energy component and blue for the high-energy component. The observations are presented schematically relative to a supposed X-type neutral line in the middle panel.

at 1703 UT. Geotail observes tailward field-aligned ion flows with negative Bz for the time interval 1705–1715 UT. Geotail is located in the Southern Hemisphere near the plasma sheet/tail lobe boundary. There are no highly accelerated electrons in this period. Electron distributions are almost isotropic until 1710 UT, after which field-aligned anisotropy appears. The electron distribution function at 1712:31 UT is presented in Plate 1 (lower right panel, TS). This distribution function is obtained in the Southern Hemisphere tailward of the magnetic reconnection site. Low-energy electrons show an excess in the earthward part, and high-energy electrons show an excess in the tailward part. Hence, accelerated electrons flow out of the magnetic reconnection site with ions, whereas low-energy electrons flow into the magnetic reconnection site.

For the first Pi2 onset at 1652 UT, Geotail observes tailward flows with a bipolar (positive and then negative) Bz signature. No evident anisotropy is seen in electron distribution functions. For the third Pi2 onset at 1719 UT, Geotail is located in the tail lobe and observes southward Bz. Tailward ion flows after 1725 UT are cold tail lobe flows. Earthward flow bursts at 1807, 1815, 1822, and 1840 UT in the recovery phase of this substorm are associated with sporadic magnetic reconnections in the distant tail and are discussed in detail in *Nagai et al.* [2002].

2.3. Current Densities

It is important to determine whether counterstreaming electrons can contribute any currents. Current densities are calculated with ion and electron distribution functions. The current density shown in Plate 1 is –9.2 nA m^{-2} for EN, –4.1 nA m^{-2} for ES, +12.0 nA m^{-2} for TN, and +35.8 nA m^{-2} for TS. Here, the negative sign is earthward and the positive sign is tailward. In the four cases in Plate 1, low-energy electrons flowing into the magnetic reconnection site dominate high-energy outflowing electrons, so the counterstreaming electrons can contribute currents flowing out of the magnetic reconnection site. These outflowing currents exist only in the outermost layer of the plasma sheet/tail lobe boundary [*Nagai et al.*, 2001]. When high-energy outflowing electrons are dominant relative to low-energy inflowing electrons for counterstreaming electrons, the current direction is into the magnetic reconnection site. Indeed, the inflowing currents are observed in the inner region, adjacent to the outflowing currents [*Nagai et al.*, 2003].

2.4. Magnetic Field Deflections

When currents exist in the GSM x, z plane, magnetic field deflections appear in the east–west magnetic field component, By. For the period 1500–1650 UT (Figure 1), Bx is approximately –20 nT and By is close to 0, except for a brief dip of Bx near 1620 UT. At 1800–2000 UT, Bx is near 0, indicating that Geotail is near the equatorial plane, and By is close to 0. Since there is no large background field in By, we take the baseline for examining By behavior to be zero in this event. It is not easy to identify By deflections properly in the presence of strong background fields.

First, we examine the magnetic field data for the period 1740–1810 UT. These data are presented in Figure 2 with a time resolution of 1/16 s. The times when counterstreaming electrons are observed are indicated with circles in the second panel. By is near –5 nT in the interval 1742–1745 UT for the earthward flows in the southern boundary region ($Bx < 0$) and near –5 nT in the interval 1749–1751 UT for the tailward flows in the northern boundary region ($Bx > 0$). By is near +5 nT in the brief time near 1754 UT for earthward flows in the northern boundary region ($Bx > 0$). These By deflections are observed mostly when electrons show counterstreaming features. There are irregular By variations in other time intervals. However, it is important to note that By returns to 0 for the first flow-reversal near 1746 UT (earthward flows, then tailward flows) and for the second flow-reversal near 1752 UT (tailward flows, then earthward flows).

Second, we examine the magnetic field data for 1700–1720 UT (Figure 3). In this tailward flow period, counterstreaming electrons are seen only after 1711 UT. The By behavior is striking. By is irregular and becomes largely negative until 1710 UT; it then becomes significantly positive after 1711 UT for the counterstreaming electrons.

These observations can be summarized as follows: By is positive (negative) in the Northern (Southern) Hemisphere earthward of the magnetic reconnection site, and By is negative (positive) in the Northern (Southern) Hemisphere tailward of the magnetic reconnection site. This magnetic field behavior is consistent with the quadrupole magnetic field structure produced by the Hall current system [*Sonnerup*, 1979].

For the 1652 UT onset, Geotail enters deep into the plasma sheet and observes tailward flows with negative Bz. The magnetic field data for the period 1650–1700 UT are presented in Figure 4. There are no counterstreaming electrons, and By becomes largely negative for $Bx < 0$. This negative By contradicts the quadrupole magnetic field structure of the Hall current system. Near 1620 UT (Figure 1), Geotail enters slightly into the inner plasma sheet (no substorm onset signatures are present on the ground), and By becomes negative for $Bx < 0$. This negative By also contradicts the quadrupole magnetic field structure of the Hall current system.

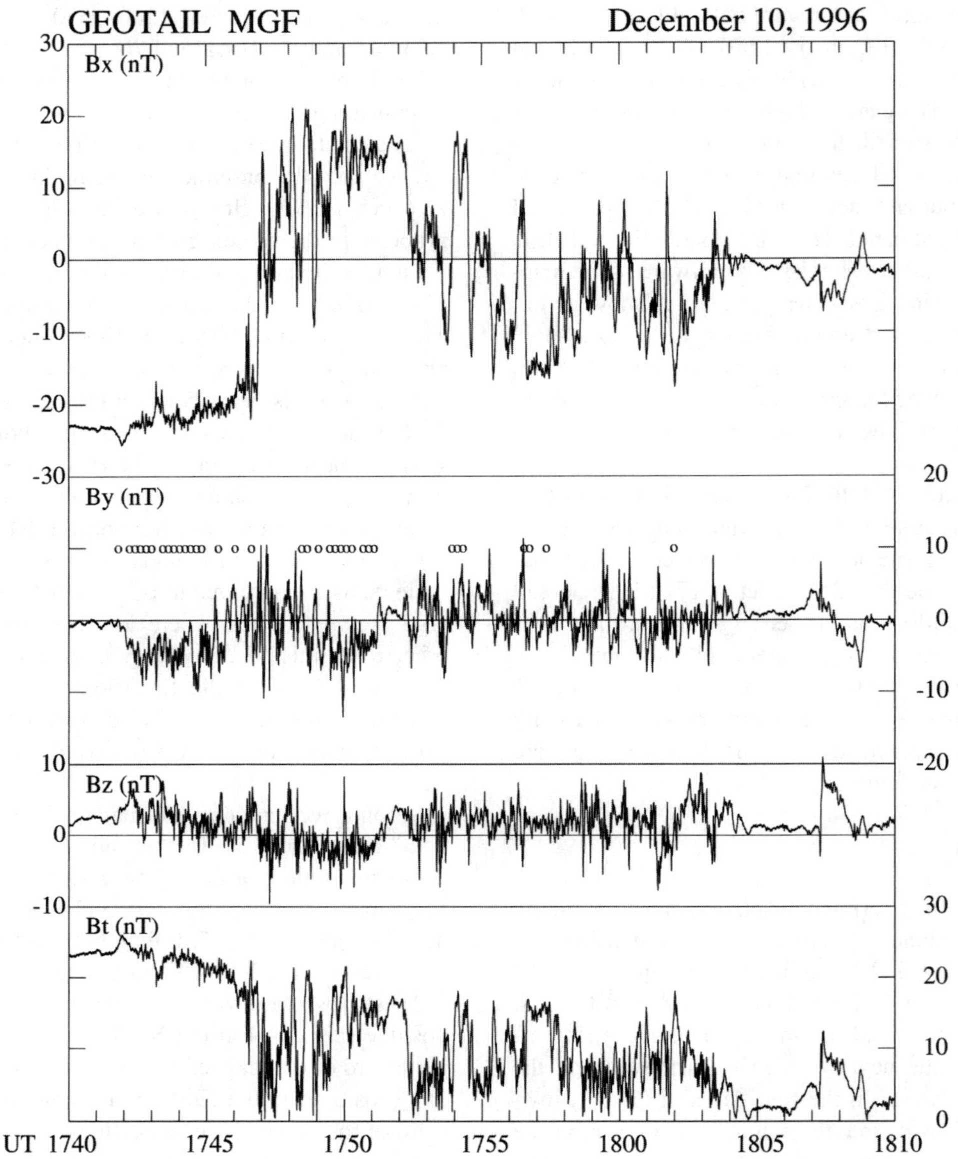

Figure 2. Magnetic field data (with a time resolution of 1/16 s) for the 1740–1810 UT period on December 10, 1996. The times when counterstreaming electrons are observed are indicated by circles.

These observations mean that the behavior of By, which is consistent with the quadrupole magnetic field structure of the Hall current system, is coupled with the counterstreaming electrons.

3. EXTENT OF THE HALL CURRENT SYSTEM

The Hall current system is generated from ion–electron decoupling, which is thought to take place at a distance on the order of the ion inertial length from the diffusion region. It is important to examine the extent of the Hall current system. For the tailward extent of the Hall current system, *Fujimoto et al.* [1997] found counterstreaming electrons and current densities of 5 nA m^{-2} in the outer boundary just after a plasmoid on January 10, 1995, at $X_{GSM} = -46.4\ R_E$. This observation implies that the Hall current system can extend beyond a distance of the ion inertial length. Since the southward magnetic field of this plasmoid is large, $Bz = -23.5$ nT [*Nagai et al.*, 1998b], this may not be a representative event. For the earthward extent of the Hall current system, *Fujimoto et al.* [2001] found tailward-flowing electrons in the outer edge of the plasma sheet during substorm-associ-

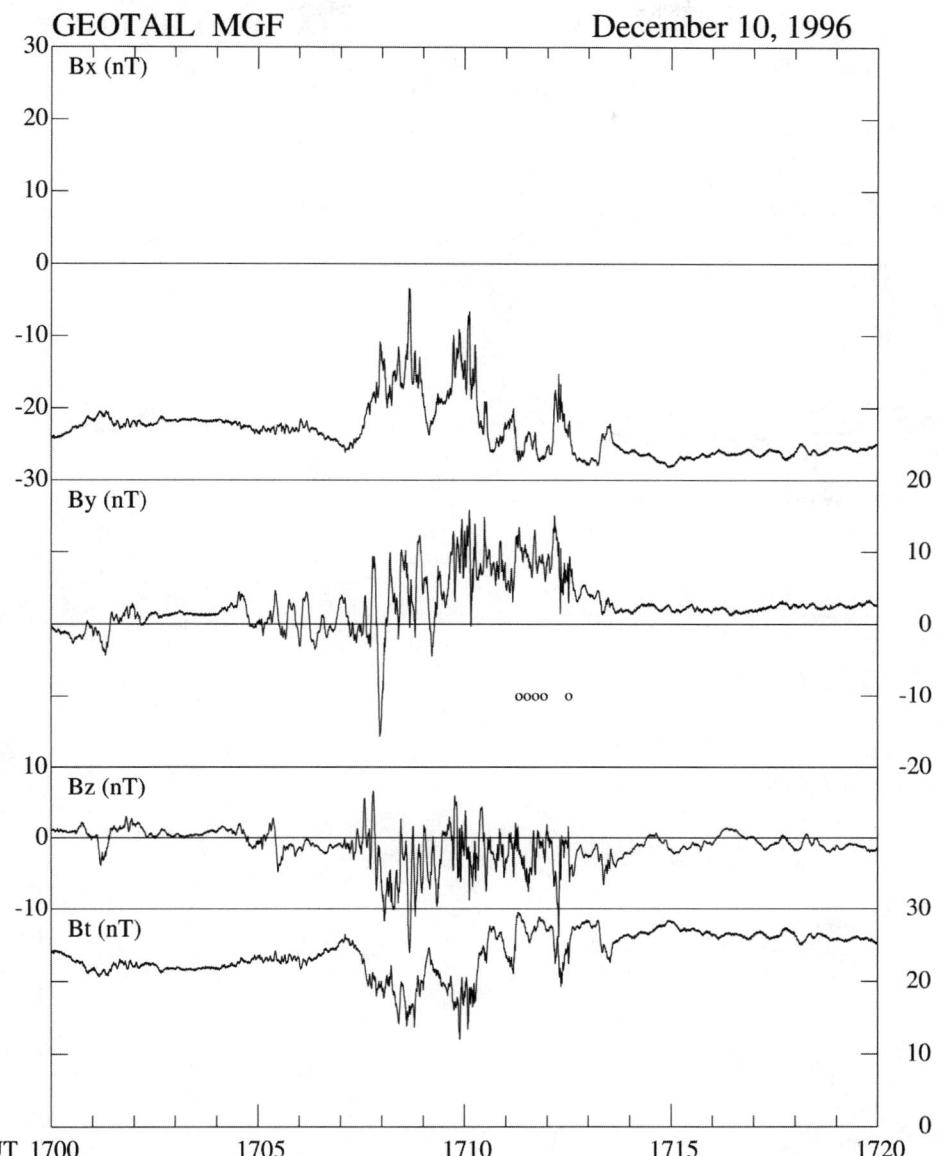

Figure 3. Magnetic field data (with a time resolution of 1/16 s) for the 1700–1720 UT period on December 10, 1996. The times when counterstreaming electrons are observed are indicated by circles.

ated dipolarization in the magnetic field. It is important to note, however, that auroral ionospheric processes may produce tailward-flowing electrons.

Since it is more unambiguous to evaluate the tailward extent of the Hall current system, we survey counterstreaming electron events for tailward flows at radial distances of 20–50 R_E in the magnetotail. We use the Geotail data, for which 3D distribution functions are available, for the period November 1994 to February 1996. The 3D distribution function data are taken only for time intervals when the data are received at the Japanese station Usuda. This limitation is unavoidable for examining counterstreaming features in electrons. We obtain most tail data at 30–50 R_E in the 1994–1995 winter season, when the apogee of Geotail was 50 R_E; in the 1995–1996 winter season, the apogee of Geotail was 30 R_E. To examine the extent of the Hall current system in the x direction, we limit the area to be examined to the magnetotail at $X_{GSM} = -20$ to -50 R_E and $Y_{GSM} = -10$ to $+10$ R_E. Here, we include the effects of the aberration for determining the survey area.

First, we select 40 tailward-flowing events for which $Vx < -500$ km/s and $Bz < 0$. We examine ground magnetic field

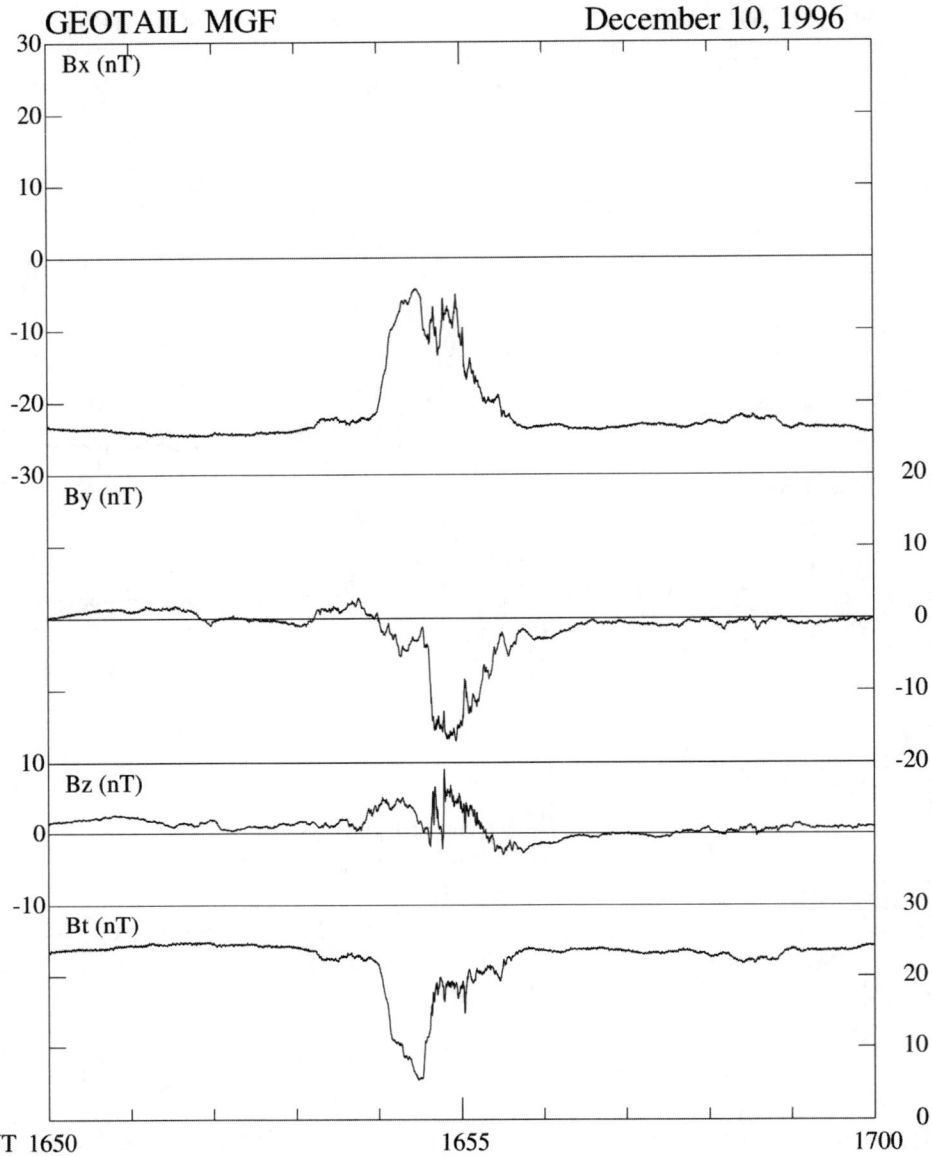

Figure 4. Magnetic field data (with a time resolution of 1/16 s) for the 1650–1700 UT period on December 10, 1996.

data and particle data at geosynchronous orbit, and confirm that these tailward flows are associated with substorm onset, pseudo-onset, or substorm intensification. Second, we determine visually whether or not counterstreaming features exist in the electron distribution functions. Third, we calculate current densities from ion and electron data for the counterstreaming electrons. Finally, we examine the By behavior for the counterstreaming electrons.

Figure 5 shows the magnetic field data for three consecutive tailward flow events on December 11, 1994. Each tailward ion flow is observed when Geotail enters the plasma sheet, which is seen as a decrease in the magnitude of Bx.

Geotail is located at $(-40.3, 8.8, -5.0\ R_E)$ at 1800 UT. These events are obtained during a large substorm, and each tailward flow event is associated with a Pi2 pulsation burst at Kakioka. Since Geotail is located mostly in the Southern Hemisphere, the Hall current–associated By deflections should be positive. For the first event, however, starting at 1756:30 UT, counterstreaming electrons are not found. There are significant negative-By changes, and By is largely negative for $Bx = 0$. Hence, these By variations are not attributable to the Hall current system. For the second event, starting at 1803:40 UT, when Geotail enters deep into the plasma sheet, By becomes negative. These By variations are

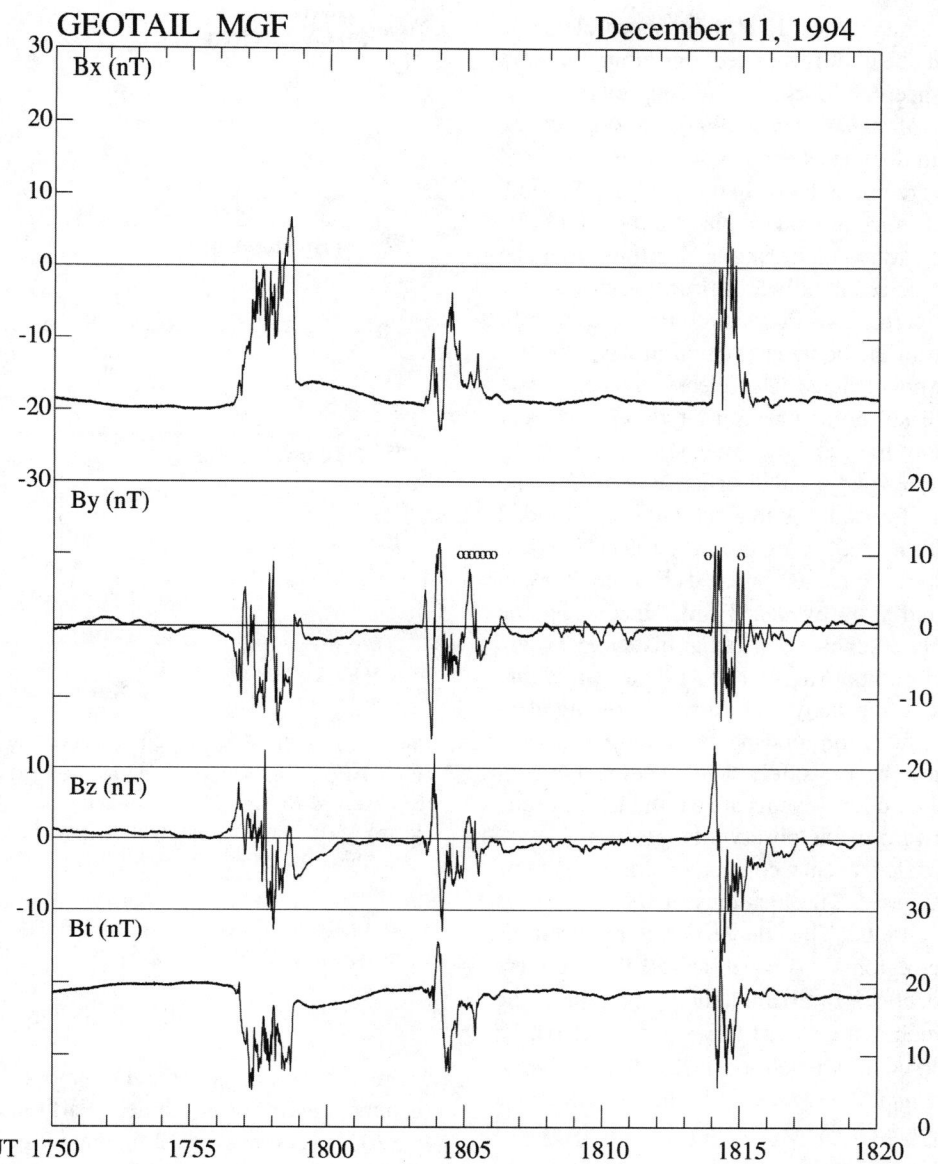

Figure 5. Magnetic field data (with a time resolution of 1/16 s) for the 1750–1820 UT period on December 11, 1994. The times when counterstreaming electrons are observed are indicated by circles.

not attributable to the Hall current system. Counterstreaming electrons are present with positive By near 1805 UT, tailward-flowing currents with a current density of 3.6 nA m^{-2} are observed. However, By becomes negative in the period of 1805:20–1806:00 UT. For the third event, starting at 1813:40 UT, Bz shows a well-defined bipolar (northward and then southward) signature, indicating a plasmoid. By becomes positive and negative for $Bx < 0$ and is largely negative for $Bx = 0$. Counterstreaming electrons are present with positive By near 1815 UT, and tailward-flowing currents with a current density of 3.5 nA m^{-2} are observed then.

These examples indicate that the magnetic field structure is complex inside tailward flow events (plasmoids). It is well known that there is a core-field By inside plasmoids [e.g., *Sibeck et al.*, 1984; *Hughes and Sibeck*, 1987]. The presence of the core field may alter the By pattern [e.g., *Karimabadi et al.*, 1999; *Pritchett*, 2001]. Any test simply using the magnetic field data is probably not reliable for examining the extent of the Hall current system [*Ueno et al.*, 2002]. It is more relevant to use the signatures of counterstreaming electrons, outflowing currents, and By deflections.

Here, we clarify three points. First, high-speed electrons can stream tailward ahead of low-speed ions along the field line in the plasma sheet/tail lobe boundary, especially ahead of plasmoids [e.g., *Machida et al.*, 1994]. In these cases, low-energy electrons do not have any excess in the earthward part, so these cases are not included in this analysis. Second, when magnetic reconnection takes place tailward of the spacecraft location, earthward-streaming electrons would be observed. For the selected tailward flow events, counterstreaming ions (earthward flows and tailward flows), which are common in the boundary region in the substorm recovery phase, are not included. The selected tailward flows are associated with substorm onsets (not in the recovery phase) on the ground. In such cases, magnetic reconnection does not take place beyond a radial distance of 50 R_E, and we observe tailward flows in the magnetotail only at radial distances of 30–100 R_E [e.g., *Nagai et al.*, 1997]. It is unlikely that magnetic reconnection proceeds beyond 50 R_E in the selected tailward flow events. Third, since magnetic reconnection usually takes place at radial distances of 20–30 R_E for substorm onsets [*Nagai et al.*, 1998a; *Nagai and Machida*, 1998], it is not likely that tailward-flowing electrons include those from the ionosphere or those mirrored near the Earth. Hence, we can safely say that counterstreaming electrons in tailward flow events are caused by magnetic reconnection earthward of the observation position.

Of the 40 tailward flow events, counterstreaming electrons are observed in 20 cases. The locations of these 20 cases are presented in Figure 6a. The counterstreaming electrons are observed in the region X_{GSM} = –20 to –50 R_E, and there is no tendency to decrease in the midtail. This region is the same as that in which fast tailward flows with negative B_z are frequently observed in association with substorms [*Nagai et al.*, 1998a; *Nagai and Machida*, 1998]. The By deflections are presented in Figure 6b. When Geotail is in the Northern Hemisphere, By is negative in 3 cases (– in Figure 6b) and positive in 1 case (X in Figure 6b). When Geotail is in the Southern Hemisphere, By is positive in 14 cases (+ in Figure 6b) and does not show any change in 2 cases (X in Figure 6b). The By behavior is generally consistent with the quadrupole magnetic field structure of the Hall current system (17 of 20 cases). However, the events are biased in the Southern Hemisphere because of the Geotail orbits. Figure 6c presents the maximum current density in each case. Of 20 cases, tailward-flowing currents are detected in 18 cases (positive in Figure 6c), and the current direction is earthward (negative in Figure 6c) in 2 cases near 40 R_E. The current density exceeds 10 nA m^{-2} inside 30 R_E, whereas it is 4–8 nA m^{-2} beyond 30 R_E. The tailward-flowing current is usually detected when Geotail exits from the plasma sheet in the late

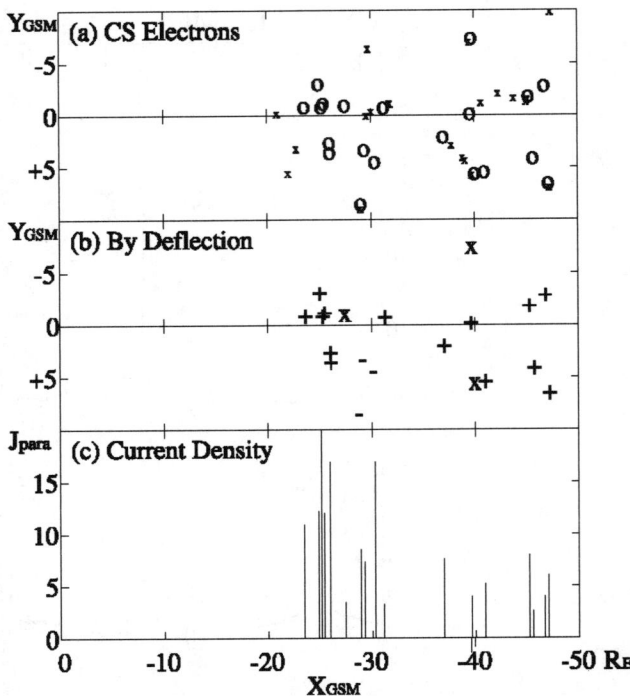

Figure 6. (a) Positions of 40 tailward-flowing events for which counterstreaming electrons are observed (o) and not observed (x); (b) observed By sign (+, –, X) for the period of counterstreaming electrons; (c) maximum current density for counterstreaming electrons (the tailward current is positive).

stage of the tailward flows, as seen in the January 10, 1995, event [*Fujimoto et al.*, 1997]

4. DISCUSSION

We have examined the December 10, 1996, magnetic reconnection event in the near-Earth magnetotail [*Nagai et al.*, 2001]. There were 12 magnetic reconnection events in 1996–1997 and 7 events in 2000–2001 (the survey was done for the data in which 3D distribution functions were available, roughly a quarter of the total tail data). There are no events in 1998–1999 and 2002. The characteristics seen in the December 10, 1996, event are representative. In the magnetic reconnection events, electrons are highly accelerated and heated, and they coexist with high-energy streaming ions. The counterstreaming electrons, for which high-energy electrons flow out of and low-energy electrons flow into the magnetic reconnection site, are common characteristics in the plasma sheet/tail lobe boundary. The low-energy inflowing electrons dominate the high-energy outflowing electrons in the outermost layer of the boundary region, resulting in outflowing currents with a current density of 10–20 nA m^{-2} in the immediate vicinity of the magnetic reconnection site. The east–west magnetic field deflection B_y observed with the counterstreaming electrons is generally consistent with the quadrupole magnetic field structure of the Hall current system. These outflowing currents are attributable to part of the Hall current system. The magnitude of the B_y deflection is 20–40 % of the background field (Figures 2–5). These values are consistent with the By magnitudes reproduced in numerical simulations for magnetic reconnection [e.g., Pritchett, 2001]. Hence, Hall physics likely plays an essential role in magnetotail magnetic reconnection. It is also evident that the Hall current system extends tailward beyond 20 R_E from the magnetic reconnection site during the substorm expansion phase.

A complementary study has been done by Øieroset et al. [2001]. In observations with the spacecraft Wind, they reported mid-tail flow reversal events (near 60 R_E) not associated with substorm activity on the ground. Although there are significant background fields, they indicated that observed magnetic field deflections are consistent with the quadrupole structure and that there are counterstreaming electrons.

In the original idea of the Hall current system for magnetic reconnection [Sonnerup, 1979], part of the Hall current loop in the inflow region was thought to be a driving force in the current circuit. In the inflow region, electrons can be transported with the magnetic field lines very close to the diffusion region, whereas ions do not approach the diffusion region because of its large inertial length. This relative motion produces the outflowing current part of the Hall current loop. Since this part of the current loop is perpendicular to the magnetic field, it results in the Hall term ($J \times B$) in the generalized Ohm's law. Other parts of the current loop are expected to be generated for charge conservation. The Hall current system is thought to form with a spatial scale on the order of the ion inertial length. The typical value of ion inertial length in the near-Earth tail is 1,000 km.

The present Geotail observations have shown that the outflowing currents are mostly field-aligned currents produced by counterstreaming electrons in the outermost region near the separatrix layer. The field-aligned current structure for the outflowing part of the Hall current system has been produced in a number of simulations [e.g., Nakamura et al., 1998; Shay et al., 1998]. The field-aligned part of the Hall current system extends beyond the ion inertial length and probably forms the substorm current system in the magnetotail; however, any closure mechanism at its end remains unknown.

Acknowledgments. The authors thank T. Mukai and S. Kokubun for their careful examination of the LEP and MGF data used here. The digital magnetic field data from Kakioka were provided by the Kakioka Magnetic Observatory. We also used ground magnetic field data supplied by WDC-C2, Kyoto University. The particle data from the Los Alamos National Laboratory geosynchronous spacecraft were obtained at http://leadbelly.lanl.gov.

REFERENCES

Birn, J., J. F. Drake, M. A. Shay, B. N. Rogers, R. E. Denton, M. Hesse, M. Kuznetsova, Z. W. Ma, A. Bhattacharjee, A. Otto, and P. L. Pritchett, Geospace Environmental Modeling (GEM) magnetic reconnection challenge, *J. Geophys. Res., 106*, 3715-3719, 2001.

Cothran, C. D., M. Landreman, M. R. Brown, and W. H. Matthaeus, Three-dimensional structure of magnetic reconnection in a laboratory plasma, *Geophys. Res. Lett., 30* (5), 1213, doi:10.1029/2002GL016497, 2003.

Fujimoto, M., M. S. Nakamura, I. Shinohara, T. Nagai, T. Mukai, Y. Saito, T. Yamamoto, and S. Kokubun, Observations of earthward streaming electrons at the trailing boundary of a plasmoid, *Geophys. Res. Lett., 24*, 2893-2896, 1997.

Fujimoto, M., T. Nagai, N. Yokokawa, Y. Yamade, T. Mukai, Y. Saito, and S. Kokubun, Tailward electrons at the lobe-plasma sheet interface detected upon dipolarizations, *J. Geophys. Res., 106*, 21255-21262, 2001.

Huba, J. D., and L. I. Rudakov, Three-dimensional Hall magnetic reconnection, *Phys. Plasmas, 9*, 4435-4438, 2002.

Huba, J. D., and L. I. Rudakov, Hall magnetohydrodynamics of neutral layers, *Phys. Plasmas, 10*, 3139-3150, 2003.

Hughes, W. J., and D. G. Sibeck, On the 3-dimensional structure of plasmoids, *Geophys. Res. Lett., 14*, 636-639, 1987.

Karimabadi, H., D. Krauss-Varban, N. Omidi, and H. X. Vu, Magnetic structure of the reconnection layer and core field generation in plasmoids, *J. Geophys. Res., 104*, 12313-12326, 1999.

Kokubun, S., T. Yamamoto, M. H. Acuña, K. Hayashi, K. Shiokawa, and H. Kawano, The GEOTAIL magnetic field experiment, *J. Geomagn. Geoelectr., 46*, 7-21, 1994.

Machida, S., T. Mukai, Y. Saito, T. Obara, T. Yamamoto, A. Nishida, M. Hirahara, T. Terasawa, and S. Kokubun, GEOTAIL low energy particle and magnetic field observations of a plasmoid at X_{GSM} = -142 R_E, *Geophys. Res. Lett., 21*, 2995-2998, 1994.

Mukai, T., S. Machida, Y. Saito, M. Hirahara, T. Terasawa, N. Kaya, T. Obara, M. Ejiri, and A. Nishida, The low energy particle (LEP) experiment onboard the GEOTAIL satellite, *J. Geomagn. Geoelectr., 46*, 669-692, 1994.

Nagai, T., R. Nakamura, T. Mukai, T. Yamamoto, A. Nishida, and S. Kokubun, Substorms, tail flows, and plasmoids, *Adv. Space Res., 20*, 961-971, 1997.

Nagai, T., and S. Machida, Magnetic reconnection in the near-Earth magnetotail, in *New Perspectives on the Earth's Magnetotail*, edited by A. Nishida, D. N. Baker, and S. W. H. Cowley, pp. 211-224, AGU, Washington, D. C., 1998.

Nagai, T., M. Fujimoto, Y. Saito, S. Machida, T. Terasawa, R. Nakamura, T. Yamamoto, T. Mukai, A. Nishida, and S. Kokubun, Structure and dynamics of magnetic reconnection for substorm onsets with Geotail observations, *J. Geophys. Res., 103*, 4419-4440, 1998a.

Nagai, T., M. Fujimoto, M. S. Nakamura, R. Nakamura, Y. Saito, T. Mukai, T. Yamamoto, A. Nishida, and S. Kokubun, A large southward magnetic field of -23.5 nT in the January 10, 1995, plasmoid, *J. Geophys. Res., 103*, 4441-4451, 1998b.

Nagai, T., I. Shinohara, M. Fujimoto, M. Hoshino, Y. Saito, S. Machida, and T. Mukai, Geotail observations of the Hall currents system: Evidence of magnetic reconnection in the magnetotail, *J. Geophys. Res., 106*, 25929-25949, 2001.

Nagai, T., M. S. Nakamura, I. Shinohara, M. Fujimoto, Y. Saito, and T. Mukai, Counterstreaming ions as evidence of magnetic reconnection in the recovery phase of substorms at the kinetic level, *Phys. Plasmas, 9*, 3705-3711, 2002.

Nagai, T., I. Shinohara, M. Fujimoto, S. Machida, R. Nakamura, Y. Saito, and T. Mukai, Structure of the Hall current system in the vicinity of the magnetic reconnection site, *J. Geophys. Res., 108* (A10), 1357, doi:10.1029/2003JA009900, 2003.

Nakamura, M. S., M. Fujimoto, and K. Maezawa, Ion dynamics and resultant velocity space distributions in the course of magnetic reconnection, *J. Geophys. Res., 103*, 4531-4546, 1998.

Nishida, A., The GEOTAIL Mission, *Geophys. Res. Lett., 21*, 2871-2873, 1994.

Øieroset, M., T. D. Phan, M. Fujimoto, R. P. Lin, and R. P. Lepping, In situ detection of collisionless reconnection in the Earth's magnetotail, *Nature, 412*, 414-417, 2001.

Pritchett, P. L., Geospace Environment Modeling magnetic reconnection challenge: Simulations with a full particle electromagnetic code, *J. Geophys. Res., 106*, 3783-3798, 2001.

Shay, M. A., J. F. Drake, R. E. Denton, and D. Biskamp, Structure of the dissipation region during collisionless magnetic reconnection, *J. Geophys. Res., 103*, 9165-9176, 1998.

Sibeck, D. G., G. L. Siscoe, J. A. Slavin, E. J. Smith, S. J. Bame, and F. L. Scarf, Magnetotail flux ropes, *Geophys. Res. Lett., 11*, 1090-1093, 1984.

Smith, D., S. Ghosh, P. Dmitruk, and W. H. Matthaeus, Hall and turbulence effects on magnetic reconnection, *Geophys. Res. Lett., 31*, L02805, doi:10.1029/2003GL018689, 2004.

Sonnerup, B. U. Ö., Magnetic field reconnection, in *Solar System Plasma Physics*, vol. III, edited by L. T. Lanzerotti, C. F. Kennel, and E. N. Parker, pp. 45-108, North-Holland, New York, 1979.

Ueno, G., S.-I. Ohtani, Y. Saito, and T. Mukai, Field-aligned currents in the outermost plasma sheet boundary layer with Geotail observation, *J. Geophys. Res., 107*(A11), 1399, doi:10.1029/2002JA009367, 2002.

M. Fujimoto and T. Nagai, Department of Earth and Planetary Sciences, Tokyo Institute of Technology, Tokyo 152-8551, Japan. (nagai@geo.titech.ac.jp)

Plasma Acceleration due to Transition Region Reconnection

J. Büchner and B. Nikutowski

Max-Planck-Institut für Sonnensystemforschung, Katlenburg-Lindau, Germany

A. Otto

Geophysical Institute, University of Alaska, Fairbanks, Alaska

Modern observations allow plasma acceleration in the solar atmosphere to be diagnosed. A probable energy source for these acceleration processes is the photospheric motion, powered by the sub-photospheric plasma convection. Among others, magnetic reconnection is a major possible mechanism for the observed plasma acceleration. In order to understand where and how reconnection can accelerate solar plasma, we developed a fully three-dimensional dissipative magnetohydrodynamic (MHD) simulation approach based on observed photospheric magnetic fields and plasma motion. Our model starts with a force-free extrapolation of the longitudinal photospheric magnetic fields and with a VAL-type stratified equilibrium distribution of the chromospheric-coronal plasma. Heat conduction, radiation cooling and gravity are neglected in the present version of the model. The applied photospheric boundary conditions do not consider emerging from or submerging to the solar interior magnetic flux. The horizontal photospheric plasma motion drives the chromospheric and coronal plasma dynamics. The resulting currents are generated not only parallel but also perpendicular to the magnetic field. Using plasma-microphysical results we consider an anomalous resistivity that is switched on as soon as the local current density exceeds a critical value. As a result, we obtain reconnection mainly in the transition region between chromosphere and corona, where the plasma density decreases and plasma collisions become less efficient. The resulting plasma velocities are consistent with the observed accelerated plasma flows.

1. INTRODUCTION

Modern solar observations identified plasma acceleration as an important energy release process in the solar atmosphere. Typical examples are explosive events, first described by *Brueckner & Bartoe* [1983]. Explosive events have been intensively investigated by the SOHO spacecraft mission (*Dere et al.* [1989], *Innes et al.* [1997], *Erdelyi et al.* [1997]). One possible mechanism of the plasma acceleration in explosive events is magnetic reconnection (e.g. *Dere et al.* [1991]). However, fundamental unresolved questions concern the location, onset conditions, and consequences of three-dimensional magnetic reconnection in the solar atmosphere (see, e.g. *Priest & Forbes* [2000], *Van Driel* [2003]).

There is no doubt that magnetic fields link the solar interior with the atmosphere of the Sun. They provide a major energy source and determine the structure, dynamics, and energetics of the chromosphere and corona. Direct high-resolution observations of the solar magnetic field are restricted mainly to the photosphere. Hence, the coronal field is obtained by extrapolation, assuming it is force-free, i.e. allowing electric currents to flow only parallel to the magnetic field. Force-free, or potential, field configurations (no currents at all, they represent the lowest energy state of a magneto-plasma), provide only a very approximate description of the coronal field structure. Coronal acceleration processes as a response of the corona upon the energy input from the solar interior cannot be addressed by potential field models. Simulations using typical spectra of surface motion have shown that the corresponding energy input in combination with magnetic reconnection can explain the observed coronal heating up to millions of degrees (*Gudiksen & Nordlund* [2002]), although, at present, there is no clear observational evidence that the corona is heated by magnetic reconnection. Investigations of the distribution of flare energy with several instruments, for example, have shown that the flaring rate extrapolated to the nanoflare energy range results in an insufficient heating rate to explain coronal temperatures (*Aschwanden & Parnell* [2002]). Plasma acceleration by reconnection has been considered by directly comparing numerical simulation results with observations. *Innes & Tóth* [1999], for example, used a two-dimensional magnetohydrodynamic (MHD) model to demonstrate the formation of jets by reconnection between newly emerging flux and the preexisting coronal field, comparing their results with observed jets. *Roussev et al.* [2001] (see also references therein) extended these simulations by taking into account the particular physical environment, typical for "quiet" sun explosive events, including radiative losses, volumetric heating and thermal conduction. This allowed them to compare their synthetized radiation lines with observations. *Galsgaard et al.* [1999] simulated the distribution of the energy release by flux braiding in coronal loops. They obtained peaks of energization near the footpoints and the summit.

In contrast to the previous work, which considers model magnetic field configurations, we developed a three-dimensional numerical simulation model based directly on the observation of the photospheric magnetic field and plasma motion. Our goal is to find the cause, the onset conditions and the location of reconnection leading to solar plasma acceleration as a result of the particular photospheric magnetic field and plasma motion. For this purpose we start with an extrapolated force-free magnetic field and line-symmetric periodic boundary conditions in the horizontal directions, compatible with the MHD equations. As a boundary condition for the simulation, we take into account the cross-field plasma motion in the photosphere that is obtained by using sequential magnetic field observations (section 2). We demonstrate the capabilities of our model using observed initial and boundary conditions of the photospheric magnetic field from SOHO MDI on October 17, 1996. Due to the photospheric motion, the coronal magnetic field loses its initially force-free character. Perpendicular sheet-current concentrations and magnetic tension build up mainly in the transition region, where the plasma density drops while these currents dissipate immediately in the highly collisional chromosphere. Motivated by kinetic plasma considerations beyond the scope of this paper, we assume anomalous resistivity to be switched on when and where the current density exceeds the threshold of an ion-acoustic instability such that the enhanced magnetic stress can relax via reconnection. In contrast to previous investigations, our model reveals the relation between the structure of the photospheric magnetic field and plasma motion and the location of reconnection. This way we located the most favorable conditions for reconnection and strong plasma acceleration in the transition region between chromosphere and corona. We present the results of our simulation in section 3. In section 4 we summarize our findings and discuss the applicability of our model as well as its limits that determine the directions of its further development.

2. SIMULATION MODEL

Our modelling approach consists of three parts. First, we use the observed longitudinal (line of sight) photospheric magnetic field to obtain an initial three-dimensional force-free magnetic field configuration, which satisfies equation

$$\nabla \times \mathbf{B} = \alpha \mathbf{B} \qquad (1)$$

with a constant coefficient α, i.e. allowing only electric currents parallel to the magnetic field (section 2.1).

We use a VAL-type (after *Vernazza et al.* [1981]) hydrostatic equilibrium model for the initial height-dependence of plasma density and temperature (section 2.2). Based on the changes of the photospheric field, we derived a consistent horizontal velocity profile, which is used as a boundary condition in our model (section 2.3). We use the observed variation of the photospheric magnetic field as an input into our model. The equations and parameters used in our three-dimensional simulations are described in section 2.4.

2.1. Initial Magnetic Field

Since H.U. Schmidt's suggestion of a potential field ($\alpha = 0$) extrapolation of photospheric fields to the corona (*Schmidt* [1964]), various methods have been developed to extrapolate the coronal magnetic field from the line-of-sight magnetic field component and from vector field measurements. One of the first constant-α, force-free magnetic field extrapolations was proposed by *Nakagawa & Raadu* [1972]. *Chiu* [1977] suggested that one should use the full photospheric magnetic field vector information as a boundary condition. *Seehafer*

[1978] demonstrated that a force-free magnetic field with constant $\alpha \neq 0$ does not have a finite energy content and, therefore, cannot be determined uniquely from only one magnetic field component given at the photosphere. Subsequent work extended the extrapolation to non-constant values of α (*Sakurai* [1981]), to the incorporation of vector-magnetic field observations (*Aly* [1989], *Cuperman et al.* [1989]), to magnetic fields, observed outside of the center of the disk (*Cuperman et al.* [1990a]) to moderately nonlinear force-free fields (*Cuperman et al.* [1990b], or introduced regularization methods to increase the height to be reached (*Amari* [1998]). The use of the force-free condition (1) is a lowest-order approach to the solar coronal magnetic field. It controls the solar atmospheric plasma filling, because in the chromosphere and corona the plasma pressure is small compared to the magnetic pressure. This does not necessarily cause a force-free field with a constant α, but the choice of a constant α is most straightforward for the sake of constructing a simple extrapolation method. The particular numerical value of α can be chosen by comparing observed bright loops with extrapolated magnetic field structures or put to zero, if there is no such information available.

For our purposes we currently incorporated a method similar to the *Seehafer* [1978] Green's function solution, because of its simplicity and efficiency. More sophisticated methods may be considered at a later time; however, our results demonstrate that the photospheric fields are never in an exact equilibrium state such that the precise initial conditions, be it a potential ($\alpha = 0$) or finite α force-free field, have only a limited impact on the dynamics, particularly at later times of the evolution.

To apply extrapolated fields in simulations where plasma boundary conditions play a crucial role, we derived a Green's function extrapolation with the horizontal components (B_x and B_y) being related to the vertical component (B_z) through fulfilling the force-free condition (1), and at the same time, $\nabla \cdot \mathbf{B} = 0$). Such an approach needs periodicity in the horizontal (X and Y) directions. *Seehafer* [1978] chose a plane symmetry of B_z across the X and Y boundaries of the system. This allows a solution in a non-periodic region which is a quarter of a fully periodic one. For a solution in the domain $0 \leq x \leq L_x$, $0 \leq y \leq L_y$ the fully periodic region would be $-L_x \leq x \leq L_x$, $-L_y \leq y \leq L_y$. Within the latter, the total magnetic flux is balanced to a high degree. Although the *Seehafer* [1978] approach already avoids the need for periodicity in the domain $0 \leq x \leq L_x$, $0 \leq y \leq L_y$, his solution is not suitable as an initial condition for plasma simulation. MHD simulations require a locally symmetric boundary condition for the current as well as for the plasma density. Using a corresponding local symmetry for the Seehafer solution generates a discontinuity in the field-aligned current and is, therefore, not suitable for an MHD simulation. We instead use a modified Green's function approach, which satisfies line-symmetric boundary conditions, consistent with the MHD demands (*Otto et al.* [2004]). Using this extrapolation technique, we can initialize our simulation model with an initial photospheric magnetic field based on MDI observations of the longitudinal components of the solar magnetic field. Since we lack additional information, we assume $\alpha = 0$ at $t = 0$; i.e. we start with a potential field extrapolation. A particular extrapolated magnetic field configuration above a region of interest is shown in Figure 1 for a SOHO-MDI observation made at 17:02 UT on October 17, 1996.

2.2. Initial Plasma Conditions

Because the plasma-β is small throughout the chromosphere and corona, the physics of the solar atmosphere is determined by the magnetic field and not by the plasma pressure. Hence, we can just "fill" the extrapolated magnetic field configuration with plasma. We choose to start with an initial density and temperature stratified with height, following the equilibrium-VAL-model of *Vernazza et al.* [1981] in its version "C". Deviating from VAL, however, we assume somewhat smaller values for the plasma density in the lower chromosphere, limiting it to 100 times the coronal density (see Figure 2). This is done partly for numerical reasons because larger densities correspond to larger collision frequencies which would require significantly smaller simulation time steps than desirable for large-scale three-dimensional simulations. This limitation of the density is justified since we do not draw special attention to the processes in the lower chromosphere, we just want to ensure a sufficiently strong coupling between the neutral gas and plasma. Further, to resolve the transition region well (by at least 10 grid points, see section 2.4), we enhanced the width of the transition region artificially to about 1500 km. In Figure 2 we compare the used equilibrium density–height profile (solid line) with the VAL model C, depicted by a dashed line. Figure 3 shows the corresponding initial temperature–height profile used to initialize our simulation model.

2.3. Boundary Conditions

For full compatibility, we assume the same line symmetry across the X and Y boundaries for plasma and neutral gas density as well as for the velocities and currents. At the lower (photospheric) boundary the tangential velocity is defined by strongly coupled plasma and neutral gas motion, and the boundary normal velocity is set to 0 (no emerging or submerging flux). The normal magnetic field uses $\nabla \cdot \mathbf{B} = 0$ and the horizontal components are computed from $\nabla \times \mathbf{B} =$

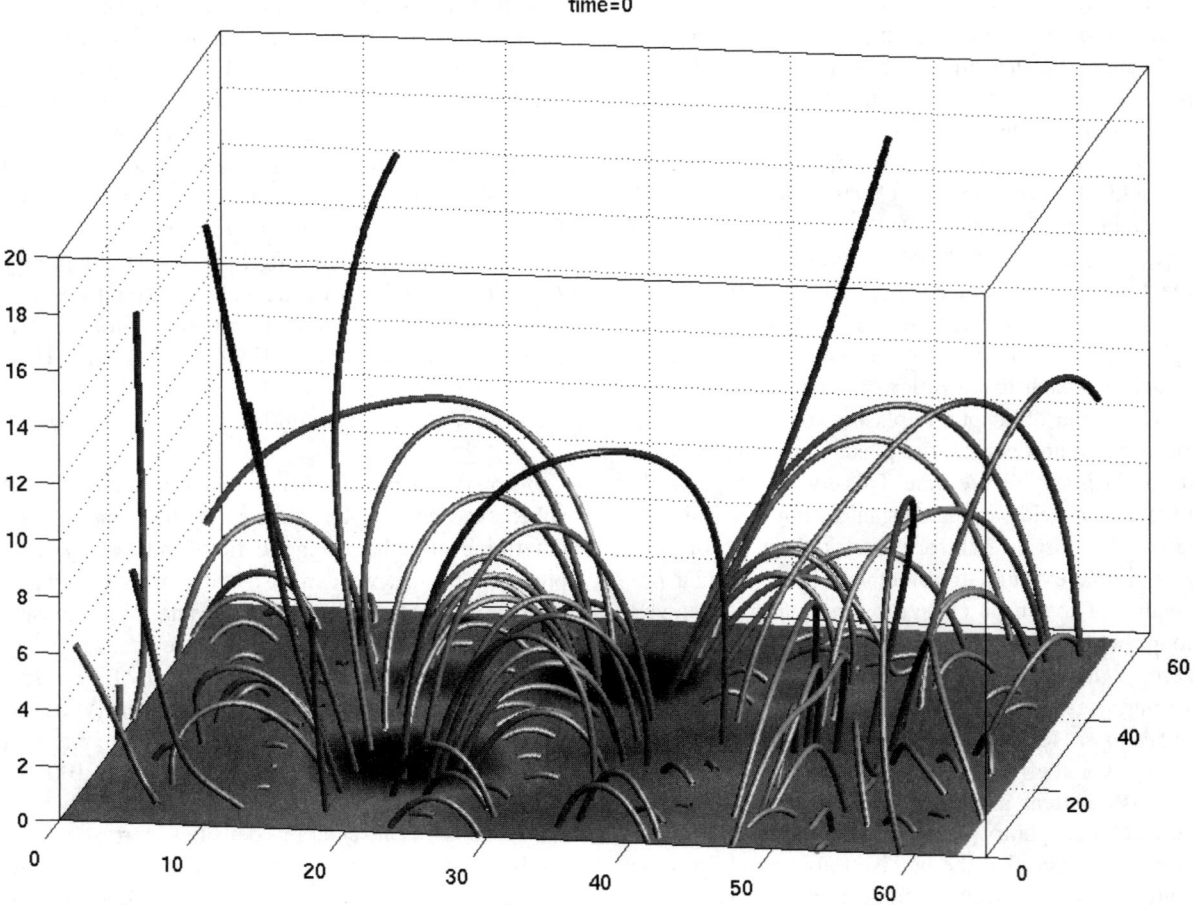

Figure 1. Potential magnetic field, extrapolated from the photospheric line-of-sight magnetic field observed by SOHO-MDI in the region of interest at 17:02 UT on October 17, 1996.

$\alpha \mathbf{B}$. Density and pressure are assumed symmetric (zero normal derivative). At the top (coronal) boundary, symmetric conditions (zero normal derivatives) are assumed.

From the variation of the photospheric magnetic field between 17:02 UT and 18:02 UT on October 17, 1996, three flow vortices can be derived, shown in Figure 4, which are used to mimic the observed chromospheric plasma and neutral gas motion. The vortex motion satisfies $\nabla \cdot \mathbf{u}_n = 0$, where u_n is the neutral gas velocity, such that density is conserved. It is contained in the X, Y plane so it can be expressed as a potential function U, i.e. $\mathbf{u}_n = \nabla \times (U\mathbf{e}_z)$. The contour lines of the function U are streamlines of the flow. This way U can be taken directly from observations. The motion is imposed for the neutral gas throughout the whole simulation box. However, it effectively couples to the plasma only in the chromosphere, where the collision frequency is sufficiently large.

2.4. Equations and Parameters

We simulate the solar atmospheric plasma and neutral dynamics solving the following set of MHD equations:

$$\frac{\partial \rho}{\partial t} = -\nabla \cdot \rho \mathbf{u} - \mu (\rho - \rho_e) \quad (2)$$

$$\frac{\partial \rho \mathbf{u}}{\partial t} = -\nabla \cdot \left[\rho \mathbf{u}\mathbf{u} + \frac{1}{2}(p + B^2)\underline{\underline{1}} - \mathbf{B}\mathbf{B} \right]$$
$$- \mu \rho (\mathbf{u} - \mathbf{u}_e) \quad (3)$$

$$\frac{\partial \mathbf{B}}{\partial t} = \nabla \times (\mathbf{u} \times \mathbf{B} - \eta \mathbf{j}) \quad (4)$$

$$\frac{\partial h}{\partial t} = -\nabla \cdot h\mathbf{u} - \frac{(\gamma - 1)}{\gamma} h^{1-\gamma} (2(\gamma - 1)\eta \mathbf{j}^2)$$
$$+ \frac{(\gamma - 1)}{\gamma} h^{1-\gamma} \mu(h - h_e) \quad (5)$$

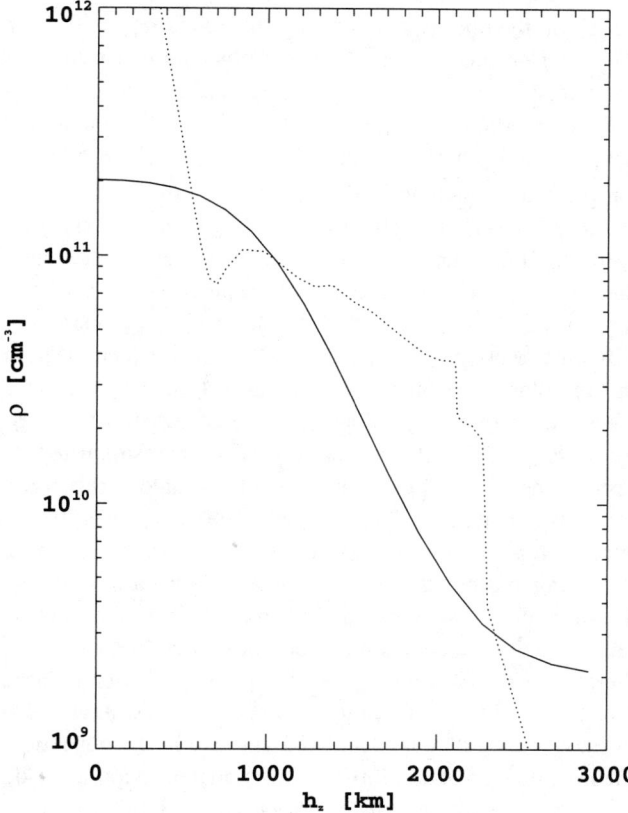

Figure 2. Equilibrium density–height profile in accordance with the VAL model C (see *Vernazza et al.* [1981], dashed line) and density–height distribution used as initial condition for our simulation (solid line).

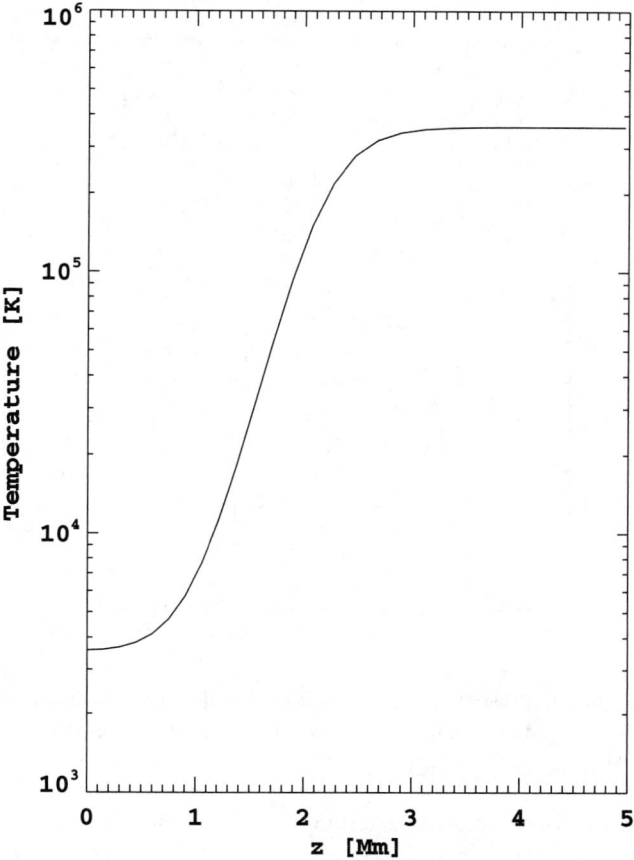

Figure 3. Initial temperature–height profile in equilibrium with the density profile shown in Figure 2.

The subscript e denotes the equilibrium variables. The pressure p is substituted by the internal energy variable $h = (p/2)^{1/\gamma}$, which yields a continuity equation for h in the absence of energy sources and sinks. For the polytropic index we use the adiabatic law $\gamma = 5/3$. As already mentioned an important consideration for modelling the photospheric-chromospheric interaction is the strong collisional coupling between the plasma and the neutral gas. Collisions not only cause friction between plasma and neutral gas and energy exchange, they also lead to ionization and recombination. In equations (2)–(5) we use the same effective frequencies for the ionization, recombination, friction and energy-exchange processes, because the sole purpose of the introduction of these collisions is to maintain a strong coupling between plasma and the neutral gas below the transition region and to let the coupling vanish in the transition to the corona, thereby maintaining the transition region itself. In this sense and for this purpose, the exact values of these frequencies do not matter as long as they are sufficiently large ($\mu \gg 1$; for the definition of the normalized μ, see equation 6). So far we did not include radiative cooling and thermal conduction processes. A fully dynamical model of the transition region requires that all its relevant collisional processes, radiation and thermal conduction are incorporated. This is beyond the scope of the present paper, where we focus on the influence of the photospheric boundary condition and the presence of the transition region on the dynamic processes in the corona.

All quantities in equations (2)–(5) are normalized, the magnetic field to $B_0 = 1$ G $= 10^{-4}$ T and the density to $n_0 = 2 \cdot 10^{15}$ m^{-3}, proton density in the corona just above the transition region (the actual proton number density may be a bit smaller, but the presence of heavier ions lets us choose this number). These values yield a normalizing Alfvén speed of $v_{A0} = \sqrt{B_0^2 / (\mu_0 mn)} = 47$ km/s. Using a length scale $L_0 = 470$ km (about the size of one MDI pixel) reveals a normalizing (Alfvén) time of $\tau_{A0} = L_0 / v_{A0} = 10$ s. The density in the chromosphere is larger by a factor of 100. Pressure is normalized to twice the typical magnetic pressure $p_0 = B_0^2 / (\mu_0) = 10^{-8}$ T$^2 / (4\pi \cdot 10^{-7}$ Vs /Am$) = 8.0 \cdot 10^{-3}$ Pa. The

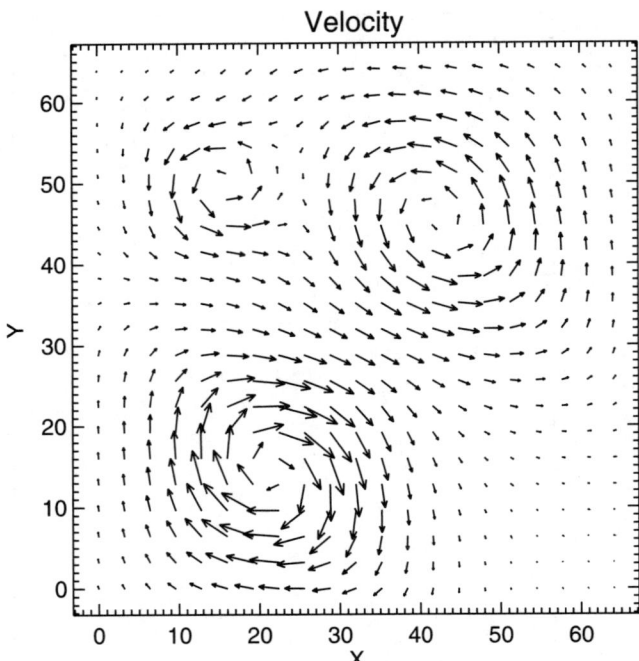

Figure 4. Streamlines of the photospheric flow velocity imposed as a boundary condition. The maximal velocity corresponds to $0.075\ v_A$.

normalizing pressure determines the normalizing temperature to be $T_0 = p_0 / (2n_0 k_B) = 1.4 \cdot 10^5$ K (2 here because of electrons and ions). This yields a typical thermal velocity of $v_{th0} = \sqrt{2 k_B T_0 / m} = \sqrt{B_0^2 / (\mu_0 m n)} = v_{A0} = 47$ km/s.

The ion-neutral collision frequency is $\nu_{in} = n_n \sigma_n v_{th}$, where the cross-section σ_n depends on the type of collision and of the colliding particles. A typical value for elastic (momentum exchange) collisions is $\sigma_n \approx 10^{-19}$ m^2. As pointed out above, we use the same frequencies for ionization, recombination, friction and energy exchange because we use these collisions mainly to maintain the transition region. The normalized ion-neutral collision frequency is

$$\mu = \nu \dot{\tau}_{A0} = 10^2 n \sqrt{p/\rho} \qquad (6)$$

where n is the plasma or neutral gas number density normalized to n_0, p is the normalized plasma pressure and ρ the normalized plasma mass density. Since $n \gg 1$ below the transition region and on the order of 1 above the transition region, we have $\nu \gg 1$ below the transition region and $\nu \ll 1$ above the transition region, in the corona where, as discussed before, the normalized ion-neutral collision frequency is negligible. The (normalized) ion-neutral collision frequency is very large below the transition region, getting even larger closer to the solar surface.

Since electron–ion collision frequencies are about $\nu_{ei} \approx 10^7$ s^{-1} in the photosphere/chromosphere and $\nu_{ei} \approx 30$ s^{-1} in the lower corona, the normalized collisional resistivity can be expressed as $\eta = \tau_A / \tau_D$, where the diffusion time τ_D is given by $\tau_D = L_0^2 / (\lambda_i^2 \nu_{ei})$, where $\lambda = c / \omega_{pi}$ is the ion inertial length, c being the speed of light and $\omega_{pi} = \sqrt{n_i e^2 / M_i}$ the ion plasma frequency. The latter can be estimated as $2.5 \cdot 10^4$ s in the chromosphere and 10^8 s in the corona. The corresponding (normalized) collisional resistivities are, therefore, $\eta \approx 4 \cdot 10^{-4}$ in the chromosphere and $\eta \approx 10^{-7}$ in the corona. Hence, the collisional resistivity is too small to explain the observed plasma acceleration by reconnection in the solar atmosphere. In our simulation, therefore, we take into account anomalously enhanced resistivity caused by wave–particle interactions. For two-dimensional reconnection the physical consequences of anomalous resistivity were investigated by *Roussev et al.* [2002], who adopted an explicit form of the anomalous resistivity motivated by a streaming plasma instability. These authors assumed that the anomalous resistivity switches on as soon as the electron drift velocity exceeds some fraction of the mean electron thermal speed. As a result, they found that sufficiently large threshold values and scaling constants of the anomalous resistivity would cause localized and patchy reconnection. Meanwhile *Watt et al.* [2002] have found that an ion-acoustic instability provides anomalous resistivity two orders of magnitude stronger than predicted by the quasilinear theory. Taking all these results into account, we switch the resistivity on when the current density exceeds the threshold of an ion-acoustic instability and scale the initial anomalous resistivity with the value, provided by *Watt et al.* [2002], enhancing it further proportionally to the current density squared.

We solve the partial differential MHD equations (2)–(5) by means of a finite difference approximation. We apply a leapfrog scheme, which is second-order in accuracy and very low in numerical dissipation. As usual in schemes based on the flux-corrected transport (FCT) paradigm, un-physical oscillations are damped on the grid scale. Second-order derivatives are treated with a Dufort–Frankel scheme, which allows the consideration of small resistivities. We choose a non-uniform grid to meet the requirement of a high plasma-β (higher plasma/gas pressure rather than magnetic pressure) at the lower boundary, reaching the much lower β corona via the low-β chromosphere. We also choose a non-uniform grid to be able to spatially resolve the chromosphere and the steep gradients in the transition region adequately. Our grid has 49 points in the vertical direction (Z), and 131 points in each of the horizontal (periodic) X and Y directions. In all directions we add two "ghost zones", additional mathematically motivated grid planes, to allow the calculation of gradients across the physical boundaries. In this paper

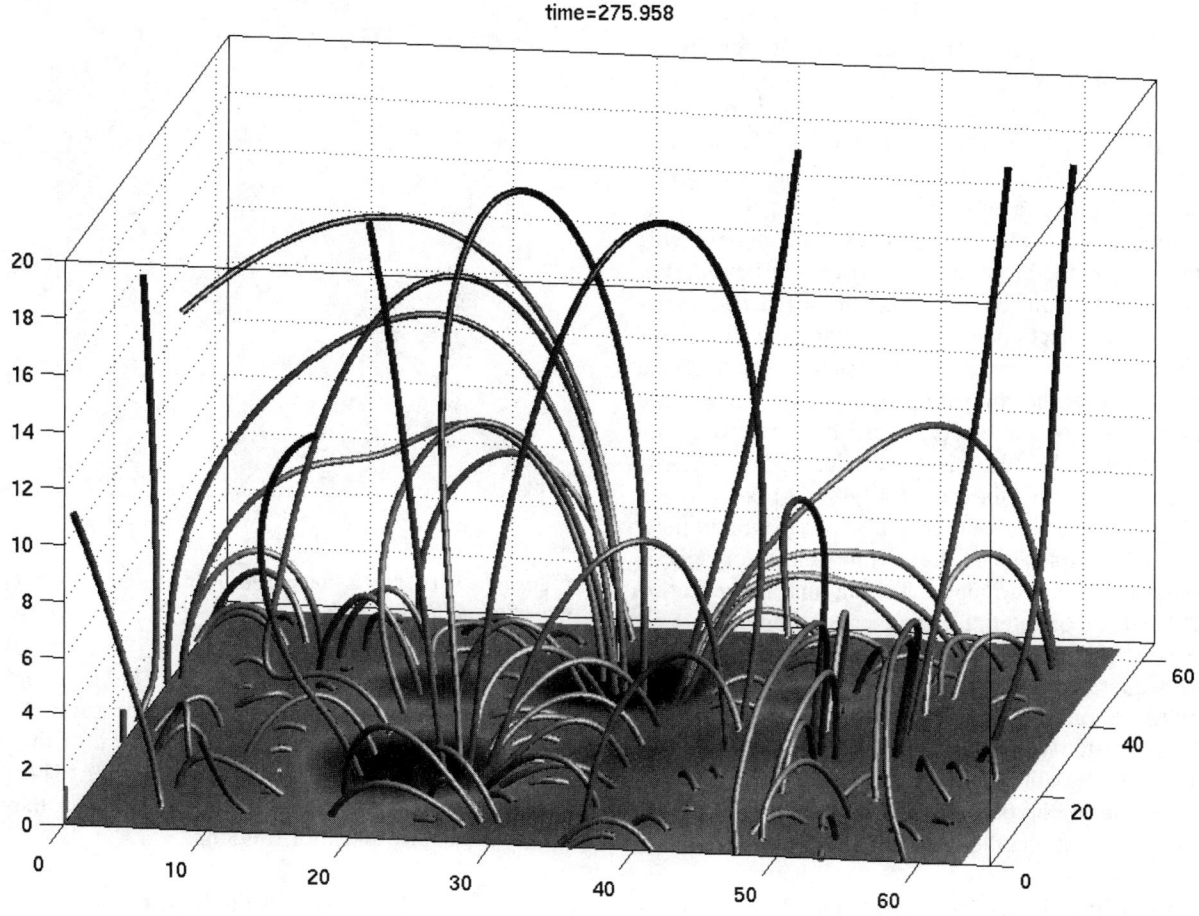

Figure 5. Simulated magnetic field at 17:48 UT, i.e. 46 min after the simulation has started.

we report the simulation of a three-dimensional rectangular shape cut out of the solar atmosphere, which extends 14 Mm in the vertical direction above a 30 Mm × 30 Mm photospheric region (horizontal direction). Our horizontal grid resolution is, therefore (the ghost zones are located outside the physical box), 230 km. In the vertical direction the grid resolution varies smoothly, starting with 140 km between the photosphere and the lower corona. Above $Z \approx$ 7.5 Mm the grid size increases to about 200 km, to 300 km above $Z \approx 9$ Mm and, finally, in the corona from $Z \approx 11$ Mm to the upper boundary of the box at $z = 14.1$ Mm, it becomes 380 km. We start the simulation with a Fourier expansion of the observed longitudinal photospheric magnetic field. To ensure a small discretization error for the finite difference scheme, we choose to resolve one wavelength of the shortest Fourier mode by eight (horizontal) grid points. For 129 grid points in each horizontal direction we can utilize, therefore, up to 16 Fourier modes per direction.

Higher Fourier modes have a much smaller scale height. Our approach includes all Fourier modes, which significantly contributes to the magnetic flux and energy at coronal heights.

3. RESULTS

As already mentioned, we have applied our simulation approach to SOHO-MDI observations of the photospheric magnetic field starting at 17:02 UT on October 17, 1996. We have chosen a region within which the magnetic field contains a dipolar magnetic structure as well as a number of smaller side-polarities. The evolution of the photospheric magnetic field is modelled by using a corresponding neutral gas velocity profile at this boundary, which, due to collisional coupling, drives the plasma below the transition region, thus causing the the plasma dynamics in the solar atmosphere above the chosen domain. For the initial and boundary conditions, as well as the simulation parameters discussed in section 2, the code runs are stable for several hours of real solar time.

After we start with a potential (force-free with $\alpha = 0$) initial configuration, our simulations demonstrate not only the generation of parallel, force-free currents but also current sheets oriented perpendicular to the magnetic field in the upper chromosphere and corona. The dynamically evolving magnetic field is, therefore, not force-free any more. After the threshold of the ion acoustic instability is reached, three-dimensional reconnection starts, and parallel electric fields accelerate the plasma. The anomalous resistivity is triggered a few minutes after the simulation starts. Reconnection and plasma acceleration take place preferably in the transition region. A result of the dynamical evolution of the magnetic configuration after 46 min (i.e. at 17:48 UT) is shown in Figure 5, which illustrates the reconfigured magnetic field. More flux is now present in the corona, and the magnetic field connectivity is rearranged. To examine reconnection in more detail and to determine the location of reconnection sites, one should identify regions with finite electric field components parallel to the magnetic field (*Hesse & Schindler* [1988]). Parallel electric fields appear as soon as the anomalous resistivity switches on, within minutes (solar time) after the simulation is started. They are located mainly in the transition region between chromosphere and corona. Figure 6 depicts the parallel electric field in the transition region at 17:48 UT.

At the same time the z-component of the plasma velocity

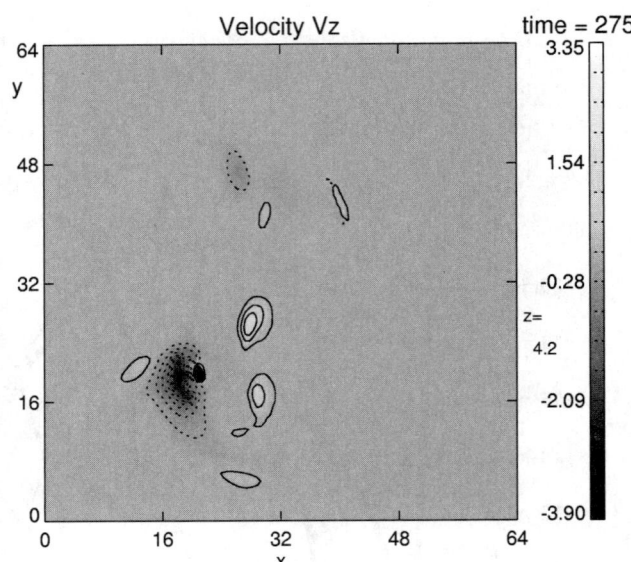

Figure 7. Simulated Z-component of the plasma velocity in the transition region at 17:48 UT.

also peaks in the transition region. The flow velocities are maximum in the same position as the parallel electric fields are. Figure 7 depicts the z-component of the plasma velocity out of the transition region proper. The plasma velocities reach the order of 100 km s^{-1}, as they were observed in the transition region.

4. SUMMARY AND DISCUSSION

Our model includes a number of assumptions that limit its applicability to the consideration of the electrodynamics of the solar atmosphere. Nevertheless it provides a major step toward a more realistic model, because it computes the three-dimensional magnetic evolution based on observed photospheric magnetic field and plasma motion. In particular our model describes the generation of non-force-free perpendicular electric currents, causing Lorentz forces, as well as parallel electric fields due to reconnection as a consequence of the photospheric forcing. Starting with particular SOHO observations of the photospheric magnetic field and plasma motion, we found that three-dimensional reconnection in the solar atmosphere preferably takes place in the transition region, where it causes up- and downward-directed plasma flows.

Reconnection operates mostly close to the transition region. Strong currents develop also in the chromosphere, but there, the strong coupling to the highly inertial neutral gas prohibits plasma acceleration. In the transition region, however, neutral gas and plasma decouple. This allows reconnection as soon as the current density exceeds a critical value and

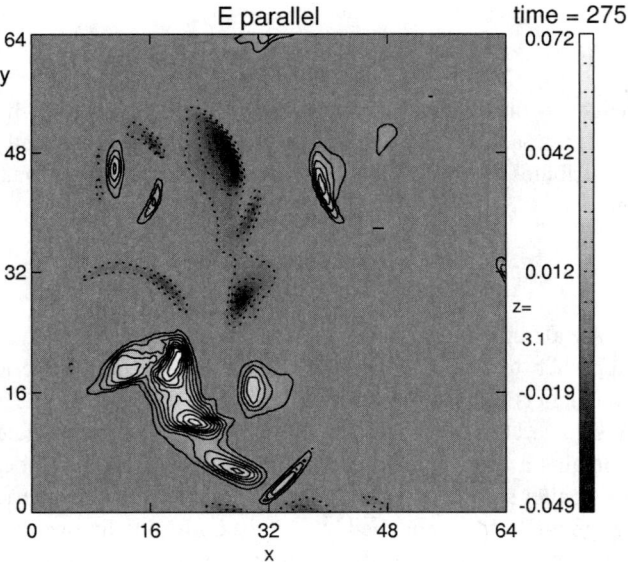

Figure 6. Simulated electric field component directed parallel to the magnetic field and diagnosed in the transition region at 17:48 UT.

anomalous resistivity is switched on. Finally, the magnetic strength decreases in the corona such that the corresponding current densities are smaller than in the transition region and chromosphere.

The limited applicability of the current version of our model is limited to describing the electrodynamical aspects due to radiation cooling and heat conduction. These latter are needed to determine the correct plasma temperatures and densities for comparison with spectral line observations. Also we did not take into account emerging or submerging magnetic flux because such fluxes were not observed during the considered event. Using a two-dimensional model *Galsgaard & Roussev* [2002] found that gravitation may decrease the reconnection rate slightly and that a new transitional initial phase arises. In the course of our simulations of the dynamic evolution, the density and temperature height profiles basically stay close to their initial equilibrium values, becoming modified just locally by the ohmic heating and plasma acceleration due to the parallel electric fields of three-dimensional reconnection. In our simulation without gravity the pressure balance is due to the constancy of the total pressure over height, such that the coronal pressure is slightly overestimated while the chromospheric one is slightly underestimated, acting, somehow, like the gravitation of the Sun. The physics of the transition region with its jump in temperature and density is described qualitatively correctly. Although the solar corona is not collisionless, the collisional scales such as the inverse magnetic Lundquist number (ratio of magnetic diffusion to Alfvén time scale) are insufficient to explain reconnection. In space, plasma dissipation usually occurs on the plasma scales such as Larmor radii or inertial lengths. The spatial scales resolved in large-scale three-dimensional MHD simulations are far too large to describe the formation of such thin current sheets. On the other hand, MHD simulations provide reasonable predictions of the evolution of the global solar atmospheric system. Our model can, therefore, be used to investigate the stability of solar atmospheric structures and to identify the critical regions where current sheets are formed and where reconnection and acceleration take place. Our model is expected to benefit considerably from the expected observational progress. For example, better information about the sub-photospheric motion from local helio-seismology can considerably improve the photospheric boundary conditions and the predictive power of our model. Also, in the future, observations of chromospheric emission lines such as He-I or Si-IV may allow direct identification of configurations that deviate from ideal force-free configurations and thus may provide a direct comparison to simulation results (*Solanki et al.* [2003]).

Acknowledgments. J.B. thanks the Isaac Newton Institute for Mathematical Sciences for its support of the parts of work carried out in Cambridge. All of us thank E. Marsch as well as D. Innes-Markiewicz at the MPS Lindau for constructive discussions.

REFERENCES

Aly J. J., On the reconstruction of the nonlinear force-free coronal magnetic field from boundary data *Sol. Phys.* **120**, 19-48, 1989.

Amari T., Boulmezaoud T. Z. & Maday Y., A regularization method for the extrapolation of photospheric solar magnetic field I. Linear force-free field *Astron. Astrophys.* **339**, 252-260, 1989.

Aschwanden M. J. & Parnell C. P., Nanoflare statistics from first principles: fractal geometry and temperature synthesis *Astrophys. J.* **572**, 1048-1071, 2002.

Brueckner G. E. & Bartoe J.-D. F., Observations of high-enery jets on the corona above the quiet Sun, the heating of the corona, and the acceleration of the solar wind *Astrophys. J.* **272**, 329-348, 1983.

Chiu Y. T. & Hilton H. H., Exact Greens'sfunction method of solar force-free magnetic-field computations with constant α, 1. Theory and basic test cases *Astrophys. J.* **212**, 873-885, 1977.

Cuperman S., Ofman L. & Semel M., Determination of constant-alpha force-free magnetic fields above the photosphere using three-component boundary conditions *Astron. Astroph.* **216**, 265-277, 1989.

Cuperman S., Ofman L. & Semel M., Extrapolation pf photospheric potential magnetic fields using oblique boundary vasluse: a simplified approach *Astron. Astroph.* **227**, 583-590, 1990a.

Cuperman S., Ofman L. & Semel M., Determination of force-free magnectic fields above the photosphere using three-component boundary condition: moderately non-linear case *Astron. Astroph.* **230**, 193-199, 1990b.

Dere K. P., Bartoe J.-D. F.& Brueckner G. E., Explosive events in the solar transition zone *Sol. Phys.* **123**, 41-68, 1989.

Dere K. P., Bartoe J.-D. F. & Brueckner G. E., Explosive events and magnetic reconnection in the solar atmosphere *J. Geophys. Res.* **96**, 9399-9407, 1991.

Erdelyi R., Doyle J. G., & Perez E. P., Explosive events observed by SOHO in *Fifth SOHO workshop*, **ESA-SP 404**, ed. A. Wilson, 353, 1997.

Galsgaard K., Mackay D. H., Priest E. R. & Nordlund A., On the location of energy release and temperature profiles along coronal loops *Sol. Phys.* **189**, 95-108, 1999.

Galsgaard K. & Roussev I., Magnetic reconnection in 2D stratified atmospheres I. Dynamical consequences *Astron. Astroph.* **383**, 685-696, 2002.

Gudiksen, B. V. & Nordlund, A., Bulk heating and slender magnetic loops in the solar corona *Astrophys. J.* **572**, L113-L116, 2002.

Hesse, M. & Schindler, K., A theoretical foundation of general magnetic reconnection *J. Geophys. Res.* **93**, 5559-5567, 1988.

Innes D. E., Inhester B., Axford W. I. & Wilhelm K., Bi-directional plasma jets produced by magnetic reconnection on the Sun *Nature* **386**, 811-813, 1997.

Innes D. E. & Tóth, G., Simulations of small-scale explosive events on the Sun *Sol. Phys.* **185**, 127–141, 1999.

Nakagawa Y. & Raadu M. A., On practical repesentation of magnetic field *Sol. Phys.* **25**, 127-135, 1972.

Otto A., Büchner J. & Nikutowski B., Force-free extrapolation of photospheric magnetic fields appopriate for numerical simulations of the solar atmosphere *Astron. Astroph.* **192**, in preparation, 2004.

Priest E. R. & Forbes, T. G., Magnetic Reconnection: MHD Theory and Applications, Cambridge Univ. Press, 2000.

Roussev I., Doyle J. G., Galsgaard K. & Erdelyi R., Modelling of solar explosive events in 2D environments III. Observable consequences *Astron. Astroph.* **380**, 719-726, 2001.

Roussev I., Galsgaard K. & Judge P. G., Physical consequences of the inclusion of anomalous resistivity in the dynamics of 2D magnetic reconnection *Astron. Astroph.* **382**, 639-649, 2002.

Sakurai T., Calculation of force-free magnetic field with non-constant α *Sol. Phys.* **69**, 343-359, 1981.

Schmidt H. U., In *Physics of Solar Flares NASA* **SP 50**, 107, 1964.

Seehafer N., Determination of constant α force-free solar magnetic fields from magnetograph data *Sol. Phys.* **58**, 215-223, 1978.

Solanki S. K., Lagg, A., Woch, J., Krupp N. & Collados, M., Three-dimensional magnetic field topology in a region of solar coronal heating *Nature* **425**, 692-695, 2003.

van Driel-Gesztelyi L., Observational signatures of magnetic reconnection in *NATO Proceedings on "Turbulence, Waves and Instabilities in the Solar Plasma"*, ed.R. Erdelyi, K. Petrovay, B. Roberts and M. Aschwanden, Kluwer, Nato Science Series, **124**, 281-304, 2003.

Vernazza J. E., E. H. Avrett & R. Loeser, Structure of the solar chromosphere. III - Models of the EUV brightness components of the quiet-sun *Astrophys. J. Suppl.* **45**, 635-725, 1981.

Watt C. E. J., Horne, J. B. & Freeman, M.P., Ion-acoustic resistivity in plasmas with similar ion and electron temperatures *Geophys. Res. Lett.* **29**, 1004, 2002.

J. Büchner and B. Nikutowski, Max-Planck-Institut für Sonnensystemforschung, Max-Planck-Str. 2, 37191 Katlenburg-Lindau, Germany. (buechner@linmpi.mpg.de)

A. Otto, Geophysical Institute, University of Alaska, Fairbanks, Alaska 99775-7320. (ao@how.gi.alaska.edu)

RHESSI Observations of Particle Acceleration in Solar Flares

R. P. Lin

Physics Department and Space Sciences Laboratory, University of California, Berkeley, California

The Sun is the most energetic particle accelerator in the solar system, producing ions up to tens of GeV and electrons to hundreds of MeV in solar flares and fast coronal mass ejections. The primary objective of the NASA Reuven Ramaty High Energy Solar Spectroscopic Imager (RHESSI) Small Explorer spacecraft (launched 5 February 2002) is to investigate particle acceleration and energy release in solar flares through imaging spectroscopy of flare hard X-ray/gamma-ray continuum and gamma-ray lines emitted by energetic electrons and ions, respectively. Here I present preliminary results on the hard X-ray imaging spectroscopy of a flare, including the spatial variation with energy of the hard X-ray sources, the energy spectra and timing of the individual sources, and spectral features and total energy content. RHESSI observes continuous solar emission in the 3-10 keV energy range, with many microflares from many active regions. These microflares have a non-thermal power-law component similar to normal flares but with much steeper spectral slopes. I also present the first high resolution gamma-ray line spectrum and the first imaging of gamma-ray lines, from the large X4.8 flare of 23 July 2002.

1. INTRODUCTION

The primary scientific objective of RHESSI (renamed after launch to honor Dr. Reuven Ramaty, who developed much of the theoretical framework for solar gamma-ray line spectroscopy) is to investigate particle acceleration and explosive energy release in the magnetized plasmas at the Sun. The Sun accelerates ions up to tens of GeV and electrons to hundreds of MeV in solar flares and in fast coronal mass ejections (CMEs). Solar flares are the most powerful explosions in the solar system, releasing up to 10^{32}–10^{33} ergs in 10^2–10^3 s. The flare-accelerated ~10-100 keV electrons (and sometimes >~1 MeV/nucleon ions) appear to contain a significant fraction, ~10–50, of this energy, indicating that the particle acceleration and energy release processes are intimately linked.

High-energy emissions are the most direct signature of the acceleration of electrons, protons and heavier ions in solar flares. Accelerated electrons colliding with the ambient solar atmosphere produce bremsstrahlung hard X-ray and gamma-ray continuum emission, while nuclear collisions of energetic ions result in a complex spectrum of narrow and broad gamma-ray lines. Hot (million-degrees) thermal flare plasmas also emit bremsstrahlung X-rays. RHESSI utilizes a single instrument (*Lin et al.*, 2002) consisting of an Imaging System and a Spectrometer (Figure 1), to provide the high resolution imaging (~2.3 arcsec) spectroscopy (~1 keV FWHM) of solar emissions from soft X-rays (3 keV) to gamma-rays (17 MeV).

Since focusing optics, commonly used for soft X-rays/ EUV and longer wavelength emissions, are not feasible for hard X-rays and gamma-rays, a novel Fourier transform imaging system is employed (see *Hurford et al.*, 2002, for description). This consists of nine rotating modulation collimators (RMCs), each made up of a pair of widely separated grids mounted on the rotating spacecraft. Each grid consists of a planar array of equally-spaced, X-ray-opaque slats separated by transparent slits. The two grids of each RMC are

Particle Acceleration in Astrophysical Plasmas
Geophysical Monograph Series 156
Copyright 2005 by the American Geophysical Union
10.1029/156GM20

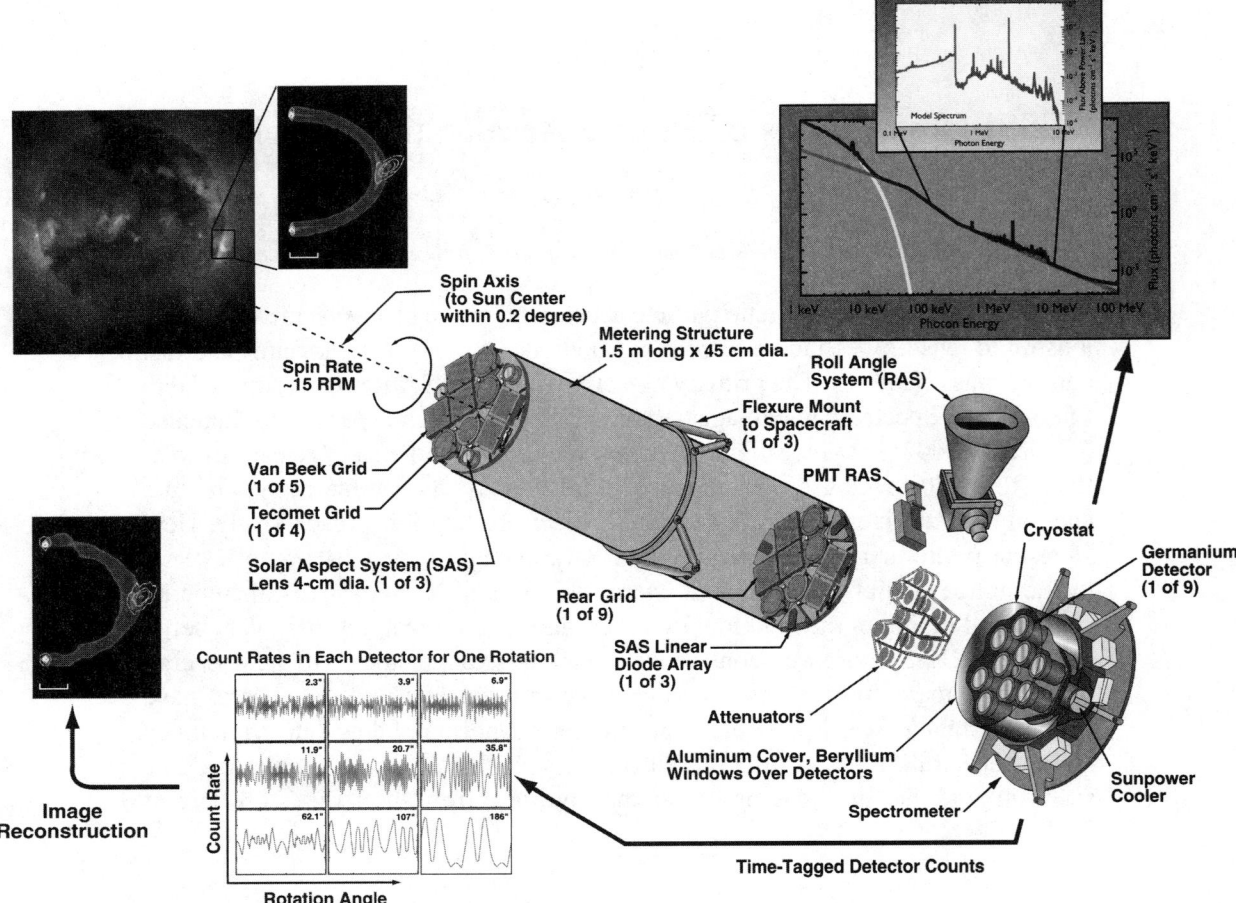

Figure 1. Schematic of the RHESSI instrument X-rays and gamma-rays from the Sun (upper left) pass through the slits of the front and rear grids of each of the nine grid pairs to reach the germanium detector. As the spacecraft rotates, the detector count rates are temporally modulated (lower left). These modulations can be analyzed to reconstruct the image. The germanium detectors are cryogenically cooled to provide high spectral resolution capable of resolving narrow gamma-ray lines and steep solar continuum spectra (upper right). The attenuators are inserted automatically when the count rate approaches saturation. The SAS, RAS, and PMTRAS provide solar pointing and roll aspect information.

aligned parallel to each other; the transmission through the grid pair depends on the direction of the incident X-rays. As the spacecraft rotates, the RMCs convert the spatial information of the source into temporal modulation of the photon-counting rates of the segmented germanium detectors (GeDs) behind each RMC.

The GeDs are segmented into a thin front planar segment and a thick rear segment. The intense hard (~3–250 keV) X-ray fluxes that usually accompany large flares are absorbed by the front segment, so the rear segment will always count ~0.3–17 MeV gamma-rays at moderate rates, essential for line measurements where optimal spectral resolution and low dead time are desired. The GeDs are cooled to <~75 K by a space-qualified long-life mechanical cryocooler, to achieve high spectral resolution (~1 keV).

To accommodate the large dynamic range (~10^7) in soft X-ray flux from microflares to very large flares, two sets of aluminum disk attenuators automatically move in front of the GeDs to absorb low energy (<~10 keV or <~20 keV) photons.

The energy and arrival time of every photon, together with pointing information, are recorded in the spacecraft's on-board 4-Gbyte solid-state memory (sized to hold all the data from the largest flare) and automatically telemetered within 48 hours. With these data, the X-ray/gamma-ray images can be reconstructed on the ground (see Figure 3). The instrument's ~1° field of view is much wider than the ~0.5° solar diameter, so all flares are detected, and pointing can be automated.

RHESSI was launched on 5 February 2002 into a near

circular, 600 km altitude orbit at 38° inclination. Observations began a week later, after the GeDs were cooled to ~75 K. Since then, RHESSI has been operating continuously; through the end of August 2002, it had detected over 1900 flares above 12 keV and over 600 above 25 keV. Here we present some of the preliminary results; the first hard X-ray imaging spectroscopy of solar flares; the first detection of 3–10 keV hard X-ray microflares; the first high resolution spectrum of solar gamma-ray lines and the first images of energetic ions and neutrons from gamma-ray line emission; and the first imaging spectroscopy of a flare/coronal mass ejection (CME).

2. 20 FEBRUARY FLARE

Bursts of bremsstrahlung hard X-rays (>~20 keV), emitted by accelerated electrons colliding with the ambient solar atmosphere, are the most common signature of the impulsive phase of a solar flare. Provided the electron energy E_e is much greater than the average thermal energy, kT, of the ambient gas, essentially all of the electron energy will be lost to Coulomb collisions, with only a tiny fraction (~10^{-5}) lost to bremsstrahlung. For this non-thermal situation, the hard X-ray fluxes observed in many flares indicate that the energy in accelerated >20 keV electrons must be comparable to the total flare radiative and mechanical output (*Lin and Hudson*, 1976). Thus, the acceleration of electrons to tens of keV may be the most direct consequence of the basic flare-energy release process.

During the flare impulsive phase, a double footpoint structure is often observed with the two footpoints brightening simultaneously to within a fraction of a second (Sakao et al., 1994, 1996). Very rapid transport of energy from the release site down to the footpoint interaction regions is required, which can only be achieved by fast electrons streaming down the loop and depositing their energy in lower coronal and chromospheric footpoints.

The 20 February, 11:05 UT flare (GOES class C7.7) observed by RHESSI (Figure 2) shows a single, simple HXR peak (*Krucker and Lin*, 2002; *Aschwanden et al.*, 2002; *Sui et al.*, 2002; *Vilmer et al.*, 2002). At ~10 arcsec resolution (comparable to Yohkoh HXT) RHESSI images (Figure 3) in energy bands as narrow as 2 keV reveal a transition from an elongated single source at 12–14 keV to two footpoints at 30–80 keV. RHESSI's imaging spectroscopy—spatially resolved photon spectra—show essentially identical power-law spectra in the two footpoints (Figure 4), suggesting that the energetic electrons producing the X-rays come from the same source, with a much steeper spectrum between the footpoints.

Surprisingly, however, the maxima of the two footpoint sources did not occur simultaneously (Figure 5, contrary to the general behavior seen by *Sakao et al.* (1996). The southern source peaked about ~8 seconds later than the northern source. However, imaging at resolution below 4" and with a cadence of 1 second (with an integration time over ~4 s, one spin period) showed three sources: source 1 to the north, source 3 to the south, and a weaker, earlier-peaking source 2 in between but much closer to source 3 (Figure 6). With this third source, the observed temporal evolution could be made consistent with simultaneous brightening of connected footpoints in two loops; one loop connecting sources 1 and 2, the second connecting source 1 and 3. In the beginning of the HXR peak (around 11:06:08), sources 1 and 2 are seen brightening simultaneously, reaching only moderate fluxes. Then, source 3 appears when source 1 peaks, while source 2 is slowly disappearing. In a third and last phase, source 3 peaks with a weaker simultaneous brightening of source 1.

Next to the three main sources, there may be a fourth source that might just be slightly above the EIT loop (*Sui et al.*, 2002), similar to the weak hard X-ray sources detected in the corona by Yohkoh HXT (*Masuda et al.*, 1994; *Alexander and Metcalf*, 1997) above the soft X-ray loop linking the hard X-ray footpoints. Such sources have been interpreted as evidence for energy release by magnetic reconnection in a region above the soft X-ray loop. The time variation of this possible loop top source is shown in Figure 6 (bottom) in black.

RHESSI's high spectral resolution (~1 keV FWHM) allows the transition between the emission of hot thermal flare plasma at low energies and the emission of the non-thermal, power-law energetic electrons at higher energies to be accurately defined, for the first time. For the 20 February flare and many others detected by RHESSI, the power-law component, integrated over the image, is observed to extend down to ~10 keV (Figure 7). This implies significantly more energy in the accelerated electrons than derived from previous hard X-ray measurements, which were limited to low energy cutoff of >20 or >25 keV (previous detectors required a thick entrance window to prevent contamination by thermal emission because of their broad spectral response). For the hard X-ray power-law index of ~3 observed for this flare below the break, the energy in electrons is ~5–10 times more above 10 keV than above 20 or 25 keV.

The spectrum of the non-thermal component in Figure 7 fits well to a double power-law shape with a relatively sharp downward break, similar to the first high resolution flare hard X-ray spectrum, obtained from a balloon-borne germanium spectrometer (*Lin and Schwartz* 1987). This is commonly observed in many RHESSI flares. *Kontor et al.* (2002) showed that a single power-law spectrum of accelerated electrons could produce such a break in the X-ray spectrum due the penetration of the higher energy electrons into the predominantly neutral chromosphere, where collision losses

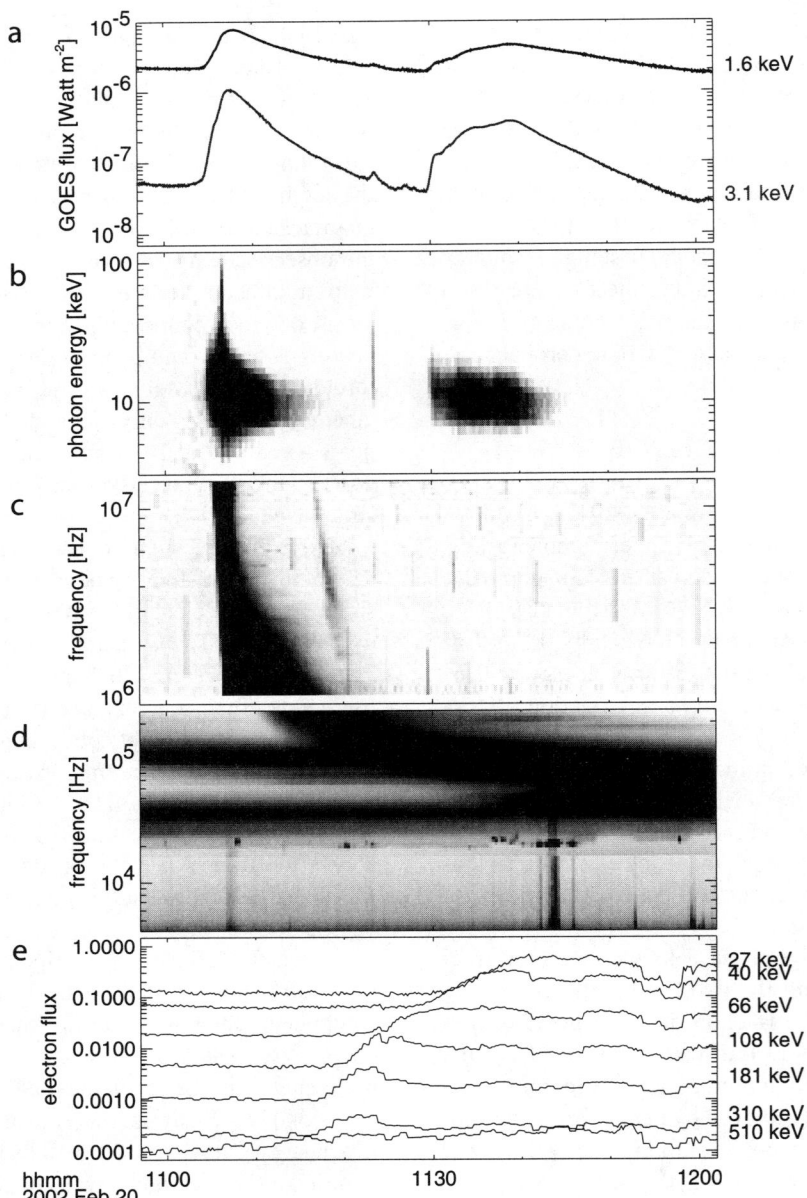

Figure 2. Example of a solar flare event versus time as seen by: (a) GOES in soft X-rays; (b) RHESSI from 3 to ~250 keV in a spectrogram, where intensity is gray-scale; (c) and (d) radio spectrogram (gray-scale) from 14 MHz to 5 kHz from the Wind spacecraft WAVES experiment (Bougeret et al., 1995), showing the type III solar radio burst emitted by energetic electrons escaping to the interplanetary medium; and (e) observations of energetic electrons near the Earth at 1 AU from the 3-D plasma and energetic particles experiment (Lin et al., 1995) on Wind, showing the velocity dispersion (faster electrons arriving earlier).

are significantly lower than in the fully ionized corona where the low-energy electrons stop. On the other hand, the electrons that escape into the interplanetary medium from impulsive flares typically have a double power-law spectrum (including the 20 February event of Figure 4), suggesting that the break may be a characteristic of the source.

3. HARD X-RAY MICROFLARES

The RHESSI instrument provides uniquely high sensitivity in the energy range from ~3 to 15 keV, together with ~1 keV FWHM spectral resolution and simultaneously imaging down to 2.3 arcsec. For comparison the Hard X-ray Imaging

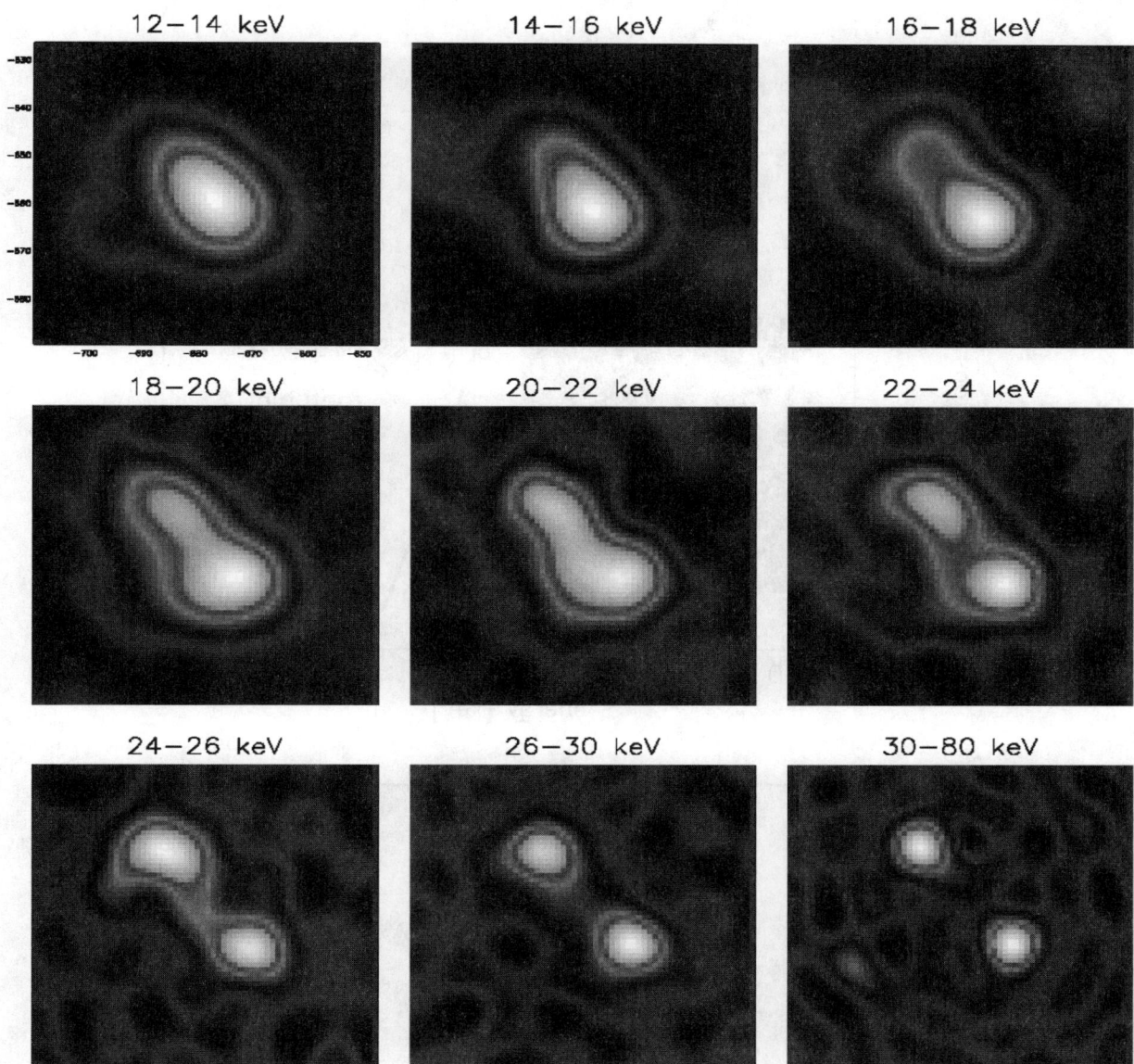

Figure 3. RHESSI images at ~10 arcsec spatial resolution in 2 keV energy bands from 12 to 26 keV, plus broader bands up to 80 keV, at the peak (1106–1106:40 UT) of the 20 February 2002 flare. The flare exhibits an elongated source at 12–14 keV changing to two footpoints at 30–80 keV. The box is 64 arcsec on a side.

Spectrometer (HXIS) on the Solar Maximum Mission provided effective areas of ~0.7 mm^2 at 3.5 keV rising to ~7–26 mm^2 at ~10–15 keV (*Van Beek et al.*, 1980). RHESSI, with shutters out, has an effective area of ~10 mm^2 at 3.5 keV, rising rapidly to ~250 mm^2 at 5 keV and ~3500 mm^2 at 10–15 keV—a factor of ~14 to >130 times larger.

With the shutter out, RHESSI detects solar flux in the ~3–10 keV band essentially all the time, with frequent small bursts evident, e.g., microflare activity. In one orbit of RHESSI data (2 May 2002, 01:40–02:40 UT) at a time of GOES soft X-ray background level around B6, at least 7 microflares (marked with numbers in Figure 8) could be easily identified (*Krucker et al.*, 2002; *Benz and Grigis*, 2002). A number of even smaller transient increases—e.g.,

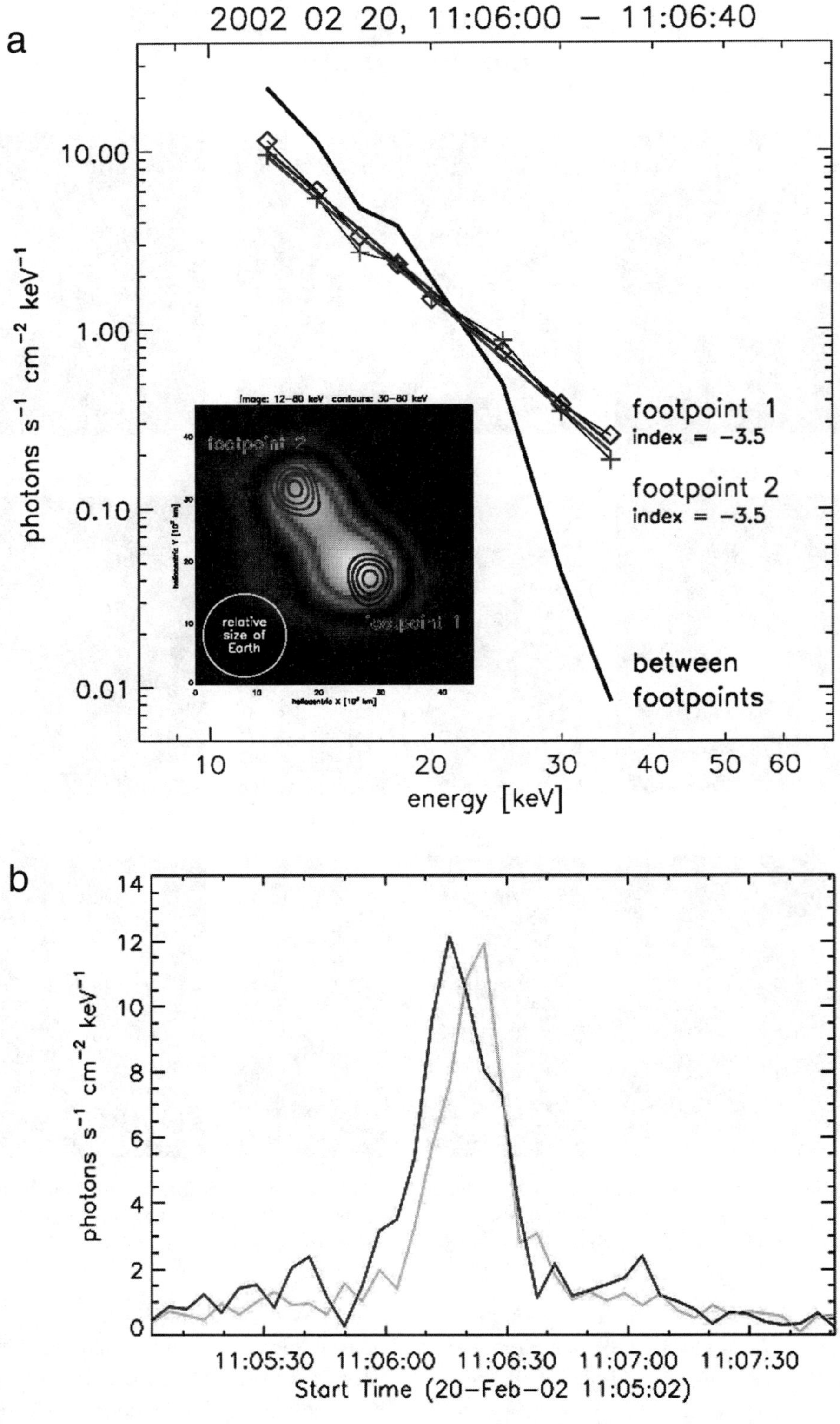

Figure 4. (a) At the peak (1106–1106:40 UT) of the 20 February 2002 flare, the spectra of the two footpoints appear to have similar power-law slopes, while the emission in between the footpoints has a much steeper spectrum. (b) The timing of the emission from the two footpoints (see Figure 4), showing a ~8 second delay between the peaks.

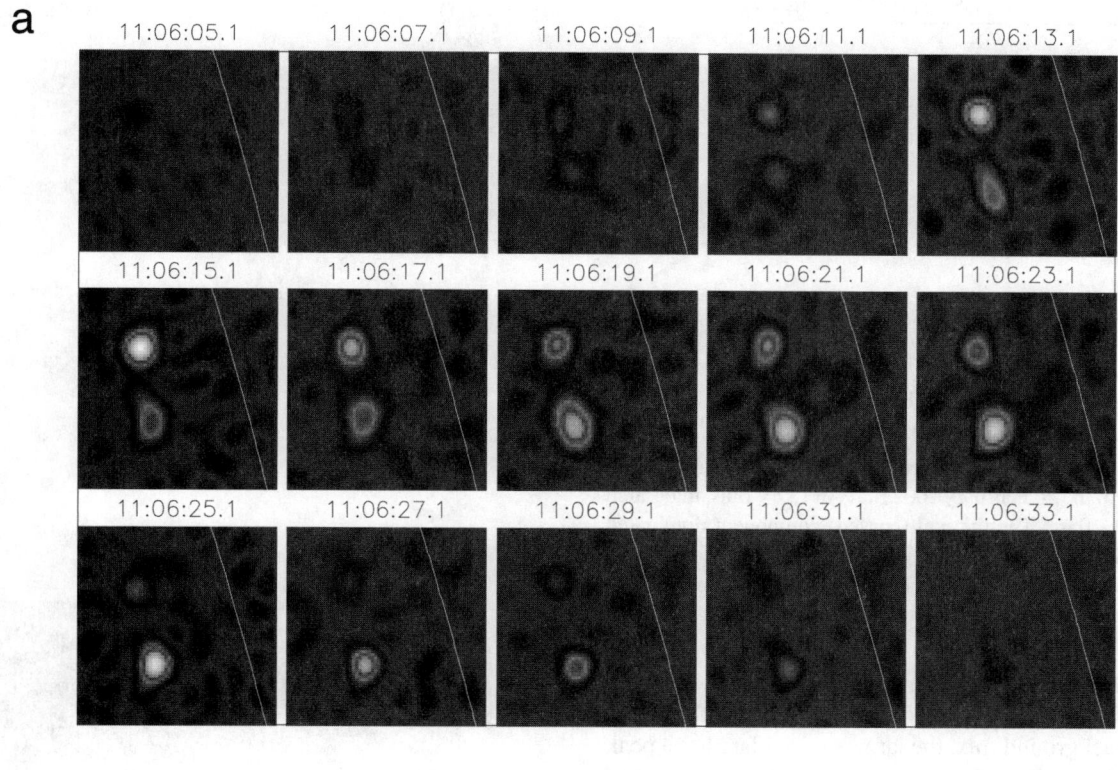

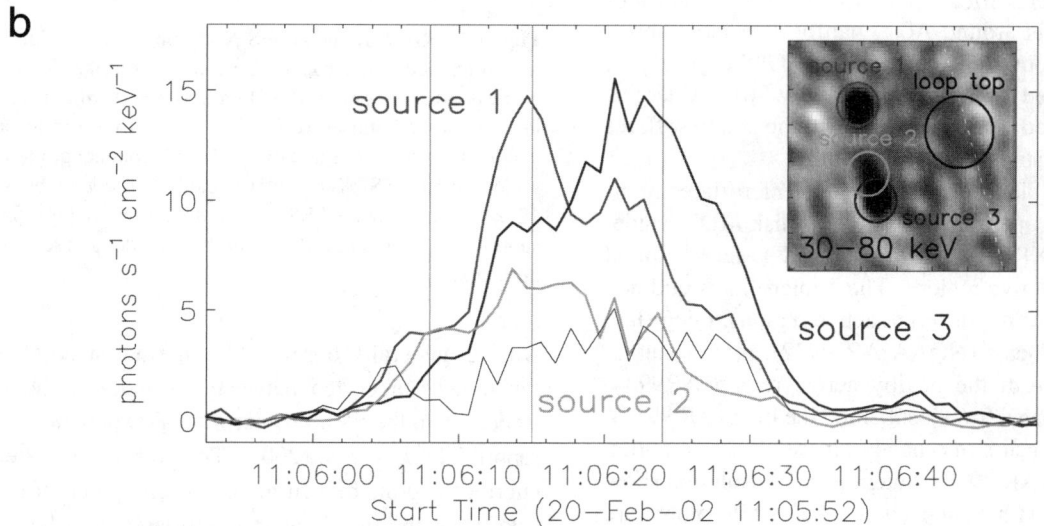

Figure 5. (a) HXR images (30-80 keV, 7″ resolution) taken every 2 s with 4-s integration time of the 20 February flare are shown. The field-of-view is 64″ by 64″, the white lines give the location of the solar limb. (b) The 3 different HXR sources appearing in the 20 February flare are shown in the top 3 figures. The images show the thermal emission (6–9 keV) with the 30–80 keV HXR contours overlaid (levels are 30, 50, 70, and 90%). Later in the event EIT 195 Å difference images show the post-flare loops (second row of figures). To outline the position of the earlier-occurring HXR sources, contours of the time that average 30–80 keV emission are overplotted. The bottom panel shows the temporal evolution of the different sources, including a possible source above the EIT loop top (black). The circles in the insert mark the source locations, but do not represent the source sizes. The three gray vertical lines give the center time of the interval of the above shown images; time intervals are ~8 s, ~8 s, and ~16 s, respectively.

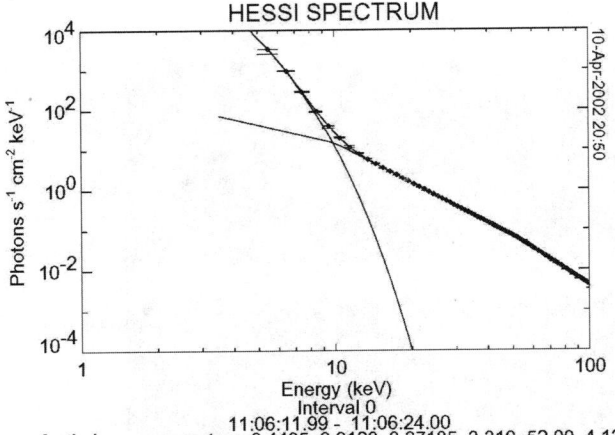

Figure 6. The photon energy spectrum near the peak (1106:12–1106:24 UT) of the 20 February 2002 flare, in 1 keV bins, integrated over the entire flare. The thermal (curved) component dominated below ~10 keV, but the power-law (non-thermal) component clearly extends down to ~10 keV. The non-thermal component fits to a double power-law with a relatively sharp break at ~52 keV.

at 02:04, 02:19, 02:27, 02:32, and 02:36 UT—appear to be significant in the 3–12 keV count rates. After subtracting the GOES background flux, the largest microflare has a peak GOES soft X-ray classification of A6, while the smallest of the 7 microflares is around A0.1, hardly detectable above the GOES background. Compared to the GOES survey of flares sizes reported by *Feldman et al.* (1996), the largest microflare presented here corresponds to the smallest flares presented in their study.

Figure 9 shows cleaned images of the microflares numbered in Figure 1 superimposed on a full-disk MDI magnetogram. All the HXR microflares analyzed so far are found to originate from active regions. The 7 microflares outlined in Figure 8 come from 4 different active regions: microflare 1 is from the southeast (NOAA AR 9932); microflares 2, 4, 5, and 6 are from the nearby active region AR9934; microflare 3 originates from just behind the limb (AR 9915); and microflare 7 occurs in a newly emerging active region near the east limb (AR 9938). Imaging at 7″ resolution show some microflares with elongated sources, while others are spatially not resolved.

As a preliminary analysis, spectral fits are performed separately to the impulsive phase and the decay phase of the microflares. Examples of spectra of the first three microflares are shown in Figure 10. A thermal fit alone cannot represent the data of microflares 1 and 2 during the impulsive phase, but a thermal plus a non-thermal power-law produces a good fit. The non-thermal power-law fit reveals steep spectra for the 2 May 01:40–02:40 UT time interval with exponents ranging from −5 to −8. These are much steeper values than

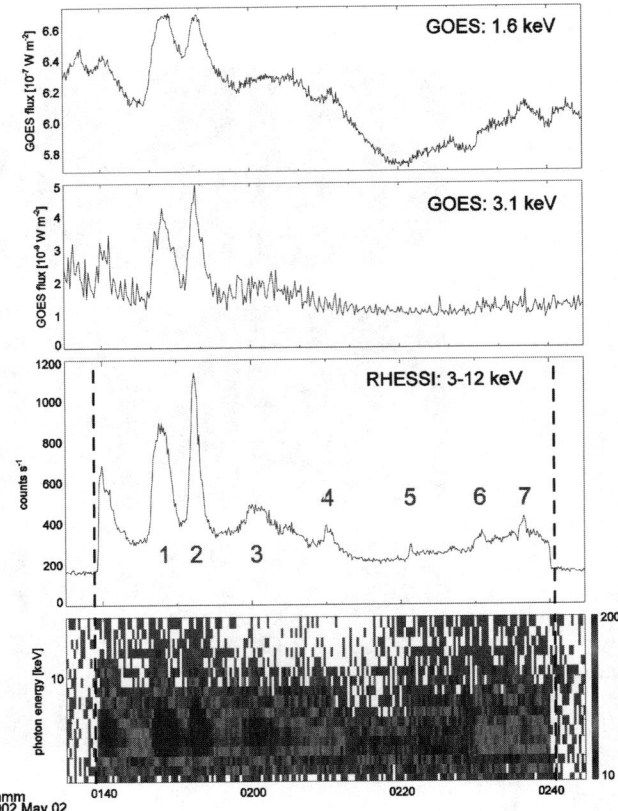

Figure 7. RHESSI and GOES X-ray observations during times of low solar activity. Around one hour of observations are shown corresponding to one RHESSI day. The vertical dashed lines give the sunrise and sunset, respectively. From the top to bottom, the panels show the softer and harder GOES channel plotted in a linear scale, the RHESSI X-ray total uncalibrated counts between 3 and 12 keV, and the RHESSI count spectrogram plot (background subtracted). The microflares studied in detail are marked with numbers.

what is generally observed for normal sized flares where the indices are often between −3 and −4, and somewhat steeper than the >8 keV microflares (exponents of −3 to −7) reported by *Lin et al.* (2001). The power laws extend to low energies and the best fit gives low energy cutoffs of around 7 keV for the two biggest microflares with best statistics. The smaller microflare 3, which occurs just behind the limb, does not show a clear non-thermal component in the spectra and is best fitted with a thermal spectra only, suggesting that the non-thermal emission is occulted.

The spectra of the decay phases do not show a picture as clear as the impulsive phase. Some events, like microflare 1, are moderately well fitted with a thermal emission only; others, like microflare 2, are better represented with a thermal plus a non-thermal fit (Figure 10).

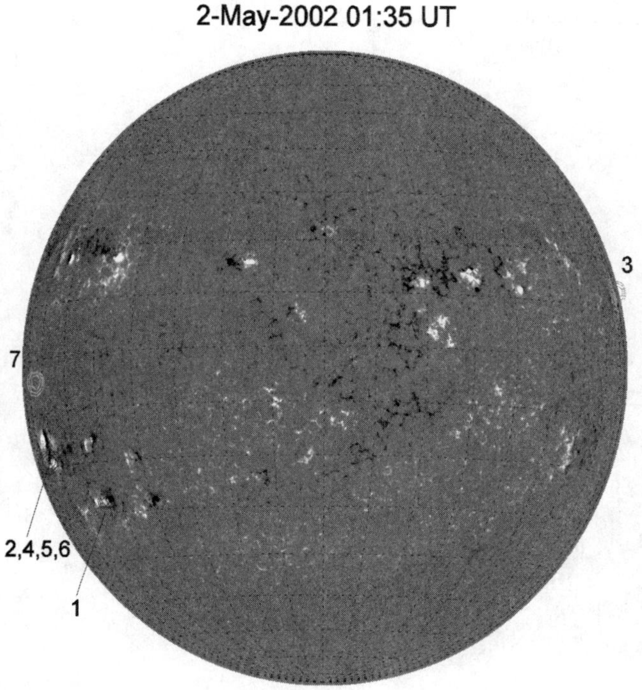

4. 23 JULY 2002 GAMMA-RAY LINE FLARE

On 23 July 2002 a GOES X4.8 flare located at S13E72 was observed by RHESSI (Figure 11) up to ~10 MeV. Figure 12a shows the hard X-ray sources were compact and grouped within ~30 arcsec of each other. A soft source to the East dominates at energies below ~20 keV, but by ~30 keV three compact sources appear to the west of the soft source. These are seen up to 100–200 keV and higher. Figure 12b shows that the north and south sources have closely similar double power-law spectra while the spectrum of the middle source is closer to a single power-law, suggesting the north and south sources are magnetically connected to the same source of accelerated electrons.

Figure 13 shows the gamma-ray spectrum for this flare from 0.3 to 10 MeV. This is the first high-resolution spectrum of a gamma-ray flare ever obtained. Since the expected gamma-ray lines are predominantly at energies below ~7 MeV, the sharp drop seen there indicates that the line emission dominates over the electron bremsstrahlung continuum in the MeV range.

The strongest and narrowest line at 2.223 MeV (expected width <~0.1 keV) is emitted by deuterium formed in an excited state by the capture of neutrons (produced by high

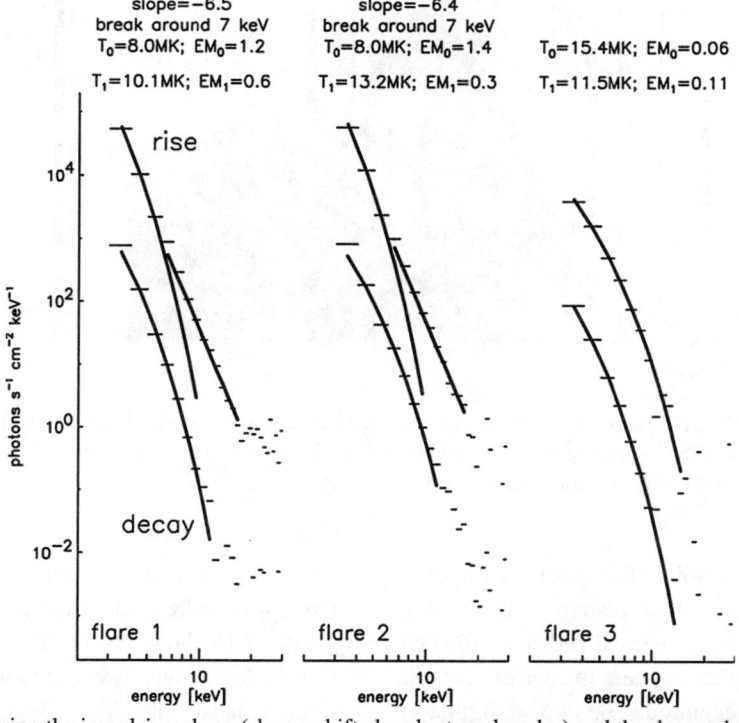

Figure 8. The locations of the 7 microflares marked in Figure 7 are shown on an MDI magnetogram.

Figure 9. Spectra during the impulsive phase (shown shifted up by two decades) and the decay phase of the microflares labeled 1 to 3 in Figure 1. The impulsive phase is fitted with both a thermal and a non-thermal component, the decay phase with a thermal component only. For the behind-the-limb microflare (flare 3), a thermal alone fits the data well enough. The curves shown give the range fitted; values above ~15 keV are dominated by noise.

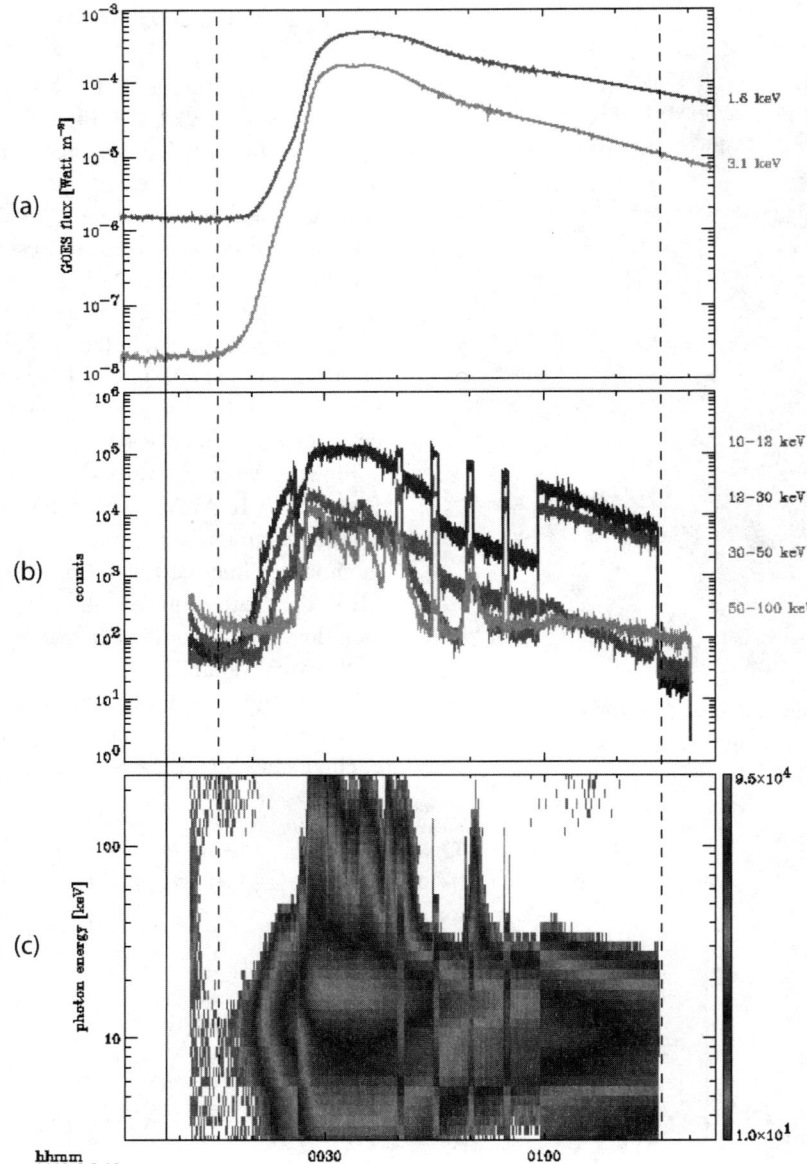

Figure 10. The 23 July 2002 flare (GOES X4.8), as observed by: (a) GOES in soft X-rays; (b) RHESSI in four energy bands; and (c) on a RHESSI spectrogram (see explanation in Figure 2). The sharp vertical features are due to the attenuator moving in and out as the count rate increases or decreases.

energy ions colliding with the solar atmosphere) by hydrogen. This line is delayed (by ~100 s) relative to the prompt lines since the neutrons must be thermalized by collisions with hydrogen in the photosphere (where the density is high enough) before they can be captured. The line strength and thermalization time depend on the density of ^{3}He in the photosphere, because the neutrons can also be captured on ^{3}He without radiation. Figure 14 shows the 2.223 MeV line profile, uncorrected for instrument response. The measured FWHM is ~4 keV, close to the intrinsic instrumental width.

Figure 15 shows the narrow Mg, Si, Ne lines in the ~1.2 to 1.8 MeV range. These are the first measurements able to resolve these lines. Their line profiles are determined by Doppler broadening; the observed widths appear consistent with a downward isotropic distribution of energetic protons. The line centroids appear to show a small red shift, although

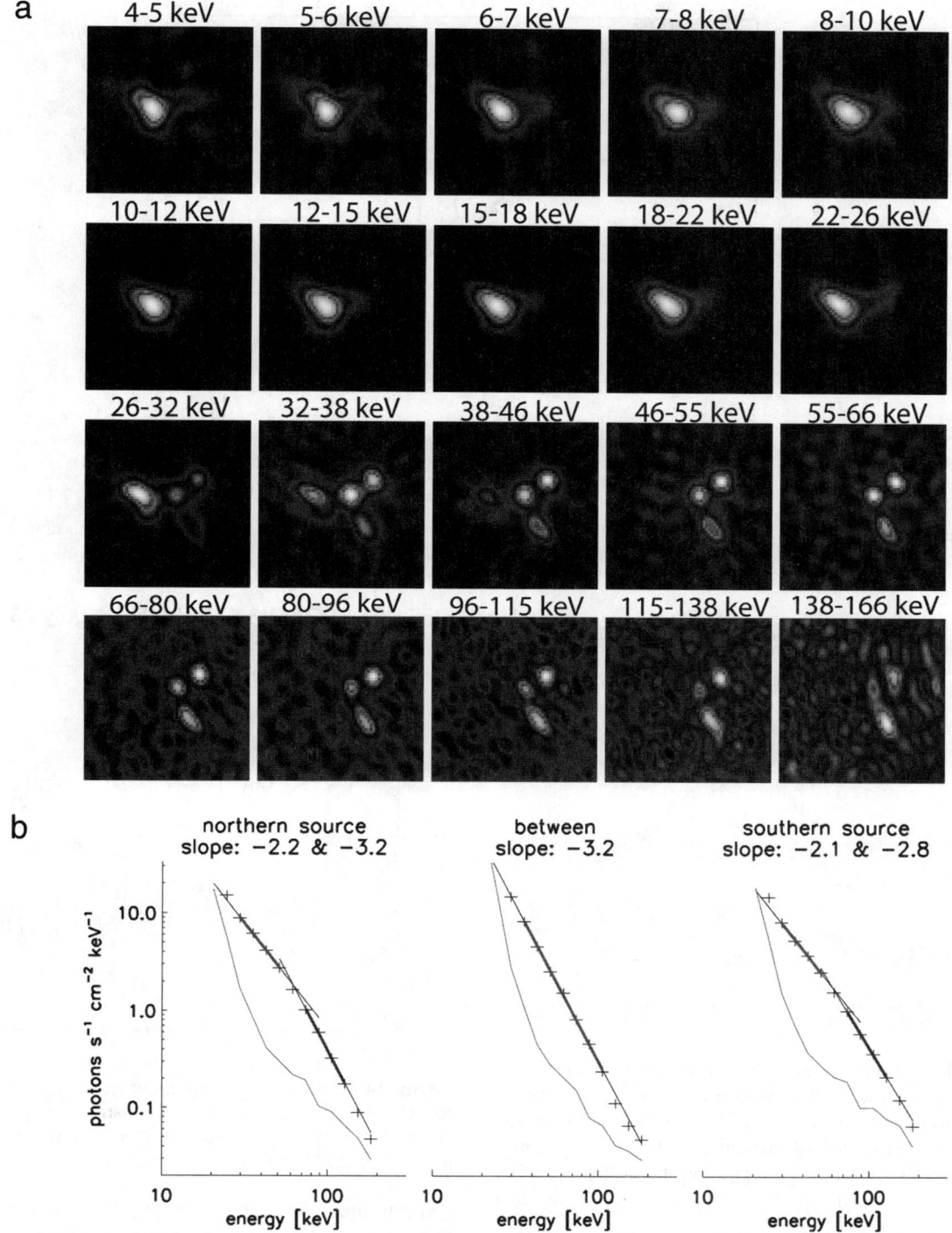

Figure 11. (a) Imaging of the 23 July 2002, X4.8 solar flare in 20 energy bands, from 1-keV-wide bins at 4 keV, to 28-keV-wide bins (for enough counts to image) at 138 keV, illustrating the changes in sources as a function of energy, from a single dominant elongated source at energies below ~30 keV to three sources above ~40 keV. The images are 64 arcsec on a side; the lower left corner is just at the southeast limb of the Sun. b) The energy spectra of the three dominant sources at energies above ~40 keV, showing that the spectra are similar for the north and south sources but both are quite different from the source in between. The dashed lines indicate background.

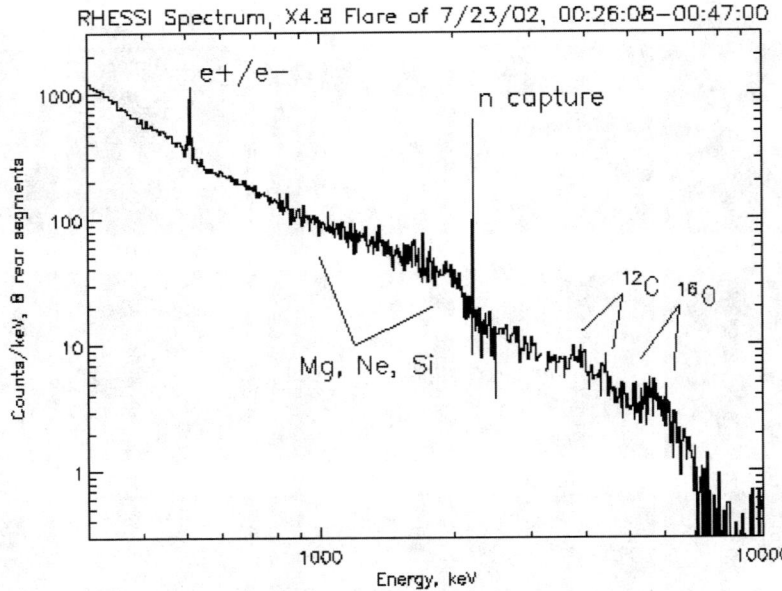

Figure 12. The gamma-ray spectrum from 0.3 to 10 MeV, integrated over the 23 July 2002 flare. The emission from ~1 to 7 MeV is dominated by broad and narrow gamma-ray line emission.

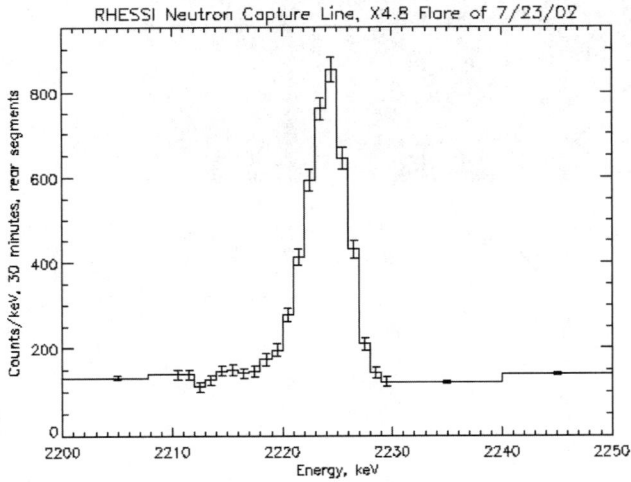

Figure 13. The (preliminary, uncalibrated) spectrum of the intrinsically extremely narrow 2.223 MeV line for the 23 July 2002 flare. This line is produced by the capture of neutrons (from spallation reactions) by hydrogen to form deuterium. The observed line width of ~4 keV is nearly the same as the instrumental width.

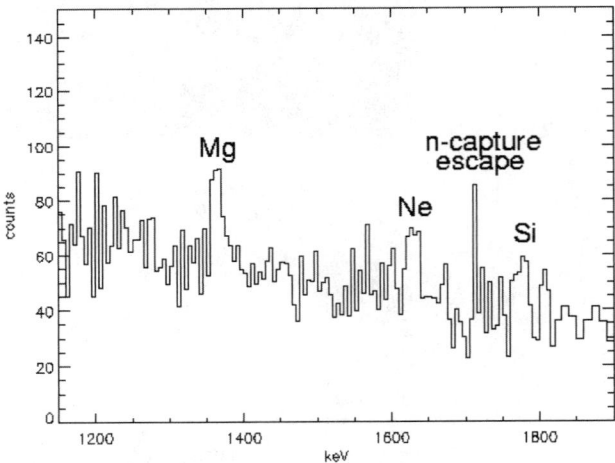

Figure 14. The gamma-ray spectrum from ~1.2 to 1.9 MeV for the 23 July 2002 flare, showing the narrow lines from Mg, Ne, and Si. The very narrow feature at ~1.7 MeV is the single-escape peak of the 2.223 MeV line.

this will require more analysis to confirm. Since the flare is located near the east limb, we would not expect a large red shift.

Up to now, there has been no information on the location of the energetic ions produced in flares. *Vestrand and Forest* [1993] reported the detection of the 2.223 MeV line from an over the limb flare by SMM. Since this line is formed by neutron capture in the photosphere, this observation suggests that the energetic ions were able to travel far away from the flare. Two of the RHESSI grids are thick enough (1.6 and 3 cm thick tungsten) to image this line, at 36 arcsec and 3 arcmin resolution, respectively. Since the line is narrow, continuum background can be almost completely rejected by limiting the energy range to <10 keV. Figure 16 shows the image of this line.

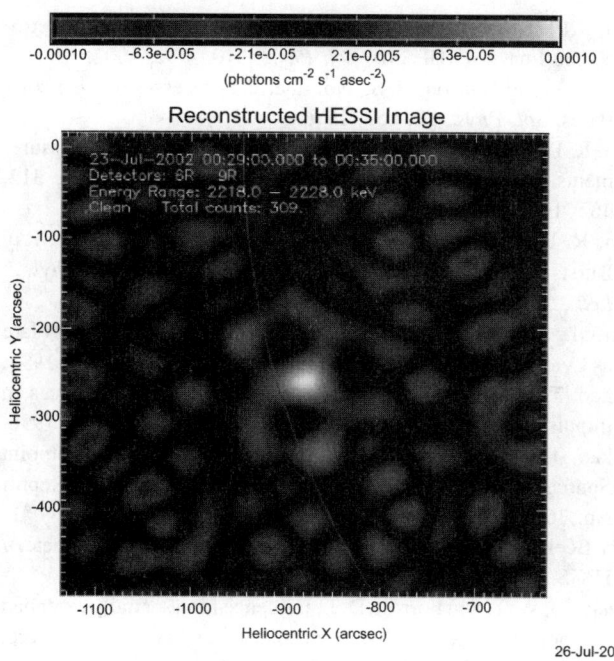

Figure 15. The first gamma-ray line (the 2.223 MeV line) image of a flare, at ~36 arcsec resolution. The solar limb is indicated by the curved line.

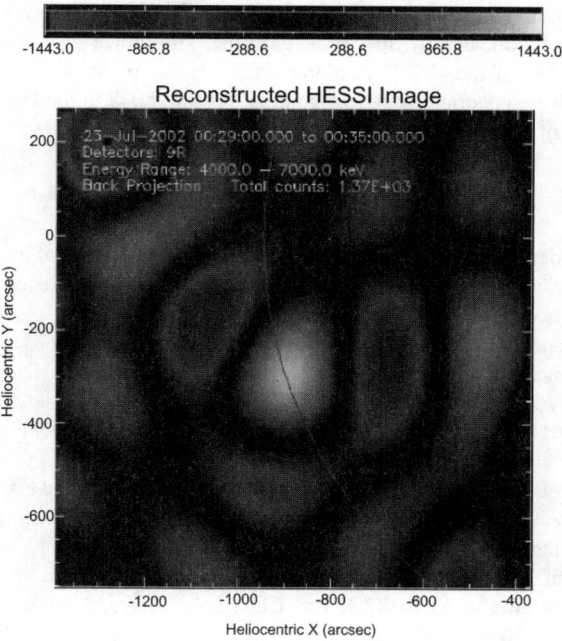

Figure 16. The gamma-ray image in the 4–7 MeV band, which is dominated by broad and narrow prompt lines of carbon and oxygen. The narrow lines are due to energetic protons and alphas colliding with ambient carbon and oxygen while the broad lines which form the continuum in this energy range are due to energetic carbon and oxygen colliding with ambient hydrogen and helium. Thus this image shows where the energetic ions are interacting with the solar atmosphere. The first gamma-ray line (the 2.223 MeV line) image of a flare, at ~30 arcsec resolution. The solar limb is indicated by the curved line (Bougeret et al., 1995).

As mentioned earlier, the nuclear line emission dominates over the electron bremsstrahlung continuum, especially in the 4–7 MeV range. Besides the narrow lines produced by energetic protons and alphas colliding with ambient carbon and oxygen nuclei, the underlying broad "continuum" is primarily due to energetic carbon and oxygen colliding with ambient hydrogen and helium. Thus, the image of the 4–7 MeV range (Figure 16) shows the location of the energetic ions for the first time.

It's clear that most of the ions are colliding with the solar atmosphere in the vicinity of the flare and hard X-ray sources, although we cannot tell at present whether the energetic electrons and ions are cospatial. Further analysis is required.

5. SUMMARY

As indicated by the preliminary results presented here, RHESSI is already providing many exciting new results. However, the power of RHESSI lies in its capability for detailed quantitative probing of the particle acceleration and energy release mechanism in flares and related phenomena. That will require careful, comprehensive analysis of the RHESSI data, together with the context measurements from other spacecraft and from the ground.

Acknowledgments. I wish to acknowledge the RHESSI team, which consists of Co-Investigators Gordon Hurford, Hugh Hudson, and Norman Madden at Berkeley; Brian Dennis, Carol Crannell, Gordon Holman, Reuven Ramaty and Tycho von Rosenvinge at Goddard Space Flight Center (GSFC); Alex Zehnder at the Paul Scherrer Institute in Switzerland; Frank van Beek at Delft University in The Netherlands; Patricia Bornman at NOAA; Richard Canfield at Montana State University; Gordon Emslie at University of Alabama, Huntsville; Arnold Benz at Institute of Astronomy, Zurich, Switzerland; John Brown at University of Glasgow in Scotland; Shinzo Enome at NAO, Japan; Takeo Kosugi at ISAS, Japan; and Nicole Vilmer at Observatoire de Paris, Meudon, France. Associated Scientists on the HESSI team are David Smith, Jim McTiernan, Isabel Hawkins, Said Slassi-Sennou, Andre Csillaghy, George Fisher, Chris Johns-Krull (now at Rice University) at Berkeley; Richard Schwartz, Larry Orwig, Dominic Zarro at GSFC; Ed Schmahl at University of Maryland; and Markus Aschwanden at Lockheed-Martin. The engineering team at Berkeley is led by Peter Harvey, Dave Curtis, and Dave Pankow; at GSFC, by Dave Clark and Rob Boyle; at PSI, by Reinhold Henneck, Akilo Michedlishvili, and Knut Thomsen. The HESSI spacecraft is being developed and fabricated by Spectrum Astro Inc. The project manager for HESSI in the GSFC Explorer Office is Frank Snow. This research is supported by NASA contract NAS5-98033 to the University of California, Berkeley.

REFERENCES

Alexander, D. and Metcalf, T. R., A Spectral Analysis of the Masuda Flare using *Yohkoh* Hard X-Ray Telescope Pixon Reconstruction, *Astrophys. J.*, 489, 442, 1997.

Aschwanden, M. J. et al., Chromospheric Height and Density Measurements in a Solar Flare Observed with RHESSI – II. Data Analysis, *Sol. Phys.*, 210, 383-405, 2002.

Benz, A. O. and Grigis, P. C., Microflares and hot component in solar active regions, *Sol. Phys.*, 210, 431-444, 2002.

Bougeret, J.-L. et al., Waves: The Radio and Plasma Wave Investigation on the WIND Spacecraft, The Global Geospace Mission, reprinted from *Space Sci. Rev.*, 71, 231-263, 1995.

Feldman, U. et al., Electron Temperature, Emission Measure, and X-Ray Flux in A2 to X2 X-Ray Class Solar Flares, *Astrophys. J.*, 460, 1034, 1996.

Hurdford, G. J. et al., The RHESSI Imaging Concept, *Sol. Phys.*, 210, 61-86, 2002.

Kontar, E. P. et al., Nonuniform Target Ionization and Fitting Thick Target Electron Injection Spectra to RHESSI Data, *Sol. Phys.*, 210, 419-429, 2002.

Krucker, S. et al., Hard X-ray Microflares down to 3 keV, *Sol. Phys.*, 2002.

Krucker, S. and Lin, R. P., Relative Timing and Spectra of Solar Flare Hard X-ray Sources, *Sol. Phys.*, 2002.

Lin, R.P. et al., The Reuven Ramaty High-Energy Solar Spectroscopic Imager (RHESSI), *Sol. Phys.*, 210, 3-32, 2002.

Lin, R. P. and Hudson, H. S., Non-thermal processes in large solar flares, *Sol. Phys.*, 50, 153, 1976.

Lin, R. P. and Schwartz, R. A., High spectral resolution measurements of a solar flare hard X-ray burst, *Astrophys. J.*, 312, 462, 1987.

Lin, R. P., P. T. Feffer and R. A. Schwartz, Solar Hard X-Ray Bursts and Electron Acceleration Down to 8 keV, *Astrophys. J. Lett.*, 557, L125, 2001.

Masuda, S. et al., A Loop-Top Hard X-Ray in a Compact Solar Flare as Evidence for Magnetic Reconnection, *Nature*, 371, 495, 1994.

Sakao, T. et al., Characteristics of hard X-ray double sources in impulsive solar flares, *Adv. Space Res.*, 17, No. 4-5, 60, 1996.

Sakao, T. et al., Hard X-ray Imaging Observations of Footpoint Sources in Impulsive Solar Flares, *Kofu Symp.*, NRO Report No. 360, 169, 1994.

van Beek, H. F. et al., The Hard X-ray Imaging Spectrometer / HXIS/, *Sol. Phys.*, 65, 39, 1980.

Vestrand, W. T. and Forrest, D. J., Evidence for a spatially extended component of gamma rays from solar flares, *Astrophys. J. Lett.*, 409, L69, 1993.

Vilmer, N. et al., Hard x-ray and Metric/Decimetric Radio Observations of the 20 February 2002 Solar Flare, *Sol Phys.*, 210, 261-272, 2002.

R. P. Lin, Space Sciences Laboratory, University of California, Berkeley, California 94720-7450. (rlin@ssl.berkeley.edu)

Magnetohydrodynamic Analysis of January 20, 2001, CME-CME Interaction Event

A. H. Wang and S. T. Wu

Center for Space Plasma and Aeronomic Research, University of Alabama in Huntsville, Alabama

N. Gopalswamy

NASA/Goddard Space Flight Center, Greenbelt, Maryland

To understand the physical mechanism of the CME-CME interaction recently observed by SOHO and WIND, we have constructed a numerical resistive MHD model by adding magnetic dissipation term into the global streamer model with bi-modal solar wind [*Wang et al.*, 1998], and emerging flux-ropes [*Wu and Guo*, 1997] to simulate these physical processes. Observations show that when two or more CMEs interact with each other, they merge and deflect each other, and can result in more complex ejecta. To compare our simulation with observations, we use the January 20, 2001 CME-CME interaction event close to the Sun recorded by SOHO/LASCO/C2/C3. The simulation results show that the acceleration and deceleration of the two CMEs are caused by the background solar wind and the deflection of each other due to the interaction. The CME cannibalization is caused by the magnetic reconnection.

1. INTRODUCTION

Solar eruptions drive the variability of geospace (i.e., near-Earth environment). These eruptions are well-known eruptive flares and coronal mass ejections. Recent observations have reinforced the view that CMEs are a key causal link to the major interplanetary disturbances and geomagnetic storms [*Kahler*, 1992; *Gosling et al.* 1991]. Thus, studies of CMEs, merged successive CMEs and CME-induced shock propagation are important for our understanding and prediction of interplanetary disturbances and geomagnetic storms.

Early Helios observations show that CMEs can interact with each other, with shocks, and co-rotating streamers. Recently, SOHO and WIND observations have revealed that CMEs cannibalize and deflect one another [*Gopalswamy et al.*, 2001; 2002]. CMEs are also accelerated and decelerated due to their interaction with the solar wind [*Gopalswamy et al.*, 2001; *Andrews and Howard*, 2001]. These CME interactions result in different solar wind signatures as compared to isolated CME events. *Burlaga et al.* [2001], using solar wind observations from the Advanced Composition Explorer (ACE) at 1 AU, studied fast ejecta, defined as transient, non-corotating flows that move past Earth during a day or more, with a maximum speed >600 km/s. These fast ejecta are important candidates for intense geomagnetic storms. *Burlaga et al.* [2001] identified two classes of fast ejecta: (1) magnetic clouds, whose local magnetic structure is a flux rope and (2) "complex ejecta," which have disordered magnetic fields. They found evidence that some of the complex ejecta could have been produced by the interaction of two or more successive CMEs. Later *Burlaga et al.* [2002] used observations from LASCO and ACE to identify three sets of successive halo CMEs directed toward Earth (two or more CMEs observed within 1–4 days in the corona) and corresponding flows and magnetic fields at 1 AU. Each set of successive CMEs merged between the Sun and 1 AU to form complex ejecta during propagation.

Particle Acceleration in Astrophysical Plasmas
Geophysical Monograph Series 156
Copyright 2005 by the American Geophysical Union
10.1029/156GM21

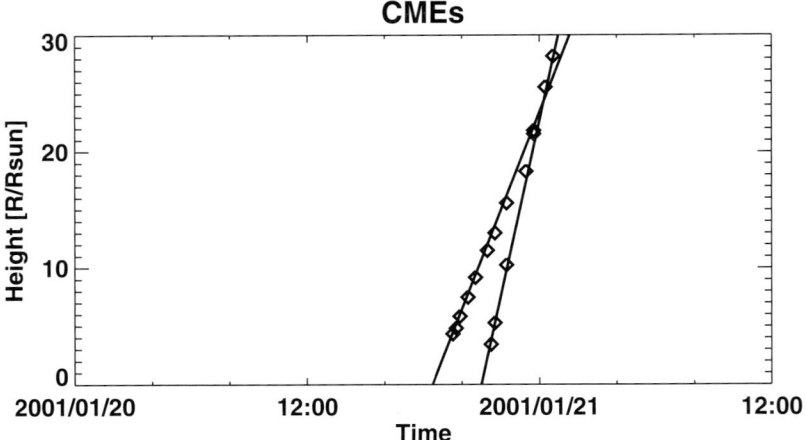

Figure 1. Height vs. UT time for the January 20–21, 2001, CMEs.

To understand the physical mechanism of these dynamical evolutionary processes, we have constructed a numerical resistive MHD model by adding magnetic dissipation term into the global streamer model with bi-modal solar wind [*Wang et al.*, 1998], and emerging flux-ropes to destabilize a streamer and launch CMEs [*Wu and Guo*, 1997]. In bi-modal solar wind, the background solar wind consists of a slow speed component about 300 km/sec near the solar equatorial region, and a fast speed component about 700 km/sec in the solar polar region. To investigate CME-CME interactions, the magnetic resistivity is included in the model for the magnetic reconnection process. The January 20, 2001, event is in the corona, and is used to compare and guide this MHD numerical simulation study. In section 2, the observations of the January 20, 2001, event used for the study are briefly described. Section 3 presents the MHD model and the numerical method. Numerical simulation results and comparison with observations are discussed in section 4 and the concluding remarks are given in section 5.

2. DESCRIPTION OF THE OBSERVATION

On January 20, 2001, SOHO/LASCO recorded the interaction between two fast CMEs in the range of 3.44 Rs (solar radius) to 30 Rs. These two CMEs originated from the same active region (AR 9313 at S07E40 and S07E46) and were separated from each other by only about 2.5 hours. Both were halo CMEs with an average speed of 840 km/sec and 1500 km/sec, respectively.

Before these two CMEs came into the field-of-view of the LASCO/C2/C3 ranges, the observations also showed occurrences of flares and radio emissions, such as Metric type II and IP type II, in correspondence with these two CMEs. Figure 1 shows two CMEs height as a function of time. The diamonds are the observation data for these two CMEs.

The height–time trajectories and corresponding velocity–height trajectories of these two CMEs are shown in Figure 2. The corresponding solar flares and radio emissions induced by these two CMEs are also shown in Figure 2a. These two CMEs merged into one at time 23:42 UT. Other observed important characteristics for these two CMEs are the acceleration of CME 1 and the deceleration of CME 2 from Figure 2. The general properties of these two CMEs are listed in Table 1.

3. NUMERICAL MHD MODEL

To reveal the related physical processes from January 20, 2001, event, we suggest that these two CMEs are formed by the streamer blow-out due to the emergence of two consecutive flux-ropes as suggested by Howard et al. [1985]. We employed our new 2D resistive MHD streamer model with bi-modal solar wind and emerging flux-ropes from the lower boundary to simulate this event. In our pre-event coronal model we assume an axisymmetric, time-dependent MHD flow of a single-fluid, polytropic, and fully ionized plasma. In order to obtain a bimodal solar wind, we added volumetric heating, momentum addition, and thermal conduction. The flow is calculated in a meridional plane (from the north pole to the equator). The governing equations for this model are represented by a set of conservation laws of mass, momentum, and energy. In addition, the induction equation is used to describe the dynamical interaction between plasma flow and the magnetic field with finite magnetic resistivity (η). This set of MHD equations is:

$$\frac{\partial \rho}{\partial t} = -\frac{1}{r^2}\frac{\partial}{\partial r}\left(r^2 \rho v_r\right) - \frac{1}{r\sin\theta}\frac{\partial}{\partial \theta}\left(\rho v_\theta \sin\theta\right), \quad (1)$$

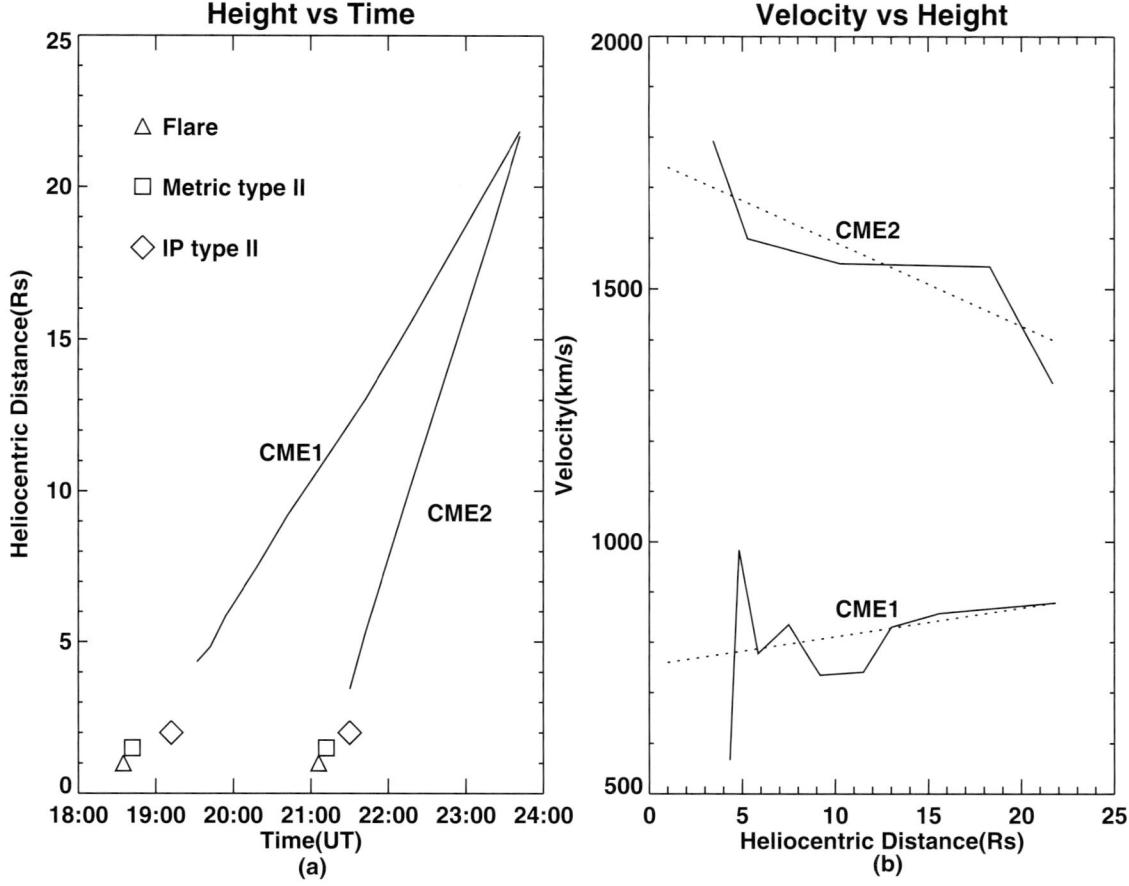

Figure 2. The observed characteristics for the January 20, 2001, event: (a) height–time curves and (b) velocity–distance curves for both CMEs. The solid lines are observed data, and the dotted lines are linear fits to the data. They exhibit acceleration of CME 1 and deceleration of CME 2. Flare and radio emissions are also shown in (a).

Table 1. Properties of Two CMEs and Their Related Sources, Flare, and Radio Emissions

Property	CME 1	CME 2
CME onset (1Rs)	18:33:51	21:08:19
Speed (km/s)	839	1507
Acceleration (m/s/s)	15.4	-41.1
Width	Halo	Halo
Central position angle	64°	71°
Solar Source	S07E40	S07E46
Flare	18:33-18:59 M1.2/2F	21:06-21:32 M7.7/2B
Metric type II	18:42-18:53 (180-30 MHz)	21:12-21:49 (180-25 MHz)
IP type II	19:12-19:16 (14-6 MHz)	21:30-24:00 (14-0.5 MHz)

the heat engine that produces both the waves and the particles? Note again, we need not consider the kinematics of wave generation, so long as we can calculate the entropy ratio of output to input.

Now the cost of coherent waves becomes clear. The input lacks both accelerated particles and coherent waves, whereas the output has both. Thus any mechanism that proposes coherent wave acceleration must include the entropy cost of providing for an even larger number of "accelerated pseudo-particles". That is, acceleration by incoherent waves is to be preferred to coherent waves because they do not change the entropy bookkeeping of the heat engine, where the difference in entropy cost can be estimated as the proportion of phase space occupied by the coherent waves in contrast to the incoherent waves.

The "cost" of coherency is similar to the calculation of the "Ockham factor" in Bayesian treatments of statistics [Sivia, 1996]. That is, if a data set can be fit by a theory with either two or three adjustable parameters, the cost of one more variable can be calculated by estimating the volume of phase space introduced by the additional parameter. It is in essence the same entropic calculation, where we look at the range of available values for the additional parameter compared to the total volume introduced by the additional parameter.

"Wave" is merely an abstract term here, which can refer to any quantized energy source, dynamic or static. The reflection from a moving wall in Fermi acceleration, may be considered a "wave", where the continuous compression of approaching walls, Fermi-I, would be a "coherent" wave (each energizing step is in the same direction), and the motion of a particle between randomly moving walls, Fermi-II, would be an "incoherent" wave (diffusive energy gain). Thus Fermi-II is to be preferred over Fermi-I, all other things being equal.

The final division is the distinction between standing waves and travelling waves, or from the particle perspective, between a trap and free streaming. The difference is whether a given wave or particle interacts more than once in the acceleration process. If the waves and/or particles are spread uniformly through the entire thermodynamic volume, then the initial energy Q_1 must be much larger to produce the same output acceleration, W, and hence a reduction in the efficiency, η. Conversely, if we keep the energy constant by spreading the same number of waves/particles over a larger volume, then the dilution reduces the probability of an accelerating interaction and greatly increases the acceleration time, possibly invalidating our stationary state hypothesis. Thus it should be clear that the highest power heat engines are those that concentrate their work.

Therefore the most likely and highest power mechanisms

Table 1. Comparison of Three Traps.

Feature	Dipole	Fermi	Quadrupole
1) Stochasticity	**Poor**	Moderate	Good
$\tau_1 : \tau_2 : \tau_3$	$10^{-3}:1:10^3$	$10^{-3}:10^3:10^4$	$10^{-1}:1:10^1$
2) Process	**Poor**	Moderate	Good
Flows from ...	Rim > ctr	End > side	Ctr > rim
And exit is ...	Blocked	By diffusion	Easy
3) Wave coupling	**Poor**	Moderate	Good
varies w/energy	Inversely	Constant	Directly
4) Trapping for	Moderate	**Poor**	Good
acceleration	Traps	Detraps	Both
5) Diffusion and	**Poor**	Moderate	Good
acceleration	Needed	Helpful	Neutral
6) Adiabatic heat	good	moderate	good
P.A.D.	2D oblate	1D prolate	2D oblate
7) Energy source	Moderate	Moderate	Good
SW vs internal	Compress	Alfvenic	Both+intern
8a) Electron E_{max}	Good	**Poor**	Moderate
MeV @ Re	900 @ 10	1.8 @ 0.1	280 @ 3
8b) Electron E_{min}	**Poor**	Good	Moderate
keV	<45	2.5	~30
9a) Trap volume	Good	**Poor**	Moderate
log(m^3)	24	20	22
9b) Trap lifetime	Good	**Poor**	Moderate
log(seconds)	>13	4	lo:hi 9:5
9c) Trap accel t	**Poor**	Good	Moderate
log(seconds)	>5.2	3.8	4.2
9d) Trap power	Good	**Poor**	Moderate
log(watts)	<8.3	6	7.3

are those that are multi-step, incoherent, and occur in a trap. For our application, three such traps have been considered: the dipole trap itself, the Fermi trap, and the quadrupole cusp trap.

3. TRAP CHARACTERISTICS

In our thermodynamic heat engine analogy for acceleration, the entire argument depends on the approach to quasi-equilibrium. If any part of the process takes too long, then the entire method is invalidated. Therefore we must consider other temporal bottlenecks beyond entropic considerations that can restrict the power of the output. For example, a restricted source of particles might throttle the power, or a rapidly diminishing probability for multiple steps may produce too soft a spectrum to explain the data. At this point, we must leave behind generalizations and focus on the particulars of the traps as accelerators. In Table 1, we compare the Earth's dipole trap and its energization through radial diffusion (e.g., *Schulz and Lanzerotti* [1974]) with both the Fermi trap at the bow-shock (e.g., *Ellison et al.* [1990]) and the quadrupole trap found in the Earth's outer cusp [*Sheldon et al.*, 1998; *Sheldon*, 2000]. The quadrupolar acceleration mechanism is discussed later, but invokes the same solar wind (SW) compressive events used by others (e.g., *Mead and Beard* [1964]; *Fälthammar* [1965]; *Fillius and McIlwain* [1967]) to produce dipole energization.

As a chain is only as strong as its weakest link, so an acceleration process can be throttled by a single step. As the comparison shows, quadrupoles may be the most robust of the three processes. One could view it as a generalization of Fermi acceleration to the other two spatial dimensions, or as an inside-out dipole trap. From this table, we see that the quadrupole is a very promising accelerator trap indeed, which, while not conclusive proof for the origin of RCI/ORBE, is reason enough to discuss it in more detail.

3.1.1. Stochasticity. From Hamiltonian dynamics, the symmetries of a trap produce constants of the motion, adiabatic invariants such as (μ,J,L) for a dipole and a similar triplet for a quadrupole, while a Fermi trap lacks the third. Now $\mu \propto E_\perp$, while $J \propto E_\parallel$, so in general, a particle cannot be accelerated without violating one or both of these invariants. Since the invariants are listed with increasing periods, an acceleration event that violates a particular invariant generally violates all invariants at longer periods as well, leaving shorter period invariants unchanged. This leads to the problem: Fermi acceleration violates the 2nd invariant, increasing $E_\parallel$, but not $E_\perp$. Likewise, adiabatic compressions in both dipole and quadrupole traps violate the 3rd invariant, but have no effect on the more important 1st and 2nd invariants. In order to have an efficient accelerator, the energy must be redistributed rapidly among the invariants, so that particles do not detrap in the Fermi case nor adiabatically return to their initial energy in the dipole/quadrupole case.

If we associate these invariants with a trajectory through phase space, then the conservation of the invariants maps to a closed curve in a Poincaré section, or an *n*-torus in the *n*-dimensional subset of 6D phase space (the Kolmogorov–Arnol'd–Moser theorem; *Arnol'd* [1964]). As fluctuations, scattering, and chaos cause these tori to "blur", Arnol'd argues that the tori can overlap and form a "web" that permits rapid stochastic transport through phase space, thereby redistributing the energy. Arnol'd shows that the formation of a web is greatly enhanced when the invariants have similar periods. The dipole trap, with three orders of magnitude separating the invariants, has no such stochastic transport, whereas the quadrupole trap is an ideal location.

3.1.2. Process. A second consideration is the flow of particles through the heat engine. The dipole has a large supply, the entire plasmasheet and geotail, but the acceleration process brings the output into the radiation belts close to the Earth, where they scatter in the upper atmosphere and are lost. All dipoles have this problem; that is, the exit is filled with the magnet that makes the dipole in the first place. Thus the dipole may be a bright source of energetic neutral atoms, but not energetic ions. The Fermi process is more dynamic, the trap forming whenever the SW magnetic field is radial and vanishing abruptly as the field wanders. Thus the exit is not so much a flow as a sudden release. Although the Fermi process isn't throttled, neither is it very continuous. In contrast the quadrupole process begins with plasma flowing into the center of the trap, where it is scattered and trapped. As it diffuses outward, it is accelerated until it finally escapes at the rim. This process is limited not by the exit but by the supply of sufficient source particles at the center.

3.1.3. Coupling. The strength of the coupling between "waves" (power in disturbances to the trap) and the particles affects the power. In the dipole trap, the more energetic particles are also found much further in, and at much higher *B*-field strength, such that the strength of magnetic disturbances, $\Delta B/B$, as well as pitchangle scattering, decreases as energy increases, resulting in higher energy particles becoming more and more decoupled from the waves. In contrast, the Fermi accelerator imparts the same relative energy to all particles that rebound from a moving wall, independently of their energy. In absolute energy terms, the more energetic ones are actually favored. But the quadrupole trap, being an inside-out dipole trap, puts its most energetic particles at the periphery. So not only does a compression have all the beneficial characteristics of a Fermi trap, but also the energetic particles are more likely to be in the largest $\Delta B/B$ region of the trap.

3.1.4. Diffusion. Diffusion, being second-order, is generally slower than first-order direct acceleration, which is why the cosmic ray community has preferred Fermi-I at shocks to the slower Fermi-II, despite the lower entropy considerations of second-order methods. In addition, diffusion is most effective when the gradients are largest, so that diffusively dominated acceleration is a victim of its own success, becoming less efficient as it erases gradients. Thus the dependence and the rate of diffusion are both critical to understanding the efficiency or power of a proposed mechanism. Dipole acceleration depends completely on diffusion to adiabatically energize, whose rate is a high power of radial distance, making it increasingly slower at higher energies. Diffusion is of secondary importance for Fermi acceleration, providing a way for energetic particles to escape the trap, or pitchangle scatter their $E_\parallel$ into $E_\perp$ so as to achieve higher energization. But in a quadrupole, diffusion is almost irrelevant. The low field re-

Table 2. Pressure Pulse Efficiency vs Trap Dimension.

D	γ	W	1.01 P	1.1 P	10 P	38 P	100 P
1	3	$0.5k^{-3}P^{0.6}$	0.003	0.03	1.82	5.16	10.3
2	2	$1.0k^{-5}P^{0.5}$	0.005	0.05	2.16	5.16	9.00
3	5/3	$1.5k^{-6}P^{0.4}$	0.006	0.06	2.27	4.90	7.96
1*	3	Normed	1.00	1.00	1.00	1.00	1.00
2*	2	Normed	1.50	1.49	1.19	1.00	0.87
3*	5/3	Normed	1.80	1.78	1.24	0.95	0.77

gion at the center of the cusp acts as a built-in scattering mechanism such that ordinary diffusion is of limited importance in redistributing the energy. Likewise, particles migrate outward in the trap under adiabatic, not diffusive forces, making the process flow independent of diffusion. This independence makes the quadrupole much faster than the dipole and gives it a slight edge over Fermi in processing speed.

3.1.5. Adiabaticity. If we assume an adiabatic compression using a polytropic equation of state, $PV^\gamma = k$, where $\gamma = (m + 2)/m$ is given by the number of degrees of freedom, m, then we can calculate how effectively a pressure pulse converts to work (acceleration).

$$-dW = PdV = (k/V^\gamma)dV = (k^{1/\gamma}/\gamma P^{-1/\gamma})dP$$
$$W = k/(1 - \gamma) V^{1-\gamma} = PV/(1 - \gamma) = k^{1/\gamma}/(1 - \gamma) P^{\gamma-1/\gamma}$$

Setting $k = 1$ for all systems, assuming a square-wave pressure pulse a factor n greater than the initial pressure, gives the results shown in Table 2.

We see that for small pressure pulses ($n = 1.01$), the 2D quadrupole trap has a 50% greater acceleration efficiency than the 1D Fermi trap, which holds true until the pressure pulse is roughly 38 times the initial pressure. Since small pulses are more common than large pulses, the quadrupole trap has the potential to be a more efficient accelerator than the Fermi trap, depending on k.

3.1.6. Energy sources. The energy source for the Fermi trap comes from the *B*-field enhancement that reflects the streaming ions. Such an enhancement might come from Alfvén waves, or compressional waves in the SW, convecting toward the bow-shock. Very occasionally, it might be actual shock fronts propagating in front of a coronal mass ejection or magnetic cloud. Likewise, for the standard dipole compression, very similar disturbances in the SW are usually invoked. In terms of cross-sectional area presented to the SW, the dipole is largest, followed by the quadrupole, and lastly the Fermi trap. In addition to these SW energy sources, the quadrupole can also absorb internal sources of waves, such as dipolarizations due to substorms. As [*Hassam*, 1995] points out, the cusp has a very low *Q*-value and is therefore a great absorber of wave power. While this source is available for the dipole as well, the $\Delta B/B$ is much smaller in the dipole than in the quadrupole, reducing its importance in the former.

3.1.7. Energy cutoffs. The total energy in the accelerated spectra can be integrated over all energies, which for both Fermi and quadrupole acceleration have power-law tails. Depending on the exact power-law, the cutoffs at both low and high energy have an effect on the total. In addition, the data constrain the models to very precise cutoff energies. The Fermi trap has an upper energy cutoff that occurs when the gyroradius of a particle exceeds the ~1° requirement for a quasi-parallel shock, whereas the cutoff for the dipole/quadrupole traps occur when the gyroradius is the same radius as the trap. Using 10 nT for the Fermi trap, and 50 nT for the dipole/quadrupole trap outer boundary, we estimate the maximum cutoff energy at trap radii of 10 Re, 0.1 Re, and 3 Re, respectively.

Likewise, the low-energy cutoff can occur when $E \times B$ drift is comparable or greater than the trapping ∇B drift. Using the same estimate for the radius of the trap as above, and estimating the voltage from $V = r^* v \times B$ of the SW, we get 2.5 keV for the Fermi trap, 360 keV for the quadrupole, and 1.2 MeV for the dipole. Clearly this is an overestimate, perhaps because both the dipole and quadrupole traps have boundary layers that short out much of the potential that develops from $v \times B$. Using satellite electric field probes, we can put more realistic limits of $E_{min} < 45$ keV for the dipole and perhaps E_{min} ~30 keV for the quadrupole.

Now the lower limit of the Fermi trap is just above the thermal energy of the SW, so that much of the SW particle distribution is available for acceleration, whereas lower cutoffs of both the dipole and the quadrupole are greatly above the thermalized SW energy, which can starve the input of both these traps. This is the essential difference between "low" and "high" quadrupole states, where we propose the high state has modified cutoffs due to topological changes in the trap. That is, diamagnetic cavities in the cusp (CDC) increase the radial magnetic gradient, which strengthens ∇B-drift and effectively lowers E_{min} while simultaneously raising E_{max}. In doing so, this increases the average power, produces CEP particles, and taps into the high fluxes available at lower energy.

3.1.8. Power. The last four entries are an attempt to estimate the average power of the proposed mechanisms using constant SW input. The trap volume is estimated for a dipole of radius 10 Re, for a Fermi trap with a 1°-wide region of a bow shock with a 12 Re radius of curvature extending

100 Re upstream, and for a quadrupole of 3 Re radius and approximately 3 Re depth. The trap lifetime estimated for a dipole is the 10^6-year flipping of the Earth's internal dipole field; for a Fermi trap, a 3-h persistence for a particular vector direction of the magnetic field; and for a quadrupole trap, two separate persistence times—the first given by the dipole + SW persistence time, the second given by the proposed "high" metastable state of the quadrupole (estimated from rise times of ORBE during high-speed SW conditions).

In calculating the acceleration time, it is not just the time for which the trap exists, but also the time to accelerate a particle to the appropriate cutoff energy. Or conversely, if the trap exists for insufficient time, the energy cutoff will be correspondingly lower. We give an estimate for the time to go from 1 keV to 1 MeV in all three traps. Estimating this for the dipole trap is difficult, since one pass from L = 10 plasmasheet to L = 5 ORBE is insufficient to explain the spectrum, and no theory of multipass (e.g., Nishida recirculation) is currently accepted or understood. Nevertheless, we optimistically estimate that four circulations of 1 day each can provide the energy. For the Fermi trap, we estimate the time required to bounce 100 Re between barriers, receiving a 400 km/s kick at one end, or 32 kicks, or $t = (800,000 \text{ km}/400 \text{ km s}^{-1}) \Sigma 1/n = 2000(4.05) = 8000$ s. For a quadrupole we assume a pulse of a 30% increase in pressure, which from Table 2 gives a 14% increase in energy, or 53 kicks, which, if occurring every 8 min (a typical peak in a power spectrum of SW pressure pulses), integrates to 25,000 s.

Clearly the dipole trap exists longer than the (uncertain) acceleration time, so power is limited by the acceleration time. In contrast, the Fermi trap has an acceleration time comparable to the trap lifetime, which means the power is limited mainly by the lifetime of the trap. The quadrupole trap is a little less clear. The "low" state exists much longer than the acceleration time, whereas the "high" state is again comparable. Thus in the high state, the power may again be limited by the trap lifetime.

The acceleration power is proportional to the energy density, ε, and volume, V, divided by the time, t, as given by $P_a = \varepsilon V/t$. We don't know the energy density well, since it can depend on waves as well as particles, but it should be comparable to some fraction of the SW energy density that can be extracted by the trap, e.g., 10% of the SW kinetic energy. So assuming the energy density has a constant value of $\varepsilon = 10^{-10}$ J/m^3 for all, we arrive at a power estimate of the three traps.

4. CONCLUSIONS

We have made qualitative thermodynamic arguments for the superiority of traps in accelerating particles. Greater rigor could be obtained by evaluating actual distributions without any change in the argument, but at the risk of losing the forest for the trees. We then consider two well-known traps, and the lesser known quadrupole trap. By estimating the average power of the three traps from many perspectives, we showed that the quadrupolar trap has the potential to outperform the others, both in the magnetosphere and in astrophysical magnetospheres. In a subsequent paper, we will refine the model for the Earth's cusp, showing the effect of CDC-entrained plasma on the topology and energy cutoffs, which may account for the "low" (without CDC) and "high" (with CDC) states of the quadrupole accelerator.

Acknowledgments. We acknowledge fruitful conversations with colleagues at NASA/MSFC/NSSTC and NASA grants NAG-5 2578, NAG-5 7677, and NAG-5 1197 at Boston University.

REFERENCES

Alfvén, H. *Phys. Rev. 75*, 1732, 1949.

Alfvén, H.*Phys. Rev.77*, 375, 1950.

Arnol'd, V. I., *Dokl. Akad. Nauk. SSR,156*, 9,1964.

Baker, D.N., T.I. Pulkkinen, X.Li, S.G. Kanekal, J.B. Blake, R.S. Selesnick, M.G. Henderson, G.D. Reeves, H.E. Spence, and G.Rostoker. "Coronal mass ejections, magnetic clouds, and relativistic magnetospheric electron events: Istp." *J. Geophys. Res., 103*(A8), 1998

Baker, D. N., J. B. Blake, R. W. Klebesadel, and P. R. Higbie, "Highly relativistic electrons in the earth's outer magnetosphere 1. lifetimes and temporal history 1979-1984," *J. Geophys. Res., 91*, 4265-4276, 1986.

Chang, S.-W. "Cusp energetic ions: a bow shock source." *Geophys. Res. Lett., 25*, 3729–3732, 1998.

Chen, J., T. A. Fritz, R. B. Sheldon, H. E. Spence, W. N. Spjeldvik, J. F. Fennell, and S. Livi, "A new temporarily confined population in the polar cap," *Geophys. Res. Lett, 24*, 1447-1450, 1997.

Chen, J., T. A. Fritz, R. B. Sheldon, H. E. Spence, W. N. Spjeldvik, J. F. Fennell, S. Livi, C. Russell, and D. Gurnett. "Cusp energetic particle events: Implications for a major acceleration region of the magnetosphere." *J. Geophys. Res., 103*, 69-78, 1998a.

Chen, J. and T.A. Fritz. "Correlation of cusp Mev Helium with turbulent ULF power spectra and its implications." *Geophys. Res. Lett., 25*, 4113-4116, 1998b.

Chen, J. and T.A. Fritz. "Origins of energetic ions in CEP events and their implications." *Int. J. Geomagn. Aeron.2*, 31, 2000.

Chen, J. and T.A. Fritz. "Energetic oxygen ions of ionospheric origin observed in the cusp" *Geophys. Res. Lett., 28*(8), 1459-1462, 2001a.

Chen, J., T.A. Fritz, R.B. Sheldon, J.S. Pickett, and C.T. Russell. "The discovery of a new acceleration and possible trapping region of the magnetosphere." *Adv. Space Res. 27*(8), 1417-1422, 2001b.

Chen, J. and T.A. Fritz. "The global significance of the CEP events." In H.N. Wang and R.L. Xu, editors, *Solar-Terrestrial Magnetic Activity and Space Environment*, volume Cospar Colloq. Ser 14, pages 239–249, 2002.

Christon, S.P., D.J. Williams, D.G. Mitchell, L.A. Frank, and C.Y. Huang. "Spectral characteristics of plasma sheet ion and electron populations during undisturbed geomagnetic conditions." *J. Geophys. Res.*, 94, 13409–13424, 1989.

Collier, M. R., "On generating kappa-like distribution functions using velocity space Lévy flights," *Geophys. Res. Lett.*, 20, 1531-1535, 1993.

Delcourt, D.C. and J.A. Savaud. "Populating of the cusp and boundary layers by energetic (hundreds of kev) equatorial particles." *J. Geophys. Res.*, 104, 22,635, 1999.

Elkington, S.R., M.K. Hudson, and A.A. Chan, Acceleration of relativistic electrons via drift-resonant interaction with toroidal-mode pc-5 ulf oscillations, *Geophys. Res. Lett.*, 26, 3273–3276, 1999.

Ellison, D. C. et al., "Particle injection and acceleration at Earth's bow shock: Comparison of upstream and downstream events," *Astrophys. J.*, 352, 1990.

Fälthammar, C.-G., Effects of time-dependent electric fields on geomagnetically trapped radiation, *J. Geophys. Res*, 70, 2503-2516, 1965.

Farrugia, C.J., J.D. Scudder, M.P. Freeman, L.Janoo, G.Lu, J.M. Quinn, R.L. Arnoldy, R.B. Torbert, L.F. Burlaga, K.W. Ogilvie, R.P. Lepping, A.J. Lazarus, J.T. Steinberg, F.T. Gratton, and G.Rostoker. "Geoeffectiveness of three Wind magnetic clouds: A comparative study." *J. Geophys. Res.*, 103(A8), 17,261-17,278, 1998.

Fermi, E. *Phys. Rev.*, 75, 1169, 1949.

Fillius, R.W. and C.E. McIlwain. "Adiabatic betatron acceleration by a geomagnetic storm." *J. Geophys. Res.*, 72, 4011-4015, 1967.

Fritz, T. A., and J.-S. Chen, "The cusp as a source of magnetospheric particles," *Radiat. Meas.*, 30, 1999.

Fritz, T. A., J.-S. Chen, and R. B. Sheldon. "The role of the cusp as a source for magnetospheric particles: A new paradigm?" *Adv. Space Res.*, 25(7-8), 1445-1457, 2000.

Fritz, T. A., J.-S. Chen, and G. L. Siscoe. "Energetic ions, large diamagnetic cavities, and {Chapman-Ferraro} cusp." *J. Geophys. Res.*, 108(A1), doi:10.1029/2002JA009476, 2003a.

Fritz, T. A., T. H. Zurbuchen, G. Gloeckler, S. Hefti, and J.-S. Chen. "The use of iron charge state changes as a tracer for solar wind entry and energization within the magnetosphere." *Ann. Geophys.*, 21, 2155-2164, 2003b.

Hassam, A. B. "Dynamics and dissipation of compressional Alfven waves near magnetic nulls." *Phys. Plasmas*, 2(12), 4662-4664, 1995.

Hess, W. N. *The Radiation Belt and Magnetosphere*. Blaisdell Pub. Co., Waltham, MA, 1968.

Ingraham, J. C., R. D. Belian, T. E. Cayton, M. M. Meier, and G. D. Reeves. "March 24, 1991, geomagnetic storm: Could substorms be contributing to relativistic electron flux buildup at geosychronous altitude?" *Eos Supplement*, 80, S294, 1999.

Krall, N. A., and A. W. Trivelpiece, *Principles of Plasma Physics*, San Francisco Press, Inc., San Francisco, 1986.

Li, X., I.Roth, M.Temerin, J.R. Wygant, M.K. Hudson, and J.B. Blake. "Simulation of the prompt energization and transport of radiation belt particles during the March 24, 1991 SSC." *Geophys. Res. Lett.*, 20, 2423–2427, 1993.

McIlwain, C.E., Processes acting upon outer zone electrons, in *Radiation Belts Models and Standards*, edited by J. F. L. et. al, AGU, Washington DC, 1996.

Mead, G. D., and D. B. Beard, "Shape of the geomagnetic field solar wind boundary," *J. Geophys. Res.*, 69, 1181, 1964.

Nishida, A. "Outward diffusion of energetic particles from the Jovian radiation belt." *J. Geophys. Res.*, 81, 1171, 1976.

Paulikas, G.A. and J.B. Blake. "Effects of the solar wind on magnetospheric dynamics: Energetic electrons at geosynchronous orbit." In W.P. Olson, editor, *Quantitative Modelling of Magnetospheric Processes, Geophys. Monogr. Ser.*, volume 21, page 180, Washington, D.C., 1979. AGU.

Reeves, G.D., R.H.W. Friedel, R.D. Belian, M.M. Meier, M.G. Henderson, T.Onsager, H.J. Singer, D.N. Baker, X.Li, and J.B. Blake. "The relativistic electron response at geosynchronous orbit during the January 1997 magnetic storm." *J. Geophys. Res.*, 103(A8), 1998.

Schulz, M. and L.J. Lanzerotti. *Particle Diffusion in the Radiation Belts*. Springer-Verlag, New York, 1974.

Sheldon, R. B., H. E. Spence, J. D. Sullivan, T. A. Fritz, and J.-S. Chen. "The discovery of trapped energetic electrons in the outer cusp." *Geophys. Res. Lett.* 25, 1825-1828, 1998.

Sheldon, R. B., J.-S. Chen, and T. A. Fritz. "Comment on "Origins of energetic ions in the cusp" by K. J. Trattner et al.'." *J. Geophys. Res.*, 108(A7), doi:10.1029/2002JA009575, 2003.

Sheldon, R.B. "The bimodal magnetosphere and radiation belt, ring current and tail transducers." *Adv. Space Res.*, 25, 2347-2356, 2000.

Sivia, D. S., *Data Analysis: A Bayesian Tutorial*, Clarendon Press, Oxford, 1996.

Trattner, K.J., S.A. Fuselier, W.K. Peterson, S.-W. Chang, R.Friedel, and M.R. Aellig. "Origins of energetic ions in the cusp." *J. Geophys. Res.*, 106, 5967–5976, 2001.

Jiasheng Chen and Theodore Fritz, Center for Space Physics, Boston University, 725 Commonwealth Avenue, Boston, Massachusetts 02215.

Robert Sheldon, NSSTC/SD50, 370 Sparkman Drive, Huntsville, Alabama 35805. (Rob.Sheldon@msfc.nasa.gov)

Electron Phasespace Density Analysis Based on Test-Particle Simulations of Magnetospheric Compression Events

Jennifer L. Gannon and Xinlin Li

Laboratory for Atmospheric and Space Physics, Boulder, Colorado

The sudden appearance of a new electron radiation belt at approximately 2.5 R_e, as observed by CRRES (Combined Radiation and Release Experiment Satellite) on March 24, 1991, has been modelled by *Li et al.*, (1993), by means of a test-particle simulation. They reproduced the observed flux levels using a guiding-center code which conserves the particles' first adiabatic invariants. The current work presented in this paper uses the output of a very similar particle tracing simulation to produce a phasespace density profile versus radial distance for the event. This work confirms the previous work and provides more detailed information about the initial particle distribution. In addition, the source population used in the *Li et al.*, (1993) model is replaced by a flat initial phasespace density profile versus radial distance. This is done in order to test the model and to study the contribution of the initial particle distribution to the results. Under this initial distribution, with no changes to the field model, the results show that the overall flat form of the original profile is retained. As another test of the model and in order to study the contribution of the field, the model field is replaced with a symmetric (no local time dependence) electric field with full reflection, with no modifications to the source population. This leaves the phasespace density profile unchanged. We also present an argument that the third adiabatic invariant is conserved for all time during the compression, as long as the model field is azimuthally symmetric.

INTRODUCTION

On March 22, 1991, an optical flare was observed on the Sun (*Blake et al.*, 1995). Twenty-eight hours later, at 3:41 UT on March 24, 1991 the effects of a strong shock impact were seen by the CRRES satellite (see Figure 1), which was fortuitously located at approximately 2.5 R_e, during its inbound pass on the nightside near the equatorial plane. During orbit 587, March 24, 1991, CRRES observed a very rapid (< 10 seconds) four order of magnitude increase in electron flux, a unipolar magnetic field enhancement, and a bipolar electric field (see Figure 2). This is believed to be associated with an interplanetary shock from the Sun impacting the Earth's magnetosphere (*Blake et al.*, 1995). The new highly energetic electron radiation belt produced by this event greatly affected the Earth's space environment and the intensity can be seen in Figure 1 during the remainder of the mission (about 6 months). The new radiation belt, while decaying slowly, persisted much longer, and was observed by the SAMPEX satellite for almost ten years (*Li and Temerin*, 2001).

Li et al (1993) modelled the March 24, 1991 event by representing the electric field due to magnetospheric compression and relaxation as two gaussian pulses, one incoming and one reflected, in the magnetosphere. The perturbed magnetic field was calculated from the specified electric field

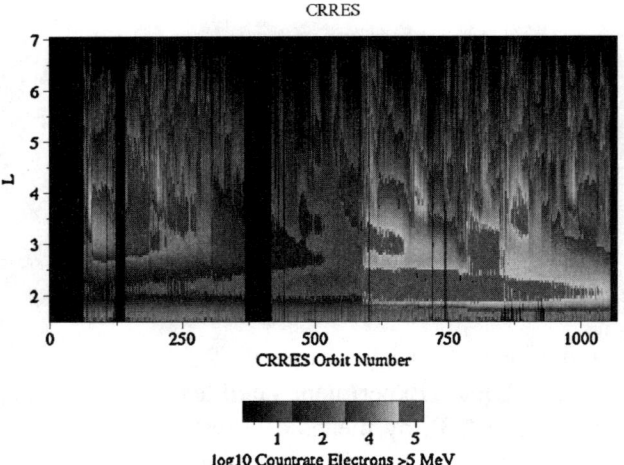

Figure 1. (Courtesy of J.B. Blake) Electron countrate measured by CRRES (> 6 *MeV*) for the whole CRRES mission: July 1990 to October 1991. The data are binned in 0.1 L-values and the range $L = 1.0$ to 7.0 is shown. The CRRES spacecraft was in an 18°-inclination geotransfer orbit (300 km x 5.2 R_e). The event of March 24, 1991 occurred at orbit 587.

using Faraday's law. Figure 2a shows the integral flux data and electric and magnetic field measurements from CRRES. Figure 2b shows the results of a particle tracing code under the modelled electric field, as well as the modelled fields for comparison to the data. The sawtooth shape in both integral flux plots are particle drift-echoes, which occur as electrons drift around the Earth and return to the CRRES detector. The period of the drift echoes, about 150 seconds, approximately corresponds to the drift period of a 15 MeV electron at $L = 2.5$, the detector position. L corresponds to the radial distance in units of Earth radii at the equator, if the Earth's magnetic field is approximately a dipole.

Li et al (1993) were able to reproduce the drift echoes, and flux magnitudes for the first 900 seconds of the event. Other work has been done on this event, such as MHD simulations by *Hudson et al.* (1995) and *Elkington et al.* (2002). These works consolidated the idea of fast acceleration of radiation belt electrons by a travelling electric field associated with a strong interplanetary shock impact on the magnetosphere.

Because of the availability of the CRRES data, the original simulation focused on reproducing the integral flux measurements. However, a phasespace density profile versus L is subject to greater physical restraints, and because of this, would allow us further insight into the processes involved in shock induced radial transport. Phasespace density is based on the canonical coordinates of the system, from which the adiabatic invariants can be derived. Liouville's theorem states that, in the absence of sources and losses, phasespace density, f, is conserved along the trajectory of a particle, or $f = constant$ (*Walt*, 1996). Because of the implications of Liouville's theorem and conservation of the adiabatic invariants, phasespace density is a more physically meaningful parameter than flux or count, which are more natural for the particle detector. Because no phasespace density profile is available observationally, we must reconstruct it from the information we do have.

In order to calculate a phasespace density profile versus radial distance, we require knowledge of differential flux. CRRES satellite observations are limited to integral channels in the energy range of interest, requiring us to reconstruct the differential flux from the available models. We use a model very similar to *Li et al* (1993), differing only in the exclusion of detector geometric factors and satellite motion. We retain the parameters and source population that they found to best reproduce the features of the March 24, 1991 event.

In this work, we determine the phasespace density profile versus L before and after the shock on March 24, 1991 using a particle tracing code. In addition, as a test to the code, and in order to independently study the contributions of the source population and field model to the result, we look at the effects of the *Li et al* (1993) model field on an initially flat source population versus L and a field representing a symmetric, local time-independant compression on the original source population.

MODEL DESCRIPTION

The model used by *Li et al* (1993) consists of an electric field model of the following form, and a magnetic field derived from it, using Faraday's law:

$$E(r,t) = -\hat{e}_\phi E_0 (1 + c_1 \cos(\phi - \phi_0))(e^{-\xi^2} - c_2 e^{-\eta^2}) \quad (1)$$

$$\eta = [r - v_0(t - t_{ph} + t_d)]/d$$

$$\xi = [r + v_0(t - t_{ph})]/d$$

$$t_{ph} = t_i + (\frac{c_3 R_e}{v0})[1 - \cos(\phi - \phi_0)]$$

where $\hat{e}_\phi$ is a unit vector in the azimuthal direction, positive eastward.

The exponential terms represent the compression and relaxation of the magnetosphere as oppositely-directed gaussian pulses. The exponents of the gaussians include a time delay. The exponential form is multiplied by a local time modulation, with the strongest point of the field corresponding with the point of impact of the impulse. The electric field looks like a wave, propagating inward and azimuthally at a constant speed, and partially reflecting at the ionosphere.

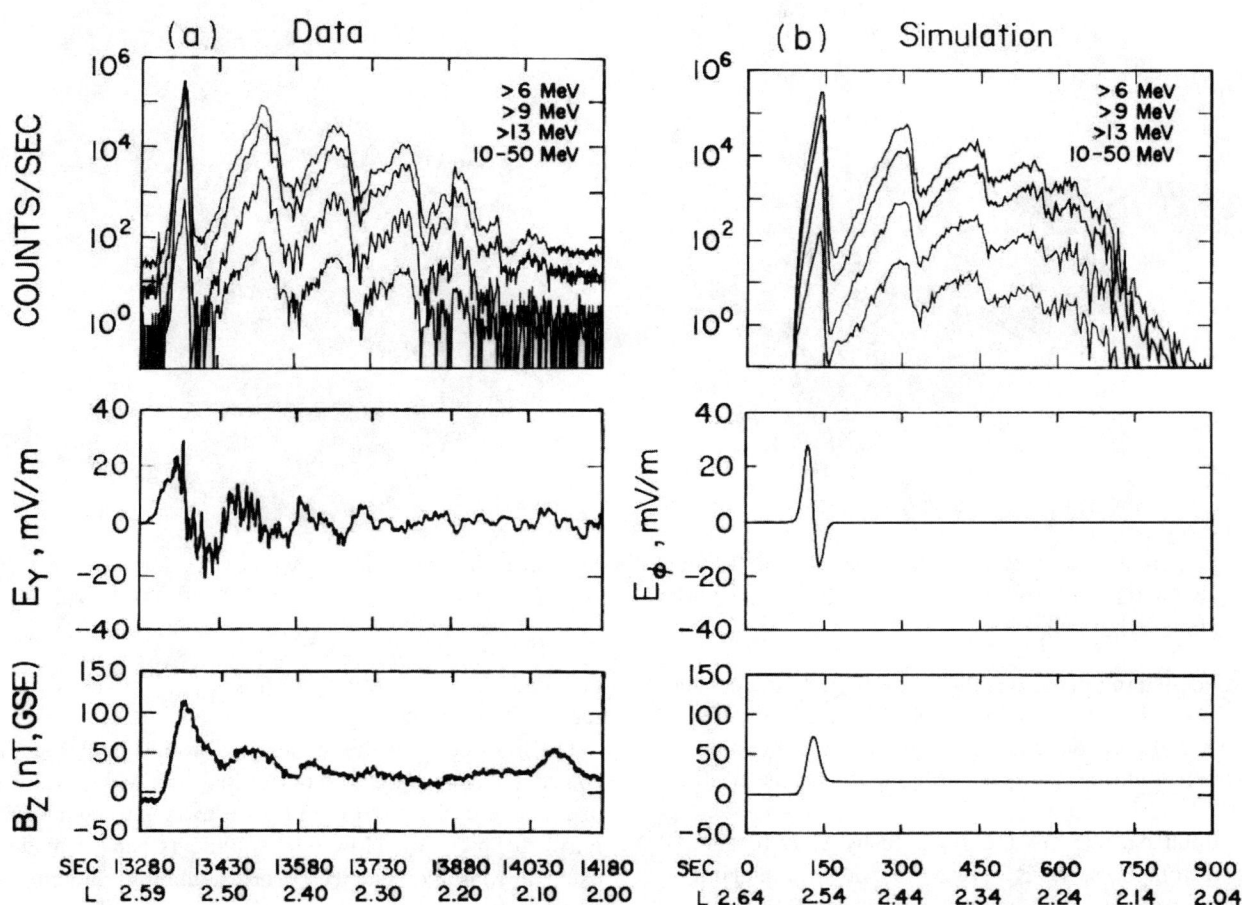

Figure 2. From Figure 1 of Li et al. (1993). (a) Count rate and field data taken from the CRRES satellite during the March 24, 1991, SSC. Top panel shows count rates as a function of time from four energetic electron channels measuring integral counts above 6, 9, and 13 *MeV* and also between 10–50 *MeV* [Blake et al. (1992)]. Middle and bottom panels show the measured electric field E_y in a co-rotational frame and the B_z magnetic field component with a model magnetic field subtracted, in GSE coordinates over the same time interval [Wygant et al. (1994)]. (b) Simulation results of the March 24, 1991, event by *Li et al.* (1993), in the same format as (a).

The calculated magnetic field pulse is added to a simple dipole background field. The parameter $c_1 = 0.8$ affects the local time dependence, $c_2 = 0.8$ determines the amount of reflection, $c_3 = 8.0$ represents the magnitude of the propagation delay in the azimuthal direction, $t_d = 2.06\, R_e/v_0$ indicates the location of the reflection at $r = 1.03\, R_e$, and $t_i = 81$ sec is the initial reference time. The parameters found to mimic the unipolar electric field and bipolar magnetic field observed by CRRES are $v_0 = 2000$ *km/s*, $E_0 = 240$ *mV/m*, $\phi_0 = 45$ deg and $d = 30,000$ *km*. The parameters t, ϕ, and r are the time, azimuthal and radial position of the particle. Figure 3 shows several snapshots of the electric field pulse at different points in time. The pulse travels in, E pointing westward, and is reflected, E pointing eastward. The field magnitude is strongest at the point of impact, ϕ_0.

Li et al (1993) applied the field model to a test particle code following about 300,000 equatorially-mirroring electrons under the *Northrop* (1963) guiding center equations:

$$\dot{W} = e\dot{\mathbf{R}}_\perp \cdot \mathbf{E} + \frac{\mu}{\gamma}\frac{\partial B}{\partial t} \qquad (2)$$

$$\dot{\mathbf{R}}_\perp = \frac{\hat{e}_1}{B} \times (-c\mathbf{E} + \frac{\mu c}{e\gamma}\nabla B) \qquad (3)$$

In the above equations, W is particle energy, E is the electric

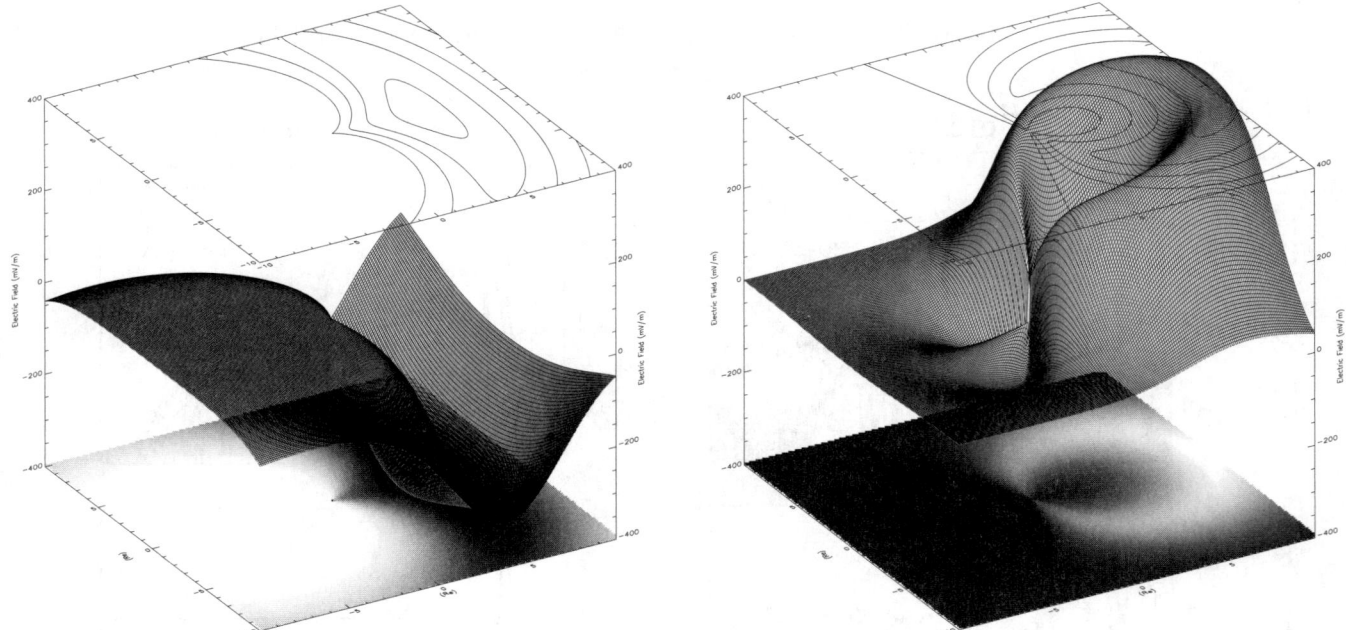

Figure 3. Snapshots at 2 times of the electric field model used by *Li et al* (1993) and the current work. The Sun is to the right at a positive radial distance. Left: The incoming pulse, $E_\phi < 0$, at t = 60 sec. Right: The reflected pulse, $E_\phi > 0$, at t = 120 sec.

field (modelled impulse), *B* is the magnetic field (calculated from the electric field pulse, plus a background dipole field), *R* is the radial position, $\hat{e}$ is a unit vector in the direction of the magnetic field, *c* is the speed of light, and *e* is the electron charge.

The initial particle distribution used by *Li et al* (1993) extended from 3.0 to 9.0 R_e, every three degrees in azimuth and every 10% in initial energy from 1.0 to 9.0 MeV. A weighting was included in order to simulate a realistic electron distribution. It included a strong power law in energy (many more low energy particles than high energy particles), as well as a parabolic weighting in radial position, in order to reproduce the observation:

$$weighting = G(L) * L^2 * \sigma * W^{-7} * \frac{v}{v_d} \qquad (4)$$

$$G(L) = 1 - \frac{(L - L_0)^2}{a_0}.$$

In the above, $a_0 = 7.5$ and $L_0 = 10$, and the L^2 term allowed for the fact that the area over which the source was spread increased with radial distance. The ratio of actual velocity, *v*, to drift velocity, v_d corrects for the fact that the guiding centers of the electrons are traced rather than the complete motion. *Li et al* (1993) used σ to represent the detector response, modelling the detector geometric factors. This is not used in the current work. In addition, the original work included satellite motion. In this work, we consider the satellite to be stationary.

The limitations of the model used in this study include a constant magnetospheric shock propagation velocity, the use of a dipole background field and restriction to equatorially-mirroring particles. However, arguments can be made for using a constant velocity of propagation if the impulse, which is believed to be very sharp and fast, does not travel via fast magnetosonic modes, but instead as a propagating discontinuity (J. Lyon, private communication, 2004), which would not slow down as much in the region of interest. The use of a dipole is a reasonable simplification for the inner magnetosphere and allows us to make the arguments in the source population and model field modification sections. Tracing particles of pitch angles other than 90 degrees would be worthwhile, but is not computationally feasible at this point.

Figure 2b, previously mentioned, shows the simulation results of the original *Li et al* (1993) work. Figure 4 shows results based on the output of the very similar simulation we used in this work. Figure 4a shows integral flux values similar to the original reproduction and data in Figure 2. Figure 4b shows a count rate versus L for particles in the simulation. This includes no detector geometric factor. It simply shows the particles seen above three energy thresholds (6, 9, 13 *MeV*) from simulation output. Figure 4c depicts the count rates versus energy of particles seen at a particular

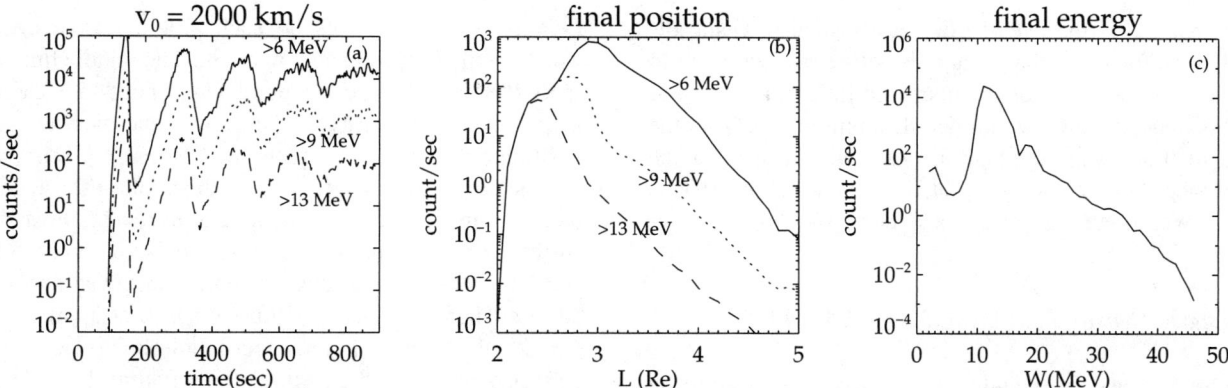

Figure 4. Simulation results for values used to simulate March 24, 1991, event. a) log(count) at detector location. b) final radial position distribution of count rate for each of the four CRRES energy channels. c) final energy spectrum at detector position.

radial location, $L = 2.6$. The peak energy is about 12–13 *MeV*, which corresponds to the energy of electrons postulated to have dominated in the observed CRRES flux measurements.

PHASESPACE DENSITY CALCULATION

The original study of the event focused on reproducing the flux levels seen by the CRRES detector. In order to calculate phasespace density from simulation output, we first choose a value of the first adiabatic invariant, μ, based on the electrons' energy at $L = 2.5$, which is where we have CRRES data for comparison at the time of the March 24, 1991 event. For a relativistic electron:

$$\mu = \frac{p_\perp^2}{2m_0 B} \quad (5)$$

where $p_\perp$ is the electron's momentum perpendicular to the local magnetic field, $m_0 = 0.511$ *MeV* is the electrons rest mass, and B is the magnetic field.

From the output of the particle tracing code we select all particles within $\pm 10\%$ of the chosen μ. We then choose a time range much earlier than the shock arrival for the pre-shock analysis, and long after the shock arrival for the post-shock analysis. We want to plot phasespace density versus L, so we separate the selected particles into L-bins of 0.5, and convert the flux in each bin to phasespace density using the following relation:

$$f = \frac{j}{p^2} \quad (6)$$

where j is differential flux and p is momentum. For equatorially mirroring particles, $p_\perp = p$. Recalling equation (5), we have:

$$f = \frac{j}{2\mu m_0 B} \quad (7)$$

We use our chosen value of μ and determine the magnetic field from the dipole relation based on the middle of each L-bin.

In order to make the phasespace density levels realistic, we scale them to match the flux data available from CRRES. The equation for phasespace density, (6), requires differential flux, for which we do not have data available for direct comparison. However, the simulation results of *Li et al* (1993), incorporated with detector geometric factors, have been matched to the observed integral countrate, and thus we can use the model output to estimate what differential flux measurements CRRES would have seen. For example, we select an energy range of 14–16 MeV and, for a chosen simulated channel ($>$ 13 MeV) we count the number of particles in the 14–16 MeV range that are seen from 300–600 seconds (in the simulation, the field interacts with the particles from $t = 80 - 150\ s$), applying the weighting used by *Li et al* (1993) to obtain the countrate that a detector would have recorded:

$$countrate = \Sigma \sigma \times weight(L_0, W_0)$$

where σ represents the geometric factor and $weight(L_0, W_0)$ represents all additional weighting based on initial position and energy. This gives us a value for differential flux including detector response. Then, we divide the countrate by the average response that the channel of the detector would have to electrons of 14–16 MeV, to give the differential flux:

$$j_{real} = \frac{countrate}{\sigma_{effective}}$$

where $\sigma_{effective}$ is the average detector response. Using the specified μ for each phasespace density curve we wish to scale, the dipole value of the magnetic field at a particular radial distance, and the calibrated differential flux from the above method, we use equation (7) to calculate what the phasespace density should be at that radial distance. This process was done for phasespace densities with other μ as well.

Phasespace Density Analysis of March 24, 1991

Figure 5 shows the calculated phasespace density profile of the particle population, pre-shock and post-shock, versus

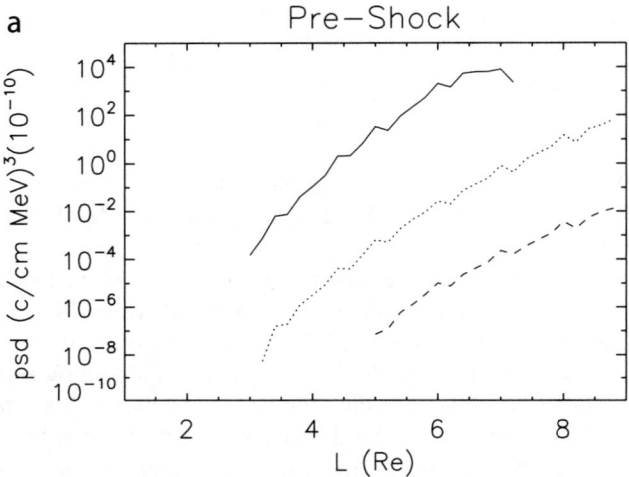

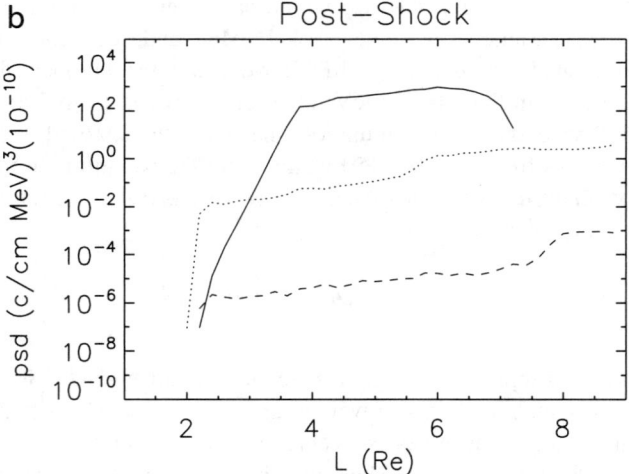

Figure 5. a) Pre-shock phasespace density profile derived from the source population used by *Li et al* (1993) to model the March 24, 1991 event. b) Post-shock phasespace density profile from the output of the Li et al. model of the March 24, 1991 event. solid: $\mu = 2077$ *MeV/G*, dotted: $\mu = 11852$ *MeV/G*, dashed: $\mu = 45898$ *MeV/G*

L-value. The three lines on each plot represent different values of the first adiabatic invariant. The middle line represents 11852 MeV/G, corresponding to about 15 MeV at 2.5 R_e, which is the energy of the bulk of the particles at that position, post-shock, as seen by CRRES and simulation results (see Figure 4c). In the pre-shock phasespace density plot (Figure 5a), there are no particles at $L = 2.5$. Post-shock, particularly in the μ range corresponding to about 15 MeV at 2.5 R_e, there are particles extending Earthwards to almost 2.0 R_e. There is a peak in flux versus L at approximately $L = 2.5$–2.6 (Figure 1) after the impulse. This peak in flux was also reproduced by simulation (Figure 4b, *>13 MeV* range), which shows them varying somewhat with L for different modelled integral channels.

The sudden energization process that produced the effects seen by CRRES is a drift-resonant process, where the resonance is determined by an electron's first invariant for a given model field. Under the model, particles of the same energy and radial position, but different local time, see different field strengths, such that particles in the right location at the right time are energized more than others, or, in other words, are resonant with the pulse. The particles of 11852 MeV/G are nearest those in the peak of the final energy spectrum (see Figure 4c), and can be considered most resonant. We see no true peaks in phasespace density, as shown in Figure 5b. Local heating (by which we mean a process that violates a particle's first adiabatic invariant) could produce a peak, as the more abundant lower energy particles are energized at a particular radial distance. However, this radial transport process, which is μ-conserving, produces no peaks in phasespace density. If a particle were to remain at a particular L and be energized, conservation of the first adiabatic invariant would be violated. Even if there is an initial existing peak, the natural tendency of such a process would be to reduce the peak. The drop-offs at the ends of each curve occur due to pre-existing phasespace density holes as a result of the initial distribution.

Li et al (1993) speculated that there was insufficient phasespace density available at $L < 6$ to produce the four order of magnitude increase seen at $L = 2.5$ by the CRRES detector while conserving the first adiabatic invariant, μ. *Li et al* (1993) determined that electrons would have to have come from $L > 6$. In the *Li et al* (1993) simulation, the bulk of the particles were accelerated from the farthest L-shells in the source population, i.e. $L = 7 - 9$. These results are reflected in the phasespace density profile versus L.

Comparison With Polar Quiescent Radiation Belt Data

In order to understand if our simulation results are of a reasonable magnitude, we compare with available known phasespace density measurements. *Selesnick et al.* (1997) show data from the Polar satellite during a quiescent period, which they describe as a steady state of particle diffusion in the magnetosphere. They look at phasespace density profiles versus L for about 100 days of 1996. We can use these data as a baseline to see if our derived phasespace densities are realistic. We cannot make a direct comparison, as their selected values of the first adiabatic invariant are lower than the values of interest in our case. However, we can show limited comparison to several features. In our profile, post-shock, between $L = 3.0$ and $L = 5.5$, we see a several order of magnitude increase in phasespace density, especially in the higher values of the first adiabatic invariant. Polar data show the same trend, especially early in the time period they observed, following a particle injection, for a first adiabatic invariant of 6000 MeV/G. Comparing actual phasespace density absolute magnitudes is more difficult, as phasespace density varies significantly with time and we are looking at only one instance in time. In the Polar data, for a μ of 6000 MeV/G, the maximum phasespace density reached was approximately 100 $(10^{-10})(cm\ MeV/c)^{-3}$. Under our simulation for a μ of 11852 *MeV/G*, a higher value of μ, we expect a lower phasespace density. The maximum phasespace density reached in our simulation was about 10 $(10^{-10})(cm\ MeV/c)^{-3}$, which is not inconsistent because of the different values of μ.

Initial Distribution Modifications—Flat Profile Versus L

As a test to the model, and in order to study the contribution of the source population, we consider the effect of the same model field on an initially flat phasespace density profile. Because of the continually evolving nature of phasespace density, a flat profile is not typically seen in nature, but it is not unphysical, especially on small scales. In this case, the initial electron distribution was forced to be flat in phasespace density versus L by adjusting the weighting. This was achieved by modifying the dependence on radial position and the energy weighting. The parabolic form based on position, G(L) in equation (4), was replaced with a power law, L^{-2}. The weighting in energy was removed by adding an additional factor W^1 in order to negate the weighting implicitly included in the 10% increase in initial energy distribution from 1.0 to 9.0 MeV. This could also be achieved by adjusting the energy spectrum from W^{-7} to $W^{1.5}$, without changing $G(L)$. Although the same results are achieved, a positive power law, in general, is unrealistic, as it implies particle numbers increasing with energy.

Figure 6 shows the post-shock curves never peak above the pre-shock curves. They remain flat except near the boundaries where there are existing phasespace density holes. Particles spread out in L, and because the total number of parti-

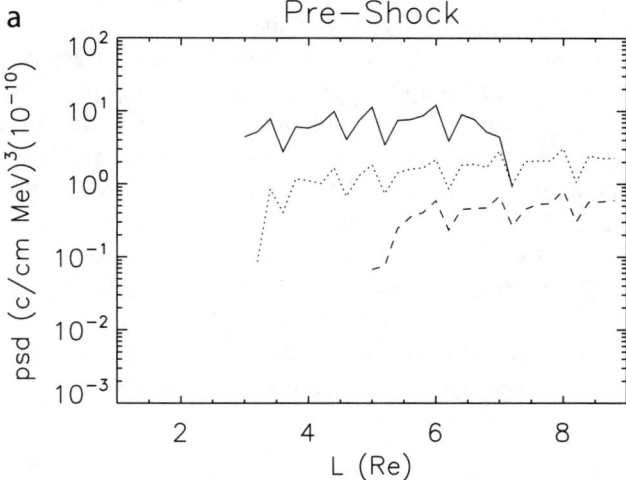

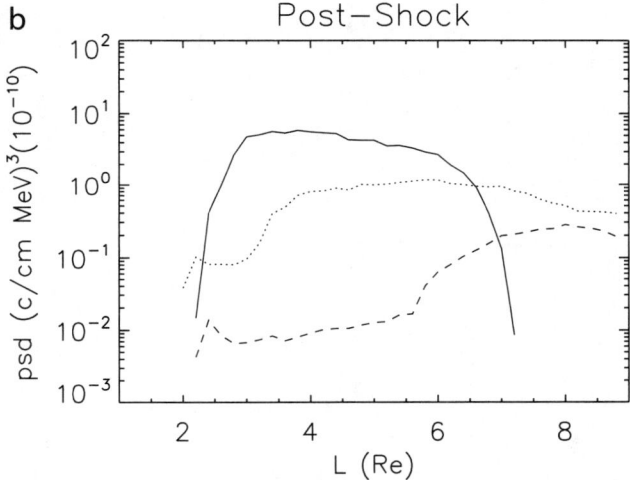

Figure 6. a) Pre-shock phasespace density profile from the output of the Li et al. model using modified source population. b) Post-shock phasespace density profile from the output of the Li et al. model using modified source population. solid: μ = 2077 *MeV/G*, dotted: μ = 11852 *MeV/G*, dashed: μ = 45898 *MeV/G*

cles is approximately conserved (less those few lost past the outer bounds of the simulation), an increase in phasespace density at one radial distance necessitates a decrease at another. This is an illustration of the idea that a μ-conserving process can not lead to increases in phasespace density versus L above the initially assumed flat phasespace density distribution. A particle cannot remain at the same radial distance and increase its energy without changing its μ.

Field Model Modifications—Azimuthally-Symmetric Compression

What happens in an azimuthally-symmetric compression case? In the *Li et al* (1993) simulation, an *80%* reflection was used to match field measurements from CRRES on the simulation time scale, and a local-time dependence of the strength of the field was included. In addition to removing the local-time dependence of the compression, we use a *100% reflection* (complete relaxation to the initial field configuration) for this part of the study. Figure 7 shows that pre- and post-shock phasespace densities are identical, apart from statistics. In the previous cases, even if there were *100% reflection*, some of the particles would remain energized even after the system has relaxed to the initial dipole configuration, a long time after the compression has passed, implying a breaking of the third adiabatic invariant. In the symmetric case, the particles return to the initial distribution—no net energization.

Is the third adiabatic invariant conserved for the symmetric case? The definition of the third adiabatic invariant is:

$$\phi_M = \int B \cdot dA \qquad (8)$$

where ϕ_M is magnetic flux through a surface A. This can be thought of as the number of magnetic field lines contained within the particle's drift path, such that the surface A is the area enclosed by that drift path. Because the field we are using is a static dipole before and after the shock, a particle that returns to its original radial distance returns to its original ϕ_M. Therefore the third adiabatic invariant is the same for the pre- and post-shock instances.

Can we say the same for every point in time in between? First, consider the motion of a cold electron during a compression. Because the electron must follow field lines, which cannot cross, the magnetic flux inside the drift path will not change as the electron moves inward (see Figure 8). The situation is no different than the beginning or end of the compression—the number of field lines in the drift path is unchanged and ϕ_M remains constant. Because the only mechanism for radial displacement under this model is $E \times B$-drift, which is energy independent, this argument can be generalized to include energetic particles, i.e., ϕ_M still remains constant. For azimuthally symmetric magnetic fields, when the guiding center approximation is valid, the third adiabatic invariant is conserved at all points along the particles' drift path. We have made no assumptions about the relative time scales of the pulse and particle drift in this argument. In our case, the time scale of the pertubation is not significantly longer than the particles' drift. A pertubation time scale longer than the drift means the corresponding invariant is conserved, but a pertubation time scale shorter than the drift does not necessarily mean the invariant is not conserved. In this azimuthally-symmetric compression example, the third adiabatic invariant is conserved, regardless of the time scale of the pertubation. Figure 9 shows the drift path, radial position and energy of a single particle. It shows that the individual particle does return to its original radial position.

In the asymetric compression case, ϕ_M is not conserved due to the local time dependence, in which the particles see a different strength electric field at different locations. In other words, the more resonant particles see more of the field for a longer amount of time. In the symmetric case, the particles see the same field independent of their velocity or position.

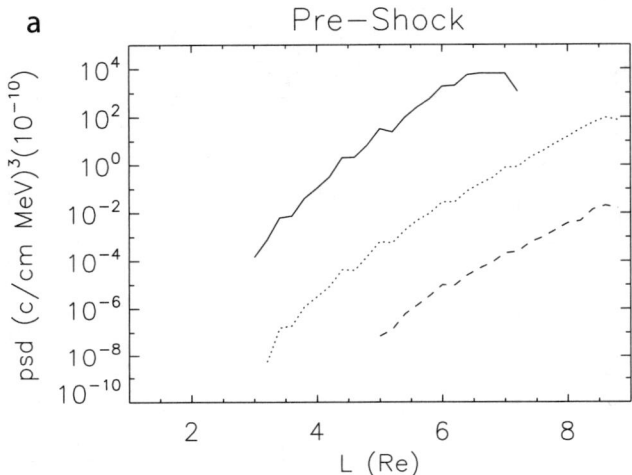

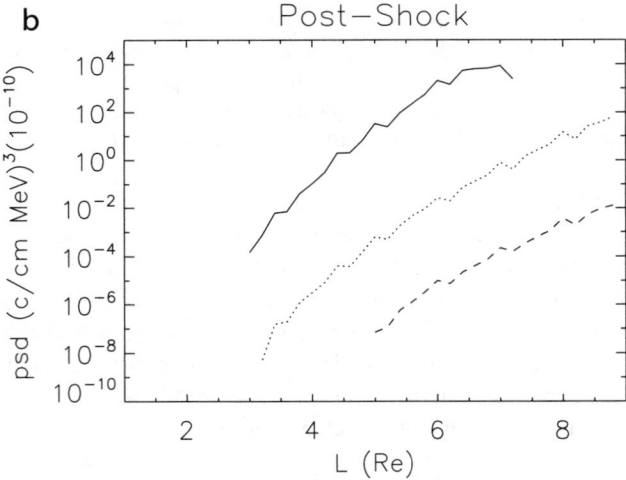

Figure 7. a) Pre-shock phasespace density profile from the output of the model using symmetric compression parameters. b) Post-shock phasespace density profile from output of the model using symmetric compression parameters. solid: $\mu = 2077\ MeV/G$, dotted: $\mu = 11852\ MeV/G$, dashed: $\mu = 45898\ MeV/G$

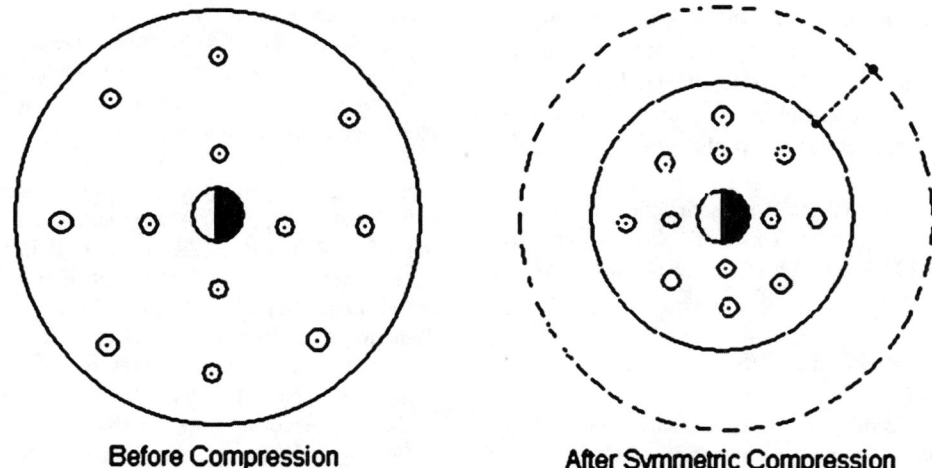

Figure 8. (Courtesy of E. Burin des Roziers) Diagram of a cold electron following a field line during a symmetric compression. Left: The solid circle indicates the initial radial position in a dipole field. Right: After the field is compressed, the electron is now at the new position (solid line). The dotted line is the original position in the left panel. Because magnetic field lines cannot cross, the flux enclosed in both solid circles is the same.

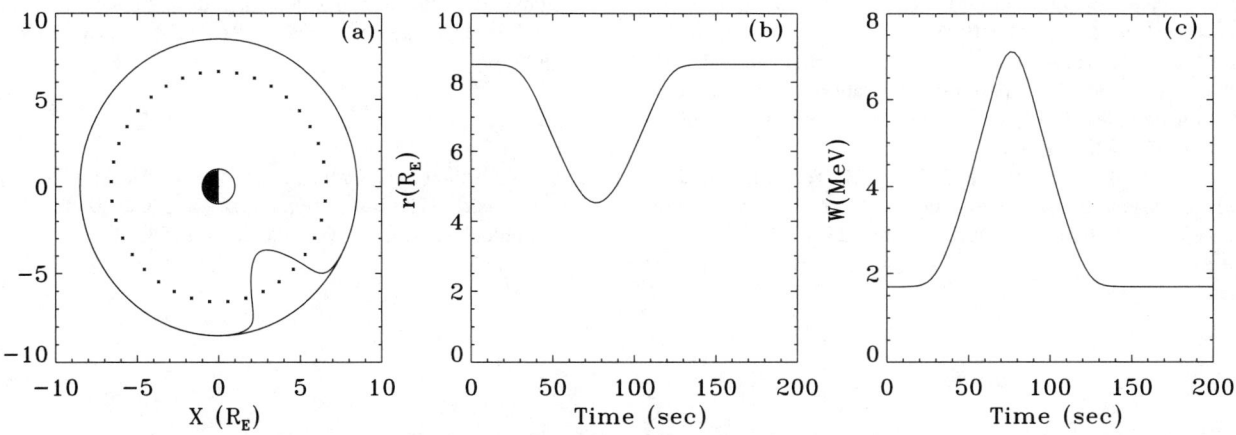

Figure 9. Single particle example under the symmetric field model. The electron begins with an energy of 1.7 MeV, a radial position of 8.5 R_e and an azimuthal position of 4 deg (near noon). Panel a) shows drift path, b) shows change in radial position, and c) shows the electrons energy versus time. Panel (b) and (c) only cover the interaction with the compression. (t = 0 is not simulation onset time.)

SUMMARY

A phasespace density profile versus L of the March 24, 1991 event was produced using output of the particle-tracing simulation by *Li et al* (1993), confirming earlier observations by CRRES which showed there was insufficient phasespace density at $L < 6$. Additional studies were done using the same electric field model, but assuming an initially flat phasespace density versus radial distance profile. This was done to study the contribution of the initial particle distribution to the results, as well as a test of the model. The phasespace density profile remains flat except near the boundaries, and the post-shock curves never surpass the pre-shock curves, illustrating that such a μ-conserving radial transport process cannot produce increases in phasespace density above an initially assumed flat phasespace density profile of a given μ versus L. As another test of the model and in order to study the contribution of the field, a symmetric (no local time dependence) electric field with full reflection is used, with no modifications to the *Li et al* (1993) source population. This yields a phasespace density profile identical to the initial phasespace density profile. Under an azimuthally-symmetric compression, particles conserve the third adiabatic invariant at all times during the compression, in addition to the first adiabatic invariant, as long as the guiding center approximation is valid. These tests, together with the phasespace density analysis, illustrate the significance of both the initial particle distribution and the actual fields associated with such compression events.

Acknowledgements. We are thankful for both reviewers for their helpful comments and suggestions. This work was supported by NASA grants (NAG5-10474 and -13518) and NSF grants (ATM-0233302 and -0101351)

REFERENCES

Blake J. B., W. A. Kolasinski, R. W. Rillius, and E. G. Mullen, Injection of electrons and protons with energies of tens of MeV into L< 3 on March 24, 1991, *Geophys. Res. Lett.*, *19*, 821, 1992.

Elkington, S.R., M.K. Hudson, M.J. Wiltberger, J.G.Lyon, MHD/particle simulations of radiation belt dynamics, *Journal of Atmospheric and Solar-Terrestrial Physics*, 64, 607-615, 2002.

Hudson, M.K., A.D. Kotelnikov, X. Li, I. Roth, M. Temerin, J. Wygant, J.B. Blake, M.S. Gussenhoven, Simulation of proton radiation belt formation during the March 24, 1991 SSC, *Geophys. Res. Lett.*, 22(3), 291, 1995.

Li, X., I. Roth, M. Temerin, J. Wygant, M. K. Hudson, and J. B. Blake, Simulation of the prompt energization and transport of radiation particles during the March 24, 1991 SSC, *Geophys. Res. Lett.*, *20*, 2423, 1993.

Li, X., M.K. Hudson, J.B.Blake, I. Roth, M. Temerin, and J.R. Wygant, Observation and simulation of the rapid formation of a new electron radiation belt during March, 24, 1991 SSC, *Workshop on the Earth's Trapped Particle Environment, AIP Conference Proceedings*, 383, 109, AIP Press, Woodbury, NY, 1996.

Li, X. and M. Temerin, The Electron Radiation Belt, *Space Science Reviews*, 95, 569-580, Kluwer Academic Publishers, 2001.

Northrop, T.G., *The Adiabatic Motion of Charged Particles*, Interscience Publishers, New York, 1963.

Schulz, M., The magnetosphere, in *Geomagnetism*, vol. 4, edited by J.A. Jacobs, p. 87, Academic, San Diego, California, 1991.

Selesnick, R.B, J.B. Blake, W.A. Kolasinski and T.A. Fritz, A quiescent state of 3 to 8 MeV radiation belt electrons, *Geophys. Res. Lett.*, 24(11), 1343-1346, 1997.

Temerin, M., I. Roth, M. K. Hudson, J. R. Wygant, New paradigm for the transport and energization of radiation belt particles, *AGU, Eos*, Nov. 1, 1994, page 538.

Walt, M., Source and Loss Processes for Radiation Belt Particles, Geophysical Monograph—Radiation Belts Models and Standards, J.F. Lemaire, D. Heyderickx, D.N. Baker, eds., 1996.

Waters, C.L., B.G. Harrold, F.W. Merik, J.C. Samson, and B.J. Fraser, Field line resonances and waveguide modes at low latitudes, *J. Geophys. Res.*, 21, A4, 7763-7774, 2000.

Wygant, J., F. Mozer, M. Temerin, J. Blake, N. Maynard, H. Singer, M. Smiddy, Large amplitude electric and magnetic field signatures in the inner magnetospher during injection of 15 MeV electron drift echoes, *Geophys. Res. Lett.*, *21*(16), 1739-1742, 1994.

Jennifer Gannon and Xinlin Li, Laboratory for Atmospheric and Space Physics, 1234 Innovation Dr.,Boulder, CO 80303. (gannonjl@colorado.edu; Xinlin.Li@lasp.colorado.edu)

Parameterization of Ring Current Adiabatic Energization

M. W. Liemohn

Atmospheric, Oceanic, and Space Sciences Department, University of Michigan, Ann Arbor, Michigan

G. V. Khazanov

National Space Science and Technology Center, NASA Marshall Space Flight Center, Huntsville, Alabama

An analysis of the factors that parameterize the net adiabatic energy gain in the inner magnetosphere during magnetic storms is presented. A single storm was considered, that of April 17, 2002. Three simulations were conducted with similar boundary conditions but with different electric field descriptions. It is concluded that the best parameter for quantifying the net adiabatic energy gain in the inner magnetosphere during storms is the instantaneous value of the product of the maximum westward electric field at the outer simulation boundary with the nightside plasma sheet density. In addition, these two quantities alone usually produced large correlation coefficients, along with several other magnetospheric quantities. An analysis is given regarding which parameters could be considered useful predictors of the net adiabatic energy gain of the ring current. Long integration times over the parameters lessen the significance of the correlation. This implies the instant response of the ring current to the particle source and convection. Finally, some significant differences exist in the correlation coefficients depending on the electric field description, and these differences are presented and discussed.

1. INTRODUCTION

Inside the Earth is a dynamo generating a large dipolar magnetic field. This field extends out in to space, creating a bubble, known as the magnetosphere, around the planet. This magnetic cavity is continuously buffeted by the solar wind, a supersonic flow of charged particles (plasma) streaming from the Sun. The solar wind pulls out magnetic field lines from the Sun, known as the interplanetary magnetic field (IMF), and together the solar wind and IMF compress the dayside of the magnetosphere and elongate the backside. This gives the magnetosphere its characteristic ovoid topology. During quiescent solar wind conditions, the dayside magnetopause distance is ~10 Earth radii (R_E) while the nightside magnetopause can extend 30–100 R_E downtail [e.g., *Dungey*, 1961; *Siscoe*, 1966].

Occasionally, the Sun emits a blast of plasma and/or magnetic field, either as continuously produced high-speed streams or as transient ejecta [see *Tsurutani and Gonzalez*, 1997, for a comprehensive review]. These fast flows form a shock wave at their leading edge, piling up the ambient solar wind and IMF into a high-density sheath region. If the flow is directed at the Earth, then the shock, sheath, and stream/ejection will slam into the magnetosphere. This causes what is known as a magnetic storm [e.g., *Gonzalez et al.*, 1994; *Rostoker et al.*, 1997; and references therein].

Magnetic storms result in numerous space weather effects. The most visible is a brightening and expansion of the auroral ovals as energetic electrons and ions cascade into the upper atmosphere [e.g., *Akasofu and Chapman*, 1961]. Intense ionospheric outflows populate the magnetosphere (along with solar wind inflow) [e.g., *Chappell et al.*, 1987]. Even though

Particle Acceleration in Astrophysical Plasmas
Geophysical Monograph Series 156
Copyright 2005 by the American Geophysical Union
10.1029/156GM24

the magnetosphere is being squeezed, there is a global depression of the magnetic field around the equator of the Earth, measured by the well-known "disturbance storm time" Dst index [*Sugiura and Chapman*, 1960]. This is largely the result of the ring current [e.g., *Gonzalez et al.*, 1994]. Particles in the magnetotail (specifically, the plasma sheet, a relatively dense blanket of plasma in the equatorial plane) are convected sunward, and during storms they are injected deep into the inner magnetosphere. Here they are adiabatically energized [e.g., *Lyons and Williams*, 1980; *Lee et al.*, 1982].

For plasma sheet-ring current ions in the 1–100 keV energy range (the particles carrying the bulk of the near-Earth plasma pressure), the gyration and bounce periods are usually much smaller than the characteristic time scales of the dominant physical processes acting on them. However, their drift periods around the Earth are usually much larger than these characteristic time scales. Therefore, it is the first and second adiabatic invariants that are being conserved, which can result in kinetic energy changes for the particles. Inward convection brings the particles into a region of higher B (magnetic field), and conservation of the first two invariants causes the particle's total kinetic energy to increase. Similarly, outward convection causes de-energization. This form of adiabatic energization is well known, as evidenced by this quote from *Alfvén and Fälthammar* [1963, p. 62], "A change in B takes place when the charged particle drifts into a region with different field strength. This drift can be produced by an electric field." It is also discussed in many recent reviews of magnetospheric physics, including *Wolf* [1995] in the widely-used textbook edited by Kivelson and Russell.

In addition to convection, they particles magnetically drift across the field lines, with electrons flowing eastward and ions flowing westward. This creates a net westward current near the Earth known as the ring current. During magnetic storms, it is largely a partial ring, becoming a symmetric ring only in the recovery phase of the storm [e.g., *Takahashi et al.*, 1990; *Chen et al.*, 1993].

The stormtime ring current is a space weather catalyst. Its inflation of the near-Earth magnetic field [e.g., *Parker and Stewart*, 1967; *Tsyganenko et al.*, 2002] alters the drift paths of the relativistic electrons in the radiation belts, causing increased precipitation into the upper atmosphere and influencing the post-storm development of the belts [e.g., *Hudson et al.*, 1998; *Green and Kivelson*, 2001]. Its modulation of the near-Earth electric field [e.g., *Jaggi and Wolf*, 1973; *Fok et al.*, 2001] alters the subauroral ionospheric wind pattern, augmenting low-density troughs and creating high-density plumes at unexpected times and places, which cause errors in GPS signal processing [e.g., *Yeh et al.*, 1991;

Foster et al., 2002]. A thorough understanding of the ring current is therefore critical to predicting space weather and mitigating space environment effects on human life here on Earth.

As was found by *Liemohn et al.* [2002] and *Khazanov et al.* [2004a], adiabatic acceleration of the ring current is a significant component of the total energy gain (and thus the total inner magnetospheric current). This paper examines the relationship of the net adiabatic energization to several solar wind and magnetospheric parameters. The analysis is based on results from numerical simulations of the stormtime ring current for the April 17, 2002, disturbance. Section 2 provides an overview of the complexity of the inner magnetospheric processes, and places this study into the context of the overall dynamics of near-Earth space. A brief description of the numerical model is provided in section 3. Section 4 is an objective presentation of the results, while section 5 is an analysis of those results. The major findings of the study are summarized in section 6.

2. COMPLEXITY OF THE INNER MAGNETOSPHERE

Figure 1 presents a schematic of the processes and interactions in the inner magnetosphere, as well as the linkages coupling all of these components. The two boxes under consideration in this study are the "Potential E-field" solver and the "Hot Ions," as it is the electric field that radially convects the particles (thus adiabatically energizing them) and it is the hot ions (rather than the electrons) that carry the majority of the plasma pressure in this region. As seen in Figure 1, however, adiabatic energy gain and loss are only parts of a much larger picture. Here we will discuss these other processes to put the adiabatic energization into perspective. For brevity, many of the references given below will be to our own work in these areas. Robust citations of previous studies by others authors are given in the reference lists of the papers cited here.

The plasma sheet is the primary source of particles for the hot electron and ion populations in the inner magnetosphere. The inflow of particles contains both electrons [e.g., *Roederer*, 1970; *Liemohn et al.*, 1998; *Khazanov et al.*, 1998] and ions [e.g., *Wolf et al.*, 1982; *Liemohn et al.*, 1999]. The flow of these particles is governed by the topology of the near-Earth magnetic and electric fields [e.g., *Alfvén and Fälthammar*, 1963; *Khazanov et al.*, 2003a]. Through a variety of processes, some of the electrons and ions are pitch-angle scattered into the loss cone, where they precipitate into the thermosphere and ionosphere [e.g., *Hultqvist et al.*, 1976; *Khazanov et al.*, 2002]. In their descent through the atmosphere, they ionize and heat the background particles, resulting in changes to the electrical conductance of the

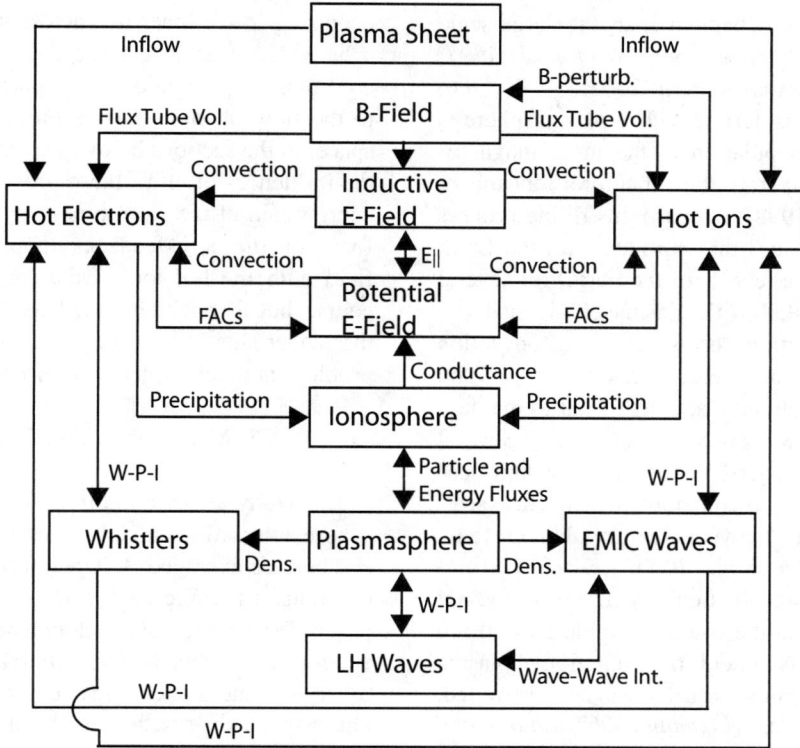

Figure 1. Schematic of the coupling between processes in the inner magnetosphere and subauroral ionosphere.

subauroral ionosphere [*Galand and Richmond*, 2001; *Khazanov et al.*, 2003a]. In addition, longitudinal asymmetries in the magnetospheric currents must be closed through field-aligned currents (FACs) into and out of the ionosphere [e.g., *Iijima and Potemra*, 1976; *Liemohn et al.*, 2001b]. The ionospheric conductance allows perpendicular currents to flow (to connect and balance the FACs), resulting in perpendicular (to the magnetic field lines) electric fields [*Goodman*, 1995; *Amm*, 1996; *Ridley and Liemohn*, 2002]. Assuming the field-aligned conductance to be infinite, the electric fields can be mapped along the field lines to the inner magnetosphere, where they will govern the subsequent flow of the hot electrons and ions. Including this feedback loop in a numerical simulation of the ring current is known as a self-consistent potential electric field calculation, because the location of the hot particles in the inner magnetosphere modifies the electric fields that determine the locations of the hot particles [e.g., *Fok et al.*, 2001, 2003; *Khazanov et al.*, 2003a; *Liemohn et al.*, 2004]. Of course, the high-latitude electric potential distribution also influences the mid-latitude electric potential pattern, and any self-consistent solution of the subauroral potential must have a high-latitude boundary condition imposed on it.

Many methods have been used to impose a non-self-consistent potential electric field within the inner magnetosphere to convect the stormtime ring current particles [e.g., *Fok et al.*, 1993; *Chen et al.*, 1993; *Liemohn et al.*, 2001a]. The choice of these various fields can lead to large differences in the resulting plasma distributions [e.g., *Jordanova et al.*, 2001; *Khazanov et al.*, 2003a], including differences in adiabatic energization [*Khazanov et al.* 2004a, b].

This convective flow is also the major loss of hot particles and energy from the inner magnetosphere. Flowout through the dayside magnetopause has been shown to be the dominant loss mechanism during the main phase of storms, and usually well into the recovery phase [*Liemohn et al.*, 1999, 2001a; *Kozyra and Liemohn*, 2003]. It is also responsible for the adiabatic de-energization of the particles while they are still close to the Earth [*Liemohn et al.*, 2002].

This central feedback loop between the hot plasma and the potential electric field is influenced by a number of other processes. One such process is the magnetic field perturbation caused by the inner magnetospheric currents [e.g., *Parker and Stewart*, 1967; *Liemoh*n, 2003]. These perturbations can be large [*Tsyganenko et al.*, 2002], altering the magnetic drift paths of the particles (another feedback loop). In addition, the magnetic field can explosively reconfigure (e.g., substorms), resulting in transient induced electric fields that can influence the particle flow. The induction electric field of substorms, however, has been shown to be a less efficient injection mechanism than the large-scale convective flow [e.g., *McPherron*, 1997; *Wolf et al.*, 1997; *Fok et al.*, 1999; *Liemohn and Kozyra*, 2002].

Another factor is the plasmasphere. The plasmasphere is the cold, dense particle population in the inner magnetosphere that is slowly filled from the subauroral ionosphere [*Lemaire and Gringauz*, 1998]. Because this filling process is slow, only those field lines that corotate with the Earth (that is, only those within a few R_E of the Earth) for several days can become populated with plasmaspheric material [e.g., *Carpenter*, 1963]. During storms, increased convection strips away the outer plasmasphere, shrinking its size and creating a large drainage plume that extends out to the dayside magnetopause [e.g., *Rasmussen et al.*, 1993]. Several processes occur in the overlap between the hot and cold particles. The first is Coulomb collisions, which cause scattering and energy transfer between the populations [e.g., *Fok et al.*, 1993; *Liemohn et al.*, 2000; *Khazanov et al.*, 2000b]. This energy transfer, particularly that from the hot O^+ to the cold electrons, is the cause of stable auroral red arcs [*Kozyra et al.*, 1997]. Another is the excitation of plasma waves. With the hot electrons, whistler waves are excited, resulting in plasmaspheric hiss [*Liemohn*, 1967; *Church and Thorne*, 1983], which can scatter the hot electrons [*Dungey*, 1963; *Liemohn et al.*, 1997] and ions [*Kozyra et al.*, 1995]. With the hot ions, electromagnetic ion cyclotron waves are excited [e.g., *Kennel and Petschek*, 1966; *Khazanov et al.*, 2002, 2003b], which can interact with (i.e., scatter) the hot ions [*Jordanova et al.*, 1997; *Khazanov et al.*, 2002, 2003b] and electrons [*Summers and Thorne*, 2003]. Given the proper plasmaspheric conditions, these low-frequency EMIC waves can excite lower hybrid waves, which are very efficient at energizing the plasmasphere [e.g., *Khazanov et al.*, 1996b, 1997, 2000a]. This is an additional energy transfer process from the hot particles to the cold plasma via wave-wave interactions. Finally, the plasmasphere can deposit this excess energy (as well as some of its particles) back into the ionosphere, influencing the subauroral conductance.

Another source of hot electrons in the inner magnetosphere is photoionization and impact ionization of the upper thermosphere. Above ~250 km, warm electrons from the ionosphere can escape along the field lines [e.g., *Takahashi*, 1973; *Khazanov and Liemohn*, 1995, 2000, 2002]. Some of these electrons are scattered and trapped in the magnetosphere, where they will drift around and eventually redeposit their energy into the ionosphere [e.g., *Khazanov et al.*, 1996a, 1998]. Because these particles can be either long or short lived in the magnetosphere, they are quite efficient at altering the energy deposition pattern into the ionosphere.

It is clear that adiabatic energization is only one component of a complex and interconnected system of processes occurring in the inner magnetosphere and subauroral ionosphere. Large-scale electric and magnetic fields, hot and cold plasma populations, and plasma waves all contribute to the flow of mass, momentum and energy in near-Earth space. In the sections below, we will not attempt to examine the influences of all of these processes on the net adiabatic energy gain of the stormtime ring current. Instead, we will focus on the core feedback loop of the potential electric field with the hot ions. Other processes are included, of course, but are not highlighted as part of this study. However, the reader should keep Figure 1 in mind, realizing that the problem is much larger than just this single interaction.

3. MATHEMATICAL FORMALISM

This study uses results from a kinetic ring current-atmosphere interaction model (RAM) that solves the gyration and bounce-averaged Boltzmann equation inside of geosynchronous orbit. See *Fok et al.* [1993] and *Jordanova et al.* [1996] for the original development of this code. The present version is described by *Liemohn et al.* [2004]. It uses second-order accurate numerical schemes to determine the hot ion phase-space distribution in the inner magnetosphere as a function of time, equatorial plane location, energy, and equatorial pitch angle. Sources are specified by geosynchronous orbit plasma data across the nightside outer boundary, splitting the composition for the E < 50 keV measurements between O^+ and H^+ according to the relations of *Young et al.* [1982] (higher energy observations are assumed to be H^+, minus a $\kappa = 5$ high-energy extension to the lower-energy O^+ distribution). Loss mechanisms include the flow of plasma out the dayside outer boundary, precipitation of particles into the upper atmosphere, pitch angle scattering and drag from Coulomb collisions with the plasmasphere, and charge exchange with the neutral hydrogen geocorona. Coupled to RAM are the plasmaspheric model of *Ober et al.* [1997] and the geocoronal model of *Rairden et al.* [1986]. A dipole magnetic field is assumed. This is reasonable in the inner magnetosphere during moderate disturbances. However, it neglects any adiabatic energy gain due to inductive forces during magnetic field reconfigurations (e.g., during substorms), which can influence the near-Earth plasma distribution [*Wolf et al.*, 1997; *Fok et al.*, 1999]. This omission is intentional, because this study focuses on the net adiabatic drift caused by the choice of electric field model.

Three choices for the inner magnetospheric potential electric field will be used. The first option is the modified *McIlwain* [1986] E5D model (as revised by *Liemohn et al.* [2001a]), which is an analytical two-cell convection pattern (hereafter the modified McIlwain field). The second option is to map the *Weimer* [1996] potential distribution into the inner magnetosphere along the dipole magnetic field (hereafter the Weimer-96 field). The third option is a self-consistent electric field description, where the inner magnetospheric electric fields are a consequence of the RAM-generated plasma distributions, the ionospheric conductance pattern, and the high-latitude boundary condition (from Weimer-96). Details of the computational scheme for this electric field choice are given by *Liemohn et al.* [2001b, 2004], *Ridley et al.* [2001, 2004], and *Ridley and Liemohn* [2002]. This electric field model is similar to that in other self-consistent inner magnetospheric electric field calculations [*Wolf et al.*, 1982, 1997; *Fok et al.*, 2001; *Khazanov et al.*, 2003a], with the primary difference being the conductance pattern.

The model includes not only drifts in configuration space (from the convection electric field, corotation, and gradient-curvature effects) but also drifts in velocity space. These latter drifts are derived from the conservation of the first and second adiabatic invariants, and result in energization (de-energization) as the particles move inward (outward) across magnetic field isocontours. Thus, it is referred to as adiabatic energization, and it is this net energy gain that will be related to other stormtime parameters in the following section.

4. RESULTS

The purpose of this study is to examine the relationship between the net adiabatic energy gain of the ring current ions during a magnetic storm and various "driver" parameters. Before addressing this topic, it is useful to quantitatively show the relationship between the ring current and the *Dst* index. Because the net current in the inner magnetosphere is westward, the resulting magnetic perturbation at the equator of the Earth is southward (opposite to the dipole field). This is the same direction as the excursion of Dst during storms (negative). *Dessler and Parker* [1959] and *Sckopke* [1966] analytically derived a formula relating the total energy content of the ring current to the magnetic perturbation at the center of the Earth, and this relationship has been upheld numerically, with some caveats [e.g., *Carovillano and Siscoe*, 1973; *Liemohn*, 2003]. Using the Dessler-Parker-Sckopke (DPS) relation, results from models such as RAM can be directly compared against *Dst* to show the relative contribution to *Dst* from the ring current. Figure 2 shows such comparisons for two storm events (for details of these magnetic storms, see *Liemohn et al.* [2001a] and *Kozyra et al.* [2002]). The agreement between model and data becomes even better when other known contributors to Dst are taken into account, such as induced currents in the Earth, the magnetopause currents, and the cross-magnetotail current. This illustrates that the ring current is a major source of magnetic perturbation in near-Earth space, with societal consequences (as mentioned in section 1). As discussed in several studies [e.g., *Lyons and Williams*, 1980; *Liemohn et al.*, 2002; *Khazanov et al.*, 2004a], adiabatic energization is a large contributor to the total energy input to the ring current ions during storms. Thus, an understanding of what controls the net adiabatic energy gain of the ring current is an essential step in understanding near-Earth space.

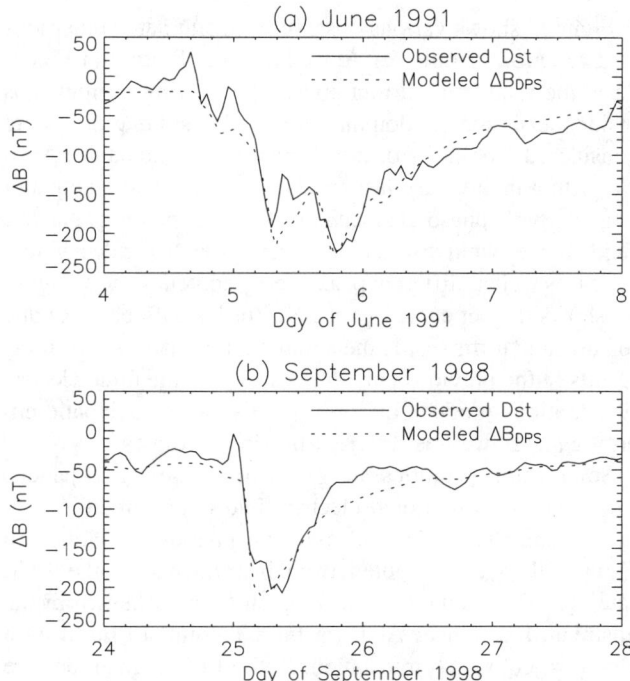

Figure 2. Observed *Dst* (solid lines) and modeled ΔB_{DPS} (dotted lines) for two magnetic storms, (a) June 4–7, 1991, and (b) September 24–27, 1998.

The storm that occurred on April 17, 2002, is chosen for the computational experiments from which we will examine the adiabatic energization. The study of *Liemohn et al.* [2004] previously analyzed results from this model for this event. They showed that several electric field options produce different levels of accuracy with respect to the observed plasmapause location (from the EUV instrument on the IMAGE spacecraft). To summarize their results, they found that a self-consistent electric field is better than a Weimer-96 electric field, which in turn is better than a modified McIlwain electric field. *Liemohn et al.* [2004] only briefly discussed the hot ion results from the numerical simulations. In the present study, one aspect of the ring current will be investigated: the net adiabatic energy gain of the hot ions in the inner magnetosphere during the storm. In this section, an objective presentation of the results is given, and these results are interpreted and discussed in section 5 below.

Figure 3 shows various observed and simulated quantities as a function of time for April 17, 2002. Plotted in Figure 3a is the total ring current energy of the ring current ions inside the simulation domain for the three simulations being considered. The storm occurs during the second half of April 17, with a main phase lasting about 6 hours and an equally long recovery phase. It is seen that the three potential electric field choices yield distinctly different growth and decay time series, as well as different peak energy content values. Figure 3b shows the net energy gain ΔE_{ad} (or loss, if negative) due to adiabatic drifts inside the simulation domain. As with the results in the previous panel, it is seen that the three electric field options produce different time series of adiabatic energy gain, as well as different maximum values.

Solar wind quantities are given in the next two panels. The solar wind motional electric field is presented in Figure 3c and the solar wind dynamic pressure is plotted in Figure 3d, both computed from instruments on the ACE satellite. These quantities have been time shifted from the upstream location of ACE (at the L1 point) to Earth by a simple $x_{GSM}/v_{x,GSM}$ translation. Both of these quantities are relatively quiet until shortly after 11:00 UT, when the shock hits the magnetosphere. There is a data gap, however, until ~8:00 UT in the solar wind plasma observations (straight line segment in Figure 3d). For the rest of the day, ACE observed the passage of the compressed sheath. It should be mentioned that the magnetic cloud (piling up the sheath material) arrived at the Earth early on April 18, causing a subsequent magnet storm on that day. Two more storms occurred over the next week as the same active region on the Sun continued to emit solar flares and coronal mass ejections.

Figure 3e is the cross polar cap potential from the Weimer-96 electric field model. This is the potential difference applied to all of the simulations. The potential jumps quickly up to nearly 200 kV, then drops back down near 50 kV during a northward interplanetary magnetic field interval (negative $E_{y,sw}$, as seen in Figure 3c), and then hovers between 150 kV and 200 kV for several hours. It then drops down below 100 kV and, with one exception just before 19:00 UT, remains low for the remainder of the day.

Each electric field option uses the cross polar cap potential difference in Figure 3e as an overall convection strength parameter, but they use it in different ways. The modified McIlwain model simply uses this quantity as a strength parameter for its analytical potential distribution. The general pattern in the inner magnetosphere is the same at all times, with variations due to this strength parameter and a shielding parameter scaled by Kp. The Weimer-96 model does not really use this potential difference in the inner magnetosphere. It is a statistical spherical harmonic fit to

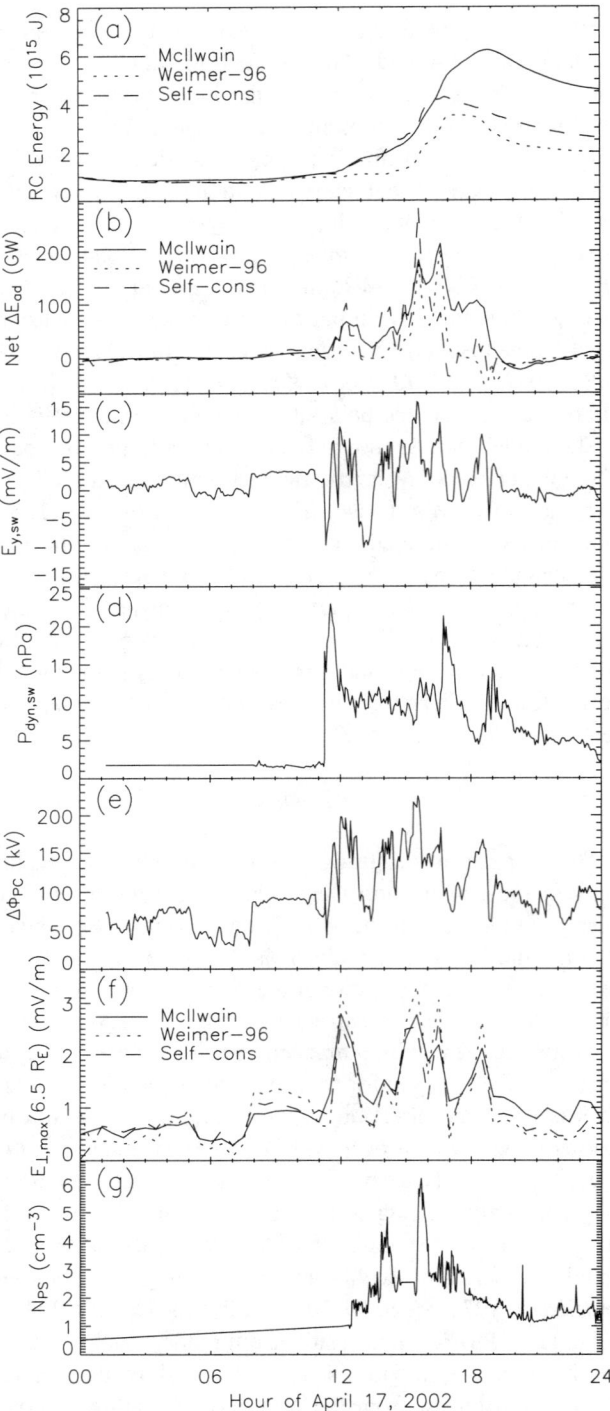

Figure 3. Time series of various simulated and observed quantities during the April 17, 2002, magnetic storm. Shown are (a) total kinetic energy of the simulated ring current for the three electric field descriptions; (b) net adiabatic energy gain from the three runs; (c) solar wind motional electric field; (d) solar wind dynamic pressure; (e) cross polar cap potential difference; (f) maximum westward electric field at 6.5 R_E; and (g) nightside plasma sheet density as observed by the LANL MPA instrument.

observed ionospheric drifts, and therefore the mid-latitude pattern is related to the total potential difference but does not have to scale with it. Similarly, the self-consistent field model only indirectly uses this quantity. Instead, it applies the Weimer-96 potential from ~72° magnetic latitude as a high-latitude boundary condition in the Poisson solver, and the potential pattern at lower latitudes is highly dependent on the location and intensity of the subauroral ionospheric conductance and field-aligned currents. Therefore, it is useful to consider another electric field quantity. Figure 3f shows the maximum westward electric field $E_{\perp,max}$ (which causes inward radial drifts) at 6.5 R_E (in the outermost ring of grid cells of the simulation). It is seen that the timing and magnitudes of the peaks of $E_{\perp,max}$ for the three simulations generally follow each other, because of the connection with the cross polar cap potential shown in Figure 3e. There are differences, however, and their impact will be discussed later.

The nightside plasma sheet density N_{PS} as observed by the magnetospheric plasma analyzer (MPA) instrument on the Los Alamos National Laboratory (LANL) geosynchronously orbiting spacecraft is shown in Figure 3g. Moments of the observed plasma distribution from 100 eV to 40 keV are included in this time series when the spacecraft is within 4 hours of local midnight. When more than one satellite is in the "nightside" window, the highest moment value was selected for this plot. Note that there is a data gap during the first half of this day (the straight line segment). The flux distributions were used as time-varying boundary conditions for the simulations. Data gaps in the time series are filled in by interpolation from the previous and next valid observations. It is seen that the near-Earth plasma sheet density was elevated above 2 cm^{-3} throughout the main phase of the storm (compared with $\leq$ 1 cm^{-3} during quiet times), with two brief intervals of densities up to 4 and 6 cm^{-3}, respectively.

Figures 4–6 present scatter plots of the net energy gain from adiabatic acceleration and deceleration (the time series shown in Figure 3b) with respect to 10 different parameters. Figure 4 shows the simulation results with the modified McIlwain electric field description, Figure 5 presents results with the Weimer-96 electric field in the model, and Figure 6 contains plots for the run with a self-consistent electric field. The 10 parameters along the x-axis of the 10 panels are, respectively: (a) solar wind motional electric field (from Figure 3c); (b) solar wind dynamic pressure (from Figure 3d); (c) nightside plasma sheet density (from Figure 3g); (d) cross polar cap potential (from Figure 3e); (e) maximum westward electric field at 6.5 R_E (from Figure 3f); (f) the average westward electric field around a circle at 6.5 R_E (summed over only those portions of the circle with westward field); (g) the multiplication of the maximum electric field and the nightside plasma sheet density; (h) like (g) but averaged over the previous hour; (i) like (g) but averaged over the previous 2 hours; and (j) like (g) but averaged over the previous 4 hours. Listed on each panel is the correlation coefficient. Each point is a ten-minute average of the net energy gain and the parameter. In addition, points are only shown for the 12-hour interval surrounding the storm peak: that is, from 11:00 UT to 23:00 UT on April 17, 2002. This also avoids the gaps in the solar wind and LANL MPA data sets. Listed on each panel is the Pearson correlation coefficient (R_P) and the Spearman rank correlation coefficient (R_S).

Several features in Figures 4–6 are worth highlighting. The first is the high level of significance for most of the scatter plots. Because there are 72 points on each panel, the 95% significance level (that the relationship is not random) is ~0.25. Thus, all of the panels in Figure 4 show statistically significant relationships, and only a few plots in Figures 5 and 6 do not meet these thresholds. The highest correlation coefficients for the modified McIlwain electric field simulation (Figure 4) are found in Figures 4g, 4h, and 4i, the instantaneous, previous-one-hour, and previous-two-hour averages (respectively) of $E_{\perp,max}$ times N_{PS}. Very high coefficients are also calculated for Figures 4c, 4e, and 4f, the MPA density, the maximum westward electric field, and the average westward electric field, respectively. Similarly, for the Weimer-96 field simulation (Figure 5), the high coefficients are found for Figure 5g and 5h. The self-consistent field simulation (Figure 6), the top coefficients are found in Figures 6d (cross polar cap potential), 6e, 6f, and 6g. Large correlation coefficients are also calculated for Figure 6a, the solar wind motional electric field. The two relationships with coefficients consistently near the bottom of the order are those between ΔE_{ad} and $P_{dyn,sw}$ and ΔE_{ad} and the previous-four-hour average of $E_{\perp,max}$ times N_{PS}.

It should be noted that the points in Figures 4–6 are not completely independent. The plasma takes some time (~hours) to flow through the inner magnetosphere, and so an injection of a blob of plasma in through the outer boundary will result in a sinusoidal time series to ΔE_{ad}. The particles will first be convected inward in radial distance, during which they will experience adiabatic energization. On the dayside, however, they will be convected outward and undergo adiabatic deceleration. In steady state, such an injection will eventually be completely lost out the dayside outer boundary and ΔE_{ad} will return to zero. Therefore, the reported statistical significance of the correlation coefficients should be treated as an upper limit, because the degrees of freedom is actually substantially lower than the 72 points in each panel. The historical influence is most pronounced in the last three panels of Figures 4–6. Because each point

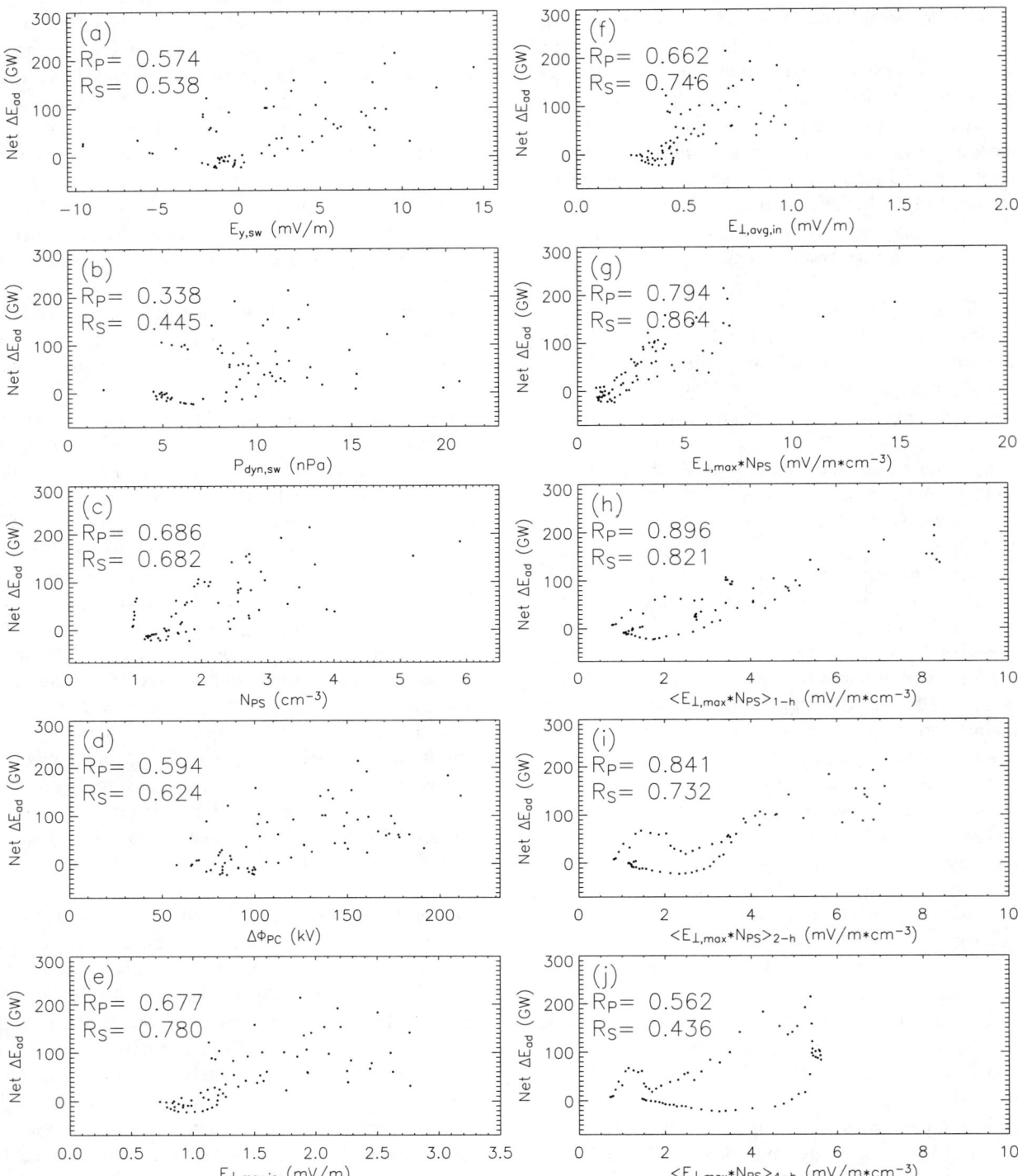

Figure 4. Scatter plots of the net adiabatic energy gain versus various quantities for the simulation using the modified McIlwain electric field model. Listed on each panel is the Pearson correlation coefficient and the Spearman rank correlation coefficient.

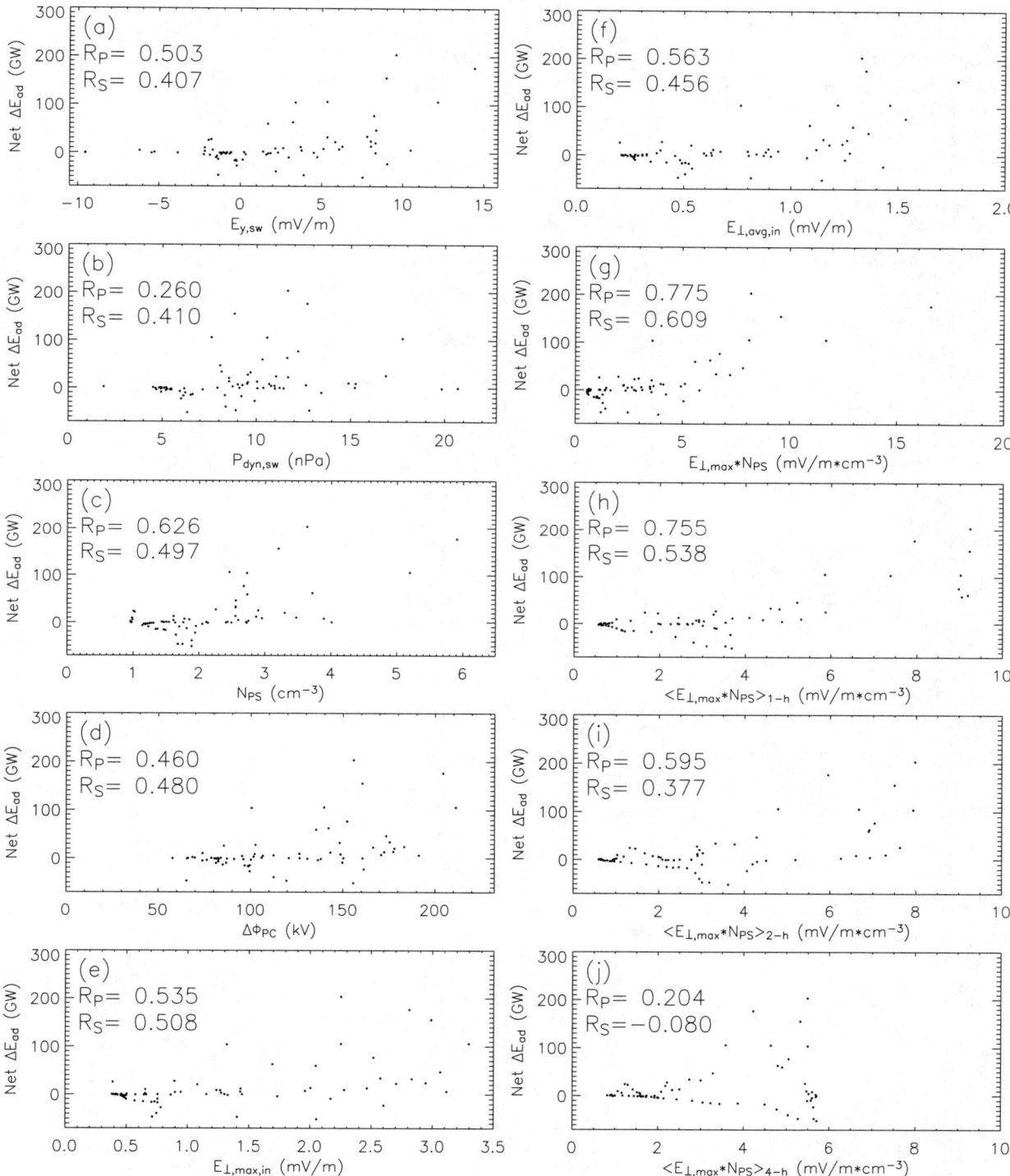

Figure 5. Same as Figure 4 except for the simulation using the Weimer-96 electric field model.

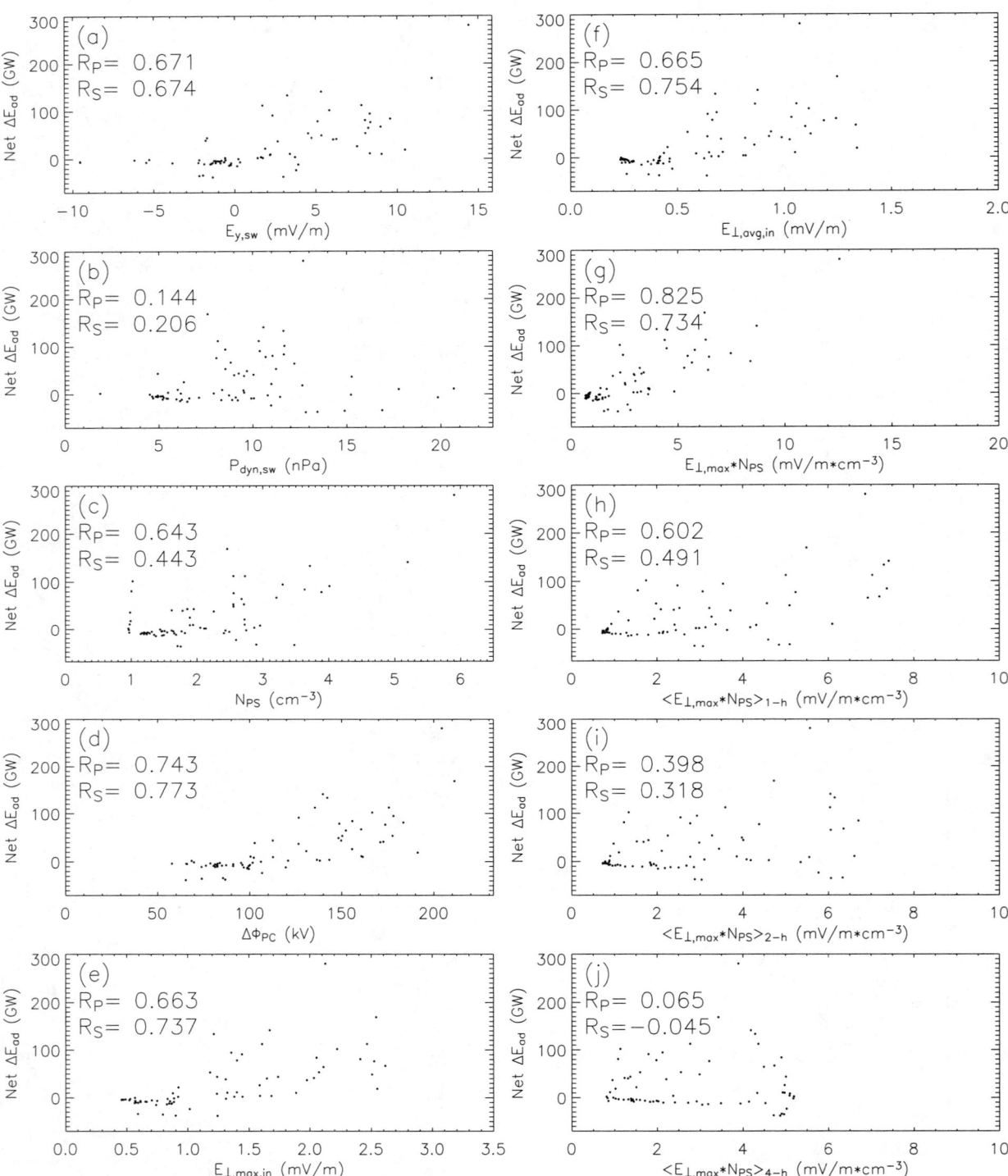

Figure 6. Same as Figure 4 except for the simulation using the self-consistent electric field model.

in the "instantaneous" panel (Figures 4g, 5g, and 6g) is a 10-minute average of the modeled ΔE_{ad} results and the abscissa quantity, the points in the last three panels are running averages over the previous 6, 12, and 24 points. Therefore, the plots resemble hodograms, because each point contains most of the data used in calculating the previous point. This trait is especially evident in the previous-four-hour average plots in Figures 4j, 5j, and 6j.

5. DISCUSSION

Let us now interpret the relationships presented in section 4. In particular, the goal of this discussion is to understand the factors influencing the adiabatic energization of ring current particles in the inner magnetosphere.

Because the solar wind and IMF are the source of the free energy in the magnetospheric system, one would expect a relationship between the solar wind parameters and the net adiabatic energy gain. Comparing Figure 3b with Figures 3c and 3d, it appears that the net adiabatic energy gain tracks the solar wind quantities. Figures 4a, 5a, and 6a reveal that the statistical significance between ΔE_{ad} and $E_{y,sw}$. The coefficients are usually not the highest among the various correlations considered, though. A strong relation is expected, because many of the Dst prediction algorithms often rely solely on $E_{y,sw}$ [Burton et al., 1975; O'Brien and McPherron, 2000a], and the ring current is a major contributor to the stormtime Dst index. $P_{dyn,sw}$ usually has an even lower correlation coefficient, as seen in Figures 4b, 5b, and 6b. For the self-consistent field results (Figure 6b), both R_P and R_S are below the 95% confidence level. The solar wind density N_{sw} (not shown) typically has even lower correlation coefficients than $P_{dyn,sw}$. There is some debate as to whether $P_{dyn,sw}$ (or N_{sw}) directly influences the stormtime ring current, with some studies finding a correlation [e.g., Fenrich and Luhmann, 1998; Borovsky et al., 1998; Smith et al., 1999] while others do not [e.g., O'Brien and McPherron, 2000b]. The present study finds a weak correlation at best.

A quantity related to the solar wind parameters is the cross polar cap potential (Figure 3e). The time series of $\Delta\Phi_{PC}$ is a convolution of that from $E_{y,sw}$ and $P_{dyn,sw}$, as derivatives of these values are included in the expansion coefficients used by the Weimer-96 model. As seen in Figures 4-6, the ΔE_{ad}–$\Delta\Phi_{PC}$ relationship usually has higher correlation coefficients than either $E_{y,sw}$ or $P_{dyn,sw}$. The significance of this relationship is reasonable because $\Delta\Phi_{PC}$ is a good measure of the cross tail convection intensity, and therefore it is a good indicator of the rate at which plasma is delivered to (and convected through) the inner magnetosphere.

A more specific quantity to consider is the westward electric field $E_\perp$ on the outer boundary of the simulation domain. Figure 3f plots the maximum value of this quantity for the three simulations, and the time series closely follow the trends in $\Delta\Phi_{PC}$. Differences exist among the simulations, though, and so it is useful to consider $E_\perp$ as well as $\Delta\Phi_{PC}$. In the scatter plots, Figures 4e, 5e, and 6e show the ΔE_{ad}–$E_{\perp,max}$ relationship is highly correlative. This is expected because $E_{\perp,max}$ is a direct measure of the convective flow (and thus the adiabatic energization) in the inner magnetosphere. Similarly, Figures 4f, 5f, and 6f, which show the ΔE_{ad}–$E_{\perp,av}$ relationship, also have very high correlation coefficients.

Another factor directly influencing the amount of adiabatic energization in the inner magnetosphere is the amount of plasma being convected into the region. Figure 3g shows ΔE_{ad} against these values, and a comparison with the lines in Figure 3b reveal a qualitative match between the time series. Figures 4c, 5c, and 6c show the quantitative relationship between ΔE_{ad} and N_{PS}. The correlations are often large, but typically not as large as those of $E_\perp$. The lower coefficients are most likely because of cumulative effect of N_{PS} on ΔE_{ad}. That is, the particles take some time to drift through the inner magnetosphere, and so a comparison against the instantaneous nightside plasma sheet density may not be the most appropriate choice. In addition, the particles must be inwardly convected by a westward electric field, and so the presence of a dense plasma sheet does not guarantee large amounts of energy gain.

A more suitable parameter with which to compare is the product of $E_\perp$ and N_{PS}. This is a measure of the energy flux into the system, and therefore should be a good indicator of ΔE_{ad}. The scatter plots for $E_{\perp,max}*N_{PS}$ are shown in Figures 4g, 5g, and 6g, revealing a very high correlation coefficient that is often larger than either $E_{\perp,max}$ or N_{PS} alone. It appears that the correct parameter has been found. However, this high correlation is for the product of the instantaneous values. Integrating over some time interval into the past might prove even better. Plots of ΔE_{ad} versus the previous-one-hour average of $E_{\perp,max}*N_{PS}$ are shown in Figures 4h, 5h, and 6h. The coefficients for the modified McIlwain electric field (Figure 4h) and the Weimer-96 electric field (Figure 5h) are similar to the instantaneous value coefficients. For the self-consistent electric field (Figure 6h), the coefficients are significantly less than those for Figure 6g. Because the particles take several hours to drift through the simulation domain, perhaps a longer integration time is needed. The final two panels of Figures 4–6 show the scatter plots for ΔE_{ad} against the previous-two-hour average of $E_{\perp,max}*N_{PS}$ and the previous-four-hour average of $E_{\perp,max}*N_{PS}$, respectively. The correlation coefficients for all three electric field descriptions drop as the integration time is increased. Figures 4i and 4j are still significant numbers, but R_P and R_S for the other two field models are quite low. It is low because the electric field from 4 hours ago only marginally influences the present-time ΔE_{ad} (by delivering the N_{PS} from 4 hours ago into the inner magnetosphere). In addition, after some time the relationship reverses (seen most clearly in Figures 4j, 5j, and 6j), and the delivery of a large blob of plasma into the inner magnetosphere eventually causes a large net decrease in energy instead of a net increase.

6. CONCLUSION

This paper presented an analysis of the factors that parameterize the net adiabatic energy gain in the inner magnetosphere during magnetic storms. Adiabatic energization (de-energization) occurs when the plasma sheet particles move radially inward (outward) to form the stormtime ring current. A single storm was considered, that of April 17, 2002. This is an "average" storm, with a Dst minimum near –100 nT, solar wind values that were not too large, and a modest duration for the main phase and recovery phase. Therefore, a generalization of these results to magnetic storms in general is possible but not robust.

It is concluded that the best parameter for quantifying the net adiabatic energy gain in the inner magnetosphere during storms is the instantaneous value of the product of the maximum westward electric field at the outer simulation boundary and the nightside near-Earth plasma sheet density. An average of this quantity's values over the previous hour yields a close second in the overall ranking of the correlation coefficients. However, all of the instantaneous magnetospheric quantities considered in this study produced large correlation coefficients. Therefore, they all could be considered useful predictors of ΔE_{ad}. Long integration times over the parameters lessen the significance of the correlation.

Differences in the correlations were noted between the three potential electric field choices. The simplest field model, the modified McIlwain description, in general yielded the highest correlation coefficients for any given parameter (the average of the coefficients was 0.67). Even the time-integrated parameters produced statistically significant correlations. This field choice also produced the highest single correlation coefficient of 0.90 (Figure 4h). The Weimer-96 field model produced, in general, the lowest correlation coefficients for any given parameter (the average of the coefficients was 0.46). For this field there was a distinct break between the top two parameters (those mentioned in the previous paragraph) and the rest. The self-consistent field produced the middle value, in general (the average of the coefficients was 0.53). The instantaneous electric field parameters (magnetospheric and solar wind) were quite good at quantifying ΔE_{ad} for this field choice.

However, it also produced the two lowest correlation coefficients of the analysis (Figure 6j), and its time-integrated parameters were the lowest of the three field choices. Finally, all three electric field choices had poor correlation coefficients when ΔE_{ad} was compared against $P_{dyn,sw}$.

As a final note, the reader is reminded of the complexity in Figure 1, and that many of the processes listed there were not considered in this study. The flow of energy through the inner magnetosphere is a nontrivial issue, with many factors contributing to and modifying the details. Furthermore, each magnetic storm is different, with the various processes waxing and waning in importance for numerous reasons. This study considered only the primary driving parameters of the net adiabatic energy gain of the stormtime ring current ions. It is believed that this is a rather general result with widespread applicability. However, much work still needs to be done to fully understand the factors influencing the development and dynamics of the inner magnetospheric hot plasma during magnetic storms.

Acknowledgments. The authors thank the organizers of the Huntsville 2002 Workshop for inviting G.V.K. to speak at the meeting. The authors would like to thank the sources of funding for this study: NASA grants NAG5-10297 and NAG-10850 and NSF grant ATM-0090165. The authors would also like to thank all of their data providers who made the ring current simulations possible, especially M. F. Thomsen and G. D. Reeves at the Los Alamos National Laboratory for the DANL data and CDAWeb for allowing access to the plasma and magnetic field data of the ACE spacecraft.

REFERENCES

Akasofu, S.-I., and S. Chapman, The ring current, geomagnetic disturbance and the Van Allen radiation belts, *J. Geophys. Res., 66*, 1321, 1961.

Alfvén, H., and C.-G. Fälthammar, *Cosmical Electrodynamics*, Oxford University Press, London, 1963.

Amm, O., Comment on "A three-dimensional, iterative mapping procedure for the implementation of an ionosphere-magnetosphere anisotropic Ohm's law boundary condition in global magnetohydrodynamic simulations" by Michael L. Goodman, *Ann. Geophys., 14*, 773, 1996.

Borovsky, J. E., M. F. Thomsen, and R. C. Elphic, The driving of the plasma sheet by the solar wind, *J. Geophys. Res., 103*, 17,617, 1998.

Burton, R. K., R. L. McPherron, and C. T. Russell, An empirical relationship between interplanetary conditions and Dst, *J. Geophys. Res., 80*, 4204, 1975.

Carovillano, R. L., and G. L. Siscoe, Energy and momentum theorems in magnetospheric processes, *Rev. Geophys. Space Phys., 11*, 289, 1973.

Carpenter, D. L., Whistler evidence of a "knee" in the magnetospheric ionization density profile, *J. Geophys. Res.*, *68*, 1675, 1963.

Chappell, C. R., T. E. Moore, and J. H. Waite, Jr., The ionosphere as a fully adequate source of plasma for the Earth's ionosphere, *J. Geophys. Res.*, *92*, 5896, 1987.

Chen, M. W., M. Schulz, L. R. Lyons, and D. J. Gorney, Stormtime transport of ring current and radiation belt ions, *J. Geophys. Res.*, *98*, 3835, 1993.

Church, S. R., and R. M. Throne, On the origin of plasmaspheric hiss: Ray path integrated amplification, *J. Geophys. Res.*, *88*, 7941, 1983.

Dessler, A. J., and E. N. Parker, Hydromagnetic theory of geomagnetic storms, *J. Geophys. Res.*, *64*, 2239, 1959.

Dungey, J. W., Interplanetary magnetic field and the auroral zones, *Phys. Rev. Lett.*, *6*, 47, 1961.

Dungey, J. W., The loss of Van Allen electrons due to whistlers, *Planet. Space Sci.*, *11*, 591, 1963.

Fenrich, F. R., and J. G. Luhmann, Geomagnetic response to magnetic clouds of different polarity, *Geophys. Res. Lett.*, *25*, 2999, 1998.

Fok, M-C., J. U. Kozyra, A. F. Nagy, C. E. Rasmussen, and G. V. Khazanov, A decay model of equatorial ring current and the associated aeronomical consequences, *J. Geophys. Res.*, *98*, 19,381, 1993.

Fok, M.-C., T. E. Moore, and D. C. Delcourt, Modeling of inner plasma sheet and ring current during substorms, *J. Geophys. Res.*, *104*, 14,557, 1999.

Fok, M.-C., R. A. Wolf, R. W. Spiro, and T. E. Moore, Comprehensive computational model of the earth's ring current, *J. Geophys. Res.*, *106*, 8417, 2001.

Fok, M.-C., et al., Global ENA image simulations, *Space Sci. Rev.*, *109*, 77, 2003.

Foster, J. C., P. J. Erickson, A. J. Coster, J. Goldstein, and F. J. Rich, Ionospheric signatures of plasmaspheric tails, *Geophys. Res. Lett.*, *29*(13), doi:10.1029/2002GL015067, 2002.

Galand, M. and A. D. Richmond, Ionospheric electrical conductances produced by auroral proton precipitation, *J. Geophys. Res.*, *106*, 117, 2001.

Gonzalez, W. D., J. A. Joselyn, Y. Kamide, H. W. Kroehl, G. Rostoker, B. T. Tsurutani, and V. M. Vasyliunas, What is a geomagnetic storm?, *J. Geophys. Res.*, *99*, 5771, 1994.

Goodman, M. L., A three-dimensional, iterative mapping procedure for the implementation of an ionosphere-magnetosphere anisotropic Ohm's law boundary condition in global magnetohydrodynamic simulations, *Ann. Geophys.*, *13*, 843, 1995.

Green, J. C., and M. G. Kivelson, A tale of two theories: How the adiabatic response and ULF waves affect relativistic electrons, *J. Geophys. Res.*, *106*, 25,777, 2001.

Hudson, M. K., V. A. Marchenko, I. Roth, M. Temerin, J. B. Blake, and M. S. Gussenhoven, Radiation belt formation during storm sudden commencements and loss during main phase, *Adv. Space Res.*, *21* (4), 597-607, 1998.

Hultqvist, B. W., W. Riedler, and H. Borg, Ring current protons in the upper atmosphere within the plasmasphere, *Planet. Space Sci.*, *24*, 783, 1976.

Iijima, T., and T. A. Potemra, The amplitude distribution of field-aligned currents at northern latitudes observed by TRIAD, *J. Geophys. Res.*, *81*, 2165, 1976.

Jaggi, R. K., and R. A. Wolf, Self-consistent calculation of the motion of a sheet of ions in the magnetosphere, *J. Geophys. Res.*, *78*, 2852, 1973.

Jordanova, V. K., L. M. Kistler, J. U. Kozyra, G. V. Khazanov, and A. F. Nagy, Collisional losses of ring current ions, *J. Geophys. Res.*, *101*, 111, 1996.

Jordanova, V. K., J. U. Kozyra, A. F. Nagy, and G. V. Khazanov, Kinetic model of the ring current-atmosphere interactions, *J. Geophys. Res.*, *102*, 14,279, 1997.

Jordanova, V. K., L. M. Kistler, C. J. Farrugia, and R. B. Torbert, Effects of inner magnetospheric convection on ring current dynamics: March 10-12, 1998, *J. Geophys. Res.*, *106*, 29,705, 2001.

Kennel, C. F., and H. E. Petschek, Limit on stably trapped particle fluxes, *J. Geophys. Res.*, *71*, 1, 1966.

Khazanov, G. V., and M. W. Liemohn, Nonsteady state ionosphere-plasmasphere coupling of superthermal electrons, *J. Geophys. Res.*, *100*, 9669, 1995.

Khazanov, G. V., and M. W. Liemohn, Kinetic theory of superthermal electron transport, in *Recent Research Developments in Geophysics*, vol. 3 (part 2), edited by S. G. Pandalai, pp. 181-201, Research Signpost, Trivandrum, India, 2000.

Khazanov, G. V., and M. W. Liemohn, Transport of photoelectrons in the nightside magnetosphere, *J. Geophys. Res.*, *107*(A5), 1064, doi: 10.1029/2001JA000163, 2002.

Khazanov, G. V., T. E. Moore, M. W. Liemohn, V. K. Jordanova, and M.-C. Fok, Global collisional model of high-energy photoelectrons, *Geophys. Res. Lett.*, *23*, 331, 1996a.

Khazanov, G. V., T. E. Moore, E. N. Krivorutsky, J. L. Horwitz, and M. W. Liemohn, Lower hybrid turbulence and ponderomotive force effects in space plasmas subjected to large-amplitude low-frequency waves, *Geophys. Res. Lett.*, *23*, 797, 1996b.

Khazanov, G. V., E. N. Krivorutsky, T. E. Moore, M. W. Liemohn, and J. L. Horwitz, Lower hybrid oscillations in multicomponent space plasmas subjected to ion cyclotron waves, *J. Geophys. Res.*, *102*, 175, 1997.

Khazanov, G. V., M. W. Liemohn, J. U. Kozyra, and T. E. Moore, Inner magnetospheric superthermal electron transport: Photoelectron and plasma sheet electron sources, *J. Geophys. Res.*, *103*, 23,485, 1998.

Khazanov, G. V., K. V. Gamayunov, and M. W. Liemohn, Alfvén waves as a source of lower hybrid activity in the ring current region, *J. Geophys. Res.*, *105*, 5403, 2000a

Khazanov, G. V., M. W. Liemohn, J. U. Kozyra, and D. L. Gallagher, Global energy deposition to the topside ionosphere from superthermal electrons, *J. Atmos. Solar-Terr. Physics*, *62*, 947, 2000b.

Khazanov, G. V., K. V. Gamayunov, V. K. Jordanova, and E. N. Krivorutsky, A self-consistent model of the interacting ring current ions and electromagnetic ion cyclotron waves, initial results: Waves and precipitating fluxes, *J. Geophys. Res.*, *107* (A6), 1085, doi: 10.1029/2001JA000180, 2002.

Khazanov, G. V., M. W. Liemohn, T. S. Newman, M.-C. Fok, and

R. W. Spiro, Self-consistent magnetosphere-ionosphere coupling: Theoretical studies, *J. Geophys. Res., 108*(A3), 1122, doi:10.1029/2002JA009624, 2003a.

Khazanov, G.V., K. V. Gamayunov, V. K. Jordanova, Self-consistent model of magnetospheric ring current and electromagnetic ion cyclotron waves: The May2-7, 1998, storm, *J. Geophys. Res., 108*(A12), 1419, doi: 10.1029/ 2003JA009833, 2003b.

Khazanov, G. V., M. W. Liemohn, T. S. Newman, M.-C. Fok, and A. J. Ridley, Magnetospheric convection electric field dynamics and stormtime particle energization: Case study of the magnetic storm of 4 May 1998, *Ann Geophys., 22*, 497, 2004a.

Khazanov, G. V., M. W. Liemohn, M.-C. Fok, T. S. Newman, and A. J. Ridley, Stormtime particle energization with AMIE potentials, *J. Geophys. Res., 109*, A05209, doi: 10.1029/ 2003JA010186, 2004b.

Kozyra, J. U., and M. W. Liemohn, Ring current energy input and decay, *Space Sci. Rev, 109*, 105, 2003.

Kozyra, J. U., C. E. Rasmussen, R. H. Miller, and E. Villalon, Interaction of ring current and radiation belt protons with ducted plasmaspheric hiss, 2, Time evolution of the distribution function, *J. Geophys. Res., 100*, 21,911, 1995.

Kozyra, J. U., A. F. Nagy, and D. W. Slater, High-altitude energy source(s) for stable auroral red arcs, *Rev. Geophys., 35*, 155-190, 1997.

Kozyra, J. U., M. W. Liemohn, C. R. Clauer, A. J. Ridley, M. F. Thomsen, J. E. Borovsky, J. L. Roeder, and V. K. Jordanova, Two-step Dst development and ring current composition changes during the 4-6 June 1991 magnetic storm, *J. Geophys. Res., 107*(A8), doi: 10.1029/2001JA000023, 2002.

Lee, L. C., J. R. Kan, and S.-I. Akasofu, Ring current energy injection rate and solar wind-magnetosphere energy coupling, *Planet. Space Sci., 30*, 627, 1982.

Lemaire, J. F., and K. I. Gringauz, *The Earth's Plasmasphere*, Cambridge University Press, New York, 1998.

Liemohn, H. B., Cyclotron-resonance amplification of VLF and ULF whistlers, *J. Geophys. Res., 72*, 39, 1967.

Liemohn, M. W., Yet another caveat to the Dessler-Parker-Sckopke relation, *J. Geophys. Res., 108*(A6), 1251, doi: 10.1029/ 2003JA009839, 2003.

Liemohn, M. W., and J. U. Kozyra, Assessing the importance of convective and inductive electric fields in forming the stormtime ring current, in *Sixth International Conference on Substorms*, edited by R. M. Winglee, Univ. Washington, Seattle, p.456, 2002.

Liemohn, M. W., G. V. Khazanov, and J. U. Kozyra, Guided plasmaspheric hiss interactions with superthermal electrons, 1, Resonance curves and timescales, *J. Geophys. Res., 102*, 11,619, 1997.

Liemohn, M. W., G. V. Khazanov, and J. U. Kozyra, Banded electron structure formation in the inner magnetosphere, *Geophys. Res. Lett., 25*, 877, 1998.

Liemohn, M. W., G. V. Khazanov, P. D. Craven, and J. U. Kozyra, Nonlinear kinetic modeling of early stage plasmaspheric refilling, *J. Geophys. Res., 104*, 10,295, 1999.

Liemohn, M. W., J. U. Kozyra, P. G. Richards, G. V. Khazanov, M. J. Buonsanto, and V. K. Jordanova, Ring current heating of the thermal electrons at solar maximum, *J. Geophys. Res., 105*, 27,767, 2000.

Liemohn, M. W., J. U. Kozyra, M. F. Thomsen, J. L. Roeder, G. Lu, J. E. Borovsky, and T. E. Cayton, Dominant role of the asymmetric ring current in producing the stormtime Dst*, *J. Geophys. Res., 106*, 10,883, 2001a.

Liemohn, M. W., J. U. Kozyra, C. R. Clauer, and A. J. Ridley, Computational analysis of the near-Earth magnetospheric current system, *J. Geophys. Res., 106*, 29,531, 2001b.

Liemohn, M. W., J. U. Kozyra, C. R. Clauer, G. V. Khazanov, and M. F. Thomsen, Adiabatic energization in the ring current and its relation to other source and loss terms, *J. Geophys. Res., 107*(A4), 1045, doi: 10.1029/2001JA000243, 2002.

Liemohn, M. W., A. J. Ridley, D. L. Gallagher, D. M. Ober, and J. U. Kozyra, Dependence of plasmaspheric morphology on the electric field description during the recovery phase of the April 17, 2002 magnetic storm, *J. Geophys. Res., 109*, A03209, doi: 10.1029/2003JA010304, 2004.

Lyons, L. R., and D. J. Williams, A source for the geomagnetic storm main phase ring current, *J. Geophys. Res., 85*, 523, 1980.

McIlwain, C. E., A Kp dependent equatorial electric field model, *Adv. Space Res., 6(3)*, 187, 1986.

McPherron, R. L., The role of substorms in the generation of magnetic storms, in *Magnetic Storms, Geophys. Monogr. Ser., 98*, edited by B. T. Tsurutani, W. D. Gonzalez, Y. Kamide and J. K. Arballo, p. 131, AGU, Washington, D. C., 1997.

Ober, D. M., J. L. Horwitz, and D. L. Gallagher, Formation of density troughs embedded in the outer plasmasphere by subauroral ion drift events, *J. Geophys. Res. 102*, 14,595, 1997.

O'Brien, T. P., and R. L. McPherron, An empirical phase space analysis of ring current dynamics: Solar wind control of injection and decay, *J. Geophys. Res., 105*, 7707, 2000a.

O'Brien, T. P. and R. L. McPherron, Evidence against an independent solar wind density driver of the terrestrial ring current, *Geophys. Res. Lett., 27*, 3797, 2000b.

Parker, E. N., and H. A. Stewart, Nonlinear inflation of a magnetic dipole, *J. Geophys. Res., 72*, 5287, 1967.

Rairden, R. L., L. A. Frank, and J. D. Craven, Geocoronal imaging with Dynamics Explorer, *J. Geophys. Res., 91*, 13,613, 1986.

Rasmussen, C. E., S. M. Guiter, and S. G. Thomas, Two-dimensional model of the plasmasphere: refilling time constants, *Planet. Space Sci., 41*, 35-42, 1993.

Ridley, A. J., and M. W. Liemohn, A model-derived description of the penetration electric field, *J. Geophys. Res., 107*(A8), 1151, doi: 10.1029/2001JA000051, 2002.

Ridley, A. J., D. L. De Zeeuw, T. I. Gombosi, and K. G. Powell, Using steady-state MHD results to predict the global state of the magnetosphere-ionosphere system, *J. Geophys. Res., 106*, 30,067, 2001.

Ridley, A. J., T. I. Gombosi, and D. L. De Zeeuw, Ionospheric control of the magnetosphere: Conductance, *Ann. Geophys., 22*, 567, 2004.

Roederer, J. G., *Dynamics of Geomagnetically Trapped Radiation*, Springer-Verlag, New York, 1970.

Rostoker, G., E. Friedrich, and M. Dobbs, Physics of magnetic storms, in *Magnetic Storms, Geophys. Monogr. Ser., 98*, edited by B. T. Tsurutani, W. D. Gonzalez, Y. Kamide and J. K. Arballo, p. 149, AGU, Washington, D. C., 1997.

Sckopke, N., A general relation between the energy of trapped particles and the disturbance field near the Earth, *J. Geophys. Res., 71*, 3125, 1966.

Siscoe, G. L., A unified treatment of magnetospheric dynamics, *Planet. Space Sci., 14*, 947, 1966.

Smith, J. P., M. F. Thomsen, J. E. Borovsky, and M. Collier, Solar wind density as a driver for the ring current in mild storms, *Geophys. Res. Lett., 26*, 1797-800, 1999.

Sugiura, M., and S. Chapman, The average morphology of geomagnetic storms with sudden commencement, *Abh. Akad. Wiss. Göttingen. Math. Phys. Kl., 4*, 1-53, 1960.

Summers, D., and R. M. Thorne, Relativistic electron pitch-angle scattering by electromagnetic ion cyclotron waves during geomagnetic storms, *J. Geophys. Res., 108*(A4), 1143, doi: 10.1029/2002JA009489, 2003.

Takahashi, T., Energy degradation and transport of photoelectrons escaping from the upper ionosphere. *Rept. Ionos. and Space Res. Jap., 27*, No.1, 79, 1973.

Takahashi, S., T. Iyemori, and M. Takeda, A simulation of the storm-time ring current, *Planet. Space Sci., 38*, 1133-1141, 1990.

Tsurutani, B. T., and W. D. Gonzalez, The interplanetary causes of magnetic storms: A review, in *Magnetic Storms, Geophys. Monogr. Ser., 98*, edited by B. T. Tsurutani, W. D. Gonzalez, Y. Kamide and J. K. Arballo, p. 77, AGU, Washington, D. C., 1997.

Tsyganenko, N. A., A model of the near magnetosphere with a dawn-dusk asymmetry: 2. Parameterization and fitting to observations, *J. Geophys. Res., 107*(A8), 1176, doi: 10.1029/2001JA000220, 2002.

Weimer, D. R., A flexible, IMF dependent model of high-latitude electric potentials having "space weather" applications, *Geophys. Res. Lett., 23*, 2549, 1996.

Wolf, R. A., Magnetospheric configuration, in *Introduction to Space Physics*, edited by M. G. Kivelson and C. T. Russell, p. 288, Cambridge University Press, New York, 1995.

Wolf, R. A., et al., Computer simulations of inner magnetospheric dynamics for the magnetic storm of July 29, 1977, *J. Geophys. Res., 87*, 5949, 1982.

Wolf, R. A., J. W. Freeman, Jr., B. A. Hausman, R. W. Spiro, R. V. Hilmer, and R. L. Lambour, Modeling convection effects in magnetic storms, in *Magnetic Storms, Geophys. Monogr. Ser., 98*, edited by B. T. Tsurutani, W. D. Gonzalez, Y. Kamide, and J. K. Arballo, American Geophysical Union, p. 161, 1997.

Yeh, H.-C., J. C. Foster, F. J. Rich, and W. Swider, Storm time electric field penetration observed at midlatitude, *J. Geophys. Res., 96*, 5707, 1991.

Young, D. T., H. Balsiger, and J. Geiss, Correlations of magnetospheric ion composition with geomagnetic and solar activity, *J. Geophys. Res., 87*, 9077, 1982.

G. V. Khazanov, National Space Science and Technology Center, NASA Marshall Space Flight Center, Huntsville, Alabama 35899. (george.v.khazanov@nasa.gov)

M. W. Liemohn, Atmospheric, Oceanic, and Space Sciences Department, University of Michigan, 2455 Hayward St., Ann Arbor, Michigan 48109-2143. (liemohn@umich.edu)

Interrelation Among Double Layers, Parallel Electric Fields, and Density Depletions

Nagendra Singh

Department of Electrical and Computer Engineering, University of Alabama, Huntsville, Alabama

Using numerical simulations of voltage- and current-driven plasmas, we demonstrate an intimate relationship between electric fields parallel to the ambient magnetic field and density depletions (cavities) in collisionless magnetized plasmas. Simulations of double layers (DLs) in plasmas driven by applied potential drops are reviewed, showing that a strong DLs form in density cavities. The dependence of scale length of the DL and the associated density structure on the potential drop and plasma density is discussed and compared with observations. Simulations of current-driven plasmas also show that a density cavity in current-carrying plasmas could charge to large potentials forming DLs. Simulations of plasmas in an auroral-type diverging flux tube show that the DLs form in the region of the lowest density in the form of a cavity extending over a few Debye lengths in the local plasma. We also review the formation of two-dimensional U- or V-shaped double layers with deep density depletions in which the parallel fields dominate. The transverse size of the narrow density depletions and the DL structures is also discussed. This review reveals that voltages, currents, and density cavities in collisionless plasmas are highly coupled; when the plasma is driven by either a voltage drop or a current, they all appear in the plasma and collectively affect the plasma electrodynamics.

1. INTRODUCTION

Mechanisms for supporting parallel electric fields in collisionless space plasma have been debated over several decades now without a clear-cut resolution [e.g. see *Shawhan*, 1978]. On the basis of laboratory experiment findings, Alfven predicted that electric fields parallel to the ambient magnetic field in collisionless plasmas could be supported by double layers (DLs). Theory [*Carlqvist*, 1972; *Block*, 1977] and simulations [*Singh*, 1980, 1982, 2000, 2003; *Borovsky*, 1983; *Newman* et al., 2001; *Singh and Khazanov*, 2003] have also clearly indicated that parallel electric fields localize as DLs in density depletions, or cavities. Only recently have satellite observations begun to reveal that parallel fields do indeed exist in density depletions [*Mozer and Hull*, 2001; *Hull et al.*, 2003; *Ergun et al.*, 2002a,b]. The purpose of this paper is to present a systematic review of simulations dealing with the relationship of plasma depletions and DLs.

When a potential difference is applied across uniform plasmas, as in simulations or in laboratory experiments, the important question is, where does the potential drop occur in the plasma—in potential sheaths near electrodes (boundaries), as shown by curves 1 and 3 in Figure 1, or elsewhere in the midst of the plasma, as shown by curve 2?

Another question pertaining to this topic is the role of density fluctuations/depletions in generating parallel electric fields in current-carrying plasmas; the fluctuations and the depletions of interest here could be self-consistently generated as a part of the plasma turbulence inherent to the state of the plasma. Recent simulations have shown that the density depletions and cavities act as seeds for DL formation in current-driven plasmas [*Newman et al.*, 2001; *Singh*, 2002]. Since a voltage drive in a plasma generates current, the role of density depletions in nucleation of DLs in a voltage-driven plasma [*Singh*, 1982, 2000, 2003] is essentially the

Particle Acceleration in Astrophysical Plasmas
Geophysical Monograph Series 156
Copyright 2005 by the American Geophysical Union
10.1029/156GM25

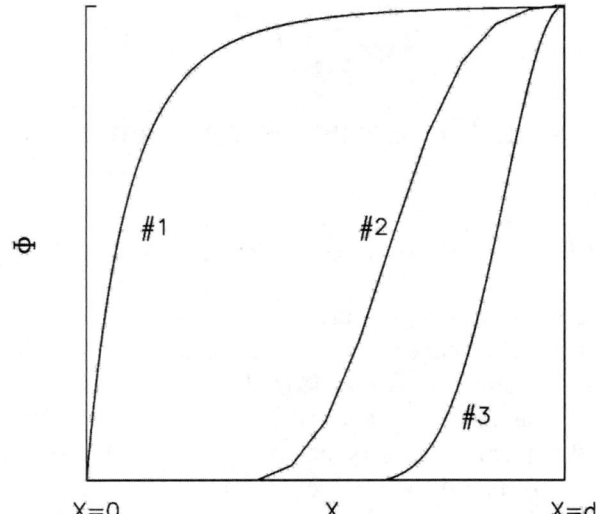

Figure 1. Schematic diagram showing possible locations of a potential drop in a collisionless plasma. What *does* determine the location of the potential drop?

same as in the current–cavity interaction mentioned here. The recurring formation of DLs seen in voltage-driven simulations [*Singh and Schunk*, 1982a] and laboratory experiments [*Iizuka et al.*, 1984] are attributable to the process of current–cavity interactions. In the older studies the process of recurring formation was called potential-relaxation instability. More recently *Singh* [2003] called the same process a cavitational instability due to the fact that cavity grows with the DL when the current exceeds the thermal current in the plasma [*Carlqvist*, 1972].

Since observations in space are now reporting a close connection between parallel electric fields and density depletions (cavities) in the auroral plasma [*Mozer and Hull*, 2001; *Ergun et al.*, 2002a,b; *Hull et al.* 2003], it is timely to develop a systematic knowledge of this connection. Therefore, we present here a brief summary of our studies on this topic.

2. DOUBLE LAYER FORMATION IN UNIFORM PLASMAS DRIVEN BY A POTENTIAL DIFFERENCE

Figures 2a and 2b show examples of DLs formed in a nearly uniform plasma as seen in a one-dimensional (1-D) simulation using a Vlasov code [*Singh*, 1980]. In the simulations described here as well as in later sections we use the following normalizations:

distance $\bar{x} = x/\lambda_{do}$, velocity $\bar{V} = V/V_{te}$, time $\bar{t} = t\omega_{po}$, electric potential $\bar{\Phi} = \Phi/\Phi_n$, electric field $\bar{E} = E/E_n$, density $\bar{n} = n/n_o$, and current $\bar{J} = J/J_{th}$, where $V_{te} = (k_B T_o/m)^{1/2}$, $\omega^2_{po} = n_o e^2/m\varepsilon_o$, $\lambda_{do} = V_{te}/\omega_{po}$, $J_{th} = n_o e V_{te}$, $\Phi_n = k_B T_o/e$, and $E_n = \Phi_n/\lambda_{do}$, k_B is the Boltzmann constant, m is the

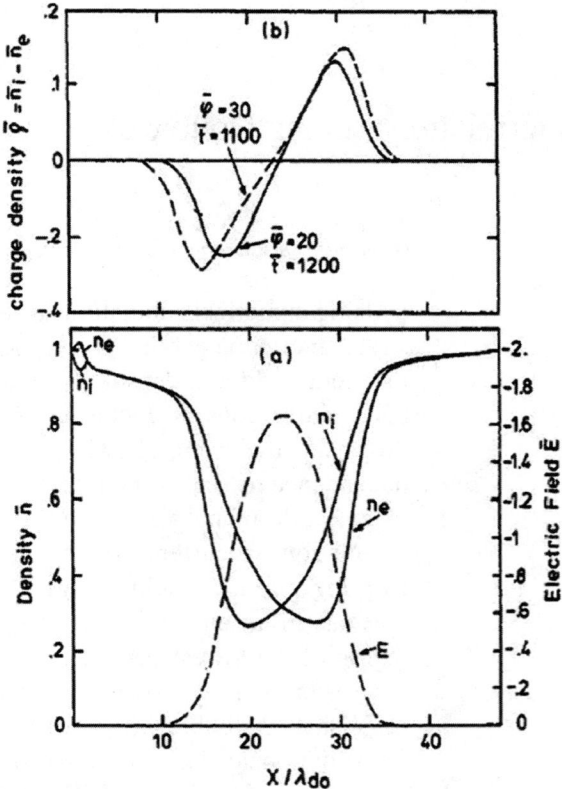

Figure 2. (a) Electron and ion density profiles and the electric field for $\Phi_0 = 20$. (b) Density of charge separation for $\Phi_0 = 20$ and 30, as labeled.

electronic mass, T_o is the electron and ion temperatures in the initially uniform plasma, and e is the magnitude of the electronic charge. Unless otherwise stated, in the following discussion we have used these normalizations and taken the liberty to remove the bars on top of the symbols. Note that in 1-D simulations the x coordinate is along the ambient magnetic field if the plasma is magnetized.

The length of the simulated plasma is $d = 50\lambda_{do}$, where λ_{do} is the Debye length in the initial plasma. This rather short length of the simulated plasma serves the purpose of suppressing any plasma instabilities generated by the electron and ion beams accelerated by the DL. Therefore, it is possible to have a laminar and stable DL. For much longer lengths, the instabilities on the high-potential side of the plasma cause plasma heating, evacuation of the heated plasma, motion of the DL, and its recurring formation and decay [*Singh*, 1982; *Singh and Schunk*, 1983a]. Thus the simulation described here serves the purpose of describing a strong laminar DL without the complications of plasma waves. Figure 2a shows the electron (n_e) and ion (n_i) densities, and electric field profiles at $t = 1200\omega_{eo}^{-1}$ when a near steady-state has been reached in the plasma after a uniform electric field $E_o = 0.4E_n$ was applied to the plasma by applying a potential difference of $\Phi_o = dE_o = 20\Phi_n$.

In Figure 2b we have plotted the space charge distributions associated with the DL. Note there are two curves for the space charge, one for $\Phi_o = 30$ and the other for $\Phi_o = 20$; only for the latter value are density and the electric field structures shown in panel (a). First we note that the electric field, which was initially uniformly distributed over $0 < x < d$, is eventually localized in a pulse confined inside a density cavity, which is charged like a capacitor, with negative and positive space charges peaking near the inflection points in the electric field profile. The two layers of separated space charge give the nomenclature of the DL. The width of the cavity and hence of the DL scale with the magnitude of the potential drop is given by [*Singh*, 1980; *Joyce and Hubbard*, 1978]:

$$L \sim 5(e\Phi_o/k_B T_o)^{1/2} \lambda_{do} \qquad (1)$$

As the potential drop Φ_o increases, the width, the electric field strength, and the magnitude of the space charge all increase and show the dependence on the square root of Φ_o. Since $\lambda_{do} = (k_B T_o/m_e)^{1/2}/\omega_{po}$, the above scaling becomes

$$L \sim 3.5(2e\Phi_o/m_e)^{1/2}/\omega_{pe} \propto n^{-1/2}, \qquad (2)$$

where n is the plasma density. In a plasma with a DL, electrons and ions get energized to energies $\sim e\Phi_o$, enhancing the effective Debye length by a factor $|e\Phi_o/k_B T_o|^{1/2}$. The scaling in expression (1) simply reflects that the size of the DL depends on the increased Debye length, which decreases with the square root of the increasing density. This scaling seems to be quite relevant to the observations of electric fields from FAST, showing that the scale length within the DL structure varies with local Debye length depending on the density [*Ergun et al.*, 2002b]. It was found that as the density increases, the scale length decreases, giving asymmetrical DL structures in the spatial regions with large density gradients. The scaling law in expression (2) is functionally quite different from the Child–Langmuir law, which yields $L \propto \Phi_o^{3/4}$. Also note that the scale length of the DL in expression (2) does not directly depend on the current through it; however, it may indirectly depend on the current because of the dependence of Φ_o on the current. For example Φ_o may be given by Ohm's law for the upward current plasma [*Knight*, 1973; *Lyons et al.*, 1979]. *Borovsky* [1983] has shown that scaling such as that in expression (2) holds good, even for oblique DLs, without any significant effect of the ambient magnetic field. The scaling issue is further discussed in Section 6.

3. DOUBLE LAYER FORMATION IN NONUNIFORM PLASMAS DRIVEN BY A POTENTIAL DIFFERENCE

We present here results from a simulation in which the initial density profile decreases linearly from normalized density $n(x = 0)/n_o = 1$ to $n_{mls}/n_o = n(x = 2000)/n_o = 0.2$ [*Singh*, 2003]. Note that the length of the simulated plasma is $d = 2000 \lambda_{do}$; λ_{do} is the Debye length in the initial plasma at $x = 0$, where the normalized plasma density is unity. Figure 3a shows the evolution of the potential profile $\varphi(x)$ from $t = 200$ to $1800 \omega_{po}^{-1}$. The corresponding evolution of the density profile is shown in Figure 3b. Since at $t = 0$, when the simulation begins, there is no space charge in the plasma, $\varphi(x)$ increases linearly from $\varphi(x = 0) = 0$ to $\varphi(x = L_x) = \varphi_o = 30$ with increasing x. The potential profile at $t = 200$ shows modification from the initial linear potential profile, with enhanced slopes (electric field) near $x = 2000$, the location of the lowest density. The subsequent profiles shown in Figures 3a and 3b show that both the plasma density and the potential break into oscillations beginning

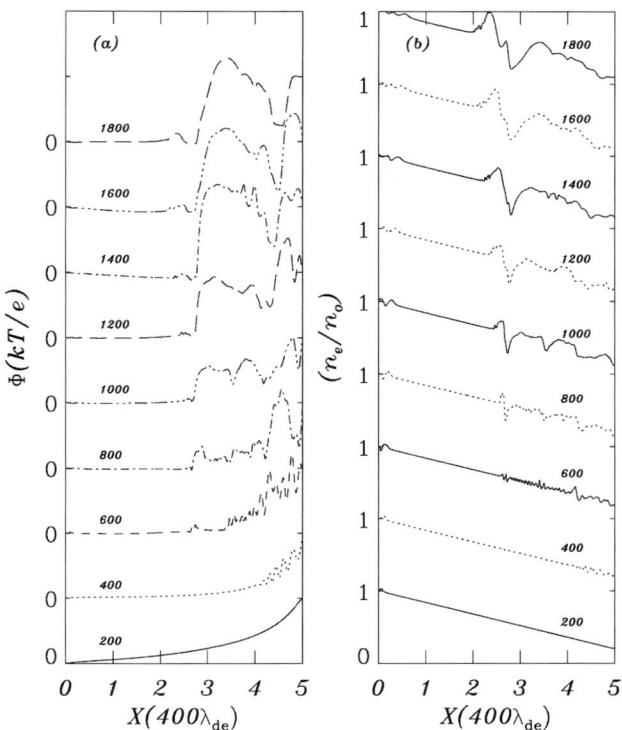

Figure 3. Evolution of DLs via the Buneman instability: (a) Potential profiles and (b) density profiles shown at the selected times labeled on the curves. The initial potential profile (not shown) is linear and the nonuniform linear density profile is like that shown for $t = 200$ in (b). Note the developing plasma oscillations, which transform into potential steps in density cavities crated by the Buneman instability.

near the boundary at $x = 2000$ and progressively extending into the plasma. These oscillations create density perturbations. The potential and density profiles for $t > 600$ show that potential steps form in the density depletions. The potential steps are DLs, across which the applied potential drops. The spatial and temporal behavior of the plasma instabilities is primarily determined by the Buneman instability [*Buneman*, 1958]. The DLs are located in the density minima. The potential drops across the DLs increase with the decreasing minimum densities.

Figures 4a and 4b show the temporal variation of plasma currents; in panel (a), electron (J_e) and ion (J_i) currents at $x = 400$ are plotted as functions of time, whereas in panel (b) only J_e ($x = 1600$) is plotted. Note that the current at $x = 1600$ is on the high-potential side, while the ones at $x = 400$ are on the low-potential side. The ion current plotted in Figure 4a is multiplied by a factor of 20 and still is nearly zero. The electron currents in the two figures, for the low- and high-potential sides, are nearly equal, demonstrating current continuity through the DLs. The electron currents build up to a large value, $J_e \sim -1.5$ at $t \sim 400$, as determined by the electron flux at the boundary $x = 0$. The large current triggers the Buneman instability and leads to the formation of multiple DLs. The instability and the DL formations repeatedly disrupt the current partially over the course of the simulation, which was run up to $t = 3000$. Figure 5a shows the evolution of the electron phase space $x-V_x$ at some selected times. These plots show that the major feature of the instability is the trapping of electrons in the potential structure. The corresponding ion phase space is shown in Figure 5b. The electron and ion phase space plots at $t = 1600$ show accelerations of the charged particles in the potential steps of the DLs as well as their trapping. The ion phase space also shows a fine-grained instability.

We saw in Section 2 that DLs formed midway in a small length of a simulated plasma (Figure 2); the processes preceding the DL formation were (i) energization of electrons and ions, (ii) formation and propagation of a giant electron hole, and (iii) plasma evacuation as a consequence of the applied electric field [*Singh*, 1980, 1982]. In this section we find that the DLs form by Buneman instability in the region of low-density plasma, where relative drifts between electrons and ions become large. This instability process also generates density cavities, in which the DLs are located. The two simple examples discussed here demonstrate different ways of DL formations in plasmas, but a common thread is that the DL formation is accompanied by cavity formation. So a question naturally arises as to the response of the plasma to an applied voltage if there is a preexisting density cavity. A well-known example of preexisting cavities in space plasmas is the auroral density cavity [*Persoon et al.*, 1988].

4. DOUBLE LAYER FORMATION IN NONUNIFORM PLASMAS WITH A PREEXISTING DENSITY CAVITY DRIVEN BY A POTENTIAL DIFFERENCE

This simulation is similar to the one described above in Section 3, except that a density cavity is initialized at $t = 0$ over $400 < x < 600$ with the minimum density $n_{cav}/n_o = 0.3$ and the density of the plasma entering the simulation region at $x = 2000$ is $n_{mls}/n_o = 0.3$ [*Singh*, 2003]. Figures 6a and 6b show the evolution of potential and density profiles from this simulation. The introduction of the cavity has a profound effect on the temporal evolution of the potential profile as shown in Figure 6a. A sharp potential step forms in the cavity very quickly and it grows in amplitude well beyond the applied potential of $\varphi_o = 30$. These features are easily seen by comparing the potential profiles at $t\omega_{peo} = 200$ and 400. The maximum potential drop across the DL at $t = 400$ ω_{peo}^{-1} is $\Delta\Phi \sim 75$, and the minimum density in the cavity is reduced from the initial value $n_{cav}/n_o = 0.3$ to n_{min}/n_o

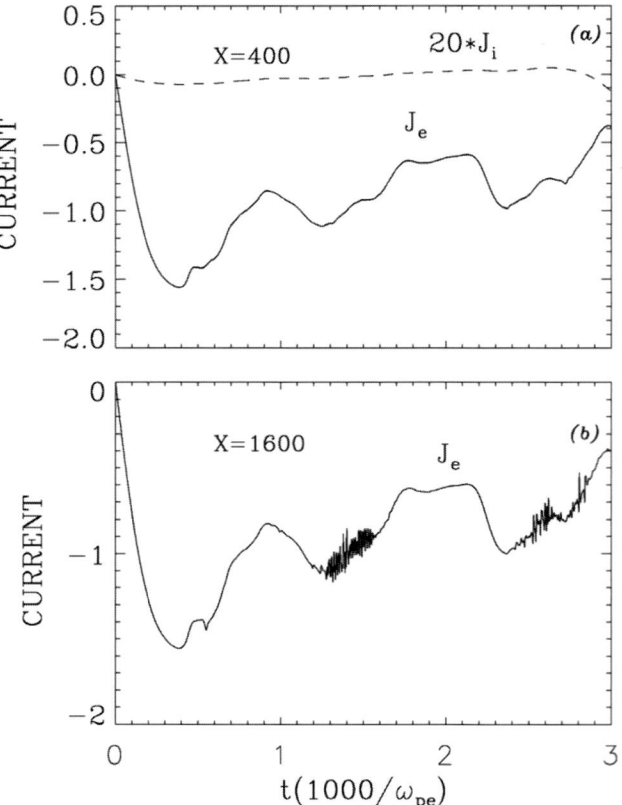

Figure 4. Evolution of plasma currents: (a) J_e ($x = 400$) and J_i ($x = 400$), (b) J_e ($x = 1600$). The dashed-line curve in (a) shows the ion current amplified by 20.

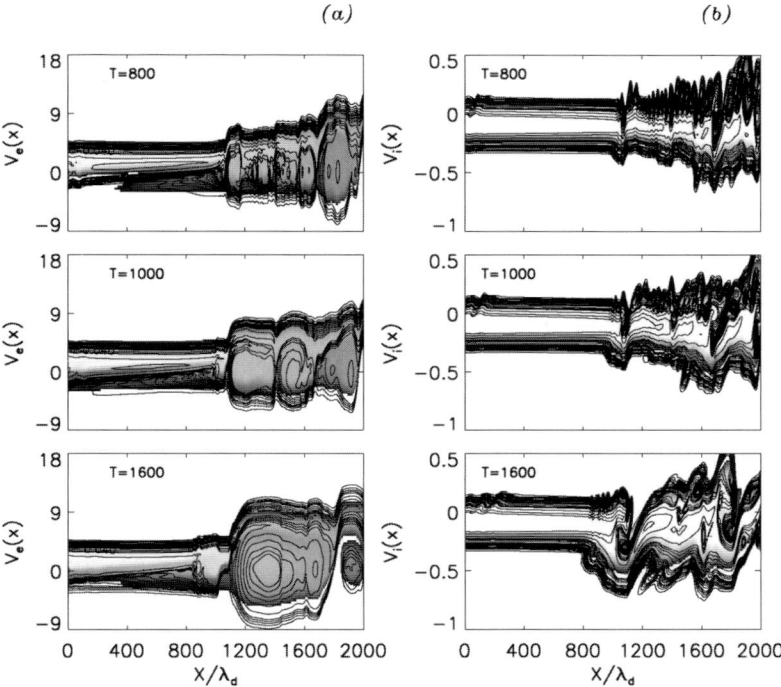

Figure 5. Phase-space: (a) electrons and (b) ions at three selected times.

~ 0.069. Thus, even when a DL forms in a density cavity, the process of DL formation involves the redeepening (recavitation) of a preexisting depletion. The cavity provides an initial "seed" for the DL formation. The redeepening of the cavity is a consequence of cavitational instability, known since the early days of research on DLs [*Carlqvist*, 1972] and further examined and verified by numerical simulations [*Singh*, 1982, 2003]. The cavitational instability occurs when the current density exceeds the thermal current density, as in the Buneman instability. The potential distributions of the type shown for $t = 400$ and its subsequent evolution generate counterstreaming of electrons on the high-potential side of the potential step formed in the cavity; the counterstreaming in turn generates electron holes [*Singh*, 2000, 2003], which are now commonly detected by satellites in the auroral plasma [*Mozer et al.*, 1997; *Ergun et al.*, 1998]. The electron density profiles evolving with the potential profiles are shown in Figure 6b. The ion density profiles are similar, except wherever charge separations occur, as in the structures of the DL itself and in the potential pulses appearing on its high-potential side. The potential pulses are electron holes; they have perturbations in the electron density but not in the ion density.

The main features of the evolving density cavity are the following. As pointed out above, the cavity first deepens near its left side, expelling plasma to the high-density side on the left. This process occurs simultaneously as the cavity charges and affects charge separation, supporting the potential step. The plasma expelled to the left creates a density enhancement, which propagates to the left and dumps the plasma from the simulation region at $x = 0$. The plasma depletion in the cavity expands toward the right, an expansion accompanied by a propagating density front as a leading edge of the cavity, spreading the density depletion until the density smoothly joins the density profile at larger distances, as seen for $t > 3600$ in Figure 6b. In addition to expanding to the right, the density cavity moves as a whole to the right, riding on expanding dense plasma from the left. The dense plasma front is the trailing edge of the evolving density depletion. The leading edge progressively accelerates as it moves into the rarer plasma on the right; its velocity ranges from $V_L \sim 0.15$ at $t = 200$ to 0.26 when $t = 4600$. The trailing edge also accelerates but moves relatively more slowly with speeds of $V_{tr} = 0.05$ at $t = 200$ and 0.14 at $t = 4400$. The DL in the cavity moves with the trailing edge. For comparison, we mention here that ion thermal speed in the initial plasma is $V_{tio} = 0.05$ and these velocities are measured in the units of the initial electron thermal speed V_{te}. Note that in the late stage of DL evolution, the DL sits near a moving density front in a narrow density minimum as seen from the density profiles for $t \geq 3600$. Such a density step could be the bottom edge of the auroral density cavity, separating the

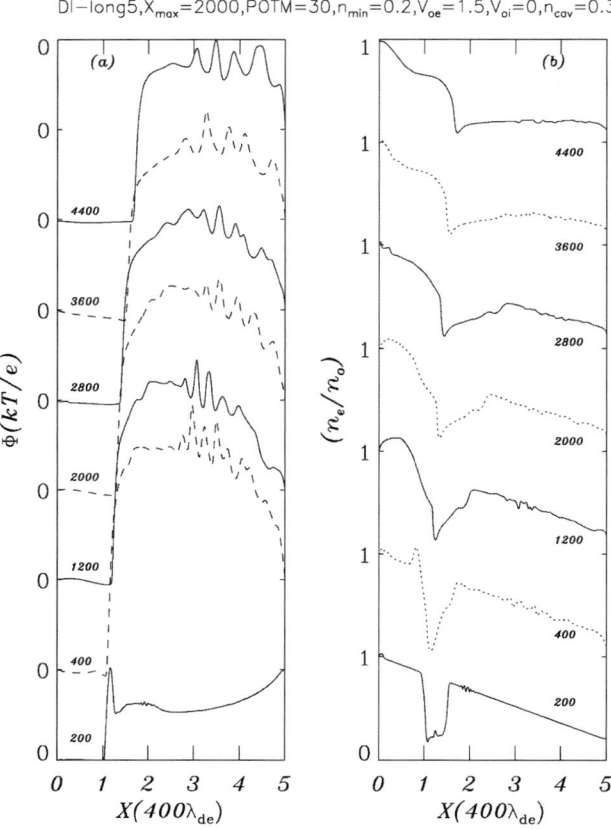

Figure 6. Nucleation of a DL in a density cavity: (a) potential profiles and (b) density profiles. The initial density cavity occupies the region $400 < x < 600$. Note the deepening of the density cavity near its left edge, where a large potential step develops with amplitude exceeding the applied potential $\varphi_o = 30$.

dense ionospheric plasma [*Temerin and Carlson*, 1998]. The motion of the step is like upward-moving polar wind.

5. DOUBLE LAYER FORMATION IN PLASMAS WITH PREEXISTING DENSITY CAVITY(IES) DRIVEN BY CURRENT

In the simulations described above, the plasma is driven by an applied potential drop. In the absence of a preexisting cavity, we found that plasma instabilities create density cavities, in which potential drops occur. In the presence of a preexisting cavity, the potential drop rather quickly localizes in the cavity, forming DL. We also saw overcharging of the cavity and its redeepening. We examine here the response of a plasma with embedded preexisting cavities when it is driven by a current [*Singh*, 2002].

As in the previous simulations, the current-driven response is studied by means of 1-D Vlasov simulations. We solve Vlasov equations for both electrons and ions, coupled with Gauss's law for the parallel electric field E. We simulate a plasma over $0 \leq x \leq d$ with the following boundary conditions on E and the electron and ion velocity distribution functions $f_\alpha(x, V_x)$. The geometry of the simulation is shown in Figure 7. For the electric field, we solve $dE/dx = \rho(x)/\varepsilon_0$ with the boundary condition $E_\parallel(x=0) = 0$ and the charge density $\rho(x) = \Sigma q_\alpha n_\alpha$ with n_α as the electronic ($\alpha = e$) and ionic ($\alpha = i$) densities given by $n_\alpha = \int_{-\infty}^{\infty} f_\alpha(x, V_x) dV_x$; $|q_\alpha| = e$, the quantum of charge, and ε_0 is the permittivity of free space. Simulations were done with current-carrying electrons and ions moving in the positive and negative x directions, respectively. We also imposed the electric potential as $\Phi(x=0) = 0$ while $\Phi(x=d)$ is floating, determining the self-consistent potential drop across the plasma. Thus these electric field boundary conditions are quite different from the ones used in the simulations described in the previous sections. We prescribe only the electron and ion velocity distribution functions for $V_x > 0$ at $x = 0$ and $V_x < 0$ at $x = d$.

For the above stated direction of electron and ion flows, the electron velocity distribution at $x = 0$ is prescribed as a Maxwellian distribution for $V_x > 0$ with a drift velocity $V_{oe} > 0$ and density n_o, and likewise ions have a drift velocity $V_{oi} < 0$ at $x = d$. The ion velocity distribution function at $x = 0$ is prescribed as a half-Maxwellian for $V_x > 0$ and likewise for the electrons the velocity distribution is half-Maxwellian for $V_x < 0$ at $x = d$. For the plasma flowing out of the simulation region, the distribution function at $x = 0$ for $V_x < 0$ and at $x = d$ for $V_x > 0$ are determined self-consistently by the processes occurring in the simulation region. Initially electron and ion temperatures are T_o, and they are kept constant at T_o for the incoming plasmas at the $x = 0$ and $x = d$ boundaries at all times. At the initial time $t = 0$, the entire region $0 \leq x \leq d$ is filled with a Maxwellian plasma with the temperature T_o. The plasma carries a current $x = 0$ determined by V_{oe} and V_{oi}. At this initial time we impose a cavity in the plasma over $x_\ell \leq x \leq x_u$ (Figure 7).

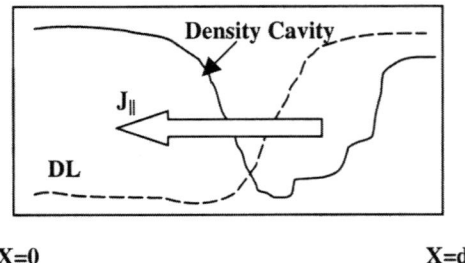

Figure 7. Schematic diagram illustrating DL formation when a field-aligned current ($J_\parallel$) flows through a density cavity. In the 1-D Vlasov simulation described here, the plasma from $x = 0$ to $x = d$ is simulated with boundary conditions mentioned in the text.

It turns out that the initial structure of the cavity does not matter much in the later evolution of the plasma and fields. Therefore, we impose the simplest structure as follows: for $x \leq x_\ell$, $n(x) = n_0$, for $x_\ell < x < x_u$, $n(x) = n_{min}$, and for $x > x_u$, $n(x) = n_0$. Simulations with a Gaussian-shaped cavity gave results similar to this simple shape as long as n_{min} was the same. The simulations were performed for an ion mass $M = 400m$, which speeds up the calculation.

We find that in an attempt to maintain the current continuity through the cavity, a DL spontaneously forms inside it. The potential drop across the DL first grows in the initial stages and then decays. Figures 8a and 8b show the temporal behavior from a simulation, in which $\overline{V}_{oe} = 0.5$, $n_{min} = 0.3$, and the field-aligned current $\overline{J}_\parallel \cong 0.5$. Figure 8a shows the growth of the potential drop $\Delta\Phi$ over the time period $100 \leq \overline{t} \leq 900$ and Figure 8b shows the subsequent decay over $1000 \leq \overline{t} \leq 1600$.

Figure 9a shows the temporal evolution of the potential

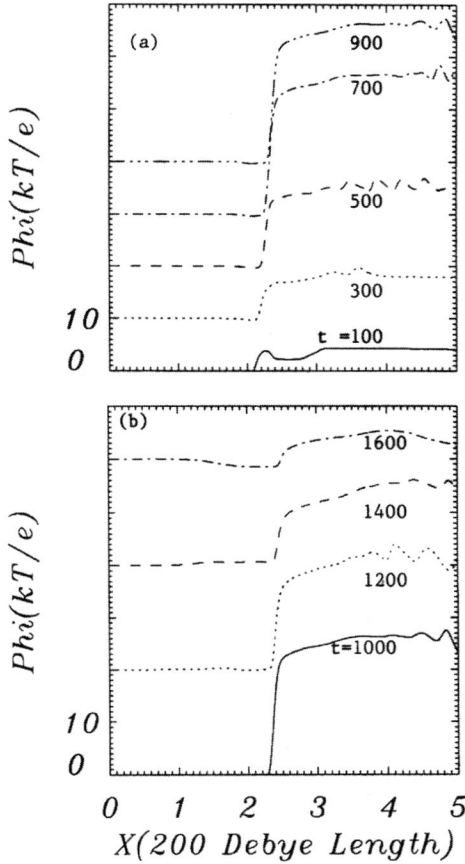

Figure 8. Temporal evolution of DL potential profile: (a) growth and (b) decay of the DL for electron drift $\overline{V}_{oe} = 0.5$ and minimum density in the cavity $\overline{n}_{min} = 0.3\ n_o$. Normalized times are shown on the curves.

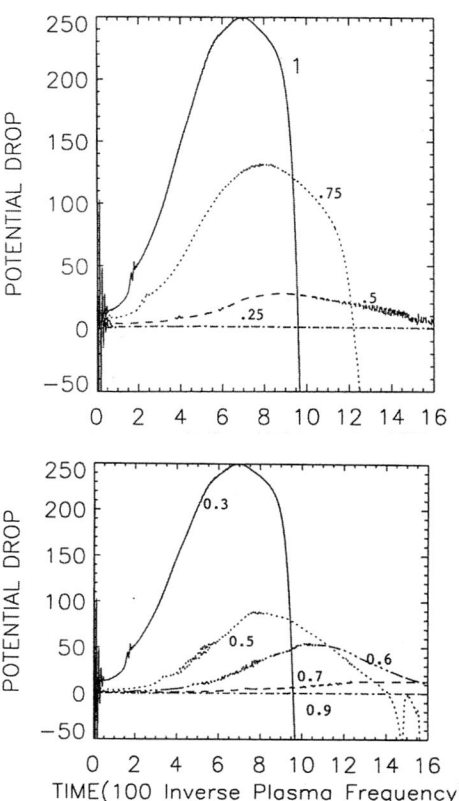

Figure 9. Temporal evolution of the maximum parallel potential drop (normalized) for a fixed value of (a) $\overline{n}_{min} = 0.3$ while $\overline{V}_{oe}$ varies from 1 to 0.25, and (b) $\overline{V}_{oe} = 1$ while $\overline{n}_{min}$ varies from 0.3 to 0.9.

drops across DLs from four simulations, in which $\overline{n}_{min}$ is fixed at 0.3 while V_{oe} is varied from $\overline{V}_{oe} = 1$ to 0.25. This figure shows that an appreciable potential drop develops only when V_{oe} is appreciably large. For $\overline{V}_{oe} \cong 0.25$, the maximum potential drop is only a few kT_o/e, but it dramatically increases with increasing V_{oe} and rises to $250\ kT_o/e$ for $\overline{V}_{oe} = 1$, which corresponds to a current $J_\parallel \cong J_{th} \cong n_o e V_{te}$.

Figure 9b shows the temporal evaluation of the potential drop for $V_{oe} = 1$ when the minimum density in the cavity is varied from $\overline{n}_{min} = 0.3$ to $\overline{n}_{min} = 0.9$. We find that the deeper the cavity, the larger the maximum potential drop in the DL.

Figure 10 shows the electron and ion phase space from considerable acceleration of both electrons and ions at $\overline{t} = 800$ when the potential drop peaks for $\overline{V}_{oe} = 1$. After acceleration by the DL, electrons appear as a beam. However, for the slow-moving accelerated ions, the temporal buildup of the DL generates a double-streaming of the ions accelerated over time; the faster ions produced by the peak potential drop across the DL take over the slower ions that were accelerated earlier by the weaker potential drops.

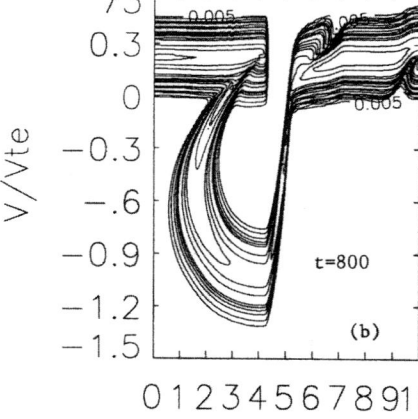

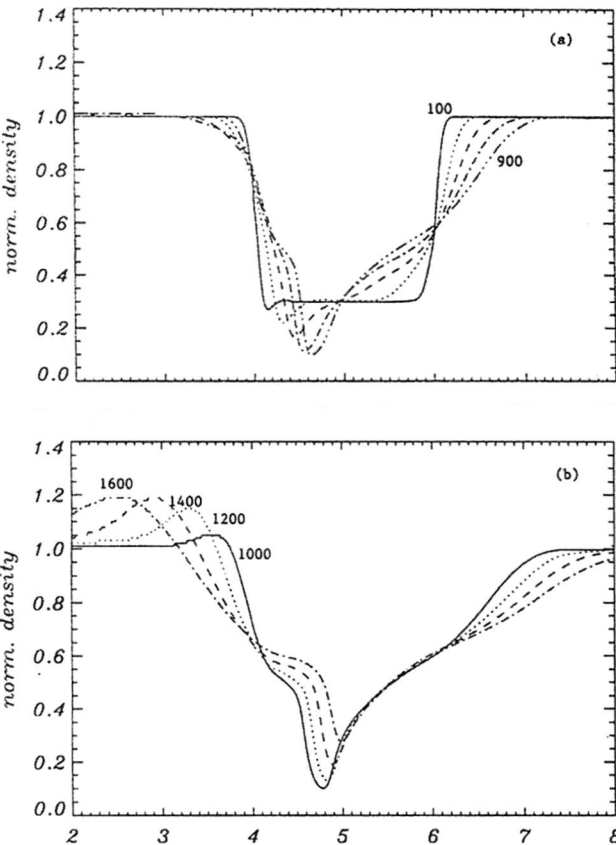

Figure 10. Electron and ion accelerations by the DL (a) electron and (b) ion x–V phase space. Contours of constant values of electron and ion velocity distribution functions are plotted.

Figure 11. Evolution of the density profile affecting the DL. Density profiles are shown corresponding to the potential profiles (a) in Figure 8a (growth phase) and (b) in Figure 8b (decay phase). The bumps in the density distributions for $x < 400\lambda_d$ in (b) are due to the slowly moving accelerated ion bunch.

To understand the evolution of the potential drop shown in Figure 8, it is useful to examine the evolution of the density profile in the cavity. Figure 11a shows the evolution of the ion density during the growth phase over $100 \leq \bar{t} \leq 900$. This figure reveals that the cavity refills by expanding plasmas from both sides. In addition to this refilling, the noteworthy feature in the region between the two expanding plasmas is a relatively localized region from which plasma is further evacuated and the minimum density continually decreases; the minimum density decreases from an initial value of $\bar{n}_i = 0.3$ to 0.1 at $\bar{t} = 900$. This evacuation process is essentially the same as noted earlier in Figure 6b in a voltage-driven simulation. The DL sits in this newly evacuated part of the density cavity and moves with it with a velocity $\bar{V}_{d\ell} \sim 0.05$, which is the ion thermal speed in the plasma with ion temperature $T_i = T_0$ and ion to electron mass ratio $M/m = 400$.

Figure 11b shows the evolution of the density profile at later times for $10^3 \leq \bar{t} \leq 1600$, showing that refilling continues but now only from the upstream side of the electron flow. This is accompanied by the motion of the DL from left to right. As the DL moves, the minimum density in the cavity increases. This increase in the minimum density leads to the progressive decay of the potential drop (Figure 8b) because less electron acceleration is needed to maintain the current.

6. DOUBLE LAYER IN A MAGNETIC FLUX TUBE WITH UPWARD CURRENT

The purpose of this section is to examine the nature of parallel electric fields in a long, diverging flux tube with cold plasma source at the bottom and hot plasma source at the top. The geometry of the flux tube is schematically shown in Figure 12. We examine the case when the bottom end of the flux tube is positive with respect to the top end,

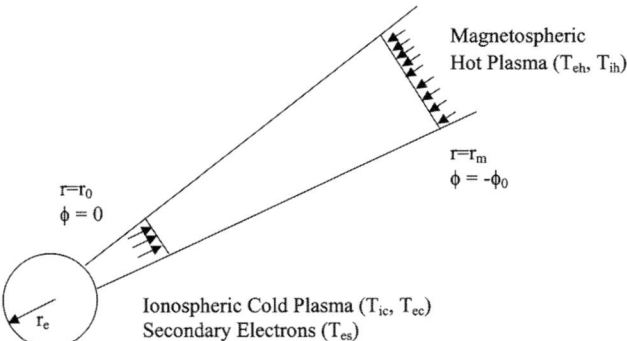

Figure 12. Geometry of the diverging flux tube model. The simulation is from geometric distance $r = r_0$ to $r = r_m$.

as is the case for upward current in an auroral flux tube. Owing to the prohibitively large computational requirement for the simulation of an actual auroral flux tube extending from an altitude of a few hundreds to thousands of kilometers, we simulate an artificially short flux tube that has all the essential physical elements, such as spatial divergence with increasing altitude, associated mirror force, and the typical particle populations of the upward current plasma. Secondary and backscattered (SBS) electron populations are also included from the bottom of the flux tube. Magnetic mirroring self-consistently generates the reflected particle populations near the bottom end. We simulate the electrodynamics of such a flux tube by using a 1-D particle-in-cell (PIC) code [*Singh and Chan*, 1993] with the resolution of local Debye length. We drive the flux tube by applying a potential difference across its length. At the bottom of the flux tube ($r = r_0$), we inject a cold, dense plasma into the flux tube. Likewise, a hot plasma is injected at the top boundary at $r = r_m$; the hot electron temperature is $T_{eh} = T_0$ and the hot ion temperature is $T_{ih} = 3T_0$. The cold electrons and ions injected at the bottom have temperatures $T_{ec} = T_{ic} = T_0/20$. We also inject at the bottom a warmer electron population having the temperature $T_{es} = 2T_0$ ($r = r_0$), which represents SBS electrons [*Evans*, 1974]. We keep the following density ratios of the various plasma populations injected at $r = r_0$ and $r = r_m$: cold ion density $n_{ico}/n_0 = 1$, cold electron density $n_{eco} = 0.95n_0$, secondary electron density $n_{eso} = 0.05n_0$, hot ion density $n_{iho} = 0.025n_0$, and hot electron density $n_{eho} = n_{iho}$. Note that these densities are relative to the normalizing density n_0 and they represent the densities of the injected Maxwellian-distributed plasma particles; the densities of the outgoing particles are determined by the processes occurring in the simulated plasma over $r_0 < r < r_m$. The ion to electron mass ratio is $M/m = 400$, instead of the real mass ratios for the typical ionospheric ions. This artificial mass ratio considerably reduces the CPU needed for completing the simulations. The radial magnetic field (B) in the flux tube is assumed to vary as follows:

$$B(r) = B_0(r_0/r)^3 \qquad (3)$$

We start the simulation at time $t = 0$ with an empty flux tube, in which plasma populations build up in time by the injections of the plasma particles as described above. We determine the electric potential distribution $\varphi(r)$ by solving the Poisson equation

$$B(r)\partial[E(r)/B(r)]/\partial r = \rho(r)/\varepsilon_0, \qquad (4)$$

where electric field $E(r) = -\partial\varphi/\partial r$, and $\rho(r)$ is the charge density at r. The above equation is solved subject to the boundary conditions $\varphi(r = r_0) = 0$ and $\varphi(r = r_m) = \varphi_0 = -60(k_B T_0/e)$, which is the parallel potential drop in the flux tube. To simulate a long flux tube, we use nonuniform grid spacing with its minimum value $\delta_0(r = r_0) = \lambda_{do}$ determined by a nominal local density n_0 at $r = r_0$ and the hot electron temperature T_0. In the simulations described here we used $r_0 = 10{,}750\,\delta_0$ and 6000 grids, giving us a simulated length $d = r_m - r_0 = 10^4\,\delta_0$ by making the grid spacing $\delta(r) = \delta_0(r/r_0)^{3/2}$. It turns out that the energized hot electrons eventually prevail all along the flux tube and therefore the grid spacing $\delta(r)$ used in the simulation is smaller than the local plasma Debye length along nearly all the length of the flux tube.

To demonstrate the relationship between a DL and a density cavity in the geometry of a flux tube, Figures 13a and 13b show examples of the potential distribution in the flux tube and the associated density structure, respectively. In Figure 13b we have plotted the total ion density, the hot ions originating at $r = r_m$ and the cold ions at $r = r_0$. Note that the horizontal axes in the figures here are $x = r - r_0$. The sharp transition in the potential distribution near $x = 6000$ is a DL structure localized in a deep density cavity with minimum density $n_{min} \sim 10^{-3}\,n_0$. On the high-potential side of the DL are some strong oscillations, not discussed here. But the growing spatial oscillations terminate in another jump in the potential, as in a DL. This jump is also associated with a sudden density jump and its maximum electric field occurs near a density minimum, although this is not clearly seen on the logarithmic scale. The thickness of the DL near $x = 6000$ is $L \cong 400\,\lambda_{do}$. Since local density $n \cong 10^{-2}\,n_0$, the local Debye length $\delta_d \cong 10\,\lambda_{do}$, giving $L \cong 40\delta_d$, whereas the scaling rule in expression (1) yields $L \cong 5 \times (60)^{1/2} \cong 39\,\delta_d$. This excellent agreement between the scale lengths of the DLs seen in quite different simulations suggests that the scaling rule in expression (1) is generally valid. According to *Borovsky* [1983], this is also valid for oblique DLs.

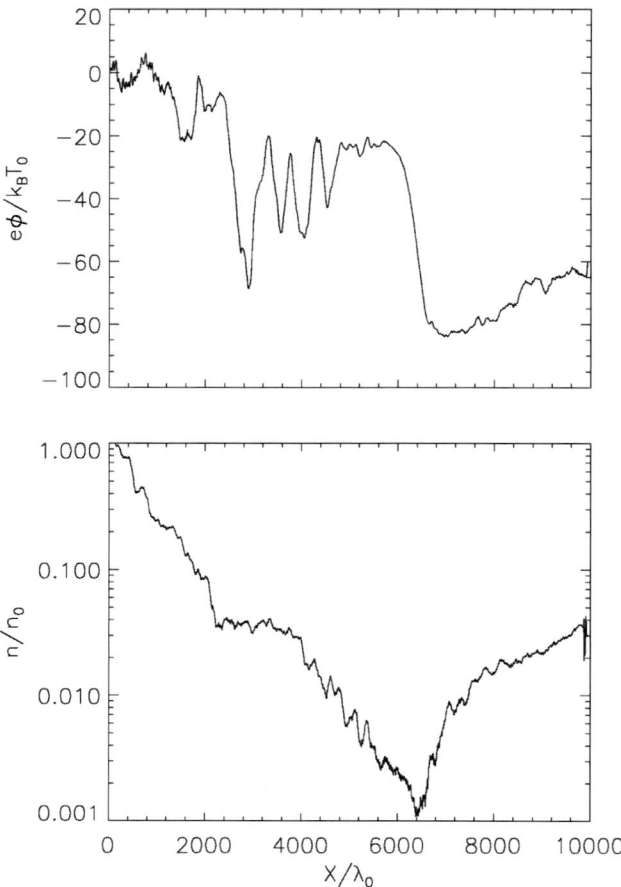

Figure 13. Example of double layers and density structures in a flux tube: (a) $\varphi(x)$ and (b) total ion density $n_i(x)$.

Figure 14 shows the phase space plots in the x–V_x plane of all the five particle species, showing the accelerations of both the hot electrons and the cold ions, originating at $r = r_m$ and $= r_o$, respectively. The formation of the hot electron beam after the acceleration by the DL near $x = 6000$ and its interactions with other particle populations are clearly seen in Figure 14b. Both cold (Figure 14a) and secondary (Figure 14c) electron populations are confined below the DL and are considerably energized by the electron–electron interactions involving the electron beam. The mirroring of the accelerated hot electrons produces an electron population with relatively quite small parallel velocities near the boundary $r = r_o$ (Figure 14b); such electrons have large perpendicular velocities.

Before acceleration by the DL, the cold ions have undergone prior accelerations by plasma expansion in the flux tube [*Singh and Schunk*, 1982c, 1983] and interactions with the energized electrons below the DL. These features are clearly seen in Figure 14d. Above the DL, the accelerated ions interact with the hot ion population (Figure 14e) and become thermalized. The vortexes seen in the ion phase space (Figure 14c) are ion holes, and they significantly contribute to the fine structures in the parallel electric field. Detailed discussion of this topic is reported elsewhere [*Singh et al.* 2005].

7. TWO-DIMENSIONAL DOUBLE LAYERS AND PLASMA DEPLETIONS

So far our discussion has been on one-dimensional features of the DLs and associated density response. We see a similar response in 2-D situations [*Singh and Khazanov*, 2003, 2005]. We performed simulations of a 2-D DL. We used a 2.5-D PIC code to model the DL structures in the x–z plane of size $L_x \times L_z$. The ambient magnetic field B_o is along z. The simulation method was described previously [*Singh et al.*, 1987; *Singh and Khazanov*, 2003, 2005]. We use an electrostatic code, solving the Poisson equation for the potential $\varphi(x,z)$ with the following boundary conditions: $\varphi(x,z=0)=0$, $\varphi(x,z=L_z)=\varphi_o$, and $\varphi(x=0,z)=\varphi(x=L_x,z)$. At the bottom ($z = 0$) and top ($z = L_z$) of the simulated plasma we inject high-density cold and low-density hot plasma, respectively, by maintaining plasma reservoirs in which plasma particles have isotropic Maxwellian velocity distribution functions. The bottom reservoir has equal electron and ion temperatures $T_{ec} = T_{ic} = T_c$; the top reservoir has the hot electron temperature $T_{eh} = T_o$, and the hot ion temperature is $T_{ih} = 3T_o$. We assume here that $T_c = T_o/10$ and that the density ratio in the two reservoirs is $n_{co}/n_{ho} = 15$, where n_{co} and n_{ho} are the densities of the cold and hot plasmas, respectively; these do not include the densities of the plasmas flowing out of the box.

At the initial time $t = 0$, the simulation region is empty. The simulation begins by filling the two plasma reservoirs and letting the plasmas in them expand into the simulation region. The plasma particles are allowed to leave the simulation region while exiting at $z = 0$ and $z = L_z$. Along x, we use a periodic boundary condition on the particles. We use the following definitions and normalizations: distances are measured in units of Debye length $\lambda_{do} = a_{eh}/\omega_{peo}$, time in $t_n = \omega_{peo}^{-1} = (n_{ho}e^2/m\varepsilon_o)^{-1}$, velocities in $a_{eh} = (k_BT_o/m)^{1/2}$, potential and electric fields in $\varphi_n = (k_BT_o/e)$ and $E_n = (k_BT_o/e)/\lambda_{do}$, respectively. The hot ions in the top reservoir at $z = L_z$ and cold electrons in the bottom reservoir at $z = 0$ have average drifts $V_{di} = -0.12a_{eh}$ and $V_{de} = 0.5a_{eh}$, respectively, to satisfy the Bohm criterion for DL formation. The simulations were performed for ion to electron mass ratio $M/m = 400$. We used a time step $\Delta t \cong 0.05\omega_{peo}^{-1}$, spatial grid spacing λ_{do} along both x and z, and number of particles per cell ranging from about 10 to 200, in keeping with the nonuniformity of the plasma.

We describe here the results from a simulation with $\Phi_o = 30$, $\Omega_e/\omega_{peo} = 3$, and $L_x = L_z = 256$. A well-formed planar DL

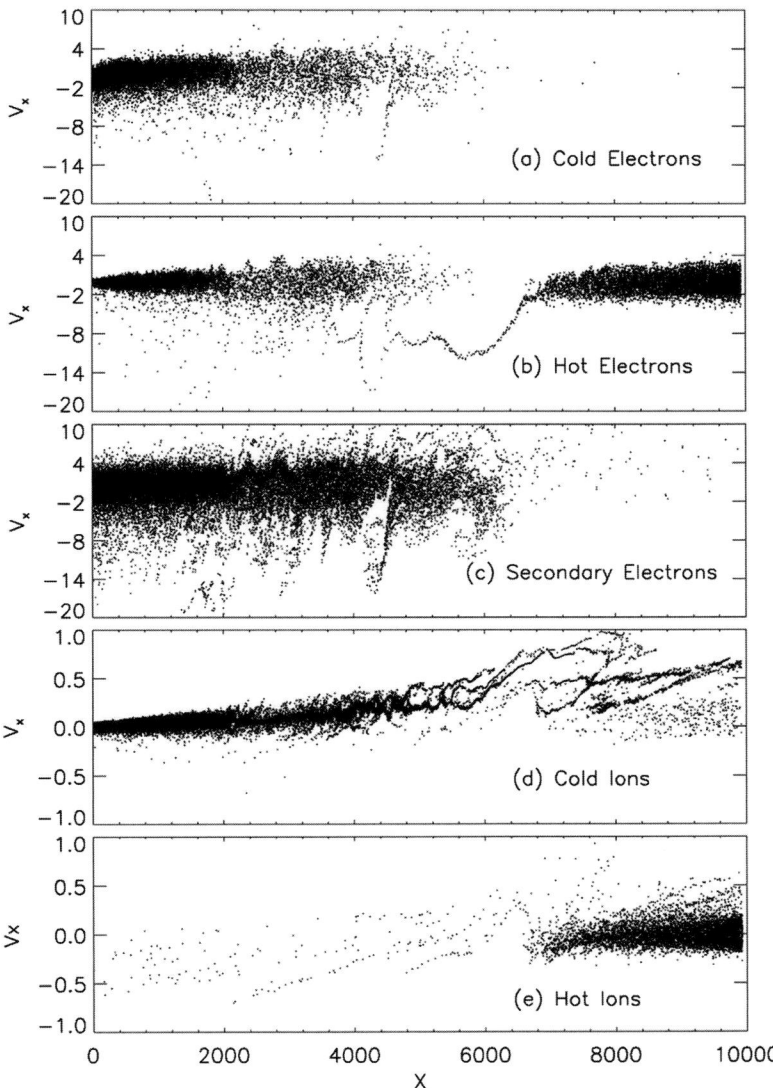

Figure 14. Phase space of various particle populations in the x–V_x plane: (a) cold electrons, (b) hot electrons, (c) secondary electrons, (d) cold ions and (e) hot ions. Velocities are normalized by the electron thermal velocity $a_{eh} = (k_B T_o / m)^{1/2}$.

appears in the simulation as early as $t \sim 500$. The potential distribution of the DL, $\varphi(z)$, shown in Figure 15a at times $t = 500$ (solid line), 600 (dashed line), and 700 (dot-dashed line), demonstrates a sharp transition from $\varphi \sim 0$ to $\varphi \sim \varphi_o = 30$ in the altitude region $64 < z < 96$. The corresponding density profiles are plotted in Figure 15b; the DL forms in a density cavity, which is continuing to fill from both bottom and top. The total plasma density below the DL reaches a maximum value of ~20; above the DL it reaches ~2. For the times shown in Figure 15a and 15b, the DL is planar in the sense that it extends nearly uniformly all over x, as shown by the equipotential contours in Plate 1a. It supports a nearly uniform strong parallel electric field E_z, which maximizes in the density cavity seen in Figure 15b.

The planar DL shown in Plate 1a is unstable. The potential structure of the DL undergoes spatial modulation along x. Plates 1b–1d reveal the salient features of the further evolution of the DL; the contours of constant potentials in the

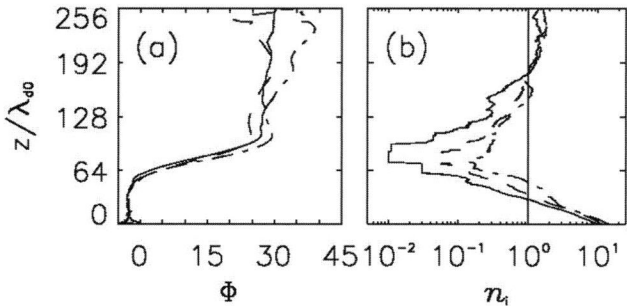

Figure 15. (a) Potential profile φ(z) at t = 500, 600, and 700 plotted in solid, dashed, and dot-dashed lines, respectively. (b) Density profile, $n_i(z)$, corresponding to φ(z) in (a); φ(z) and $n_i(z)$ do not vary much with x for the times shown.

transition region develop modulations, implying that φ(x,z) is no more constant at a fixed z as x is varied. The transverse spatial modulation is triggered by electrostatic waves below the planar DL by ion beams accelerated by it. In the nonlinear stage of the waves, the modulations transform the initial laminar DL structure near the plane y ~ 64 into jagged DL substructures (DLSSs) as amply revealed by the color equipotential plots in Plates 1a–1d. The density distributions shown in Plates 1e–1h also reveal the modulations. Plate 1 shows that the DL breaking into DLSSs is a fast process, occurring in a short time interval from t ~ 500 to t ~ 1300; once formed, however, the DLSSs undergo only slow changes, being seen at all times between the time interval t = 1288 to 5000 but continually undergoing vertical and horizontal displacements as well as distortions. A simulation extending to t = 10,000 demonstrated similar behavior. How does the DL break into DLSSs?

As the ion-beam–driven fluctuations begin to grow, their superposition on the initially uniform distribution of the DL potential φ(x) introduces perpendicular fields in the DL structure. Another consequence of the superposition is that the parallel electric field of the DL is modulated; in the locations where $E_\parallel$ weakens, the cold dense ions rise to higher heights than in the locations where $E_\parallel$ is strong, thus modulating the horizontal density structure, as is clearly visible from the color plots in Plates 1e–1h. The simulation shows the complete nonlinear evolution of this modulating process that leads to the formation of DLSSs. At t ≅ 600 the cold plasma density below the DL is nearly uniform in x (Plate 1e). At a later time, such as that at t = 1000 in Plate 1f, the horizontal distribution in the density below the DL develops fine-grained structures with a typical scale length $\delta_\perp \sim \rho_{ih}$. The density structures become progressively broader with increasing time. The substructures in the potential and density distributions show a high degree of correspondence; where the equipotential contours are locally almost horizontal and $E_\parallel$ is strong, the dense, cold plasma rises to the bottom of the DLSS. In contrast, between the DLSSs, where $E_\parallel$ is weak and $E_\perp$ is strong, the cold plasma rises as a column. The downward penetration of the DLSS into the dense cold plasma generates V- or U-shaped DLSSs (Plates 1c and 1d) that contain low-density plasma (Plates 1g and 1h). Note the presence of four well-formed V-shaped DLSSs in Plates 1c and 1d.

If a satellite were to traverse the DLSSs horizontally, it should be able to detect diverging electrostatic shocks (D-shocks) having diverging pairs of $E_\perp$ [*Carlson et al.*, 1998]. Figure 16a shows the $E_\parallel$ and $E_\perp$ along the horizontal line y = 32 in the potential structure shown in Plate 1d for t = 5000. The perpendicular field shown by the dashed line reveals two clear examples of D-shocks; such large fields occur at the borders of severe density depletions (Figure 16b). In the DLSS, cold electrons are accelerated upward by a significantly large $E_\parallel$. $E_\perp$ in the DLSS could be even larger than ~ $E_n = (k_B T_o/e)/\lambda_{do}$, the normalizing field (Figure 16a). If we assume that T_o is ~ 100–1000 eV and n_o is ~ 1 cm^{-3}, then λ_d ~ 60–200 m and $E_n \cong$ 1.6–5 V/m. Narrow D-shock structures with large $E_\perp$ (up to 1 V/m) have been measured by FAST [*Carlson et al.*, 1998]. The field structures such as those shown in Figure 16a could be seen at other times as well as other heights, but they became increasingly noisy as the perturbations created by electron beams grew and formed electron holes with increasing height in a DLSS.

The low-density plasma above the DLSS is crowded with electron holes. Plates 1b–1d show several electron hole structures appearing in orange, yellow, and white shades, depending on their amplitudes. At early times, such as at t = 1000 (Plate 1b), when the DL is laminar and not fragmented, some electron holes appear above the DL as vertically thin structures but quite elongated horizontally, spanning nearly the entire width (64 < x < 256), but they break into transversely smaller structures as they propagate upward [*Goldman et al.*, 1999]. At later times the electron holes are continually emitted from the DLSS, their transverse size being dictated by the width of the DLSS directly underneath them. The electron holes have their characteristic vortex structure.

Figures 17a and 17b show the phase space of the cold ions lying below z = 64 in the V_x–V_z plane at t = 1000 and 2000, respectively. The phase space at t = 2000 (Figure 17b) shows a dramatic change; the cold ions have undergone large perpendicular as well as parallel accelerations. The transverse acceleration could be affected by these fluctuations, which fragment the initial planar DL into DLSSs, in combination with the strong gradients in E_x in the DLSSs [*Cole*, 1976; *Singh et al.*, 1987]. The cyclotron motion in nonuniform E_x is described by

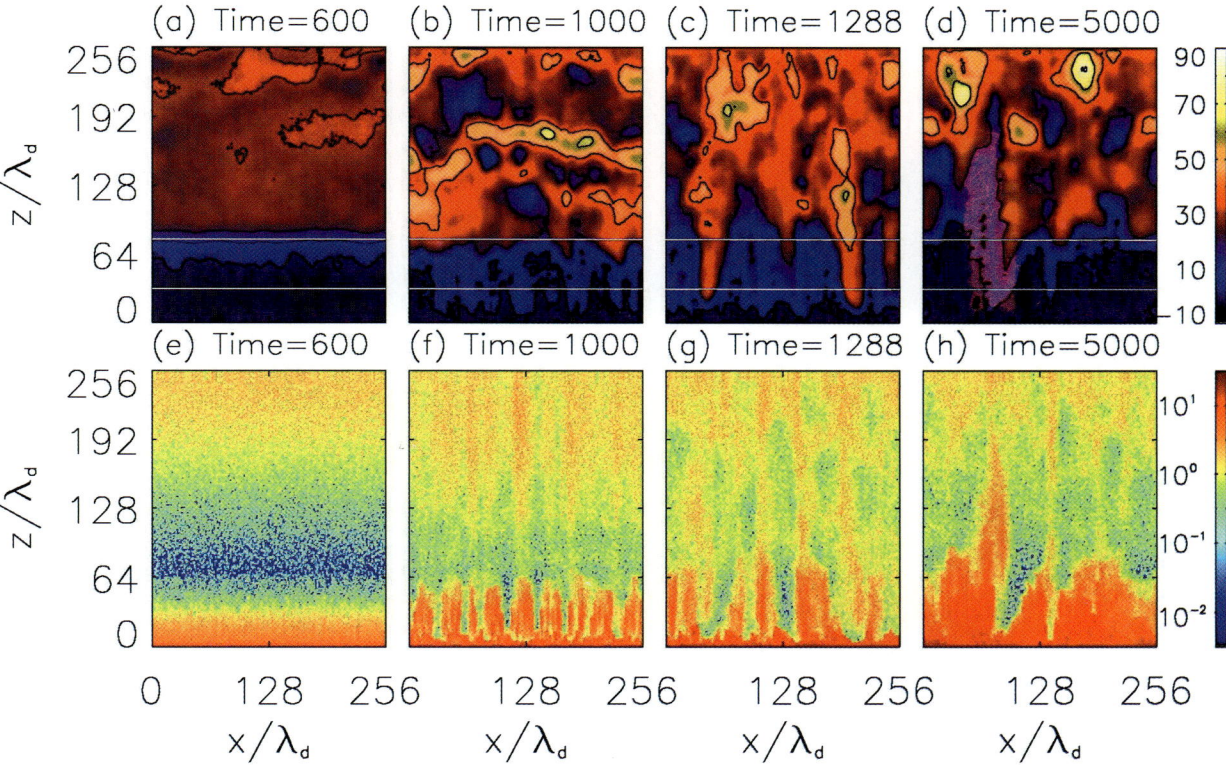

Plate 1. Evolution of the DL equipotential contours: (a) $t = 600$, (b) $t = 1000$, (c) $t = 1288$, and (d) $t = 5000$. (e)–(h) Plasma density structures, plotted on a log scale, corresponding to the potential structures in (a)–(d). Note the filamentation of the original DL into DLSSs and the presence of electron holes as localized large-potential pulses above the DL and DLSSs.

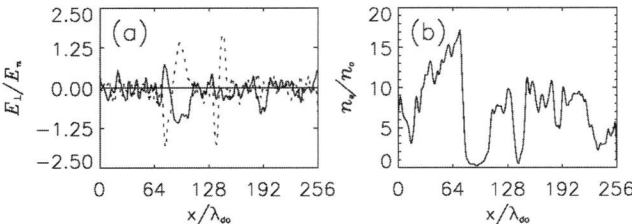

Figure 16. (a) Spatial structures of parallel ($E_z = E_\parallel$) and perpendicular ($E_x = E_\perp$) fields along the line $z = 32$ at $t = 5000$, showing two examples of clearly diverging electrostatic shock structures. (b) Plasma density structure corresponding to the field structure in (a).

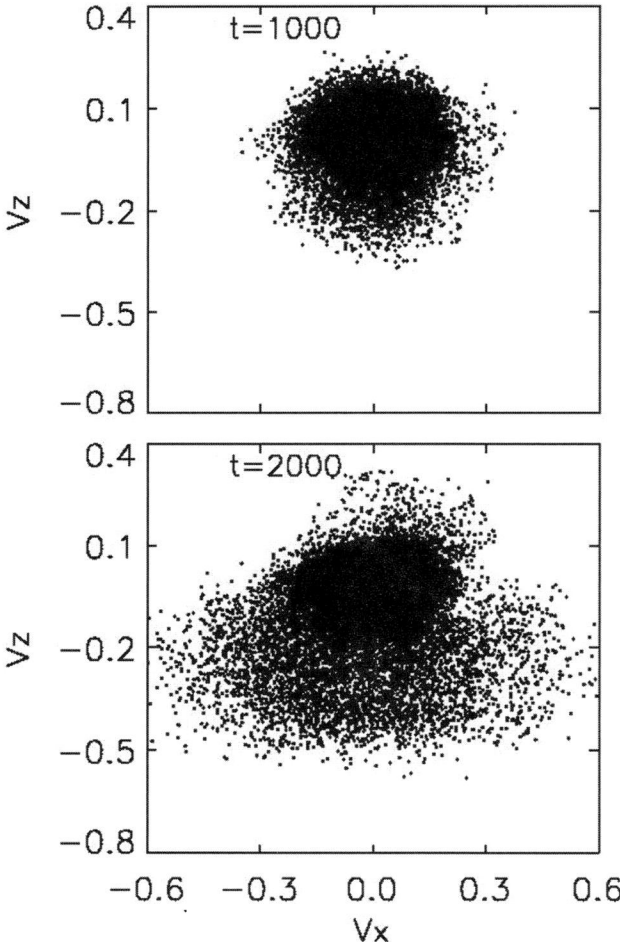

Figure 17. (a) Scatter plot in the V_x–V_z plane for the initially cold ions at $t = 1000$. (b) Same as (a) but at $t = 2000$.

$$\partial^2 V_x/\partial t^2 + [(\Omega_i/\omega_{peo})^2 - (m/M)\partial E_x/\partial x]V_x \quad (5)$$
$$= (m/M)(\partial E_x/\partial t + V_z \partial E_x/\partial z)$$

where Ω_i is the ion cyclotron frequency. For sufficiently large and positive $\partial E_x/\partial x$, the quantity in the square brackets becomes negative, destroying the cyclotron motion, and the ions demagnetized thus are freely accelerated by E_x. This happens when the transverse scale length in $E_x(x)$ is $L_\perp < L_o = (M/m)(\omega_{peo}/\Omega_e)^2 E_o \lambda_{do}$, where E_o is the magnitude of E_x [Singh et al., 1987]. For $M/m = 400$, $\omega_{peo}/\Omega_e = 1/3$ and $E_o \sim 2$ in the simulation, then $L_o \sim 90 \lambda_{do}$ and the condition $L_\perp < L_o$ is well satisfied in the DLSSs. Thus the strong gradients in E_x demagnetize the cold ions and the ions are freely accelerated by the electric fields in the evolving DLSSs.

The cold ions are accelerated up to $|V_x| \sim 0.6$, compared with their initial thermal velocity $a_{it} = 0.02$. The acceleration by a factor of ~30 in $V_\perp$ amounts to an energy gain by a factor of ~900. If this heating were to occur in the topside ionosphere, ions could be generated having transverse energy up to several hundred eV. In our simulations we do not have the mirror force, which converts such heated ions into ion conics. Note that the heated ions occur below the DL. In the presence of the mirror force, such ions are expected to be confined below the DL until the mirror force dominates the downward electric force in the DL. This confinement may lead to a large ion heating, as originally discussed by *Gorney et al.* [1985].

We have not performed a detailed analysis of the perpendicular scale length ($L_\perp$) of the DLSSs seen in the simulations. Note that density cavities are an integral part of the DLSSs. We find that when they appear for the first time by the modulation of the planar DL by the ion-beam–driven waves, $L_\perp \cong 2\pi \rho_{ih}$, where ρ_{ih} is the hot ion Larmor radius. As the ions are transversely accelerated, the transverse size of the DLSSs also increase. A full analysis of the time-dependent scale length is discussed elsewhere [*Singh and Khazanov*, 2005].

8. CONCLUSIONS

The main goal of this paper is to establish the intimate relationship between DLs and density depletions or cavities in plasmas. For this purpose we have reviewed our old and new studies on DLs using simulations. In view of recent satellite observations reporting that parallel electric fields occur in regions of reduced densities, we hope such a review will be timely and useful. The review shows that whether a plasma is driven by an applied voltage or a current, the DL always forms in a density cavity. Such density cavities could be preexisting in the plasma or could evolve as a part of the plasma turbulence or through the process of DL formation. We have found that if there is a density cavity in a plasma, a current in the plasma charges the cavity, forming a DL, which becomes the site for the acceleration of electrons and ions. When a plasma is driven by an applied voltage, current is produced and therefore the current–cavity interaction leads to a DL formation, as in the case of current-driven plasma. However, there are some important differences in the current- and voltage-driven cases. In the current-driven case, the DL is transitory and its life depends on the time taken to refill the cavity, which could be quite short. On the other hand, in the voltage-driven case, the acceleration by DLs could be long lasting, depending on the lifetime of the driving mechanism. This is achieved by the recurring formation of the DLs through the process of cavitational instability [*Carlqvist*, 1972; *Singh*, 1982, 2003; *Singh and Schunk*, 1982].

How do the cavities form in a plasma? As we saw in Section 3, cavity formation could occur through plasma instabilities. Such cavities self-consistently form when current exceeds a certain threshold. Preexisting cavities in plasma are a consequence of plasma accelerations that result in plasma evacuation. Transverse ion heating could create deep density cavities, as shown by both hydrodynamic [*Singh*, 1992] and kinetic [*Singh and Chan*, 1993; *Singh*, 1996] simulations of the transverse heating of the polar wind ions. An extended ion heating along an auroral flux tube creates an extended cavity that moves upward, riding on top of the expanding polar wind. The transition from high to low density near the ionospheric boundary could be a location for the DL formation both in downward and upward current plasmas. Such DLs move upward with the polar wind. Thin cavities, generated by spotty, localized transverse ion heating, could also act as sites for DL formation. Since localized cavities randomly evolve in the plasma near the sites of spotty ion heating, the DL location is also likely to shift from site to site, adding to the complexity of the plasma turbulence. A self-consistent study incorporating transverse ion heating [*Singh*, 1992, 1996; *Singh and Chan*, 1993] and DL formation has the promise of revealing the inherent complexity of the auroral acceleration, which probably involves many coupled processes.

Another important process for the cavity formation is the nonlinear pondermotive force associated with waves. Several authors have considered the effects of this nonlinear force as an acceleration mechanism for the ionospheric ions as well as the resulting density cavitation. *Li and Temerin* [1978] studied the pondermotive force associated with Alfven waves. *Guglielmi and Lundin* [2001] have studied the effect of low-frequency Alfven waves on the ions and resulting plasma cavitation that form filamentary density cavities in the polar region. *Singh* [1994] has compared the effectiveness of transverse ion heating and pondermotive force of nonlinear lower hybrid waves in plasma evacuation. *Staciewicz et al.* [1997] and *Singh* [1999] have considered the pondermotive force in the resonance-cone structure of large-amplitude inertial Alfven (IA) waves in creating narrowly structured density cavities.

Observations of lower hybrid density cavities (LHDC) in the auroral plasma have been extensively reported from both rocket and satellite experiments [*Kintner et al.*, 1992; *Pecseli et al.*, 1996; *Høymark et al.*, 2000]. Such cavities are at times filled with lower hybrid waves; empty cavities also are not rare. Such cavities in the regions of parallel currents could be a rich source of weak DLs that have spatial and geometrical features similar to those of the cavities. A correlative study between weak DLs [*Temerin et al.*, 1982; *Malkki et al.*, 1993] and LHDC [*Høymark et al.*, 2000] could prove very valuable for the auroral plasma physics.

The current-driven DLs in density depletions appear quite relevant to the observations of some unexplained nonlinear features of narrow-structured IA waves, namely, the large parallel electric fields ($E_\parallel$) observed from Freja [*Chust et al.*, 1998; *Stasiewicz et al.*, 1997, 1998, 2000] and FAST [*Chaston et al.*, 1999]; it is reported that ($E_\parallel$) ~ 100–200 mV/m occurs accompanied with dispersive Alfven waves (DAWs) in strong density depletions. The DAWs occur in the IA wave regime, in which the perpendicular wave number, $k_\perp \sim \omega_{pe}/c$, where ω_{pe} is the electron plasma frequency and c is the velocity of light. The reported DAW events are associated with a few hundreds of microamperes of parallel currents. On the basis of our report here, we suggest that the measured $E_\parallel$ are generated by DLs driven by the parallel currents in the deep density cavities seen with the waves [*Singh*, 2002]. *Chust et al.* [1998] had conjectured about the DL formation. It is important to point out that current-driven DLs in a uniform plasma are normally attributed to plasma instabilities such as the Buneman instability [*Singh and Schunk*, 1982b] or the ion-acoustic instability [*Sato and Okuda*, 1981]. In the presence of a cavity, the DL forms spontaneously without the mediation of any instability; essentially, the preexisting cavity charges to a potential difference so that the charged particles accelerated in it could carry the current, maintaining current continuity. Since currents are a common feature of Alfven waves having short transverse scale lengths, DL formation via the interaction of the wave current with the wave-generated density cavities or with naturally occurring ones could be a universal mechanism for dissipating Alfven waves in space and astrophysical plasmas. A transverse modulation of planar DLs by ion-beam–generated instabilities also can form structured density cavities with transverse size of the order of ion cyclotron Larmor radius [*Singh and Khazanov*, 2003, 2005].

Finally we comment that plasma turbulence in space plasmas consists of a complex set of coupled processes. Currents in plasmas, voltage drops, and density cavities in them are all mutually coupled; the presence of either a current or a voltage drop could generate the other by formation of DLs in the density cavities. Even the cavities have their origin in the acceleration processes driven by the voltages and current. Are the auroral acceleration processes driven by voltage drops or currents? In view of the coupling mentioned here, this question is probably redundant.

Acknowledgments. This work was supported by NSF grant ATM 0206669 and NASA grants NAG5-12897 and NAG5-13489. Some of the simulations were done at the Jet Propulsion Laboratory's supercomputing facility.

REFERENCES

Block, L.P., A double layer, *Rev. Astrophys. Space Sci.*, 55, 59, 1978.

Borovsky, J.E., The scaling of oblique plasma double layers, *Phys. Fluids* 26, 3273, 1983.

Buneman, O., Instability turbulence, and conductivity in current carrying plasma, *Phys. Rev. Letters, 1*, 8, 1958.

Carlqvist, P., *Cosmic Electrodynamics, 3*, 377, 1972.

Carlson, C.W., et al., Fast satellite observations in the downward current region: Energetic upgoing electron beams, parallel potential drops, and ion heating, *Geophys. Res. Lett.*, 25, 2017, 1998.

Chaston, C.C., et al., FAST observations of inertial Alfven waves in the dayside aurora, *Geophys. Res. Lett.*, 26, 647, 1999.

Chust, T., et al., Electric fields with a large parallel component observed by the Freja spacecraft: Artifacts or real signals? *J. Geophys. Res., 103*, A1, 215-224, 1998.

Cole, K.D., Effects of crossed magnetic and (spatially dependent) electric fields on charged particle motion, *Planet. Space Sci.*, 24, 515, 1976.

Ergun, R.E., et al., Fast satellite observations of electric field structures in the auroral zone, *Geophys. Res. Lett.*, 25, 2025, 1998.

Ergun, R.E., L. Anderson, C.W. Carlson, D.S. Main, J.P. McFadden, F.S. Mozer, and Y.-J. Su, Parallel electric fields in the upward current region of the aurora: Indirect and direct observations, *Physics Plasmas*, 9, 3685, 2002a.

Ergun, R.E., L. Anderson, C.W. Carlson, M.V. Goldman, D.S. Main, J.P. McFadden, F.S. Mozer, D.L. Newman, and Y.-J. Su, Parallel electric fields in the upward current region of the aurora: Numerical solutions, *Physics Plasmas*, 9, 3695, 2002b.

Ergun, R.E., et al., Fast auroral snapshot satellite observations of very low frequency saucers, *Phys. Plasmas*, 10, 454, 2003.

Evans, D.S., Precipitating electron fluxes formed by a magnetic field-aligned potential difference, *J. Geophys. Res.*, 79, 2853, 1974.

Goldman, M.V., et al., Nonlinear two-stream instabilities as an explanation for auroral bipolar wave structures, *Geophys. Res. Lett.*, 26, 1821, 1999.

Gorney, D.J., Y.T. Chiu, and D.R. Croley, Jr., Trapping of ion conics by downward electric fields, *J. Geophys. Res.*, 90, 4205, 1985.

Guglielmi, A., and R. Lundin, Pondermotive upward acceleration of ions by ion cyclotron and Alfven waves over the polar regions, *J. Geophys. Res.*, 106, 13,219, 2001.

Høymork, S.H., H.L. Pecseli, B. Lybekk, J. Trulsen, and A. Eriksson, Cavitation of lower hybrid waves in the Earth's ionosphere: A model analysis, *J. Geophys. Res., 105*, 18,519, 2000.

Hull, A.J., J.W. Bonnell, C.C. Chaston, F.S. Mozer, and J.D. Scudder, Large parallel electric fields in the upward current region of the aurora: Evidence for ambipolar effects, *J. Geophys. Res., 108*, 1265, 2003.

Iizuka, S., R. Hatakeyama, P. Michelsen, J.J. Rasmussen, K. Saeki, N. Sato, and R. Schrittwieser, Dynamics of a potential barrier formed on the tail of a moving double layer in a collisionless plasma, *Phys. Rev. Lett.*, 48, 145, 1982.

Joyce, G., and R.F. Hubbard, *J. Plasma Physics* 20, 391, 1978.

Kintner, P.M., J. Vago, S. Chesney, R.L. Arnoldy, K.A. Lynch, C.J. Pollock, and T.E. Moore, Localized lower hybrid acceleration of ionospheric plasma, *Phys. Rev. Lett.*, 68 (16), 2448, 1992.

Knight, L., Parallel electric fields, *Planet. Space Sci.*, 21, 741, 1973.

Li, X., and M. Temerin, Pondermotive effects on ion acceleration in the auroral zone, *Geophys. Res. Lett.*, 20, 13, 1993.

Lyons, L.R., D.S. Evans, and R. Lundin, An observed relation between magnetic field aligned electric fields and downward electron energy fluxes in the vicinity of auroral forms, *J. Geophys. Res.*, 84, 457, 1979.

Malkki, A., et al., A statistical survey of auroral solitary waves and weak double layers, 1, Occurrence and net voltage, *J. Geophys. Res.*, 98, 5763, 1993.

Mozer, F.S., et al., New features of time domain electric field structures in the auroral acceleration region, *Phys. Rev. Lett.*, 79, 1281, 1997.

Mozer, F.S., and A. Hull, Origin and geometry of upward parallel electric fields in the auroral acceleration region, *J. Geophys. Res.*, 106, 4205, 2001.

Newman, D.L., M.V. Goldman, R.E. Ergun, and A. Mangeney, Formation of double layers and electron holes in a current-driven space plasma, *Phys. Rev. Lett.*, 87, 2559-01, 2001.

Okuda, H., and K.-T. Nishikawa, Ion-beam-driven electrostatic hydrogen cyclotron waves on auroral field lines, *J.Geophys. Res.*, 89, 1023, 1984.

Pecseli, H.L., K. Iranpour, O. Holter, B. Lybekk, J. Holtet, J. Trulsen, A. Eriksson, and B. Holback, Lower-hybrid wave cavities detected by the Freja satellite, *J. Geophys. Res., 101*, 5299, 1996.

Persoon, A.M., et al., Electrons density depletions in the nightside auroral zone, *J. Geophys. Res.*, 93, 1871, 1988.

Sato, T., and H. Okuda, Numerical simulations of ion acoustic double layer, *J. Geophys. Res.*, 86, 3357, 1981.

Shawhan, S.D., C.-G. Fälthammar, and L.P. Block, On the nature of large auroral zone electric fields at 1-R_E altitude, *J. Geophys. Res.*, 83, 1049, 1978.

Singh, N., Computer experiments on the formation and dynamics of electric double layers, *Plasma Physics*, 22, 1, 1980.

Singh, N., Double layer formation, *Plasma Physics*, 24, 639, 1982.

Singh, N., Plasma perturbations created by transverse ion heating in the magnetosphere, *J. Geophys. Res.*, 97, 4235, 1992.

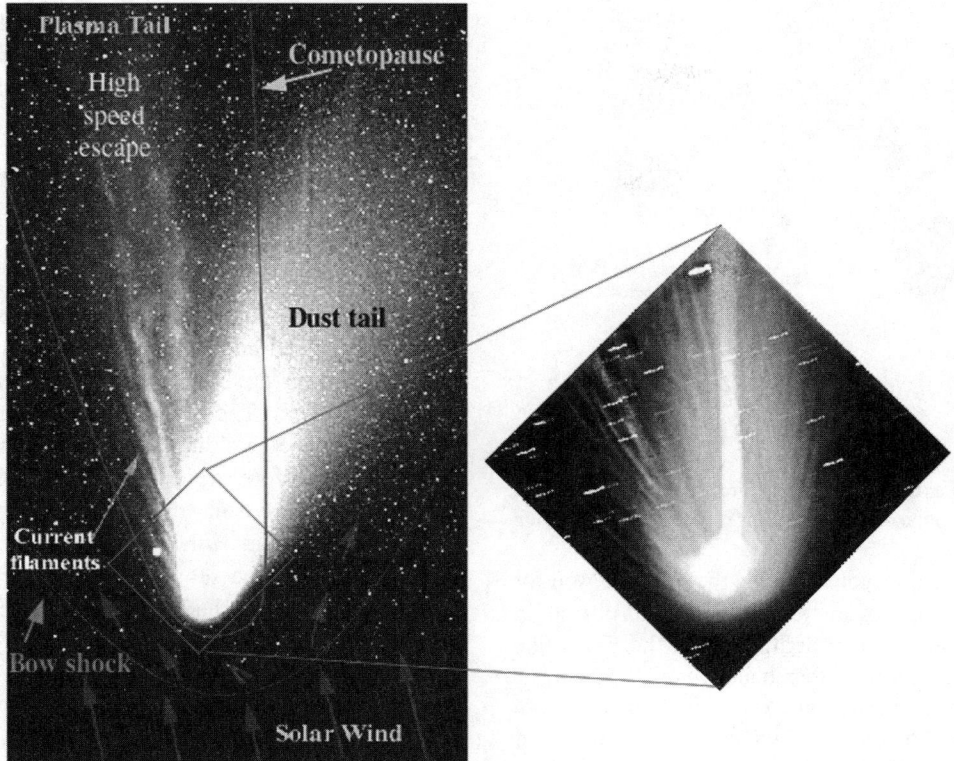

Figure 2. Example of externally driven outflow, comet Hale-Bop. Also illustrated: the structured "hairy" outflow of a cometary tail. The close-up view to the right (from the Hubble Space Telescope) illustrates the hairiness down to the small-scale features.

Comparative magnetospheric physics may be applied to more remote celestial objects out of reach of space probes (other stars, pulsars, the stellar environment, and galaxies). Magnetospheric topologies are easily identified in astrophysical objects imaged by radio telescopes and modern ground-based and space-based telescopes (e.g. the Hubble Space Telescope).

It is the purpose of this report to demonstrate that existing knowledge in magnetospheric physics may be used to understand remotely located magnetospheres of other stars and galaxies. In particular, we discuss the ejection and acceleration of plasma in the polar regions of stars and galaxies. We note that electromagnetic forcing dominates over gravitational forcing in any gaseous environment even for low ionization rates. It is well known from ionospheric research that electromagnetic forcing strongly influences the macroscopic behavior of gases for ionization rates as low as 10^{-7}. High-speed neutral winds in the upper atmosphere, caused by ion drag, are known to result from strong convection electric fields in the ionosphere [see e.g. *Rees*, 1989]. Ionization in an interstellar environment may be due to galactic radiation or UV radiation from nearby stars. The important message of this is that electromagnetic forcing is more important in the universe than is generally conceived—as is also the case for neutral gases with ionization rates as low as 10^{-7}.

Electromagnetic forces are extremely powerful in accelerating charged particles up to very high, even relativistic, velocities. Certain topologies such as those with a diverging magnetic field (e.g. magnetic dipole) are favorable sites of plasma acceleration. An example of this is the strong plasma acceleration that occurs on magnetic field lines connected to the Earth's auroral oval. The auroral acceleration process, taking place within an altitude range of 0.1–3 Earth radii near the Earth, provide plasma acceleration commonly in the range of tens of keV, in extreme cases up to some 100 keV. As will be demonstrated, the diverging topology of the Earth's dipole magnetic field (Figure 3) promotes upward ejection and focussing of the outflow into narrow beams. The latter is an important effect because it helps maintain the beam property on open field topologies. Another important region for plasma acceleration is the Earth's magnetotail, where ions and electrons may become accelerated up into the hundreds of keV range. In extreme cases, charged particles entering into the innermost part of the magnetosphere, the radiation belts [*Van Allen et al.*, 1958], become accelerated up into the MeV range.

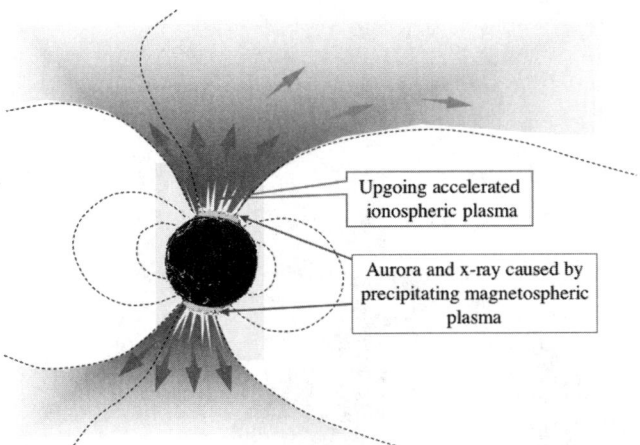

Figure 3. Effects of external forcing on a magnetized object such as the Earth, causing polar aurora and polar region plasma outflow.

External forcing by the solar wind is the prime reason for the erosion and outflow of matter from celestial objects in the inner part of our solar system, such as the Earth-like planets and comets. On the other hand, internal forcing is the prevalent cause of erosion and outflow of plasma from the Sun, giving rise to the solar wind. A combination of internal and external forcing is obtained when a star is affecting other nearby stars, such as the Trapezium in the Orion nebula (Figure 4). Notice here that newborn stars show signatures of external forcing, i.e. shock fronts and plasma tails, caused by the intense stellar wind from the central star. The individual sizes of the newborn stellar systems are similar to that of the heliosphere; i.e. the standoff distance to the interstellar wind is of the order 100–200 AU. Normally a newborn star is strongly magnetized. Therefore by analogy with the Earth's magnetosphere, the stellar magnetopause defines the standoff distance to the interstellar wind. The stellar magnetopause emerges from pressure balance conditions, i.e. a balance between internal and external total pressure (plasma and magnetic field). The stellar magnetopause marks a magnetic shielding, a cellular cavity, against the interstellar wind. The stronger the internal magnetic field, the stronger the shielding against external plasma forcing. One may therefore argue that a strong internal magnetic field is a requirement for retaining isolated structures of matter/plasma in a "windy" interstellar environment.

A number of processes are believed to be directly responsible for accelerating charged particles [see e.g. the book by *Hultqvist et al.*, 1999]. The fundamental force acting in the acceleration of charged particles is the electric (coulomb) force. The electric field may act along or perpendicular to the magnetic field. An important aspect of a fluctuating transverse (to B) electric field is that it provides an upward forcing of plasma, independent of charge, in a diverging magnetic field geometry—a ponderomotive force denoted

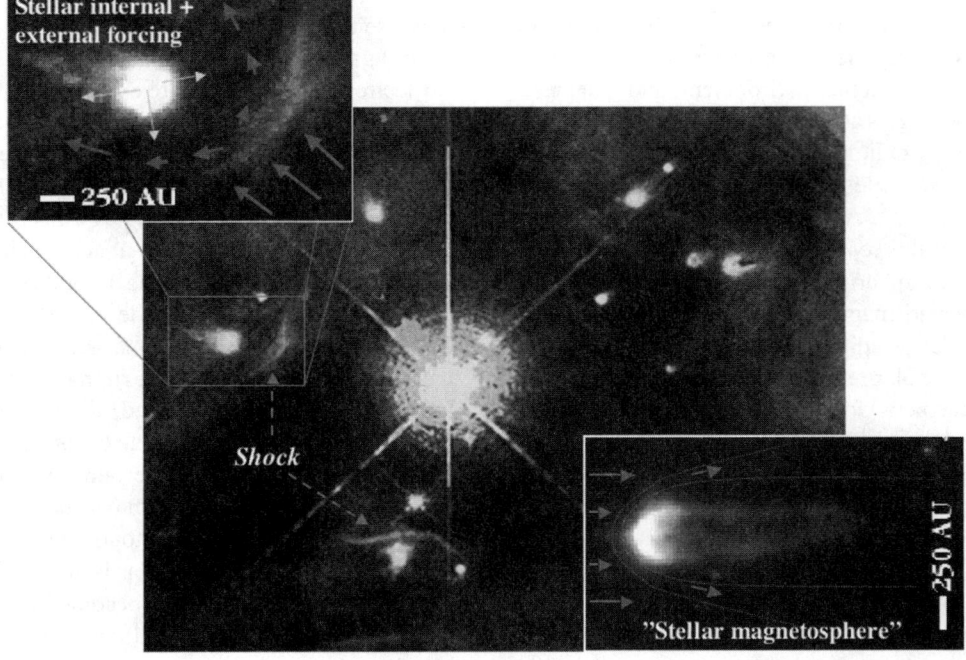

Figure 4. Combined internal and external forcing during stellar formation in the Orion nebula. The central star, Trapezium, provides external forcing on the newborn stars in the surrounding region. The newborn stars are themselves also providing internal forcing. (Source: Collage of NASA Hubble Space Telescope images.)

magnetic moment pumping [*Lundin and Hultqvist*, 1988; *Guglielmi and Lundin*, 2001]. Ponderomotive forcing as a general acceleration and outflow mechanism is described in the next section.

The influence of the magnetic field on plasma acceleration is not straightforward. The electric force is the major forcing term, the Lorenz term (v × B) providing no forcing in the sense of plasma energization. On the other hand, magnetic fields may set up spatial and temporal resonance criteria that facilitate the acceleration of charged particles. A dipole magnetic field provides a volume in space in which charged particles may energize via resonances with the gyro- and bounce motion. The relatively weak magnetic dipole of the Earth with a surface field of ≈0.5 gauss is capable of accelerating particles up into the hundreds of keV range. Jupiter, with a magnetic field some 10 times that of the Earth, is observed to sustain acceleration up to tens of MeV. Finally, the Sun, with local magnetic fields sometimes two orders of magnitudes higher than on Jupiter, may accelerate charged particles into the GeV range, i.e. well into the relativistic energies.

A general tendency for plasma acceleration near celestial objects appears to be that:

1. magnetic fields facilitate the acceleration of plasma to high energies, and
2. the maximum acceleration energy depends on the magnetic field strength.

2. PLASMA OUTFLOW AS A NON-THERMAL ESCAPE PROCESS—PHYSICAL ASPECTS

Macroscopic forces on particles near celestial bodies (volatile substances, gas, etc.) are besides gravitation also electromagnetic forces. An upward escape of particles requires forces counteracting gravitation. Escape of volatile substances from a celestial body, or neutral atmosphere escape, occurs when the net forcing results in particles acquiring velocities above escape velocity. Two mechanisms, thermal escape or non-thermal escape, may achieve this. Thermal escape implies that particles with ballistic (collisionless) orbits have thermal velocities above escape velocity.

Non-thermal escape is a generalized term for all non-Maxwellian acceleration processes, usually originating in magnetized plasma, that give rise to particle escape. The acceleration may be upward (parallel or antiparallel to the magnetic field) or transverse (to the magnetic field). In the latter case, the magnetic moment in a diverging magnetic geometry will facilitate escape. Non-thermal escape is a very efficient process because it may bring all particles into escape velocity, whereas thermal escape originates from the minor fraction of the particle thermal distribution tail that has velocities above escape velocity.

Because the exosphere of most celestial bodies are ionized and largely controlled by electromagnetic forces, the plasma formalism is more useful to describe the upward/downward motion of matter. Under the assumption that upward means parallel or antiparallel with the magnetic field vector, like in the polar region of a magnetic dipole, the motion of particles upward along the field lines of magnetized plasma is given by:

$$f_\parallel = qE_\parallel + GM_c \frac{m_o}{r^2} - \mu \frac{\partial B}{\partial s} + \frac{m_o v_\parallel}{B} \vec{V}_D \vec{\nabla}_\perp B \quad (1)$$

Where $qE_\parallel$ is the Coulomb force parallel to the magnetic field, B, $G \cdot M_C m_0/r^2$ is the gravity force term above the celestial object, $\mu \cdot \partial B/\partial s$ is the magnetic mirror force along B, and

$$\frac{m_0 v_\parallel}{B} \vec{V}_D \cdot \vec{\nabla}_\perp B$$

represents the force by gradients in the magnetic field.

Neglecting perpendicular magnetic field gradients, assuming equipotential magnetic field lines (no parallel electric field), and assuming that the first adiabatic invariant is conserved (μ is constant) we obtain for a dipole field:

$$v_\perp(r) > v_e(r) = \sqrt{\frac{2GM_c}{r}} \quad (2)$$

where $v_e(r)$ is the escape velocity in a central gravitational field. This implies that not only vertical forces, such as parallel electric fields, are required to bring particles up to escape velocities. A perpendicular/transverse forcing may be sufficient to bring charged particles to such high gyration velocities that they escape upward. A simple illustration of this is forcing by the electric drift:

$$E(r) \times B(r) > v_e(r) = \sqrt{\frac{2GM_c}{r}} \quad (3)$$

Assuming a magnetic dipole field where R_0 is the Earth's radius and B_0 is the magnetic induction on ground, the electric field E_e required to obtain escape velocity v_e at a specific altitude z is given as follows:

$$\frac{E_e}{B_o} = v_e \left(\frac{z}{R_o} \right)^{-3/2} \quad (4)$$

The above equations have the following implications:

- There is a balance between the magnetic mirror force and the gravitation force. The stronger both forces are, the longer the particles may remain in an acceleration "well" balanced by the gravitation and the magnetic mirror force, $\mu \cdot \partial B/\partial s$. Particles may eventually reach escape velocity the same was as air particles in a pressure cooker.

- The E_e/B_o ratio marks the electric drift required to achieve escape velocity at an altitude z. Acceleration to velocities above escape velocity ($v_a > v_e$) is achieved if $E_o > E_e$. The velocity will subsequently increase versus increasing normalized distance, r/R_o, by the relation:

$$v_a = \frac{E_0}{B_0}\left(\frac{r}{R_0}\right)^{3/2} \quad (5)$$

For relativistic energies, W_a,

$$W_a = m_o c^2 (\gamma - 1) \quad (6)$$

where $\gamma = \left(1 - \left(\frac{v_a}{c}\right)^2\right)^{-1/2}$.

we may from equation (5) compute the energy gain versus normalized distance r/R_o.

The above formalism describes in a qualitative way transverse electric forcing in a diverging magnetic dipole geometry. Ponderomotive forcing such as "magnetic moment pumping" [*Lundin and Hultqvist*, 1989; *Guglielmi and Lundin*, 2001] based on low-frequency Alfvén waves describe more quantitatively the velocity and energy increase versus normalized distance [equations (5) and (6)].

Magnetic moment pumping (MMP) is one of three fundamental ponderomotive forces [*Guglielmi and Lundin*, 2001]. MMP operates in an inhomogeneous ambient magnetic field with a nonzero magnetic moment force acting along B [see equation (1)]. *Guglielmi and Lundin* (2001) demonstrated that a monochromatic Alfvén wave modifies the magnetic moment such that its invariance is broken and now includes a nonlinear (quadratic) dependence on the amplitude $|E|$ of electric field oscillations. An additional wave-modified magnetic moment term replacing the third term in equation (1) may be approximated by:

$$F_P \propto E^2 \frac{\partial \ln B}{\partial z}$$

Using travelling Alfvén waves and the WKB-approximation, one obtains [*Guglielmi and Lundin*, 2001]:

$$F_P = -\left(\frac{cE}{2B}\right)^2 \frac{\partial \ln \rho}{\partial z} \quad (7)$$

where ρ is the plasma density. Equation (7) illustrates the dependence of the MMP-force versus E, B, and ρ. The parallel velocity $V(z)$ versus altitude, diagrammatically illustrated in Figure 5, may be obtained after some mathematical treatments as

$$V(z) = \frac{c}{\sqrt{2}} \frac{E(0)}{B(0)} \left[\sqrt{\frac{\rho(0)}{\rho(z)}} - 1\right]^{1/2} \quad (8)$$

Equation (8) describes the change of parallel velocity with respect to an altitude $z = 0$, where the parallel velocity is zero. Using the WKB approximation, with $r \approx B^4 \leq E(z) =$ constant, equation (8) becomes:

$$V(z) = \frac{c}{\sqrt{2}} \frac{E(0)}{B(0)} \left\{\left[\frac{B(0)}{B(z)}\right]^2 - 1\right\}^{1/2} \quad (9)$$

For a celestial body magnetic dipole field, where R_o is the radial distance to the surface and r is the radial distance, we obtain

$$V(z) = \frac{c}{\sqrt{2}} \frac{E(0)}{B(0)} \left\{\left[\frac{r}{R}\right]^6 - 1\right\}^{1/2} \quad (10)$$

Comparing equations (5) and (10), we note that the velocity increases more rapidly with distance for the MMP-forc-

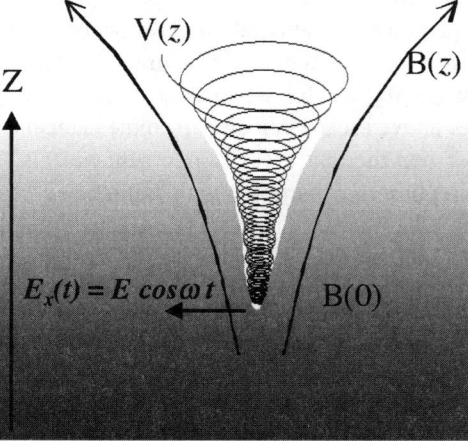

Figure 5. Diagram illustrating the ponderomotive forcing by waves in a diverging magnetic field geometry and the corresponding outflow of plasma from a density cavity.

ing. This difference is due to the fact that the electric field, E, is constant with distance in the WKB-approximation of equation (10), while we assumed the electric field to scale as "equipotentials" in equation (5). Noting that equation (5) is then a more conservative approach for modeling the upward acceleration, we use equations (5) and (6) to compute the ponderomotive upward acceleration along a dipole magnetic field model.

The diagram in Figure 6 illustrates ponderomotive acceleration of ions by Alfvén waves along the flux tubes of a dipole magnetic field. The flux tubes are considered very extended, as they are in the polar region, and thus maintain uniform divergence up to >10,000 r/R_0. The upper curve illustrates the acceleration and subsequent escape from a stellar corona (here, the Sun), where we for the sake of simplicity have assumed that all ions are gravitationally "trapped" ($r \approx R_0$) until they acquire escape velocity (620 km/s, $W \approx 2$ keV for solar protons). In the subsequent energization we assume that $v_e = E_o/B_o = 1$ is maintained [equation (6)]. We note that relativistic proton energies (≈ 100 MeV) are acquired within a relatively short range ($r/R_0 \approx 30$). The lower curves in Figure 6 illustrates the acceleration in a weaker gravitational environment, the middle curve being representative for the Earth (escape velocity ≈ 11 km/s). Clearly the acceleration in a weaker gravitational environment requires a much more extended region of a diverging magnetic field to acquire relativistic velocities. The main conclusion from Figure 6 is that a strong gravitational binding energy will enable a force balance such that particles may acquire high-to-relativistic energies in a comparatively short distance from the central body. Massive magnetized celestial objects are therefore ideal sites for the acceleration of charged particles (cosmic rays) to extreme energies.

The bar at $r/R_0 = 10$ in Figure 6 represents the modeled acceleration of H$^+$ ions by ponderomotive forcing (≈ 0.8 keV) to a distance near the dayside magnetopause. Since the acceleration process is velocity dependent [equations (5) and (10)], we note that the energy acquired for O$^+$ will be 16 times higher, i.e. 13 keV. Thus, whenever outflowing ions from the Earth are observed at equal velocities, they may have been subject to acceleration by ponderomotive forcing. Figure 7 shows data from the Cluster CIS ion experiment [*Reme et al.*, 2001] from an inbound traversal of the Earth's dayside magnetopause ($\approx 20:07$ UT). The spacecraft subsequently traversed polar cusp magnetic field lines, eventually accessing the high-altitude polar region. Throughout the entire period inside the magnetosphere, and to some extent even outside the magnetopause, one finds out-flowing ionospheric ions. The lowermost panel illustrates that parallel velocities are similar for the three ion species over an extended period of time, the major discrepancy found in the innermost part, where statistics are poor for He$^+$. Equal parallel outflow velocities for H$^+$, He$^+$, O$^+$, gradually increasing their energy/velocity with increasing altitude, is the signature of ponderomotice force acceleration.

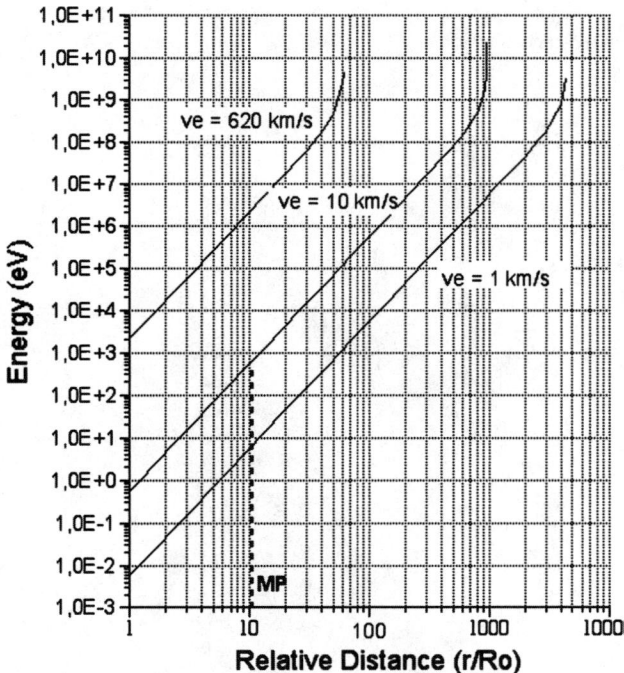

Figure 6. Diagram of the ponderomotive force acceleration of H$^+$ ions in a dipole magnetic field and with an E/B ratio of 1. The upper curve illustrates the faster acceleration curve within a high-gravity force environment (e.g. the Sun). The lower curves give the acceleration in lower gravity environments (10 km/s is near the Earth's escape velocity). The bar at 10 R_0, marked MP, corresponds to the acceleration of H$^+$ ions up to the Earth's magnetopause (Figure 7).

3. PLASMA OUTFLOW FROM STARS AND GALAXIES

In the previous section we discussed plasma outflow as a result of primarily ponderomotive forcing, neglecting other possible outward plasma forcing mechanisms from a celestial body, for instance parallel electric fields. However, the advent of the MMP-ponderomotive force is that, in principle, it affects all charged particles unidirectionally, thereby preserving charge neutrality. Moreover, the MMP-force is always in a direction of decreasing magnetic field strength, i.e. away/outward in a dipole magnetic field.

Having concluded that the structure of the plasma outflow represents an imprint of the magnetic structure (e.g. Figures 1–3), it is possible to infer the magnetic structure also from

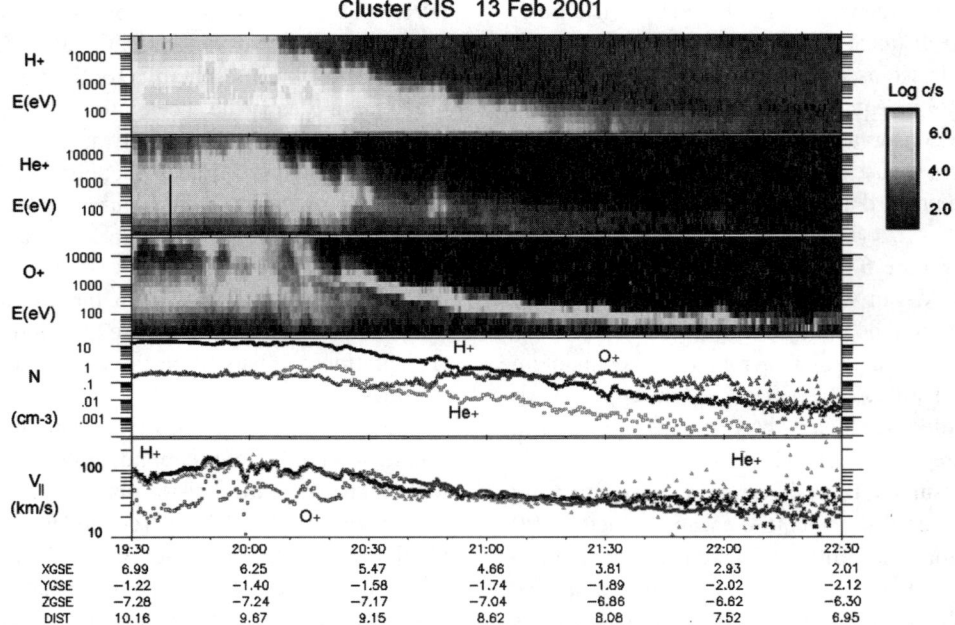

Figure 7. Cluster CIS ion data from an inbound traversal of the magnetopause (≈20:07 UT) subsequently encountering high-altitude polar ionosphere ion outflow. Upper panels show time–energy spectrogram of H^+, He^+, and O^+. The two bottom panels show density (N) and field-aligned flow velocity ($V_{\parallel}$) of H^+, He^+, and O^+, the lowest panel illustrating that upgoing ions have been subject to continuous velocity-dependent acceleration.

other, more remote celestial objects in the Universe characterized by plasma outflow. Notice that the fundamental basis for our interpretation of the objects is comparative magnetosphere physics. Although the knowledge of plasma outflow today stands on a relatively mature scientific base, in situ measurements, such a base is not possible for remote astrophysical objects. From magnetospheric physics we have learned that the "subvisual" outside remains unknown until accessed. For instance, both the Heliosphere extension into interstellar space and the nature of the effect of interstellar matter on the Heliosphere, remain open issues. Nonetheless, comparative magnetosphere physics is so far the best method to describe an environment characterized by plasma embedded in the magnetic field, intrinsic or induced, near a celestial body. In what follows, the magnetosphere analogy will be utilized on astrophysical objects.

Figure 8 (negative image) shows an example of a newborn star (IRAS 04302+ 2247), where the dark plumes extending radially outward from the central inclined white lane represent a massive outflow/loss of matter. The star itself is hidden behind a dust ring (white lane). The outflow of plasma and matter from the polar region of the star is here considered to be due to acceleration and outflow along the magnetic field lines. The magnetic field lines are tentatively inferred. An intrinsic dynamo similar to that of, e.g., the Sun and Jupiter produces the magnetic dipole field of the star. The

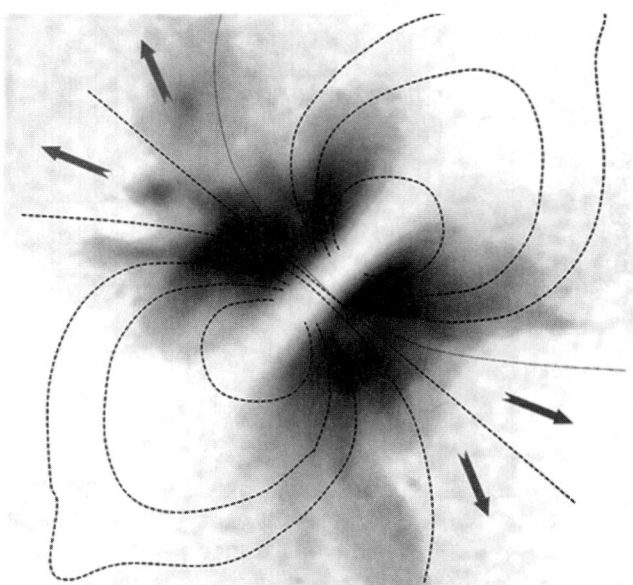

Figure 8. Outflow from a newborn star (IRAS 04302+ 2247). A dust belt orbiting the stellar system hides the visible star. The outflow of plasma and matter from the polar region suggests acceleration along the magnetic field. The magnetic field lines are tentatively drawn to mark the dipolar characteristics of the stellar magnetosphere. (Photo credit: D. Padgett, IPAC/Caltech; W. Brandner, IPAC; K. Stapelfeldt, JPL; and NASA.)

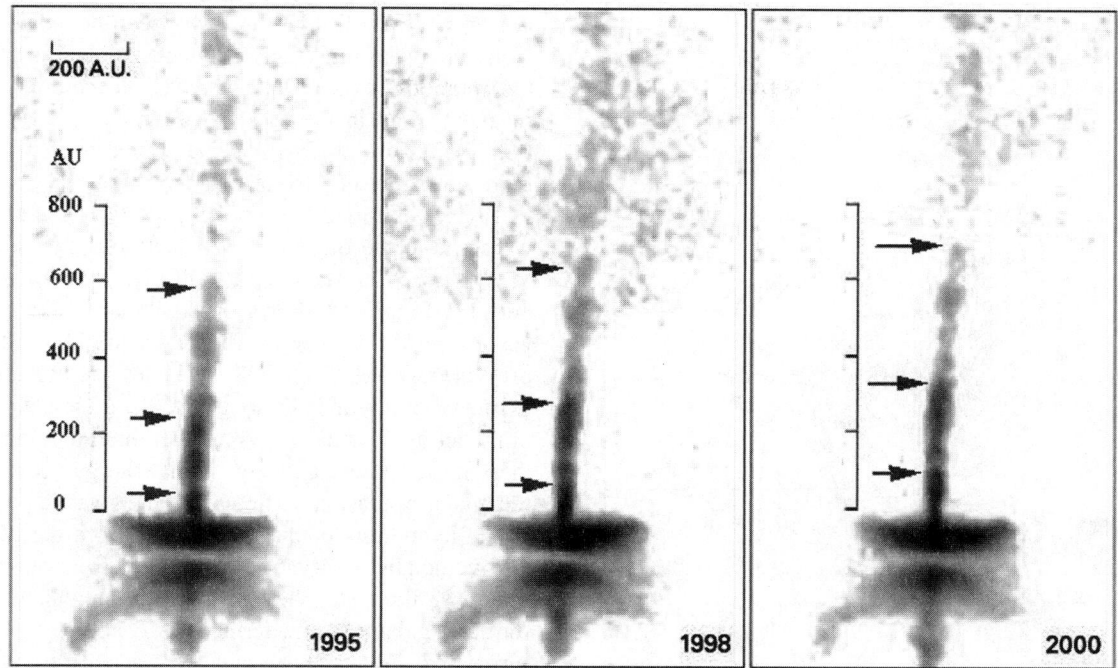

Figure 9. Interpretation of the outflow from a stellar Herbig Haro object (HH 111). A strong magnetic dipole is inferred, the outflow concentrated to a narrow region along polar magnetic field lines. Arrows indicates tracking of individual features of the plasma outflow (≈70–100 km/s)

fact that the outflow covers a wide latitude range implies a strongly perturbed or moderately strong magnetic dipole field.

Figure 9 shows another example of a newborn star, a Herbig Haro object (HH 30) characterized by narrow jets extending from the star. The difference between the HH 30 and IRAS 04302+ 2247 outflow probably reflects the intrinsic magnetic field environment. The narrow jets for the HH 30 object may be due to a stronger magnetic moment combined with the expulsion of a high-beta plasma. Whether this difference is due to mass, composition, or age is beyond the scope of this paper. Plasma jets from magnetized objects are expected to trace magnetic field lines, except when the beta ($\beta \gg 1$) or the kinetic pressure of the plasma is high [e.g. *Lindberg* 1978]. In the case of HH objects the jets are narrowly confined and straight, suggesting that the jets emerge from the magnetic poles. Magnetospheric physics tells us that a narrow region of polar plasma outflow requires the intrinsic magnetic dipole to be much stronger than the externally induced magnetic fields. For instance, the Earth's auroral region and the corresponding ionospheric plasma outflow narrows during times of low external forcing/low magnetic activity [e.g. *Yau et al.*, 1985]. Thus, one interpretation of the existence of confined bidirectional outflow jets in HH objects is that low-altitude plasma acceleration takes place in a strongly magnetized newborn star.

It is generally believed that planetary and stellar magnetic dipole fields result from a dynamo driven by the intrinsic rotation. *Russell* (1978) demonstrated that the magnetic moment is essentially proportional to the angular momentum of the magnetized planets in our solar system. In a similar manner, the strength of a potential galactic magnetic dipole is related to the angular momentum of matter in the central part of the galaxy. A simplistic view is that of a dipole field driven by the differential motion of plasma near the galactic center. In fact, based on the hetegonic nebulae theory by *Alfvén and Arrhenius* (1976) and *Alfvén* (1981), the magnetic dynamo and the transfer of angular momentum to the environment emerge before the accretional phase of the planets. *Alfvén and Arrhenius* (1976) argued that the early Sun was characterized by a strong intrinsic magnetic field (fast rotation), which eventually weakened while transferring angular momentum to planetary matter.

One may hypothesize that the Alfvén and Arrhenius scenario also applies for galaxies; i.e. galaxies are formed by initially huge plasma vortices that deliver angular momentum to star-forming matter in their neighborhood. This implies that newborn galaxies should be characterized by strong intrinsic dipole magnetic fields, and plasma outflow should be similar to that of newborn stars—only vastly different in size. Similarly, middle-aged galaxies would magnetically resemble middle-aged stars; i.e. when the spin of the galactic

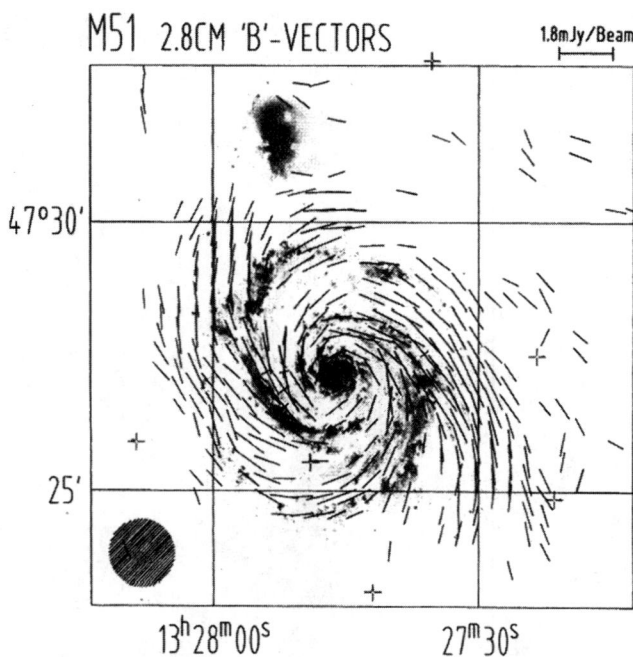

Figure 10. The spiral magnetic field within the M 51 galaxy. Notice that Parker-like spirals trace the visual galactic arms, suggesting a close relation between the two [after *Neininger*, 1992].

center slows down, the magnetic dipole weakens and the angular momentum is successively transferred to stars orbiting the galaxies.

An example of an elliptic galaxy, M 51, viewed from above the galactic pole is shown in Figure 10 [after *Neininger*, 1992]. Radio polarization measurements have established the orientation of the magnetic field, which in this case is spiraling outwards from the galactic center. The analogy with the Parker spiral in the Heliosphere is evident, although the Parker spiral is generated by the outward expansion of plasma with the solar magnetic field "frozen" into it. Notice that the magnetic field is aligned with the galactic arms, indicating a close connection between the two. Exactly why the magnetic field aligns with the galactic arms remains an intriguing quest for future research.

In a similar manner one may also obtain an equatorial perspective of the magnetic field of galaxies based on polarization measurements [e.g. *Beck*, 1990]. As expected, magnetic field vectors are predominantly directed out of the equatorial plane. Also, our own galaxy, the Milky Way, displays a large-scale dipole-like magnetic field component emerging from the center of the galaxy. Figure 11 shows the complexity of the magnetic field from the Milky Way as determined from polarization measurements [after *Matthewson and Ford*, 1970]. The direction of the magnetic field suggests a fuzzy magnetic dipole, its tentative magnetization vector being marked by the arrow near the galactice center. Notice that the inferred "magnetic dipole" is offset from the rotational axis of the galaxy. The complexity of the magnetic field in the Milky Way compared to the dipole field of individual stars may be related to that of galaxies representing ensembles of magnetized stars.

Having identified the dipole property of the Milky Way, we come to the other aspect of celestial magnetic dipoles as discussed for stars (e.g. Figures 8 and 9), the plasma outflow. Figure 12 shows observations of low-metallicity gas in the Milky Way [B. Wakker and NASA, University of Wisconsin-Madison, USA]. Their interpretation of the existence of low-metallicity gas is that it indicates accretion of matter by the galaxy. Yet another interpretation, inferring the magnetosphere analogy, is that the gas constitutes the remainder of plasma outflow from the galactic polar region.

The hypothesis of plasma outflow from the Milky Way is governed by observations of outflow from other galaxies. Figure 13 illustrates the outflow of plasma/matter from the starburst galaxy M 82, termed by *Lehnert et al.* (1999) a superwind. Dashed lines and arrows mark the anticipated (dipole) magnetic field and the corresponding plasma outflow guided by the magnetic field, respectively. The dipolar outflow of plasma and neutral gas from the central part of the galaxy is given by the dashed lines extending out from the galactic poles. The M 82 represents as mentioned a "starburst" galaxy, i.e. a galaxy consisting most probably of a large ensemble of young stars such as those described by Figures 8 and 9.

Our final example represents the outflow from an elliptic galaxy, M 84. The right-hand part of Figure 14 is an image that combines visual (center) and radio-wave (jets) measurements, the radio-wave image displaying the polar beams. The left-hand part in Figure 14 displays the apparent magnetic field vectors obtained from polarization measurements of the entire M 84 nebula [after *Laing and Bridle*, 1987]. The existence of narrow outflow jets from the center of the galaxy suggests (from a magnetosphere point of view) a strong intrinsic magnetic dipole, in which the beams (high-beta plasma?) emerge from the magnetic poles. From a cosmic plasma point of view, an elliptic galaxy in its early state may be considered a huge "Alfvén vortex". An "Alfvén vortex" may be defined as a fast-rotating plasma cloud, its shape maintained by a strong dipole magnetic field in the center [*Lundin and Marklund*, 1995]. Under the assumption that the magnetic field is "frozen" into the out-flowing plasma, the fast rotation also leads to a spiraling of the magnetic field in the outflow jets; i.e. the bipolar outflow jets constitute a magnetic flux rope. Considering the object from a three-dimensional perspective, and assuming that the bottom jet is located closer to the observer, we believe the

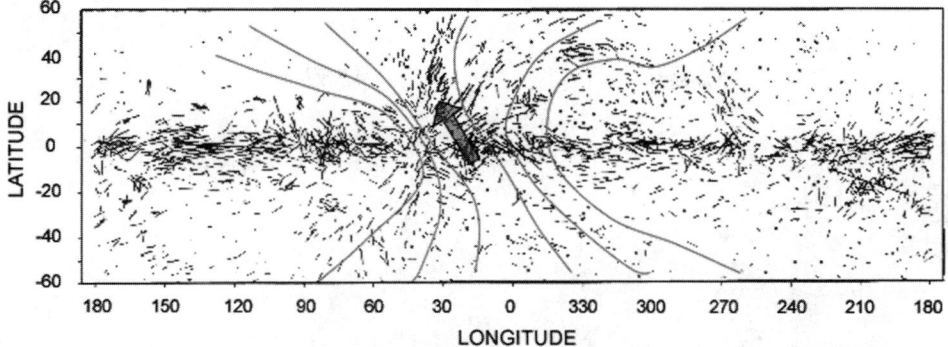

Figure 11. Polarisation measurements of the Milky Way magnetic field, the tentative magnetic dipole inferred by the arrow, and the field lines. After *Matthewson and Ford* [1970].

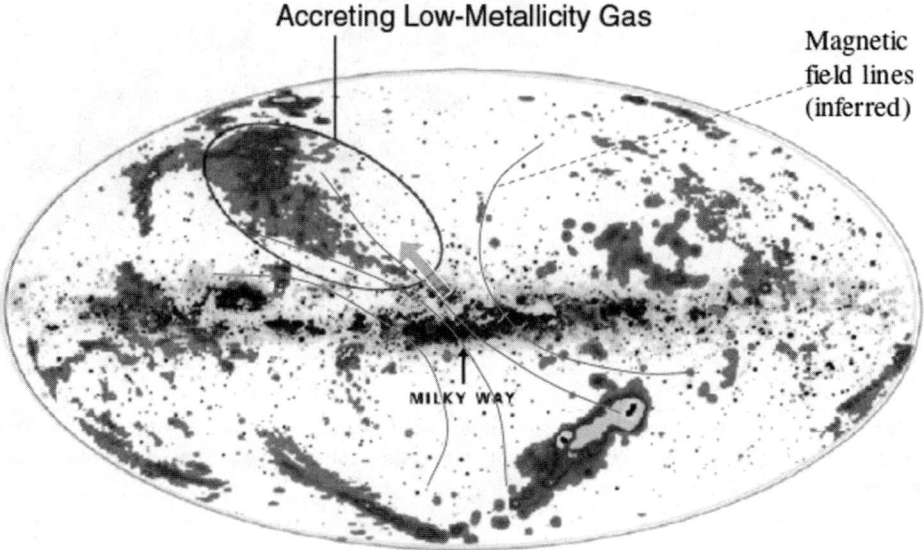

Figure 12. Observations of low-metallicity gas in the Milky Way, here interpreted as polar region plasma outflow rather than an accretion of matter. Picture credit: NASA and B. Wakker, University of Wisconsin-Madison.

magnetic field vector measurements by *Laing and Bridle* [1987] corroborate our interpretation of a magnetic field in the jet curled by the rotating central body. Notice that the outer plumes of the outflow jets make up big magnetic swirls, with the lower plume having a view directly into the beam center.

4. DISCUSSIONS AND CONCLUSIONS

The outflow of plasma and neutral gas from stars and galaxies is a topic of increasing interest and concern among the space science and astrophysics communities. Various ideas about the astrophysical jets are considered within the astrophysics community [e.g. *Reipurt and Bally*, 2001] but "traditional" astrophysical concepts seem to fall short of conceivable models for the fast and focussed ejection of matter, unless magnetic fields and plasmas are also inferred. Early theories and models of magnetic field collimation of astrophysical jets were proposed by, e.g., *Blandford* [1976] and *Lovelace* [1976], but little progress was made during the succeeding ≈20 years. To some extent this may be related to interdisciplinary knowledge transfer constraints. Today the awareness of the overall morphological consequences of instrinsic magnetic fields of celestial objects has increased [*Nishikawa et al.*, 1997; *Celotti et al.*, 1999; *Eichler*, 2003; *Celotti*, 2003]. However, the role of the magnetic field in the jet-acceleration process is less obvious. A typical example of an object where new observations and plasma modeling are expected to have a major implications on astrophysics is the pulsar [*Lyubarsky and Eichler*, 2001]. Recent plasma simulations made by *Komissarov and Lyubarsky* [2004] of the magnetized pulsar wind and the associated synchrotron

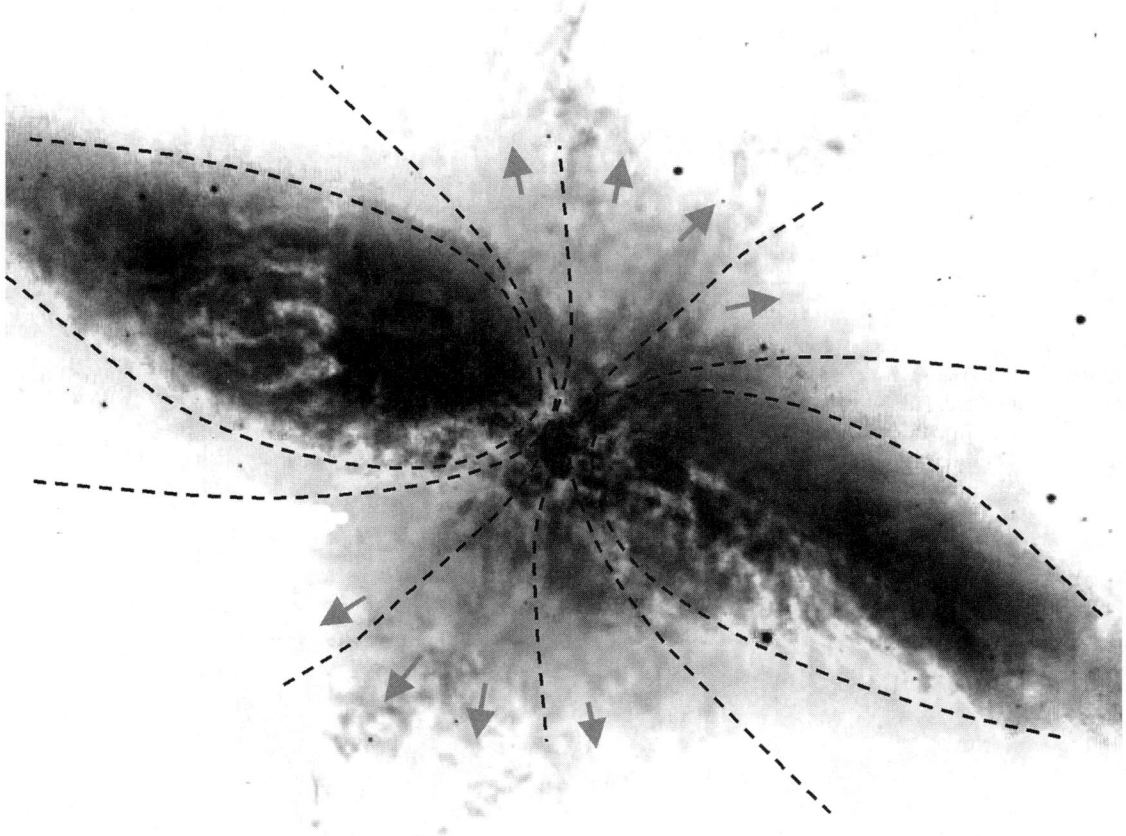

Figure 13. Image of the starburst galaxy M 82 with anticipated magnetic field and plasma outflow. The structured emissions (dashed lines) extending out from the galactic poles represent the bipolar outflow of plasma and gas from the central region of the galaxy. Source: *Lundin* [2003].

nebulae are enlightening examples of what to expect from studies and simulations of astrophysical objects. Figure 15 is an image from simulations by Komissarov and Lyubarsky, illustrating the dipole magnetic field and the narrow bipolar beams of a pulsar.

While MHD-modeling is useful in describing the overall magnetized plasma topology and flow of astrophysical objects, the acceleration of plasmas is more readily treated by kinetic theory. Non-thermal escape characterized by acceleration and outflow of plasmas from the Earth's ionosphere has been a major theme in space plasma physics for over 30 years. In this report we focused on one specific process for non-thermal plasma escape, ponderomotive forcing by electromagnetic waves (magnetic moment pumping, MMP), as a cause of outflow from stars and galaxies. Outflow of plasma by MMP is consistent with observations in the Earth's magnetosphere [e.g. *Lundin and Hultqvist*, 1989; *Hultqvist*, 1996] but should instead be considered a basis for understanding the erosion of plasma from stars and galaxies.

Qualitatively, the hot plasma in the stellar corona is likely to contain sufficient energy (electromagnetic waves) to support upward acceleration. The heating of the solar corona to $>10^6$ K is an example in itself of wave heating of plasmas. Directional (upward) acceleration is then a consequence of further heating and outflow in diverging (e.g. dipolar) magnetic fields, the acceleration being driven by a zoo of localized plasma wave processes (e.g. MHD, electrostatic, electromagnetic). MMP forcing constitutes a unidirectional parallel acceleration of plasma in the direction of the magnetic divergence, a forcing that is independent of the wave propagation direction. Also, waves propagating in the opposite direction of the plasma outflow can contribute to the acceleration of plasma in diverging magnetic fields. Plasma escape caused by MMP is therefore equally valid for internal as well as for external forcing on celestial bodies. Ponderomotive MMP may accelerate particles up to relativistic velocities in quite short distances in a magnetic dipole field, provided the gravitation field is sufficiently strong. Properly

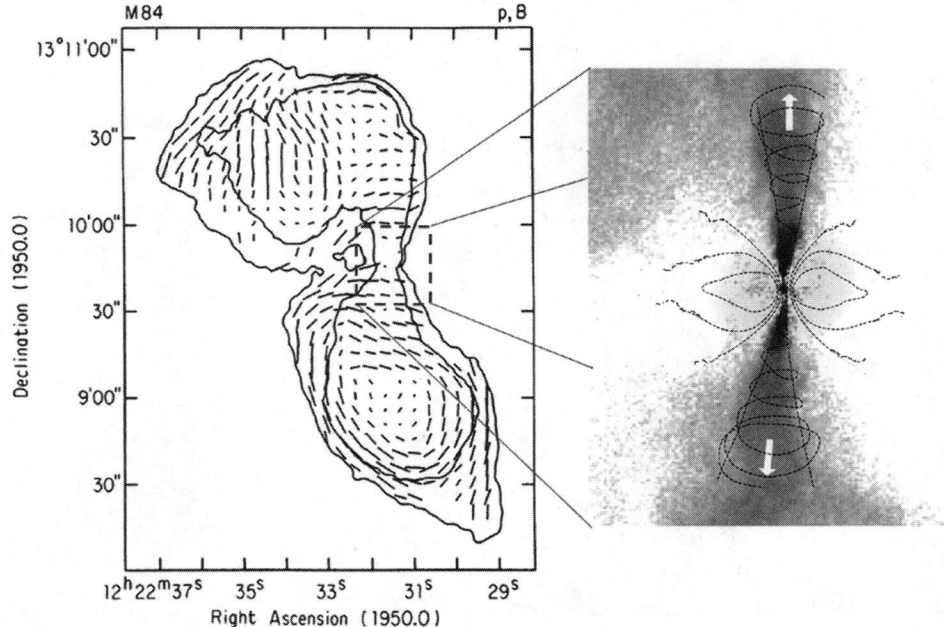

Figure 14. (Right) Image of the radio galaxy M 84 from a magnetosphere point of view with an intrinsic magnetic dipole at the center of the galaxy and bipolar plasma outflow with field-aligned current-induced perturbation (flux rope); image from NRAO/AUI, Aland Bridle. (Left) Apparent magnetic field distribution obtained from polarization measurements (*Laing and Bridle*, 1987).

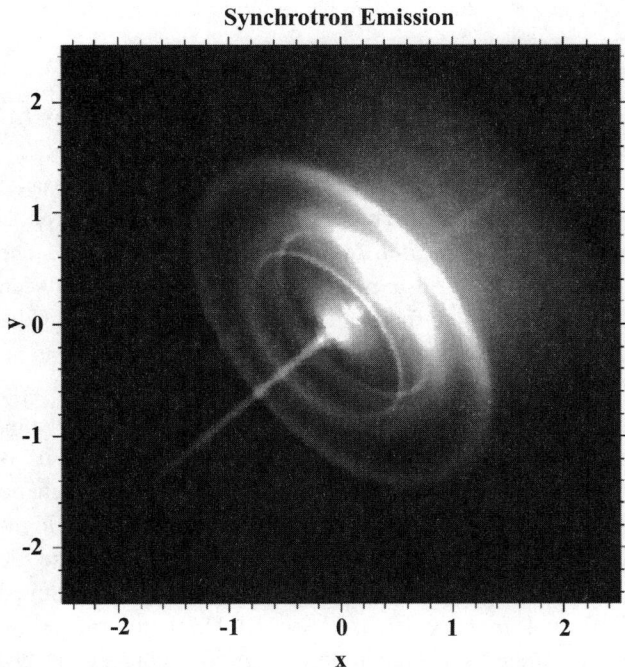

Figure 15. MHD-simulation of the Crab Nebula with magnetic field at the axis (*Komissarov and Lyubarsky*, 2004).

balanced, strong gravity and MMP-forcing provide a potential well for "pre-energization" to escape velocity, narrowing the altitude range needed to acquire relativistic velocities. In the case of the Sun, relativistic energies (≈ 1 GeV for protons) can theoretically be obtained in 100 solar radii (Figure 6).

A quantitative energy and momentum analysis of the outward acceleration of plasma is beyond the scope of this paper. However, a simple qualitative evaluation of the energy and momentum exchange, based on the assumption that wave energy is fed from below (internal forcing, Fig. 16), would be of some interest. The wave energy is assumed to successively energize the plasma, but as soon as the wave energy is consumed at a critical altitude, Z_{cp}, acceleration stops. Provided escape velocity has been reached, the accelerated plasma will propagate/expand further upwards at an approximately constant velocity. The denser the plasma and the lower the wave energy per unit volume, the lower the Z_{cp}. From this point of view, dense plasma acts to deplete wave energy in the plasma acceleration process. Besides accelerating plasma, therefore, wave energy propagating along a funnel of magnetic field lines may create a plasma density cavity, its width determined by the width if the wave duct. Because the efficiency of wave acceleration is directly proportional to the plasma density gradient [equation (8)],

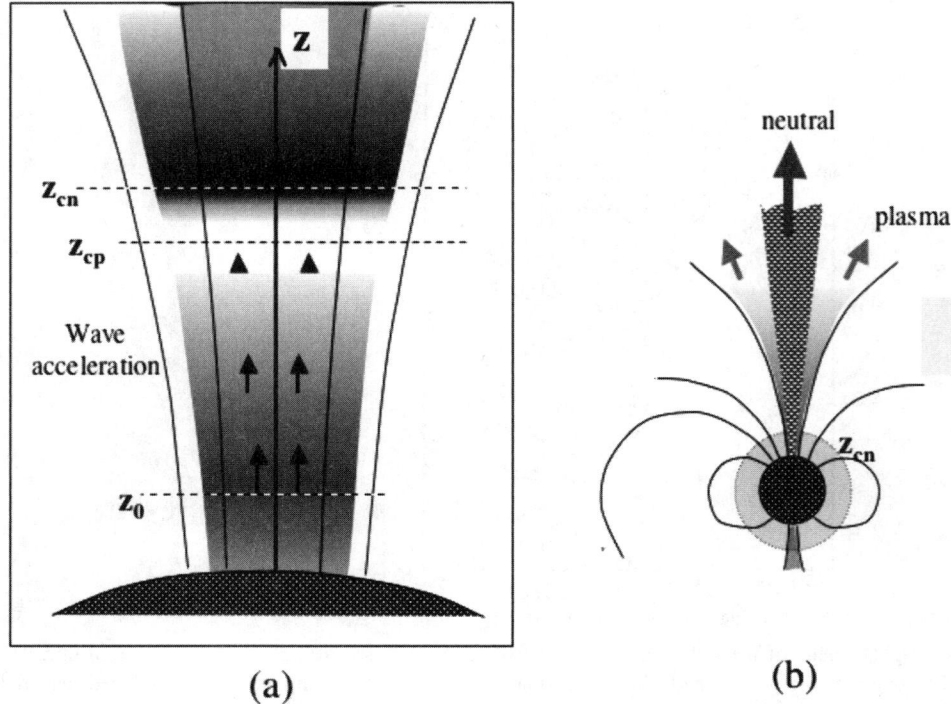

Figure 16. A simple diagrammatic illustration of polar region acceleration of plasma and neutral particles or gases due to wave energy propagating outward from a central body. (a) Closeup view of the wave-driven acceleration between Z_0 and a critical altitude Z_{cp}. Above Z_{cp} the wave energy is absorbed and no further plasma acceleration occurs. The second critical altitude, Z_{cn}, defines a momentum exchange, drag, boundary between ions and neutral particles/gases. Above Z_{cn} ions and neutral materials flow with the same velocity. (b) Illustration of how a dense central jet may be formed near the magnetic pole when the acceleration occurs at low altitudes in a strong magnetic dipole field (high density → low Z_0 → narrow beam). For a tenuous plasma the acceleration continues toward higher altitudes (high Z_c) and traces the magnetic field lines.

the ducting will enhance acceleration and a deep plasma density cavity may form, as shown in Figure 5. Such deep plasma density cavities have been reported from satellite measurements above the auroral oval [e.g. *Calvert*, 1981; *Lundin et al.*, 1991]

Another mass-loading mechanism is associated with the neutral gas and dust surrounding newborn stars. Ion-neutral collisions and charge-exchange will transfer energy and momentum from the plasma to the neutral gas. Ideally, both ions and neutrals eventually achieve the same velocity above Z_{cn}. Ion-neutral collisions in the acceleration region will also lower Z_{cp}. If the neutral jet outflow is driven by plasma acceleration, the neutral jet width will be given by the local magnetic field flaring angle—the lower the altitude of Z_{cp}, the narrower the jet. Heavy mass loading of the outflow from strongly magnetized newborn stars is therefore likely to be associated with narrow jets and relatively low outflow velocities (e.g. HH 30, Figure 9). Once cleared of "debris", the stellar wind may reach higher velocities (as in the solar wind, $V_{sw} \approx 300$–1000 km/s), and the outflow jet may become broader (Figure 8).

The magnetic field is therefore expected to play a crucial role for the acceleration processes as well as for the morphology of the plasma outflow from stars and galaxies. The strength of the intrinsic magnetic field, the relation between the intrinsic dipole field and the ambient/induced magnetic field, the plasma beta of the outflow, the neutral gas content, and the critical altitude Z_{cp} for the plasma acceleration are expected to be decisive for the outflow beam/jet focusing. Narrow and extended jets (HH objects) would, on the basis of these arguments, characterize objects with ideal magnetic field topologies and low critical altitudes. Notice that a high-density, high–kinetic energy plasma beam may oppose the magnetic field control and continue along a predetermined direction [*Lindberg*, 1978; *Alfvén*, 1981]. Therefore, narrow jets from stellar magnetic poles may also constitute high-beta plasmas. On the other hand, objects with wider escape plumes (e.g. Figure 8) may be associated with, e.g., lower intrinsic dipole fields and higher critical altitudes (for example, the Sun).

Dipole magnetic fields play an essential role in the hetegonic nebula theory of *Alfvén and Arrhenius* (1976). The stellar plasma environment is frequently controlled by strong, quasi-static dipole magnetic fields. A dipole magnetic field may play an important role for the transfer of angular momentum from fast-rotating newborn stars to planetary nebulae. The hetegonic nebula theory is based on an early ionization (critical velocity ionization) and plasma momentum exchange governed by field-aligned electric currents connecting to the central body (dynamo). Alfvén and Arrhenius also discussed the outflow (superprominiscences) coupled to field-aligned currents. Our focus in this report was on the outflow process (non-thermal escape) from a morphological point of view (polar jets) without connecting them to field-aligned electric currents. However, we also strongly believe that electric currents are important for the acceleration processes as well as for the morphology of the outflow. A good illustration of this is the outflow from the galaxy M 84.

In summary we conclude:

- Non-thermal escape is a process capable of sustaining significant outflow from celestial objects.

- The non-thermal escape process is governed by internal forcing, external forcing, or both.

- Magnetic moment pumping, MMP, is a viable cause of plasma escape to equal velocities in a diverging magnetic field geometry. MMP is independent of wave propagation direction; i.e. parallel acceleration is in the direction of a diverging magnetic field. MMP promotes escape in, e.g., a dipole magnetic field for *external* as well as *internal forcing*.

- The morphology of the outflow is governed by internal and external magnetic fields.

- Outflowing plasma may transfer energy and momentum to neutral matter (ion-neutral collisions and charge exchange), eventually producing a mixed beam of plasma and neutrals going with the same velocity. However, the neutral beam, not being controlled by the magnetic field, consequently jets away independently of the magnetic field.

- The width of neutral jets is determined by the strength of the dipole magnetic field and the critical altitude, Z_{cp}. Narrow jets require strong intrinsic dipole magnetic fields and low Z_{cp}.

Acknowledgments. This paper in comparative magnetosphere physics is the result of many years of stimulating interdisciplinary contacts and discussions. Most cross-disciplinary contacts have been with fundamental plasma physics and planetary sciences. Comparative magnetosphere physics with focus on astrophysical objects, plasma astrophysics, is bound to grow in the years to come, but it may take much longer to reach a wide acceptance for the "Plasma Universe" by Hannes Alfvén some 50 years ago. However, his disciples struggle along. We thank them for their courage and persistence. This work was funded in part by grants from the Swedish National Space Board.

REFERENCES

Alfvén, H., and G. Arrhenius, Evolution of the solar system, NASA Sp-345, 1976.

Alfvén, H., Cosmic plasma, *Astrophys. Space Sci. Library*, Vol. 82, D. Reidel. Dordrecht, Holland, 1981.

Beck, R, Magnetic fields in spiral galaxies, *IEEE Trans. Plasma Sci., 18*, 33, 1990.

Blandford, R.D., Accretion disc electrodynamics—A model for double radio sources, *Mon. Not. R. Astron. Soc., 176*, 465–481, 1976.

Calvert, W., The auroral plasma cavity, *Geophys. Res. Lett., 8*, 919–922, 1981.

Celotti, A., Astrophysical jets, in proceedings from the conference Particle Acceleration in Astrophysical Objects, Krakow, June, 2003.

Celotti, A., and R.D. Blandford, Black holes in binaries and galactic nuclei. Proceedings of the ESO Workshop held at Garching, Germany, 6-8 September 1999. L. Kaper, E.P.J. van den Heuvel, P.A. Woudt (eds.), p. 206, Springer, 2001.

Chappell, C.R., T.E. Moore, and J.H. Waite Jr., The ionosphere as a fully adequate source of plasma for the Earth's magnetosphere, *J. Geophys. Res.* 92, 5896, 1987.

Eichler, D. Plerions and polar jets, in proceedings from the conference Particle Acceleration in Astrophysical Objects, Krakow, June, 2003

Guglielmi, A., and R. Lundin, Ponderomotive upward acceleration of ions by ion-cyclotron and Alfvén waves over the polar regions, *J. Geophys. Res., 106*, 13219–13236, 2001.

Hultqvist, B., On the acceleration of positive ions by high-latitude large-amplitude electric field fluctuations, *J. Geophys. Res., 101*, 27111, 1996.

Hultqvist, B., M. Oieroset, and G. Paschmann (eds.), Magnetospheric Plasma Sources and Losses, Kluwer Academic Publishing, 1999.

Komissarov, S.S., and Y.E. Lyubarsky. Synchrotron nebulae created by anisotropic magnetized pulsar winds, *Mon. Not. R. Astron. Soc., 349*, 3, 779–792, 2004.

Laing, A.R., and A.H. Bridle, Rotation measure variation across M84, *Mon. Not. R. Astron. Soc., 228*, 557–571, 1987.

Lehnert, M.D., T.M. Heckman, and K.A. Weaver, very extended x-ray and halpha emission in M82: Implications for the superwind phenomenon, *Astrophys J., 523*, 2, 575–584, 1999.

Lindberg, L., Plasma flow in a curved magnetic field, *Astrophys. Space Sci,. 55*, 203, 1978.

Lovelace, R.V.E., Dynamo model of double radio sources Authors, *Nature, 262*, Aug. 19, 649–652, 1976.

Lundin, R., The universal importance of auroral research, in J. Moen and J. Holtet (eds.), Egeland Symposium on Auroral and Atmospheric Research, ISBN 82-91853-09-6, 55–74, 2003.

Lundin, R., and B. Hultqvist, Ionospheric plasma escape by high-altitude electric fields: Magnetic moment pumping, *J. Geophys. Res., 94*, 6665, 1989.

Lundin, R., L. Eliasson, G. Haerendel, M. Boehm, and B. Holback, Large-scale auroral plasma cavities observed by Freja, *Geophys. Res. Lett., 21*, 1903, 1995.

Lundin R., and G. Marklund, Plasma vortex structures and the evolution of the solar system—The legacy of Hannes Alfvén, *Physica Scripta, T60*, 198, 1995.

Lyubarsky, Y., and D. Eichler, The x-ray jet in the Crab Nebula: Radical implications for pulsar theory?, *Ap. J., 562*, 494–498, 2001.

Matthewson, D.S., and V.L. Ford, *Mem. R. Astron. Soc., 74*, 143, 1970.

Neininger, N., The magnetic field structure of M 51, *Astron. Astrophys. 263*, 30–36, 1992.

Nishikawa K., S. Koide, J.-I. Sakai, D. Christodoulou, H. Sol, and R.L. Mutel, Three-dimensional magnetohydrodynamic simulations of relativistic jets injected along a magnetic field, *Astrophys. J., 483*, 45–48, 1997.

Rees, M.H., Physics and Chemistry of the Upper Atmosphere, Cambridge University Press, 1989.

Reipurt, B., and J. Bally, Herbig-Haro flows: Probes of the early stellar evolution, *Annu. Rev. Astron. Astrophys., 39*, 403. 2001.

Russel, C.T., Re-evaluating Bode's law on planetary magnetism, *Nature, 272*, 147–148, 1978.

Van Allen, J.A., G.H. Ludwig, E.C. Ray, and C.E. McIlwain, Observations of high intensity radiation by satellites1958 Alpha and Gamma, *Jet Propul., 28*, 588, 1958.

Yau, A. W., E. G. Shelley, W. K. Peterson, and L. Lenchsyschyn, Energetic auroral and polar ion outflow at DE-1 altitudes: Magnitude, composition, magnetic activity dependence, and long-term variations, *J. Geophys. Res., 90*, 1985.

Stanislav Barabash and Rickard Lundin, Swedish Institute of Space Physics, Box 812, SE-981 28 Kiruna, Sweden. (stas@irf.se; rickard.lundin@irf.se)

Anatol Guglielmi, Institute of Physics of the Earth, Russian Academy of Sciences, B.Gruzinskaya, 10, 123810 Moscow, Russia. (gugl@atsmyt.comcor.ru)

Exotic Acceleration Processes and Fundamental Physics

Giovanni Amelino-Camelia

Dipartimento di Fisica, Università di Roma "La Sapienza", Roma, Italy

Gamma-ray bursts and ultra-high-energy cosmic rays provide an important testing ground for fundamental physics. A simple-minded analysis of some gamma-ray bursts would lead to a huge estimate of the overall energy emitted, and this represents a potential challenge for modelling the bursts. Some cosmic rays have been observed with extremely high energies, and it is not easy to envision mechanisms for the acceleration of particles to such high energies. Surprisingly some other aspects of the analysis of gamma-ray bursts and ultra-high-energy cosmic rays, even before reaching a full understanding of the mechanisms that generate them, can already be used to explore new ideas in fundamental physics, particularly for what concerns the structure of spacetime at short (Planckian) distance scales.

1. INTRODUCTION

Much of the work in astrophysics concerns the use of the presently accepted laws of fundamental physics in the development of consistent descriptions of the observations. In some cases the key challenge comes from an energy-balance issue. For example, for some gamma-ray bursts the overall observed fluences can be as large as 10^{-4} ergs/cm^2, which would correspond to a luminosity of 10^{52} ergs/s and higher, if one can assume isotropy. It would be difficult to describe such levels of luminosity in terms of conventional physics. We now understand [*Piran*, 2004] that most gamma-ray bursts are narrowly beamed (not at all isotropic), which lowers the luminosity estimate to a more mundane level; however, there is still no real consensus on the mechanism of emission of gamma-ray bursts.

Another energy-balance issue is encountered in the study of cosmic rays. In fact, recent data [*Takeda et al.*, 1998; *Abbasi*, 2005] suggest that some cosmic rays have energies of 10^{20} eV and higher. And some argue [*Bahcall and Waxmann*, 2003] that at least some of these cosmic rays could originate from within our galaxy. But we are unable to find within our galaxy sources which are good candidates for accelerating particles to such high energies.

In the study of these issues some "exotic acceleration mechanisms" are being considered. Usually the relevant acceleration mechanisms are deemed "exotic" not because of some role for new laws of physics, but rather because of the role played by creative ways to rely on the presently accepted laws of fundamental physics to devise mechanisms that could explain the energetics of the system of interest. Only in a few of the most speculative papers on these subjects it has been argued that new physics might be responsible for the emission mechanism.

While the possibility of new physics is at best a marginal hypothesis for the study of emission mechanisms, surprisingly these fascinating problems in astrophysics do provide, independently of the understanding of the emission mechanisms, some of the best arenas for testing ideas for new laws of fundamental physics, especially for what concerns the structure of spacetime at Planckian distance scales. We do not need to know exactly how gamma-ray bursts are emitted in order to observe that a gamma-ray burst is a very rich signal propagating over very large distances. The analysis of gamma-ray bursts is therefore a wonderful opportunity for high-sensitivity studies of the laws of propagation of signals through space. Similarly we do not need to know exactly how a 10^{20} eV cosmic ray is produced in order to observe that the collisions between such a cosmic ray and the low-energy photons it encounters on the way to our Earth laboratories provide a rare opportunity to test the nature of boost transformations. In fact, the same collisions are studied in our particle-physics laboratories but only relatively close to the center-of-mass frame.

2. ON THE EMISSION OF GAMMA-RAY BURSTS

Even taking into account the fact that most gamma-ray bursts are narrowly beamed, in some cases the inferred luminosity is as high as 10^{51} ergs/s. This is still rather large but does not represent a record-setting energy release, the total energy emitted being comparable to that of supernovae [*Piran*, 2004].

The structure of a gamma-ray burst, viewed from the perspective of signal analysis, is extremely rich, and this of course represents a challenge for emission models. We need a model capable of reproducing faithfully all aspects of that rich structure. A popular idea is the one of the "fireball internal–external shocks model" [*Piran*, 2004], but it is perhaps too early to speak of a general consensus.

An example of the issues that are being discussed in the effort to establish the emission mechanism is provided by the contribution by *Kaneko et al.* {this volume}, which stresses how important insight on the emission mechanism of gamma-ray bursts can be obtained by establishing the proper description of gamma-ray-burst spectra. On the basis of an interesting broadband spectral analysis of two spectrally hard gamma-ray bursts *Kaneko et al.* conclude that gamma-ray-burst spectra evolve on time scales that are much longer than the synchrotron cooling time. This may be consistent with some arguments discussed in the gamma-ray-burst literature (see, e.g., *Piran* [1999, 2000]), but *Kaneko et al.* argue instead that this should lead to the conclusion that the acceleration mechanism is more complicated than predicted by some popular gamma-ray-burst models.

3. GAMMA-RAY BURSTS AS A QUANTUM SPACETIME LABORATORY

Gamma-ray bursts have recently attracted much interest (see, e.g., the recent reviews by *Amelino-Camelia* [2000a, 2000b], *Sarkar* [2002], and *Mavromatos* [2002]) also from the research community exploring the hypothesis that spacetime might have to be "quantized", i.e., that there might be a quantization of spacetime observables in a way somehow analogous to the quantizations of all other observables encountered in other aspects of fundamental physics.

Some scenarios for spacetime quantization involve a sort of granularity of spacetime, and as a result one may expect some departures from the smooth laws of Lorentz symmetry [*Amelino-Camelia et al.*, 1998]. The mechanism would be analogous to the one that applies to phonons (some particle excitations encountered in certain condensed-matter systems): The law of propagation of phonons is formally relativistic at low energies, when the reticular structure of the underlying material can be ignored, but at high energies there are departures from the relativistic behaviour. If spacetime itself were granular, then some analogous effect might be present. A sort of light-cone fuzziness is essentially inevitable in the presence of spacetime granularity, and recently it was realized that some specific schemes for introducing the granularity length scale may also affect the propagation of photons by introducing a small dependence of the speed on the photon energy.

These effects on photon propagation are expected to be very small, since their magnitude is set by the ratio of the photon energy over the Planck energy scale ($\sim 10^{28}$ eV); however, as I stress in my contribution elsewhere in this volume, through the analysis of observations of gamma-ray bursts one has a chance to discover (or rule out) these Planck-scale effects. The properties of gamma-ray bursts used in this type of analysis are, as discussed in *Amelino-Camelia et al.* [1998] and *Biller et al.* [1999],

- the fact that gamma-ray bursters are often at cosmological distances,
- the fact that a typical gamma-ray-burst spectrum should extend up to the tens of MeV and higher,
- and the fact that some "microbursts" within a gamma-ray burst can have very short duration, as short as 10^{-3} s (or even 10^{-4} s).

The key point I want to stress here is that this use of gamma-ray burst for quantum spacetime studies is largely insensitive to the nature of the emission mechanism. The properties of gamma-ray bursts that are used are well established, and the analysis is largely independent of the modelling of the emission mechanism. So the convergence of interest on gamma-ray bursts from the exotic acceleration mechanisms community and the quantum spacetime community is accidental.

4. ON THE MOST ENERGETIC COSMIC RAYS

Several mysteries surround the observation of cosmic rays, particularly the "ultra-high-energy (UHE) cosmic rays", with energies higher than 10^{19} eV. The identification of the particles is problematic, since we observe them indirectly through their interactions in the atmosphere. It appears, however, that most UHE cosmic rays are protons.

Most UHE cosmic rays are believed to be of cosmological origin, but this then should imply that the so-called "GZK cutoff" [*Greisen*, 1966; *Zatsepin and Kuzman*, 1966] should be observed: the spectrum of observed cosmic rays should basically stop around $E_{gzk} \simeq 5 \cdot 10^{19}$ eV, where photons in the Cosmic Microwave Background (CMB) become viable targets for photopion production. A cosmic ray that starts its journey with energy higher than E_{gzk} should lose rather rapidly the energy in excess of E_{gzk} in the form of pions. There is great interest in recent observations of cosmic rays with energies beyond the GZK cutoff [*Takeda et al.*, 1998; *Abbasi*, 2005]. *Bahcall and Waxman* [2003] have suggested that perhaps some of the UHE cosmic rays originate from within our galaxy. This would allow them to evade the GZK cutoff, since over "short" (galactic) distances the expected energy loss through photopion production is negligible. But then we should identify within our galaxy some sources that could accelerate protons to such high energies.

So there are issues of interest for the analysis of acceleration mechanisms also in the context of cosmic-ray studies, although of a rather different nature with respect to the case of gamma-ray bursts.

5. COSMIC RAYS AS A QUANTUM SPACETIME LABORATORY

As I stress in my contribution in this volume, the fact that some cosmic-ray observatories have reported above-GZK events has also generated strong interest from the quantum spacetime research community [*Kifume*, 1999; *Aloisio et al.*, 2000; *Amelino-Camelia and Piran*, 2001]. This interest originates from the observation that spacetime granularity, besides affecting the laws of particle propagation, can also affect the energetic balance of particle physics processes. The conventional estimate of the GZK cutoff implicitly assumes that the process of photopion production would occur in an exactly smooth classical spacetime. The GZK scale is set by the minimum energy that, in classical spacetime, is required of a proton to produce a pion in a collision with a photon of CMB energy. In some quantum spacetimes this estimate of the minimum energy is shifted upward by a Planck-scale effect. Of course if the "quantum spacetime GZK scale" is higher than that estimated classically, it would be natural to expect some cosmic-ray observations that are above the classical GZK scale.

Essentially these quantum spacetime–inspired studies are using the cosmic-ray context to probe a regime of high boosts that is not accessible in laboratory experiments. The photopion production process, $p + \gamma \rightarrow p + \pi$, is well understood and studied in the laboratory at center-of-mass energies comparable to those available in collisions between a 10^{20}eV proton and a CMB photon, but in our laboratories (when the center-of-mass energies are so high) we are able to study the process only in frames that are not highly boosted with respect to the center-of-mass frame. In the observation of UHE cosmic rays we are instead observing (some consequences of) the photopion production process in a frame that is highly boosted with respect to the center-of-mass frame, involving indeed an extremely hard proton and a very soft photon. The nature of boosts is a one-to-one relation [*Amelino-Camelia, this volume* with the short-distance structure of spacetime, and therefore it is not surprising that these studies would be of interest for the quantum spacetime community.

Again, I stress that the convergence of interest on a problem (in this case the cosmic-ray spectrum problem) by the "acceleration mechanisms community" and the "quantum spacetime community" is accidental. The new physics associated with spacetime quantization is not being advocated as a way to explain the high energies reached by cosmic rays. Instead, the quantum spacetime community simply uses the experimentally established fact that some cosmic rays have huge energies. From a quantum spacetime perspective it does not matter how cosmic rays are accelerated to such high energies; it is just a wonderful opportunity that such high-energy particles are available for study.

6. INTERPLAY BETWEEN THE UNDERSTANDING OF THE EMISSION MECHANISMS AND THE QUANTUM SPACETIME STUDIES

To this point, the fact that the growing number of instances in which the interests of the "acceleration mechanisms community" and the "quantum spacetime community" converge is largely accidental needed to be stressed because readers who have not been following closely the development of these fields might quickly assume that the new physics of quantum spacetime is being advocated as a way to devise new acceleration mechanisms—which is usually not the case. However, I should also stress there are some issues that require clarification from an acceleration mechanism perspective that are quite crucial for the success of the quantum-spacetime studies.

To illustrate this point, consider a specific example that is relevant for the gamma-ray-burst studies mentioned in sections 2 and 3 above. Basically, the quantum spacetime interest in gamma-ray-burst observations originates from possible attribution by the new Planck-scale laws of particle propagation of different speeds to photons of different energies. The fact that we see some microbursts within a given burst that reach different energy channels of our detectors at the same time, within the accuracy available at our observatories, allows us to set limits on this energy dependence of the speed of photons. The Planck-scale effect could be "discovered" if future, more sensitive observatories eventually show this energy dependence. The analysis would be severely affected, of course, if at-the-source correlations between energy of the photons and time of emissionwere poorly understood. As observed recently [*Piran,* in press], one can apparently infer such an energy/time-of-emission correlation from gamma-ray-burst data. The quantum spacetime studies will be therefore confronted with a severe challenge to remove background noise. It will be crucial for the quantum spacetime analysis to have available a reliable description of the emission mechanism, which could allow removal of the undesired at-the-source effect.

7. OTHER AREAS OF FUNDAMENTAL PHYSICS

I have so far focused on the fact that some areas of interest from the acceleration mechanism perspective are also of interest for the investigation of certain quantum spacetime scenarios. However, some of the relevant phenomena are also relevant for other types of fundamental physics studies. Again, to illustrate this point, I use a specific example relevant for the cosmic-ray studies mentioned here in sections 4 and 5.

For instance, although the most energetic cosmic rays are most likely protons, this does not necessarily imply that they are emitted as protons. We infer that they are protons on the basis of the nature of their interactions in the atmosphere, but plausibly the cosmic ray might have started as some other particle, which then decays into a proton at a relatively small distance from the Earth. This would also be another way to describe the observations of cosmic rays with energies higher than the GZK scale: Such a cosmic ray might well be originally some exotic particle, one that does not lose energy through interactions with CMB photons; moreover, this particle might then decay into a proton (plus other particles) at only a relatively small distance from the Earth, when the residual time of travel is not sufficient for substantial energy loss through interactions with CMB photons.

Various new types of particles that are independently of interest from a particle physics model-building perspective have been considered in this cosmic-ray context (see, e.g., *Anchordoqui et al.* [2003]; *de Vega and Sanchez* [2003]; *Ellis et al.* [2004]). Progress in our understanding of the cosmic-ray spectrum could provide insight on these new particle physics scenarios.

REFERENCES

Abbasi, R. U., astro-ph/0407622. *Astrophys. J. 622* (2005) 910.

Aloisio, R., P. Blasi, P. L. Ghia, and A. F. Grillo, *Probing The Structure of Space-Time with Cosmic Rays*, astro-ph/0001258 *Phys. Rev. D62* (2000) 053010.

Amelino-Camelia, G., J. Ellis, N. E. Mavromatos, D. V. Nanopoulos, and S. Sarkar, astro-ph/9712103, *Nature 393* (1998) 763-765.

Amelino-Camelia, G., J. D. Bjorken, and S. E. Larsson, hep-ph/9706530, *Phys. Rev. D56* (1997) 6942-6956.

Amelino-Camelia G., and T. Piran, astro-ph/0008107, *Phys. Rev. D64* (2001) 036005.

Amelino-Camelia, G., gr-qc/9910089, *Lect. Notes Phys. 541* (2000) 1.

Amelino-Camelia, G., gr-qc/0012049, *Nature 408* (2000) 661-664.

Anchordoqui, L., T. Paul, S. Reucroft, and J. Swain, *Int. J. Mod. Phys. A18* (2003) 2229; hep-ph/0206072.

Bahcall, J. N., and E. Waxman, hep-ph/0206217, *Phys. Lett. B556* (2003) 1.

Biller, S. D., et al., *Phys. Rev. Lett. 83* (1999) 2108.

Bird, D. J., et al., *Astrophys. J. 424* (1994) 491-502.

Bird, D. J., et al., The Cosmic Ray Energy Spectrum Observed by the Fly's Eye. By HIRES (D. J. Bird et al.) (1994) 12 pp.

de Vega, H. J. and N. G. Sanchez, *Phys. Rev. D67* (2003) 125019.

Ellis, J., V. E. Mayes and D. V. Nanopoulos, hep-ph/0403144, *Phys. Rev. D70* (2004) 1025.

Greisen, K., *Phys. Rev. Lett. 16* (1966) 748.

Kifune, T. astro-ph/9904164, *Astrophys. J. Lett. 518* (1999) L21.

Mavromatos, N. E., hep-th/0210079, in Oulu 2002, Beyond the Desert (2002), 3-28.

Piran, T., astro-ph/9810256, *Phys. Rept. 314* (1999) 575.

Piran, T., astro-ph/9907392, *Phys. Rept. 333* (2000) 529.

Piran, T., astro-ph/0405503, *Rev. Mod. Phys. 76* (2004) 075015.

Piran, T., astro-ph/0407462, in *Proceedings, 40th Karpacz Winter School of Theoretical Physics,* in press.

Sarkar, S., gr-qc/0204092, *Mod. Phys. Lett. A17* (2002) 1025.

Takeda, M. et al., *Phys. Rev. Lett. 81* (1998) 1163.

Watson, A., Proc. Snowmass Workshop, p 126, 1996

Zatsepin, G. T. and V. A. Kuzmin, *Sov. Phys.-JETP Lett. 4* (1966) 78.

G. Amelino-Camelia, Dipartimento de Fisica, Università di Roma "La Sapienze" and INFN Sez. Roma 1, P. le Moro 2, 00185 Roma, Italy. (amelino@roma1.infn.it)

Improving Limits on Planck-Scale Lorentz Symmetry Test Theories

Giovanni Amelino-Camelia

Dipartimento di Fisica, Università di Roma "La Sapienza", Roma, Italy

In the recent quantum gravity literature there has been strong interest in the possibility of Planck-scale departures from Lorentz symmetry. I focus on two "minimal" test theories, a pure kinematics test theory and an effective field-theory-based test theory, that could be used in this phenomenology. Planck-scale-significant bounds on some parameters of the two test theories can be established by using observations of TeV photons from blazars. Crab Nebula synchrotron radiation analyses, for which preliminary sensitivity estimates had raised high hopes, actually do not lead to any bound on the parameters of the two test theories. Very stringent (beyond Planckian) limits could be obtained, for both test theories, if the GZK cutoff for cosmic rays is confirmed experimentally.

1. ON THE FATE OF LORENTZ SYMMETRY IN QUANTUM SPACETIME

It has been recently realized that in various approaches to the quantum gravity problem, one encounters nonclassical features of spacetime that lead to small departures from Lorentz symmetry, and a quantum gravity-motivated phenomenology of departures from Lorentz symmetry has been proposed [*Amelino-Camelia*, 1997, 1998]. The idea that Lorentz symmetry might be only an approximate symmetry has since been considered in quantum gravity models based on spacetime foam pictures [*Garay*, 1998]; in loop quantum gravity models [*Gambini and Pullin*, 1999; *Alfaro et al.*, 2000]; in noncommutative geometry models [*Amelino-Camelia and Majid*, 2000; *Bruno et al.*, 2001; *Douglas and Nekrasov*, 2001; *Kowalski-Glickman*, 2001; *Amelino-Camelia*, 2001, 2002a], including some scenarios for noncommutative geometry that are relevant in string theory [*Matusis et al.*, 2000; *Douglas and Nekrasov*, 2000]; and in other string-theory-inspired scenarios [*Kostelecky et al.*, 2000].

At a strictly phenomenological level, one can view this interest in possible Planck-scale departures from Lorentz symmetry as originating from the idea that the sought quantum gravity might involve some sort of "granularity" of spacetime ("spacetime quanta"), and on the basis of experience with certain physical systems (especially condensed-matter systems)[1], one can expect that granularity of the medium in which propagation occurs might lead to energy-dependent corrections [*Amelino-Camelia*, 1997, 1998] to the dispersion relation. At energies much larger than the particle mass but smaller than the granularity (Planckian) energy scale, the dispersion relation could be of the type

$$m^2 \simeq E^2 - \bar{p}^2 + \eta \bar{p}^2 \left(\frac{E^n}{E_p^n}\right) + O\left(\frac{E^{n+3}}{E_p^{n+1}}\right) \quad (1)$$

where $E_p \simeq 1.2 \cdot 10^{16}$ TeV is the Planck scale, η parametrizes the ratio between the Planck scale and the scale of quantization of spacetime, and the power n is a key characteristic of the magnitude of the effects to be expected.

The literature on the subject is characterized primarily by debate on whether or not one should assume that these departures from Lorentz symmetry can be introduced within the framework of effective low-energy field theory. Those who are most concerned about the reliability of this assumption opt to limit the phenomenology to contexts in which

---

[1] Some authors have also argued (see, e.g., *Garay* [1998]) that the quantum spacetime environment might act in a way that to some extent resembles that of a thermal environment. It is well established (see, e.g., *Amelino-Camelia and Pi* [1993]) that in a thermal environment the energy-momentum dispersion relations are naturally modified.

one is able to perform a pure kinematics test. Those who assume the validity of low-energy effective field theory can use it to describe some aspects of dynamics and therefore have available a wider class of limit-setting opportunities.

For these two perspectives on the problem there are correspondingly two "natural" [*Amelino-Camelia*, 2004] starting points as test theories on which to base the phenomenology. When one does not assume the validity of low-energy effective field theory, it is natural to set up a pure kinematics test theory based on the dispersion relation (Eq. (1)) and on the assumption that the energy-momentum conservation rules are not Planck-scale modified (even though there are frameworks in which instead both the dispersion relation and the rules of energy-momentum conservation would be modified; see, e.g., *Amelino-Camelia* [2001, 2002]). The pure kinematics objectives are also fully compatible with the assumption of "universality"; that is, the modification of the dispersion relation is assumed to affect all particles in the same way (same values of parameters). These hypotheses constitute the "minimal AEMNS test theory" [*Amelino-Camelia*, 2004] (where "AEMNS" refers to names of the authors who introduced [*Amelino-Camelia et al.*, 1997, 1998] the first building blocks of this test theory). While it is rather natural to get started on this pure kinematics phenomenology by using the "minimal AEMNS test theory", of course one can contemplate various generalizations ("nonminimal versions of the AEMNS test theory"), including a "nonuniversal" Planck-scale modification of the dispersion relation.

When adopting the alternative perspective, which assumes the applicability of effective low-energy field theory, one can again take the dispersion relation parametrization of Eq. (1), but consistency with the use of the field theory setup imposes an immediate default to the simplifying hypothesis of universality. In fact, the constraint of introducing the new effects within the field theory formalism, with its reference to a Lagrangian density, restricts the types of modifications one can consider. In particular it is easy to show [*Gambini and Pullin*, 1999; *Myers and Pospelov*, 2003] that the allowed terms in the Lagrangian density lead to a polarization dependence of the effect for photons: In the field theory setup it turns out that, when right-circular polarized photons satisfy the dispersion relation $E^2 \simeq p^2 + \eta_\gamma p^3 /E_p$, then necessarily left-circular polarized photons satisfy the "opposite sign" dispersion relation $E^2 \simeq p^2 - \eta_\gamma p^3 /E_p$. For spin-1/2 particles the analysis reported *Meyers and Pospelov* [2003] leads to the introduction of two independent parameters for dispersion relation deformation, one for each helicity. Since photons experience a complete correlation of the sign of the effect with polarization, it seems natural to assume (at least in the first works exploring this scenario) that for fermions as well the modification of the dispersion relation should have the same magnitude for both signs of the helicity, and the sign of the helicity should be correlated with the sign of the dispersion relation modification. This would correspond to the natural-seeming assumption that the Planck-scale effects are such that in a beam composed of randomly selected particles the average speed in the beam is still governed by ordinary special relativity (the Planck-scale effects average out when summing over polarization/helicity). These observations provide the ingredients of the "minimal GPMP test theory" [*Amelino-Camelia*, 2004] ("GPMP" from the initials of the authors of who introduced most of the ingredients of this scenario [*Gambini and Pullin*, 1999; *Myers and Pospelov*, 2003]).

In these notes I will comment on certain types of data that are being considered as opportunities to test scenarios for Planck-scale violations of Lorentz symmetry, and analyze their applicability to the two "minimal" test theories. My discussion is not of the type "status of experimental limits on the test theories"; rather, I intend to illustrate how the (apparently small) differences between the two minimal test theories can affect the phenomenology significantly. I will therefore not consider all the opportunities for testing that have been discussed in the literature but will instead focus on a few illustrative examples.

2. DERIVATION OF LIMITS FROM TIME-OF-FLIGHT ANALYSES

The most popular strategy for establishing experimental limits on Planck-scale modifications of the dispersion relation is based [*Amelino-Camelia et al.*, 1997, 1998] on the fact that in both the minimal AEMNS test theory and the minimal GPMP test theory, one expects a wavelength dependence of the speed of photons, by combining the modified dispersion relation and the relation $v = dE/dp$. At "intermediate energies" ($m < E \ll E_p$), this velocity law will take the form

$$v \simeq 1 - \frac{m^2}{2E^2} + \eta \frac{n+1}{2} \frac{E^n}{E_p^n}. \quad (2)$$

Whereas in ordinary special relativity two photons ($m = 0$) emitted simultaneously would always reach simultaneously a faraway detector, according to Eq. (2) two simultaneously emitted photons should reach the detector at different times if they carry different energy. Moreover, in the case of the GPMP test theory, even photons with the same energy would arrive at different times if they carry different polarization. In fact, while the minimal AEMNS test theory assumes universality, and therefore a formula of this type would apply to photons of any polarization, the sign of the effect in the GPMP test theory is correlated with polarization. As a result, whereas the AEMNS test theory is best tested by comparing the arrival times of particles of different energies, the GPMP test theory is best tested by considering the highest energy photons available in the data and looking for a sizeable spread in times of arrivals (which would then be attributed to the different speeds of the two polarizations).

This time-of-arrival-difference effect can be significant [*Amelino-Camelia et al.*, 1997, 1998; *Biller et al.*, 1999] in the analysis of short-duration bursts of photons that reach us from faraway sources.

In the near future an excellent opportunity to test this effect will be provided by observations of gamma-ray bursters. How long a time a gamma-ray burst travels before reaching our Earth detectors is not uncommonly on the order of $T \sim 10^{17}$ s. Given that microbursts within a burst can have very short duration, as short as 10^{-3} s (or even 10^{-4} s), this means that the photons that compose such a microburst are all emitted at the same time, up to an uncertainty of 10^{-3} s. Some of the photons in these bursts have energies that extend at least up to the GeV range. For two photons with energy difference on the order of $\Delta E \sim 1$ GeV, a $\eta \Delta E/E_p$ speed difference over a time of travel of 10^{17} s would lead to a difference in times of arrival on the order of

$$\Delta t \sim \eta T \frac{\Delta E}{E_p} \sim 10^{-2} \text{ s}, \quad (3)$$

which is significant (the time-of-arrival differences would be larger than the time-of-emission differences within a single microburst).

For the minimal AEMNS test theory, the Planck-scale–induced time-of-arrival difference could be revealed [*Amelino-Camelia et al.*, 1997, 1998; *Biller et al.*, 1999] by comparing the "average arrival time" of the gamma-ray-burst signal (or better, of a microburst within the burst) in different energy channels. The GPMP test theory would be most effectively tested by looking for a dependence of the time-spread of the microbursts that grows with energy.

Since the quality of relevant gamma-ray-burst data is still relatively poor, the present best limit was obtained by *Biller et al.* [1999], using a slightly different type of observations: The negative results of a search of time-of-arrival/energy correlations for a TeV gamma-ray short-duration flare from the Markarian 421 blazar allowed *Biller et al.* to deduce the limit $|\eta| < 3 \cdot 10^2$. For the minimal GPMP test theory, one also correspondingly concludes that $|\eta_\gamma|$ is of $O(10^2)$ or smaller.

Considering the sensitivities achievable, it appears that the next generation of gamma-ray telescopes, such as GLAST [*Norris et al.*, 1999; *de Angelis*, 2000], might be able to test

very significantly Eq. (2) in the case $n = 1$, by possibly pushing the limit on η far below 1. To achieve this level of sensitivity, however, it might be necessary to gain some understanding of certain types of potentially troublesome at-the-source effects (discussed in *Piran* [in press]).

3. LIMITS OBTAINED FROM OBSERVED ABSORPTION OF TEV PHOTONS FROM BLAZARS

In addition to a possible manifestation in time-of-arrival/energy correlations, the quantum gravity–scale modifications of the dispersion relation [*Kifuno*, 1999; *Aloisio et al.*, 2000; *Protheroe and Meyer*, 2000; *Amelino-Camelia and Piran*, 2001] could result in observably large implications for what concerns the opacity of our Universe to various types of high-energy particles. Of particular interest is the fact that, according to the conventional (classical spacetime) description, the infrared diffuse extragalactic background should give rise to strong absorption of "TeV photons" (here understood as photons with energy between 1 and 30 TeV). The relevant process is of course $\gamma\gamma \to e^+ e^-$. With a given dispersion relation and a given rule for energy-momentum conservation, one has a complete "kinematic scheme" for the analysis of the requirements for particle production in collisions (or decay processes). Both the minimal AEMNS test theory and the minimal GPMP test theory involve modified dispersion relations and unmodified laws of energy-momentum conservation (the fact that the law of energy-momentum conservation is not modified is explicitly among the ingredients of the AEMNS test theory, while in the GPMP test theory it follows from the adoption of low-energy effective field theory).

Combining a modified dispersion relation with unmodified laws of energy-momentum conservation, one naturally finds a modification of the threshold requirements for the $\gamma\gamma \to e^+ e^-$ process. Let us in particular consider the dispersion relation (Eq. 1), with $n = 1$, in the analysis of a collision between a soft photon of energy ϵ and a high-energy photon of energy E. For given soft-photon energy ϵ, the process $\gamma\gamma \to e^+ e^-$ is allowed only if E is greater than a certain threshold energy E_{th} which depends on ϵ and m_e^2. For $n = 1$, combining Eq. (1) with unmodified energy-momentum conservation, this threshold energy is estimated as (assuming $\epsilon \ll m_e \ll E_{th} \ll E_p$)

$$E_{th} \epsilon + \eta \frac{E_{th}^3}{8E_p} \simeq m_e^2. \quad (4)$$

The special relativity result $E_{th} = m_e^2/\epsilon$ corresponds of course to the $\eta \to 0$ limit of Eq. (4). For $|\eta| \sim 1$, the Planck-scale correction can be safely neglected as long as $\epsilon > (m_e^4/E_p)^{1/3}$.

But eventually, for sufficiently small values of ϵ (and correspondingly large values of E_{th}) the Planck-scale correction cannot be ignored.

In particular, if the photon of energy ϵ is part of the infrared diffuse extragalactic background and the photon emitted by a blazar is of TeV-range energy, then the prediction for absorption of the hard photon by the infrared diffuse extragalactic background is significantly modified.

The fact that the observations still give us only a preliminary picture of absorption together with the significant amount of uncertainty in phenomenological models of TeV blazars and in phenomenological models of the density of the infrared diffuse extragalactic background does not allow us to convert these observations into tight limits on departures from the classical spacetime analysis. However, even the basic fact that we see absorption of TeV γ-rays, as now suggested by several analyses, allows us to derive [*Amelino-Camelia*, 2004] the limit $\eta \geq -46$ (*i.e.* either η is positive or η is negative with an absolute value smaller than 46).

Up to this point my discussion is strictly applicable only to the minimal AEMNS test theory. For the minimal GPMP test theory, the analysis must be modified to take into account the fact that the modification of the dispersion relation carries opposite signs for the two polarizations of the photon (and for the two helicities of the electron/positron). In light of this polarization dependence in the minimal GPMP test theory, only one of the two polarizations of the photon can escape absorption. Whereas in the classical spacetime picture one would expect a cutoff behaviour to affect the entire spectrum (and in the minimal AEMNS test theory, one might find no cutoff at all), in the minimal GPMP test theory one would expect a cutoff behaviour for only a part of the spectrum, while the rest could be unaffected by the cutoff.

4. DERIVATION OF LIMITS FROM ANALYSIS OF SYNCHROTRON RADIATION

A recent series of papers [*Amelino-Camelia*, 2004; *Ellis et al.*, 2004; *Jacobson et al.*, 2002, 2003a, 2003b, 2003c; *Carroll*, 2003] has focused on the possibility of setting limits on Planck-scale–modified dispersion relations, focusing on their implications for synchrotron radiation. In *Jackson et al.* [1999], the starting point is the observation that in the conventional (Lorentz-invariant) description of synchrotron radiation, one can estimate the characteristic energy E_c of the radiation through a heuristic analysis [*Jackson*, 1999] that leads to the formula

$$E_c \simeq \frac{1}{R \cdot \delta \cdot [v_\gamma - v_e]}, \qquad (5)$$

where v_e is the speed of the electron, v_γ is the speed of the photon, δ is the emission angle for the radiation, and R is the radius of curvature of the trajectory of the electron.

Assuming that the only Planck-scale modification in this formula should come from the velocity law (described as $v = dE/dp$ in terms of the modified dispersion relation), we find that in some instances the characteristic energy of synchrotron radiation may be significantly modified by the presence of Planck-scale departures from Lorentz symmetry. As an opportunity to test such a modification of the value of the characteristic synchrotron radiation energy, we can attempt to use some relevant data (available from *Jacobson et al.* [2003b, 2003c]) on photons detected from the Crab Nebula. This must be done with caution since the observational information on synchrotron radiation emitted by the Crab Nebula is indirect: Some of the photons we observe from the Crab Nebula are attributed to synchrotron processes on the basis of a promising conjecture, and the value of the relevant magnetic fields is not directly measured.

If we assume that indeed the observational situation has been properly interpreted, and that the only modification to be taken into account is the one of the velocity law, this type of analysis has the potential to establish very stringent limits on some Lorentz symmetry-breaking parameters. However, this will of course depend on the detailed structure of the Lorentz symmetry-breaking scheme. In particular, as it turns out, the minimal test theories I am considering cannot be constrained in this way.

For what concerns the minimal AEMNS test theory it is important to realize that synchrotron radiation is due to the acceleration of the relevant electrons and therefore implicit in the derivation of Eq. (5) is a subtle role for dynamics [*Amelino-Camelia*, 2004]. From a field-theory perspective the process of synchrotron radiation emission can be described in terms of Compton scattering of the electrons with the virtual photons of the magnetic field. One would therefore be looking deep into the dynamical features of the theory.

The minimal AEMNS test theory does assume a modified dispersion relation of the type in Eq. (1), universally applied to all particles, but this is a pure kinematics framework and, since the analysis involves some aspects of dynamics, it cannot be tested by an analysis of Crab Nebula synchrotron radiation.

The GPMP test theory relies on a description of dynamics within the framework of effective low-energy theory, but, as mentioned, this in turn ends up implying the impossibility of assuming that a dispersion relation of the type in Eq. (1) universally applies to all particles. Actually, within this framework, the two polarizations of photons must satisfy different (opposite-sign Planck-scale corrections) dispersion relations. And for the description of electrons, one naturally encounters two more free parameters, which in my "minimal GPMP test theory" are also of equal magnitude and opposite sign (to preserve c as the "average speed", as discussed in section 1). As a result the "minimal GPMP test theory" automatically evades the type of constraint that could come from analysis of the Crab Nebula synchrotron radiation: The Crab Nebula constraint is significant when both helicities of the electron are affected by negative-sign-type Planck-scale modification of the dispersion relation, but in the "minimal GPMP test theory" the two helicities are affected by opposite-sign modifications of the dispersion relation.

5. DERIVATION OF LIMITS FROM ANALYSIS OF UHE COSMIC RAYS

In section 3, I discussed the implications of possible Planck-scale effects for the process $\gamma\gamma \to e^+ e^-$, but of course this is not the only process in which Planck-scale effects can be important. In particular, there has been strong interest [*Kifune*, 1999; *Aloisio et al.*, 2000; *Protheroe and Meyer*, 2000; *Amelino-Camelia and Piran*, 2001; *Ng et al.*, 2001; *Amelino-Camelia*, 2002b; *Jacobson et al.*, 2002, 2003a; *Bertolami*, 2002; *Amelino-Camelia et al.*, 2003] in "photopion production", $p\gamma \to p\pi$, where again the combination of Eq. (1) with unmodified energy-momentum conservation leads to a modification of the minimum proton energy required by the process (for given photon energy). In the case in which the photon energy is the one typical of CMBR photons, one finds that the threshold proton energy can be significantly shifted upward (for negative η), which in turn should affect at an observably large value the expected GZK cutoff for the observed cosmic-ray spectrum. Observations reported by the AGASA cosmic-ray observatory [*Takeda et al.*, 1998] provide some encouragement for the idea of such an upward shift of the GZK cutoff, but the issue must be further explored[2]. Forthcoming cosmic-ray observatories, such as Auger [*Blumer*, 2003], should be able to fully investigate this possibility [*Kifune*, 1999; *Amelino-Camelia and Piran*, 2001].

In this context comparison of the AEMNS test theory and the GPMP test theory is rather straightforward. We are in fact considering a purely kinematical effect: the shift of a threshold requirement. For the minimal AEMNS test theory, there is a clear prediction that for negative η there should be an upward shift of the GZK threshold. For the minimal GPMP test theory, where for one of the helicities of the proton the dispersion relation is of negative-η type and for the other helicity the dispersion relation is of positive-η type, one would expect roughly half of the UHE protons to evade the GZK cutoff; the cutoff would still be violated but in a softer way than in the case of the AEMNS test theory with negative η.

If the Auger data should actually show evidence of the expected GZK cutoff, then the case of negative η for the minimal AEMNS test theory would be severely constrained (both for $n = 1$ and $n = 2$), and the fermion-sector parameter of the minimal GPMP test theory would also be severely constrained. Indeed, in the minimal AEMNS test theory, violations of the GZK cutoff are predicted for negative η (although not in the positive-η case), whereas in the minimal GPMP test theory, violations of the GZK cutoff (although less numerous than expected in the minimal AEMNS test theory with negative η) are always expected, independently of the sign of the fermion-sector parameter (depending on the sign of fermion-sector parameter, the protons that violate the GZK cutoff would have a corresponding helicity).

6. CLOSING REMARKS

In this chapter I have focused on the minimal GPMP test theory and the minimal AEMNS test theory, representing the field-theory intuition and the no-field-theory intuition, respectively, in the study of Planck-scale departures from Lorentz symmetry. I found that the differences between these two test theories, although they might at first appear to be rather marginal differences (both test theories essentially adopt the same type of modification of the dispersion relation), lead to significant differences in the outcome of certain phenomenological analyses. This should be kept in mind in the relevant "quantum gravity phenomenology" [*Amelino-Camelia*, 2000, 2002c] literature. Several papers have claimed to improve limits on Planck-scale modifications of the dispersion relation, but the studies did not rely on a well-defined test theory. From outside the quantum gravity phenomenology community, these papers were perceived as a gradual improvement in the experimental bounds on the overall idea of Planck-scale departures from Lorentz symmetry, to the extent that there is now a widespread perception that, in general, departures from Lorentz symmetry are already experimentally constrained to be far beyond the Planck scale. Instead, as shown by the analysis of the two "minimal" test

---

[2] This AGASA data-based "GZK puzzle" has been very important in providing motivation for studies of Planck-scale departures from Lorentz symmetry. Even if a future improved understanding of the cosmic-ray spectrum resolves the puzzle, the lessons learned in the study of the quantum gravity problem will still be very valuable. An analogous situation has been recently encountered in the particle physics literature: Discussion of the so-called "centauro events" led to strong theoretical progress in our understanding of the possibility of "misaligned vacua" in QCD (see, e.g., *Amelino-Camelia et al.* [1997]), and this progress on the theory side remains valuable even though most authors now believe that centauro events might have been a mirage.

theories I considered, some of the experimental-limit opportunities that have generated the most excitement in the recent literature are inapplicable to some meaningful scenarios for Planck-scale departures from Lorentz symmetry.

For the objectives I was pursuing here, it was necessary to focus indeed on rather similar test theories, so that I could illustrate the fact that even relatively small differences in the structure of the test theories can significantly affect the phenomenology. Both the minimal AEMNS test theory and the minimal GPMP test theory adopt the same type of dispersion relation and both assume that Planck-scale effects would break Lorentz symmetry. Even more significant differences, from the perspective of phenomenological analyses, should be expected in test theories that explore the possibility that the Planck scale would "deform" Lorentz symmetry (in the sense of the "doubly special relativity" scenario [*Alexander and Magueijo*, hep-th/0104093; *Amelino-Camelia*, 2001, 2002a; *Bruno et al.*, 2001; *Kowalski-Glikman*, 2001; *Kowalski-Glikman and Nowak*, 2003; *Magueijo and Smolin*, 2003]. Work on the development of such a doubly special relativity test theory is in progress (see, e.g., *Amelino-Camelia et al.* [in press]), but this analysis is still at too early a stage for comment on it here.

REFERENCES

S. Alexander, and J. Magueijo: hep-th/0104093.
J. Alfaro, H.A. Morales-Tecotl and L.F. Urrutia: Phys. Rev. Lett. **84**, 2318 (2000).
R. Aloisio, P. Blasi, P.L. Ghia, and A.F. Grillo: Phys. Rev. D **62**, 053010 (2000).
G. Amelino-Camelia: "Are we at the dawn of quantum-gravity phenomenology?", gr-qc/9910089, Lect. Notes Phys. 541 (2000).
G. Amelino-Camelia: hep-th/0012238, Phys. Lett. B **510**, 255 (2001).
G. Amelino-Camelia: gr-qc/0012051, Int. J. Mod. Phys. D **11**, 35 (2002a).
G. Amelino-Camelia: gr-qc/0107086, Phys. Lett. B **528**, 181 (2002b).
G. Amelino-Camelia: gr-qc/0204051, Mod. Phys. Lett. A**17**, 899 (2002c).
G. Amelino-Camelia: gr-qc/0212002, New J. Phys. **6**, 188 (2004).
G. Amelino-Camelia and S. Majid: Int. J. Mod. Phys. A **15**, 4301 (2000).
G. Amelino-Camelia and S.-Y. Pi: hep-ph/9211211, Phys. Rev. D **47**, 2356 (1993).
G. Amelino-Camelia and T. Piran: astro-ph/0008107, Phys. Rev. D **64**, 036005 (2001).
G. Amelino-Camelia, J.D. Bjorken and S.E. Larsson: hep-ph/9706530, Phys. Rev. D **56**, 6942-6956 (1997).
G. Amelino-Camelia, J. Ellis, N.E. Mavromatos, and D.V. Nanopoulos: hep-th/9605211, Int. J. Mod. Phys. A **12**, 607 (1997).
G. Amelino-Camelia, J. Ellis, N.E. Mavromatos, D.V. Nanopoulos, and S. Sarkar: astro-ph/9712103, Nature **393**, 763 (1998).
G. Amelino-Camelia, J. Kowalski-Glikman, G. Mandanici, and A. Procaccini: gr-qc/0312124, Int. J. Modern Phys. A, in press.
G. Amelino-Camelia, Y.J. Ng, and H. van Dam: gr-qc/0204077, Astropart. Phys. **19**, 729 (2003).
O. Bertolami: hep-ph/0301191
S.D. Biller et al.: Phys. Rev. Lett. **83**, 2108 (1999).
J. Blumer: J. Phys. G **29**, 867 (2003).
R. Bruno, G. Amelino-Camelia, and J. Kowalski-Glikman: Phys. Lett. B **522**, 133 (2001).
S. Carroll: Nature **424**, 1007 (2003).
A. de Angelis: astro-ph/0009271, in Faro 2000, New Worlds in Astroparticle Physics, 140.
N.R. Douglas and N.A. Nekrasov: Rev. Mod. Phys. **73**, 977 (2001).
J. Ellis, N.E. Mavromatos, and A.S. Sakharov: astro-ph/0308403, Astroparticle Phys. **20**, 669 (2004).
R. Gambini and J. Pullin: Phys. Rev. D **59**, 124021 (1999).
L.J. Garay: Phys. Rev. Lett. **80**, 2508 (1998).
J.D. Jackson: Classical Electrodynamics, 3rd edn., J. Wiley & Sons, New York (1999).
T. Jacobson, S. Liberati, and D. Mattingly: astro-ph/0212190, Nature **424**, 1019 (2003c).
T. Jacobson, S. Liberati, and D. Mattingly: hep-ph/0112207, Phys. Rev. D **66**, 081302 (2002).
T. Jacobson, S. Liberati, and D. Mattingly: hep-ph/0209264, Phys. Rev. D **67**, 124011 (2003a).
T. Jacobson, S. Liberati, and D. Mattingly: arXiv.org/abs/astro-ph/0212190v1, Nature **424**, 1019 (2003b).
T. Jacobson, S. Liberati, and D. Mattingly: gr-qc/0303001.
T.A. Jacobson, S. Liberati, D. Mattingly, and F.W. Stecker: astro-ph/0309681, Phys. Rev. Lett. **93**, 021101 (2004).
T. Kifune: astro-ph/9904164, Astrophys. J. Lett. **518**, L21 (1999).
V.A. Kostelecky, M. Perry, and R. Potting: Phys. Rev. Lett. **84**, 4541-4544 (2000), hep-th/9912243.
J. Kowalski-Glikman: hep-th/0102098, Phys. Lett. A **286**, 391 (2001).
J. Kowalski-Glikman and S. Nowak: hep-th/0304101, Class. Quant. Grav. **20**, 4799 (2003).
J. Magueijo and L. Smolin: gr-qc/0207085, Phys. Rev. D **67**, 044017 (2003).
A. Matusis, L. Susskind, and N. Toumbas: JHEP **0012**, 002 (2000).
R.C. Myers and M. Pospelov: hep-ph/0301124, Phys. Rev. Lett. **90**, 211601 (2003).
Y.J. Ng, D.S. Lee, M.C. Oh, and H. van Dam: Phys. Lett. B **507**, 236 (2001).
J.P. Norris, J.T. Bonnell, G.F. Marani, and J.D. Scargle: astro-ph/9912136, in Salt Lake City 1999, Cosmic Ray **4**, 20.
T. Piran: astro-ph/0407462, in press.
R.J. Protheroe and H. Meyer: Phys. Lett. B **493**, 1 (2000).
M. Takeda et al.: Phys. Rev. Lett. **81**, 1163 (1998).

G. Amelino-Camelia, Dipartimento di Fisica, Universitá di Roma "La Sapienza" and INFN Sez. Roma 1, P. le Moro 2, 00185 Roma, Italy. (amelino@roma1.infn.it)

Spectral Evolution of Two High-Energy Gamma-Ray Bursts

Yuki Kaneko[1,2], Robert D. Preece[1,2], María Magdalena González[3,4], Brenda L. Dingus[4], and Michael S. Briggs[1,2]

The prompt emission of the gamma-ray bursts is found to be very energetic, releasing ~10^{51} ergs in a flash. However, their emission mechanism remains unclear and understanding their spectra is a key to determining the emission mechanism. Many GRB spectra have been analyzed in the sub-MeV energy band, and are usually well described with a smoothly broken power-law model. We present a spectral analysis of two bright bursts (GRB910503 and GRB930506), using BATSE and EGRET spectra that cover more than four decades of energy (30 keV–200 MeV). Our results show time evolutions of spectral parameters (low-energy and high-energy photon indices, and break energy) that are difficult to reconcile with a simple shock-acceleration model.

INTRODUCTION

Gamma-ray bursts (GRBs) are among the most energetic phenomena in the universe and emit a tremendous amount of energy in seconds, primarily as gamma rays. Many GRBs of duration longer than a few seconds are followed by an afterglow of longer wavelengths, lasting days to months after the burst. Despite the numerous observations of GRBs and their afterglows, their creation mechanism and origins are still unclear. GRB spectra are non-thermal and continuous from a few keV to GeV; however, the distribution of the peak energy of the emitted power (E_{peak}) is found to be a narrow lognormal with a centroid value of 250 keV [*Mallozzi et al.*, 1995; *Preece et al.*, 2000].

Many models have been suggested to explain the observed non-thermal spectra of GRBs [*Mészáros*, 2002]. The most widely accepted picture is the synchrotron shock model (SSM). When shocks are formed, electrons are accelerated by the Fermi mechanism to a power-law energy distribution, $N(E_e) \propto E_e^{-p}$, and these highly relativistic electrons radiate synchrotron radiation due to the magnetic field behind the shock, producing the GRB. However, the SSM has some difficulties when confronted with the observational data [*Preece et al.*, 2000]. Studying the broadband energy spectra of GRBs is crucial to revealing the shock acceleration and gamma-ray emission mechanisms.

Here we present the results of broadband time-resolved spectral analysis of two spectrally hard GRBs.

SPECTRAL ANALYSIS

Instruments and Data Types

The Burst and Transient Source Experiment (BATSE) on board the *Compton Gamma-Ray Observatory* (*CGRO*) observed 2704 GRBs in its 9-year lifetime (1991–2000). The BATSE observation provides the largest GRB database to date, with excellent time and energy resolution over the energy band 15 keV to ~10 MeV. Also on board the *CGRO* was the Energetic Gamma-Ray Experiment Telescope (EGRET), designed to observe gamma-ray sources in energies above 1 MeV. Some bright GRBs were observed with both BATSE and EGRET, providing spectra over a broader energy range.

BATSE was a collection of eight modules, each of which consisted of a Large Area Detector (LAD) and a Spectroscopy Detector (SD). Both are NaI(Tl) scintillation detectors coupled with photo multiplier tubes (PMTs). In this work,

---

[1]Department of Physics, University of Alabama in Huntsville, Huntsville, Alabama.
[2]National Space Science and Technology Center, Huntsville, Alabama.
[3]Department of Physics, University of Wisconsin at Madison, Madison, Wisconsin.
[4]Los Alamos National Laboratory, Los Alamos, New Mexico.

data from the brightest LAD (which varies from event to event, depending on the source direction) are used instead of the SD data due to the LAD's large effective area. There are several different data types for the LAD, of which two are used for this analysis: HERB (High Energy Resolution Burst) data and MER (Medium Energy Resolution) data. For both data types, the accumulation of the data began at the BATSE burst trigger. MER data are used when HERB data are not available or the HERB data are not complete, i.e., when the HERB data do not cover the entire duration of burst. Incomplete HERB data is common for bright events since HERB had a fixed memory space that could fill before the burst was over.

EGRET consisted of a spark chamber and a calorimeter (Total Absorption Shower Counter, TASC). The TASC was located at the bottom of EGRET and was made of a much larger NaI(Tl) scintillation crystal than were used in BATSE. Independently from the spark chamber events, the TASC observed a few dozen GRBs in its Burst mode, which was initiated by a BATSE trigger. In the Burst mode, spectra were accumulated in four commandable time intervals (normally 1, 2, 4, and 16 seconds). Each detector's characteristics are listed in Table 1.

LAD-TASC Joint Spectral Analysis

Combining the LAD and TASC data provides spectra that span 4 decades of energy (30 keV-200 MeV). In general, GRB spectra are well fit with two power-laws joined smoothly at a break energy that is uniquely related to E_{peak}. As E_{peak} approaches the upper limit of the BATSE passband, the BATSE data alone cannot adequately determine the high-energy power law index (β). Having TASC data along with BATSE data can extend the spectrum energy range up to 200 MeV, which may constrain β as well as E_{peak} values for the spectrally hard GRBs. The joint fit can also test the validity of the smoothly broken power-law model at higher energies that have been typically fitted using BATSE data.

The analysis was performed using the spectral analysis software RMFIT. To jointly fit time-resolved spectra, LAD data were binned in time to match the TASC time bins. RMFIT employs forward-fitting procedures with one or more spectral models specified by users. In the actual fitting procedure, a multiplicative Effective Area Correction term was used because of uncertainties in the calculated effective areas of each detector. The goodness of fit is determined by χ^2.

GRB Spectral Model

The photon model used in this analysis is an empirical "GRB" function, which consists of two power laws smoothly joined together [*Band et al.*, 1993]:

$$f(E) = A \left(\frac{E}{100 \text{ keV}} \right)^\alpha \exp \left(\frac{-(2+\alpha)E}{E_{peak}} \right)$$

if $E < (\alpha - \beta) [E_{peak}/(2 + \alpha)]$, and

$$f(E) = A \left(\frac{(\alpha - \beta) E_{peak}}{(2+\alpha) 100 \text{ keV}} \right)^{\alpha - \beta} \exp(\beta - \alpha) \left(\frac{E}{100 \text{ keV}} \right)^\beta$$

if $E \geq (\alpha - \beta) [E_{peak}/(2 + \alpha)]$;

Table 1. Detector Characteristics

Detector	Energy Range (MeV)	Time Resolution (s)	No. of Energy Channels
LAD	0.03–2	0.128[a] (HERB) 0.016 (MER)	128 (HERB) 16 (MER)
TASC	1–200	1, 2, 4, 16[b] (BURST)	256

[a]Minimum time resolution; increases by 64-ms increments.
[b]Commandable.

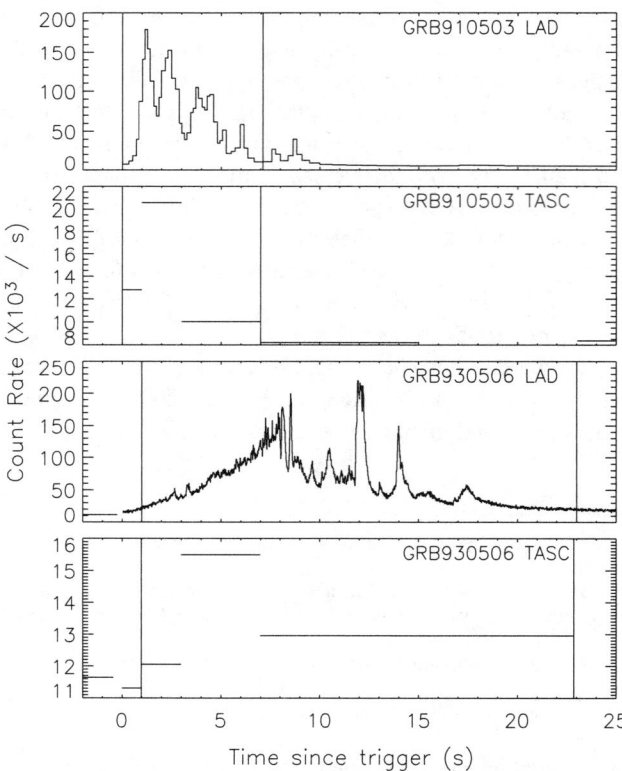

Figure 1. Time profiles of GRB 910503 (top two) and GRB 930506 (bottom two) as observed by LAD and TASC. Time intervals used in the analysis are indicated.

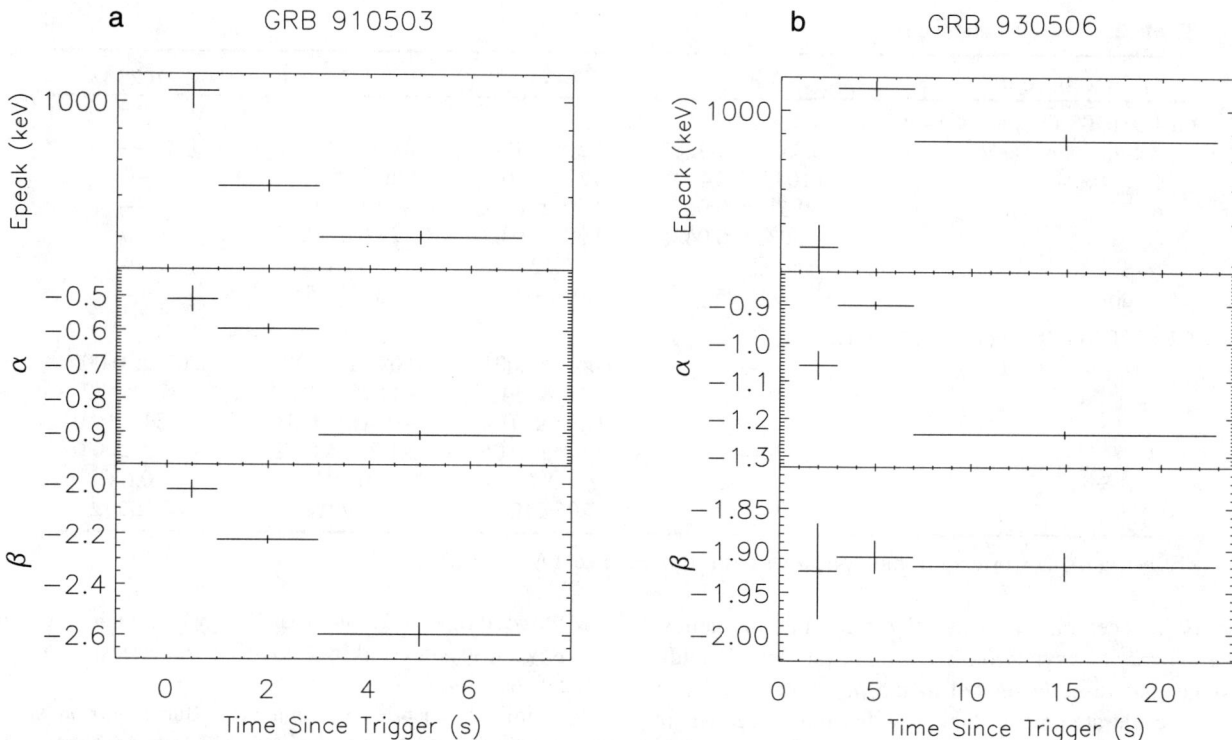

Figure 2. Spectral parameter evolutions in each burst. The values correspond to those in Table 2.

where A is the amplitude in photons $s^{-1}cm^{-2}keV^{-1}$, E_{peak} is the peak energy of the power density spectrum, α is the low-energy photon index, and β is the high-energy photon index.

The Events

Two events, GRB910503 (BATSE trigger # 143) and GRB930506 (BATSE trigger # 2329), were selected for this analysis due to their brightness and their data availability. Lightcurves of these two bursts are shown in Figure 1. Both bursts are very hard and found to have fairly high E_{peak} values, and therefore higher energy spectra are required to better constrain E_{peak} and β values.

RESULTS AND DISCUSSION

Table 2 presents the best-fit spectral parameters for each event. The results clearly show the time evolution of the deduced photon spectra in each of the two events (see Figure 2).

GRB910503 (Trigger # 143)

It is evident that the spectra evolve from hard to soft with statistically significant changes in α and β ($\Delta\alpha$ and $\Delta\beta$, respectively). Interestingly, we find $\Delta\alpha \sim \Delta\beta$. Moreover, the difference between α and β (i.e., $\alpha - \beta \equiv \Delta s$) seems to remain approximately constant throughout the burst ($\Delta s \sim 1.6$). This value of Δs is high compared with the average of ~1.4 or the most likely value of ~1.0 found by *Preece et al.* [2002] based on the analysis of 5500 time-resolved BATSE GRB spectra. For the first two time intervals, we find $\alpha > -2/3$ by about 4 σ and 2 σ respectively, which violates the synchrotron "line of death" predicted by the SSM [*Preece et al.*, 1998].

GRB930506 (Trigger # 2329)

The spectra for this event do not seem to evolve from hard to soft, but rather β stays constant while α and E_{peak} evolve soft-hard-soft. In this case, since β is clearly above -2, the fitted value for E_{peak} is actually the break energy of the spectral model (where the high energy power law begins), and not the peak energy of the corresponding power density spectrum. This requires the existence of another spectral break (and thus the true E_{peak}) at an energy above the fitted E_{peak} value.

The standard SSM involves optically-thin synchrotron radiation by energetic electrons that are left to radiate without further acceleration. The fact that the GRB spectra evolve on timescales much longer than the synchrotron cooling time may require an acceleration mechanism that is more complicated than those presumed in the SSM. Whatever

Table 2. Best Fit Parameters

Time since trigger	0–1 s	1–3 s	3–7 s	7–23 s
GRB910503 (Trigger # 143; EAC[a] = 0.66)				
A (ph s^{-1}cm^{-2}keV^{-1})	0.05 ± 0.001	0.32 ± 0.003	0.13 ± 0.001	—
E_{peak} (keV)	1040 ± 74	727 ± 16	600 ± 15	—
α	−0.51 ± 0.04	−0.60 ± 0.01	−0.91 ± 0.01	—
β	−2.03 ± 0.04	−2.22 ± 0.02	−2.60 ± 0.05	—
$\alpha-\beta = \Delta s$	1.52	1.62	1.69	—
χ^2/dof	374/325	380/325	394/325	—
GRB930506 (Trigger # 2329; EAC[a] = 0.54)				
A (ph s^{-1}cm^{-2}keV^{-1})	—	0.04 ± 0.001	0.09 ± 0.0005	0.07 ± 0.0004
E_{peak} (keV)	—	540 ± 54	1104 ± 41	871 ± 32
α	—	−1.06 ± 0.04	−0.90 ± 0.01	−1.24 ± 0.01
β	—	−1.93 ± 0.06	−1.91 ± 0.02	−1.92 ± 0.02
$\alpha-\beta = \Delta s$	—	0.87	1.01	0.68
χ^2/dof	—	205/212	293/212	272/212

[a]Effective Area Correction, multiplicative term to normalize TASC to LAD.

allows the reacceleration of the electrons must somehow balance the very fast synchrotron cooling timescale. In addition, since β is directly related to the power-law index of the shock-accelerated electron energy distribution, p, where $\beta = -(p + 1)/2$, the changes in β observed in GRB910503 may indicate change in p. Electrons accelerated by the Fermi mechanism are expected to have a power law distribution with p ~2.2–2.3 that is constant in time [*Gallant, Achterberg & Kirk*, 1999]. This also implies that $\beta < -2$ at all times, which is contradicted by the observations. The currently-standard SSM does not account for our results; therefore, the SSM needs modifications or a new shock-acceleration model of the GRB emission mechanism is required.

REFERENCES

Band, D.L., et al. BATSE Observation of Gamma-Ray Burst Spectra I – Spectral Diversity, *Astrophysical Journal*, 413, 281-292, 1993.

Gallant, Y.A., Achterberg, A. and Kirk, J.G., Particle Acceleration at Ultra-Relativistic Shocks–Gamma-Ray Burst Afterglow Spectra and UHECRs, *Astronomy & Astrophysics Supplement Series*, 138, 549-550, 1999.

Mallozzi, R.S., et al. The νF_ν Peak Energy Distributions of Gamma-Ray Bursts Observed by BATSE, *Astrophysical Journal*, 454, 597-603, 1995.

Mészáros, P., Theories of Gamma-Ray Bursts, *Annual Review of Astronomy and Astrophysics*, 40, 137-169, 2002.

Preece, R.D., et al. The Synchrotron Shock Model Confronts a "Line of Death" in the BATSE Gamma-Ray Burst Data, *Astrophysical Journal Letters*, 506, 23-26, 1998.

Preece, R.D., et al. The BATSE Gamma-Ray Burst Spectral Catalog I – High Time Resolution Spectroscopy of Bright Bursts Using High Energy Resolution Data, *Astrophysical Journal Supplement Series*, 126, 19-36, 2000.

Preece, R.D., et al. On the Consistency of Gamma-Ray Burst Spectral Indices with the Synchrotron Shock Model, *Astrophysical Journal*, 581, 1248-1255, 2002.

M.S. Briggs, Y. Kaneko, and R.D. Preece, National Space Science and Technology Center, 320 Sparkman Drive, Huntsville, Alabama 35805. (yuki.kaneko@msfc.nasa.gov)

B.L. Dingus and M.M. González, M.S. H803 P-23, Los Alamos National Laboratory, Los Alamos, New Mexico 87545.